HYPERSPECTRAL IMAGING

HYPERSPECTRAL IMAGING

TECHNIQUES FOR SPECTRAL DETECTION AND CLASSIFICATION

Chein-I Chang
Remote Sensing Signal and Image Processing Laboratory
University of Maryland, Baltimore County
Baltimore, Maryland

Kluwer Academic / Plenum Publishers
New York, Boston, Dordrecht, London, Moscow

Library of Congress Cataloging-in-Publication Data

Chang, Chein-I.
Hyperspectral imaging: techniques for spectral detection and classification/Chein-I Chang.
p. cm.
Includes bibliographical references and index.
ISBN 0-306-47483-2
1. Remote sensing. 2. Multispectral photography. 3. Image processing—Digital techniques. I. Title.

G70.4 .C46 2003
621.36′78—dc21

2002040796

ISBN 0-306-47483-2

233 Spring Street, New York, New York 10013

http://www.wkap.com

10 9 8 7 6 5 4 3 2 1

A C.I.P. record for this book is available from the Library of Congress

Printed in the United States of America

This book is dedicated to the memory of my late father, Hsun Chang (張洵) for his constant support of my career. It is also dedicated to my mother, Kuo Kuei-Lan Chang (張郭貴蘭) for her timeless support and countless encouragement during thc course of preparing this book as well as to my sister, Feng-Chu Chang (張鳳珠)'s family, my sister Mann-Li Chang (張曼莉) and my brother Chein-Chi Chang (張建祺)'s family, particularly, my beloved little youngest nephew, Yu-Rei Raymond Chang (張育瑞).

PREFACE

Hyperspectral imaging is an emerging technique in remote sensing data processing that expands and improves capability of multispectral image analysis. It takes advantage of hundreds of contiguous spectral channels to uncover materials that usually cannot be resolved by multispectral sensors. This book is an outgrowth of my research in hyperspectral image processing and personal communications in response to many people who are interested in my work previously published in various journals. At the first glimpse, this book may look like a collection of papers authored and co-authored by me. As a matter of fact, it is not the case. The book has been organized in a way that all the chapters are logically connected and can be referred back and forth one another for more details. In particular, most of computer simulations and experiments have been reworked out in order to have a consistent treatment throughout the book. The title of "Hyperspectral Imaging: Techniques for Spectral Detection and Classification" is used to reflect its focus on spectral techniques, i.e. non-literal techniques that are especially designed and developed for hyperspectral imagery rather than multispectral imagery. Although many techniques already exist in multispectral image processing, some of them may not be effective when they are directly applied to hyperspectral imagery. This book takes an opposite approach to develop techniques from a hyperspectral imagery viewpoint where noise is generally not Gaussian and interference plays a more dominant role than does noise in hyperspectral image analysis. More importantly, the detection and classification is performed and carried out by targets of interest rather than pattern classes.

A significant difference from other books is that this book explores applications of statistical signal processing techniques in hyperspectral image analysis, specifically, subpixel detection and mixed pixel classification. It includes many techniques developed in my lab with my former and current Ph.D. students, and systematically integrates these techniques in such a unified framework that readers can capture how the ideas were developed and evolved. Since many readers whose background is not engineering may find a gap in understanding the concepts presented in this book, another objective of this book is to make it self-contained so that readers can easily pick up and implement the techniques without much difficulty. In doing so, I have included detailed mathematical derivations and experiments for illustration. Nevertheless, it by no means claims to be comprehensive; rather, it can be viewed as a recipe book that offers various techniques for hyperspectral data exploitation. Some of these techniques such as OSP (Orthogonal Subspace Projection), CEM (Constrained Energy Minimization) are mature for practical implementation. They are treated in the book in great detail. In addition, many techniques developed in this book may also become handy for years to come. Due to limited scope of the book, many well-known techniques, such as linear spectral mixture analysis that can be found in numerous references, will not discussed in this book.

Instead, this book only covers the work that has been done over the years in the Remote Sensing Signal and Image Processing Laboratory (RSSIPL) at the University of Maryland, Baltimore County.

Like most books, this book owes much credit to many people who deserve my sincere gratitude. These individuals are my former Ph.D. students, Drs. Mark L.G. Althouse, Clark Brumbley, Shao-Shan Chiang, Qian Du, Joseph C. Harsanyi, Daniel Heinz, Agustin Ifarragaerri, Chien-Shun Lo, Hsuan Ren, Chuin-Mu Wang and Ching-Wen Yang as well as my current Ph.D. students, Ms. Eliza Yingzi Du, Ms. Kerri Guilfoyle and Ms. Jianwei Wang. Specifically, I would like to thank Drs. Shao-Shan Chiang and Hsuan Ren who spent so much time in helping me generate most of figures in this book. This book cannot be completed without their contributions. For the data used in this book I would like to thank the Spectral Information Technology Applications Center (SITAC) and Dr. Harsanyi who provide me with their HYDICE and AVIRIS data respectively. My sincere thanks also go to Mr. Paul Lewis and Ms. Judy Powelson who read part of this book and provided valuable suggestions.

Finally, I would like to thank Mr. James Buss, Dr. Irving W. Ginsberg, Mr. Paul Lewis and Dr. Gregory Palvin for their kind support of my research under funding received from the Office of Naval Research, the Bechtel Nevada Corporation through the Department of Energy and Spectral Information Technology Applications Center (SITAC). Last but not least, I would like to particularly thank Dr. James O. Jensen for his enthusiastic support of an NRC (National Research Council) Senior Research Associateship award that I received from the US Army Soldier and Biological Command, Edgewood Chemical and Biological Center (ECBC), Aberdeen Proving Ground, MD. This timely support allows me to take advantage of my sabbatical leave to complete this book.

CONTENTS

HYPERSPECTRAL IMAGING

1

INTRODUCTION

Hyperspectral imaging is a fast growing area in remote sensing. It expands and improves capability of multispectral image analysis. Two hyperspectral sensors currently in use and operated in airborne platform are 224-band AVIRIS (Airborne Visible/Infrared Imaging Spectrometer) and 210-band HYperspectral Digital Imagery Collection Experiment (HYDICE). They take advantage of hundreds of contiguous spectral channels to uncover materials that usually cannot be resolved by multispectral sensors. However, this advantage also comes with a price that many unknown signal sources may be also extracted by the sensors with no prior knowledge. In particular, these signal sources may include targets with size smaller than the ground sampling distance (GSD), which are generally embedded in a single pixel and cannot be identified by visual inspection. Their presence can be only measured by their spectral properties. Under these circumstances detection of such small targets cannot be accomplished by classical spatial-based image processing techniques. Instead, it must be carried out at subpixel level. Therefore, one of great challenges for hyperspectral imaging is subpixel detection, which is not treated in standard spatial-based image processing. After a target is detected, the next step is to classify detected targets according to their spatial or spectral properties. However, due to the high spectral resolution of a hyperspectral sensor and a large spatial coverage by GSD, it is often the case that more than one material substance will be present in a single pixel. In this case, a pixel may contain two or more material substances; thus it is no longer a pure pixel. To deal with such a pixel effectively, the pixel must be considered as a mixed pixel wherein several substances are present. This mixing activity further complicates image classification since traditional pure pixel-based classification techniques may not be applicable or effective even if they can be applied. Therefore, another challenging problem is to develop effective techniques for mixed pixel classification. This book is particularly written to aim at these two areas, subpixel detection and mixed pixel classification. Specifically, it is focused on problem solving techniques rather than theoretical treatment. Most techniques developed in this book are based on engineering perspectives and derived from aspects of statistical signal processing. They will be presented in a unifying framework so that readers can follow the flow, easily grasp the ideas and get hands-on algorithmic implementation.

1.1 BACKGROUND

The topics discussed in the book are related to target detection and classification rather than pattern classification as commonly treated in Duda and Hart (1973). The targets referred here are primarily man-made targets or targets whose signatures are spectrally distinct from image background. For this reason, the terminology of target classification used in this book is very different from land-cover classification generally considered in remote sensing literature. In pattern classification a classifier must classify image data into a number of pattern classes, which also include background classes. Although the background knowledge may be obtained directly from the image data in an unsupervised means, it may not be accurate. In some cases, it may not be reliable, particularly, when the targets are relatively small or when the image background is complicated due to the fact that many unknown signal sources can be uncovered by high-resolution sensors. Besides, image background generally varies and is difficult to characterize. As a result, it is nearly impossible to classify image background without complete prior knowledge. On the other hand, in target classification we are generally interested in classification of targets rather than image background. In many situations, we may have prior knowledge about targets that we would like to classify. Under this circumstance, we perform target classification with no need of background knowledge. Additionally, we also make a distinction among detection, classification, discrimination, identification and quantification. For instance, target detection does not necessarily imply target classification, neither does target classification imply target discrimination. A target detector, which achieves 100% detection may result in 0% classification since it may not be able to classify the targets it detected. One such detector is an anomaly detector, which can detect anomalous targets but may not be able to classify the detected anomalies. Similarly, a target classifier may classify targets correctly, but may fail to discriminate all the classified targets. For example, a target classifier can classify vehicles as wheeled and tracked vehicles, but it may be unable to differentiate a jeep from a truck. On the other hand, a target discriminator may discern one target from another, but does not necessarily classify these targets. However, in some applications, researchers may consider target discrimination as a follow-up operation of target classification. In this case, the target discrimination is only applied after targets are classified. A target identifier usually requires a database or spectral library to identify targets of interest. It performs more subtle functionality than detection, discrimination and classification. A target quantifier can perform quantification as a target is detected. The diagram depicted in Fig. 1.1 may help clarify these terminologies. Nevertheless, it should be noted that this differentiation is the author's preference and by no means a standard taxonomy.

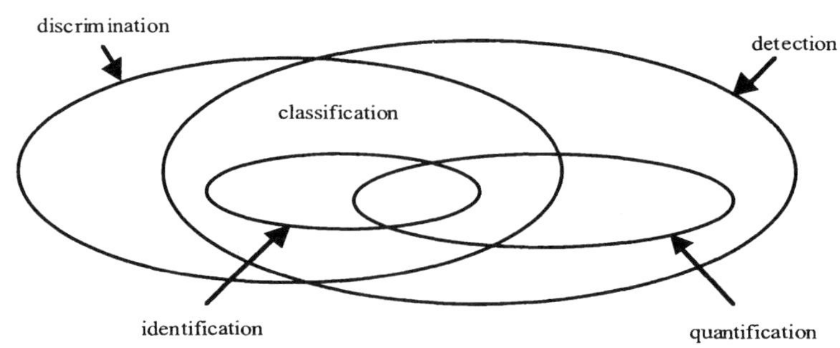

Figure 1.1. Diagram of functional taxonomy

As the book title implies, the primary goal of this book is to develop spectral techniques for subpixel detection and mixed pixel classification. In general, target detection and classification in remotely sensed images can be conducted spatially, spectrally or both. Here, we are specifically interested in non-literal (referred to as spectral) analysis techniques as opposed to literal (spatial-based) processing techniques. Non-literal exploitation is "the process of extracting non-spatial information from digital image data, automatically, semi-automatically, using nontraditional, advanced processing techniques which may employ models, measurements, signatures or other information to perform one or more of the following functions at the requisite levels of specificity: detect, geolocate, classify, discriminate, characterize, identify, quantify, point, track, predict, target or map and chart objects, emissions, activities or events of interest represented in the imagery" (HYMSMO, 1998).

1.2 OUTLINE OF THE BOOK

This book consists of 18 chapters, Chapters 2-7 on subpixel detection, Chapters 8-17 on mixed pixel classification (MPC) and Chapter 18 on conclusions and further topics which are not treated in this book. Each chapter is accompanied by extensive experiments for illustration. Chapters 2-17 are further categorized into five parts. PART I is made up of only one chapter, i.e., Chapter 2, which investigates several hyperspectral measures for spectral characterization. PART II is devoted to subpixel detection, which comprises of five chapters, Chapters 3-7, where a wide range of topics are covered; supervised unconstrained and target abundance-constrained subpixel detection in Chapter 3, supervised target signature-constrained subpixel detection in Chapter 4, unsupervised subpixel detection in Chapter 5, anomaly detection in Chapter 6 and detection sensitivity to different levels of target information and noise in Chapter 7. PART III consists of 2 chapters, Chapters 8-9 on unconstrained linear mixture analysis-based MPC. PART IV extends PART III to constrained linear mixture analysis-based MPC where 3 chapters are included, target abundance-constrained MPC in Chapter 10, target signature-constrained MPC in Chapters 11-12. PART V deals with automatic mixed pixel classification (AMPC) when no *a priori* target knowledge is available. Five chapters are covered in this part, unsupervised MPC in Chapter 13, anomaly classification in Chapter 14, linear spectral random mixture analysis in Chapter 15, projection pursuit in Chapter 16 and estimation of virtual dimensionality in Chapter 17. Finally, Chapter 18 concludes some topics that are not treated in this book.

1.2.1 Stochastic Hyperspectral Measures

A remotely sensed image is actually an image cube with the third dimension specified by spectral wavelengths. As a result, each image pixel is indeed a column vector, of which each component represents a particular spectral band. The spectral information contained in an image pixel vector generally cannot be explored by spatial-based pure pixel classification methods. The first part of this book, Chapter 2 considers some commonly used similarity measures in pattern classification and also develops several information theoretic criteria that can be used to design new hyperspectral measures. The concept of uncertainty is introduced to characterize spectral variations caused by unpredicted sources such as atmospheric and scattering effects,. With this interpretation, the well-established information theory can be readily applied to analysis of spectral properties. In particular, two information theoretic hyperspectral measures, spectral information divergence (SID) and hidden Markov model-based information divergence (HMMID), can be derived from self-information for spectral similarity. They can be viewed as stochastic measures as opposed to deterministic measures commonly used in the literature such as spectral angle mapper (SAM), Euclidean distance. A functional taxonomy of the hyperspectral measures studied in this chapter is depicted in Fig. 1.2.

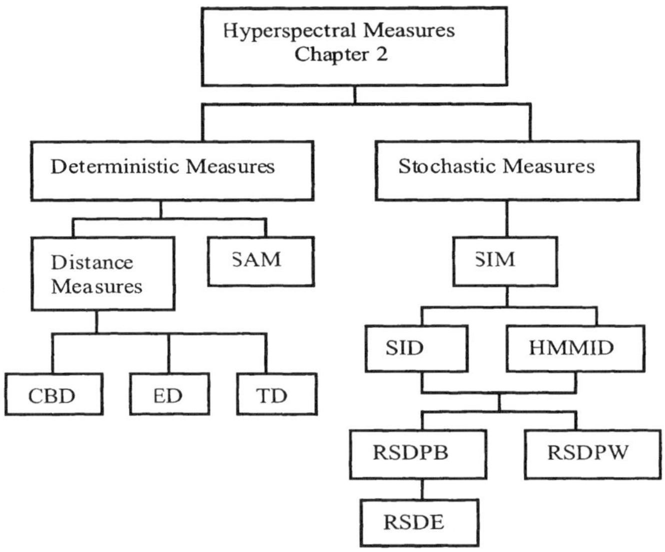

Figure 1.2: Functional taxonomy of hyperspectral measures

1.2.2 Subpixel Detection

The subpixel detection presented in this book is referred to spectral target detection, i.e. non-literal and conducted on a single pixel basis. Three types of subpixel detection are considered, supervised subpixel detection, unsupervised subpixel detection and automatic subpixel detection. Supervised subpixel detection requires *a priori* target knowledge and can be implemented with or without constraints, whereas unsupervised subpixel detection does not require any prior target information where the needed target information can be extracted directly from the image data in an unsupervised manner. Unlike supervised and unsupervised subpixel detection, automatic subpixel detection, generally referred to as anomaly detection does not need target information at all. A functional taxonomy of subpixel detection is delineated in Fig. 1.3.

In Chapter 3 we first start with the simplest subpixel detection technique, orthogonal subspace projection (OSP), which is supervised and unconstrained. From there we consider two supervised partially target abundance-constrained least-squares methods, sum-to-one constrained least-squares (SCLS) and non-negatively constrained least-squares (NCLS). Chapter 4 derives two supervised target signature-constrained methods, constrained energy minimization (CEM) and target-constrained interference-minimized filter (TCIMF). Because it is generally difficult to obtain *a priori* target knowledge for supervised subpixel detection methods, three unsupervised algorithms are introduced in Chapter 5, unsupervised vector quantization (UVQ) algorithm, unsupervised target generation process and least-squared error method. These allow us to extract necessary target information directly from the image data, referred to as *a posteriori* target information to expand ability of the supervised subpixel detection techniques. Since the *a posteriori* target information may not be accurate, another type of subpixel detection, anomaly detection, is considered in Chapter 6, which does not require *a priori* or *a posteriori* target information for subpixel detection. It extracts targets whose spectral signatures are distinct from their surrounding pixels. Since the unsupervised subpixel detection is sensitive to the used target information, whereas anomaly detection is sensitive to noise, Chapter 7 is devoted to an analysis of sensitivity to the level of used target information and noise.

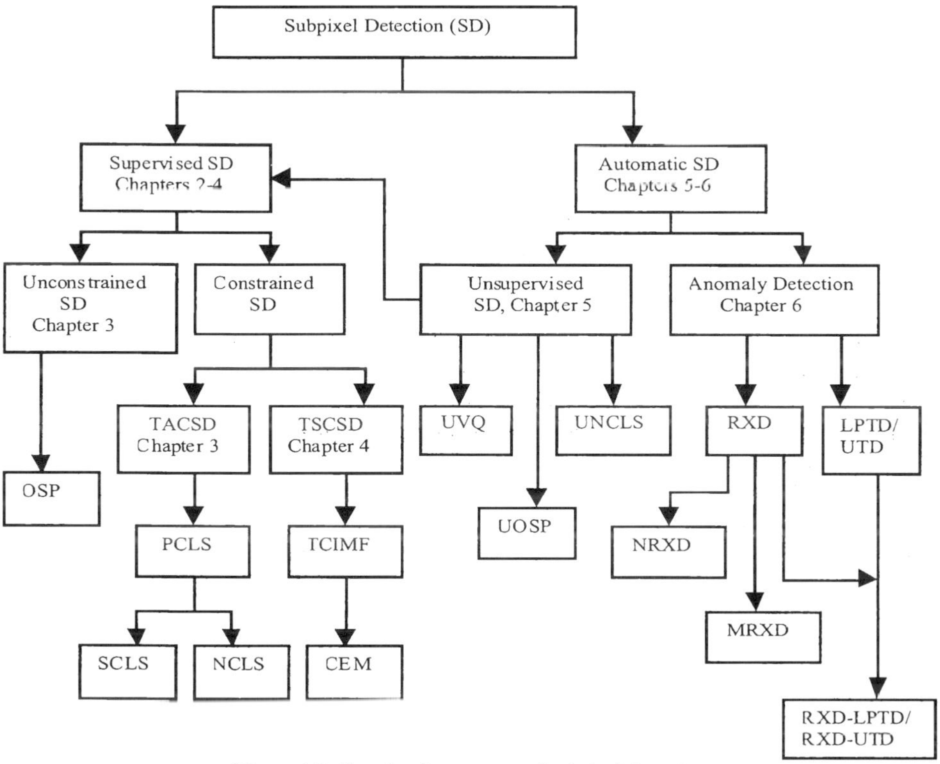

Figure 1.3. Functional taxonomy of subpixel detection

1.2.3 Mixed Pixel Classification (MPC)

Classification in standard image processing is usually performed spatially on a pure pixel basis where objects of interest are classified based on their spatial properties. In a remote sensing image scene, the image pixel is generally composed of different material substances, which cannot be analyzed by spatial-based pure pixel classification methods. In this case, using spectral properties is an effective means to detect and classify the materials present in a single pixel. Mixed pixel classification (MPC) provides such a solution. A commonly used approach to MPC is linear unmixing. It models the spectrum of each image pixel vector as a linear mixture of spectra of targets that are assumed to be present in the image data. The classification is then performed based on estimated abundance fractions of target signatures. As a result, the images produced by

MPC are generally gray scale and represent target abundance fractions. The MPC is then carried out by visual inspection of the target abundance fractional images. This abundance estimation approach is very different from the classical class-membership labeling process carried out by pure pixel classification (PPC) where each pure pixel must be assigned to one and only one class. Compared to PPC, MPC estimates the abundance fraction of each possible target and uses it as a criterion for classification. In this case, a pixel can be classified by MPC into more than one class, a situation which never occurs in PPC. Furthermore, if all the target abundance fractions are non-negative, we can normalize these fractions to unity to form a probability vector. The resulting probabilities can be viewed as *a posteriori* probabilities that provide the likelihood of each target to be assigned to the pixel. With this interpretation, MPC can be thought of as an approach which produced the likelihood of each target to be detected in each image pixel, whereas PPC can be considered a method that thresholds the estimated target abundance fractions into either 1 or 0 where a "1" indicates a target detected and a "0" for target absence. In this book, different approaches to MPC are investigated and a functional taxonomy of MPC is given in Fig. 1.4.

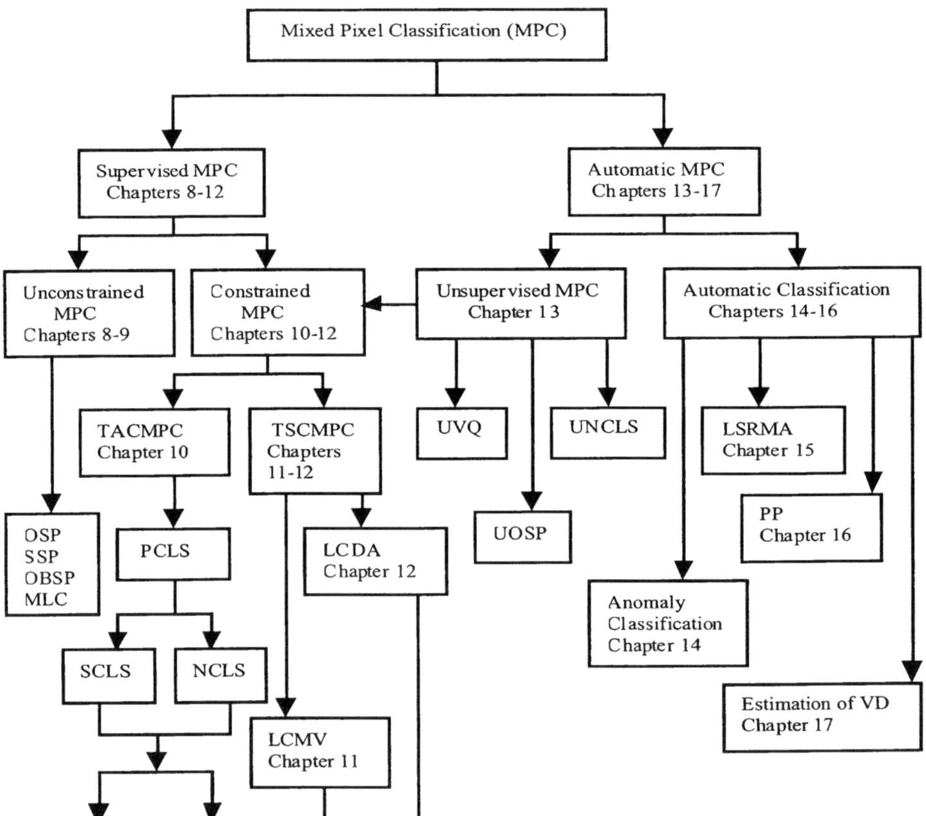

Figure 1.4: Functional taxonomy of mixed pixel classification

1.2.3.1 Unconstrained MPC

The simplest approach to MPC is linear spectral mixture analysis (LSMA), which has been widely used in the past. Due to mathematical tractability most LSMA-based techniques are supervised and unconstrained, meaning that the complete target knowledge must be given and no constraints are imposed on the linear mixture model used in LSMA. One such approach is the Gaussian maximum likelihood classification. It is an *a priori* approach, which assumes that noise in the linear mixture model is Gaussian so that a closed-form solution for MPC can be derived. Recently, an alternative approach, called orthogonal subspace projection (OSP) approach was developed and has received considerable interest in MPC. Unlike Gaussian maximum likelihood classification, the OSP only requires second-order statistics of noise and *a posteriori* target knowledge, which can be obtained from the image data. Chapter 8 considers a class of *a posteriori* OSP-based classifiers, which can be derived from the original OSP classifier developed by Harsanyi and Chang and can be considered as *a priori* OSP. Interestingly, it can be also shown that the *a priori* Gaussian maximum likelihood classifier is essentially an *a posteriori* OSP classifier. In order to evaluate the performance of MPC relative to that of PPC, Chapter 9 conducts a comparative analysis and quantitative study among several commonly used methods, specially a comparison between the OSP-based approach to MPC and Fisher's linear discriminant analysis for PPC. Since MPC basically performs abundance estimation not the way the traditional PPC does, two mixed-to-pure pixel converters (MPCV) are also suggested to convert the abundance fractional images produced by MPC to binary images so that MPC can be reduced to PPC. As a result, MPC can be compared against PPC via MPCV.

1.2.3.2 Constrained MPC

One of main reasons that we are interested in unconstrained MPC is that it has closed-form solutions and is simple and easy to implement. Constrained MPC is considered to be more challenging because constraints generally prohibit MPC from rendering analytical solutions. Two types of constraints are usually imposed. One is referred to as target abundance-constrained MPC (TACMPC), which imposes constraints on target abundance fractions to satisfy desired properties. In this case, the unconstrained LSMA considered in PART III can be extended to constrained LSMA with the abundance sum-to-one constraint (ASC) and abundance non-negativity constraint (ANC) imposed on the linear mixture model used in LSMA. Since no closed-form solutions are generally available, we must rely on numerical algorithms to generate optimal solutions. To meet this need, Chapter 10 presents a least-squares based efficient algorithm, called fully constrained least-squares (FCLS) algorithm. Because the proposed FCLS algorithm is designed for solving fully constrained linear mixing problem, it can be further used for material quantification. One major drawback of LSMA is requirement of complete target knowledge, which must be given *a priori*. This also includes image background signatures. Unfortunately, obtaining such complete knowledge, specifically, image background is very difficult if not impossible. In order to cope with this issue, another type of constrained MPC, referred to as target signature-constrained MPC (TSCMPC) is considered in Chapter 11 where the LCMV detector considered in Chapter 4 is extended to an LCMV classifier. Rather than constraining target abundance fractions as does TACMPC, TSCMPC constrains the directions of target signature vectors. In this case, only target signature vector directions are of interest. The LCMV classifier constrains a set of multiple target signature vectors so that only targets with desired constrained directions will pass through the filter while the output energy caused by signature vectors from other directions will be minimized. This approach is based on the fact that two hyperspectral image pixel vectors, which have similar spectra or point to the same direction will be considered to be in the same target class. Chapter 12 develops an another new TSCMPC approach, linearly constrained discriminant analysis (LCDA) which constrains the directions of target signature vectors in a linear mixture model to being aligned with predetermined coordinates so that the targets of interest can be classified and separated in a way that we desire. Surprisingly, the derived LCDA classifier turns out to be a constrained version of Harsanyi-Chang's OSP classifier in Chapter 8.

1.2.3.3 Automatic Mixed Pixel Classification (AMPC)

The automatic mixed pixel classification (AMPC) considered in this part includes unsupervised MPC and automatic target classification. Like automatic subpixel detection, the former requires *a posteriori* target knowledge that can be obtained directly from the image data by an unsupervised means, while the latter does not need any *a priori* or *a posteriori* target information. The unsupervised MPC can be viewed as an unsupervised extension of supervised MPC considered in PART III and PART IV. Chapter 13 presents two versions of unsupervised MPC, referred to as desired target detection and classification algorithm (DTDCA) which is developed to search specific targets, and automatic target detection and classification algorithm (ATDCA) which is designed for the purpose of surveillance and monitoring. Two types of automatic target classification are of particular interest. One type is anomaly classification in Chapter 14, which extends the ability of anomaly detection considered in Chapter 5 to anomaly classification. Another type discussed in Chapters 15-16 performs target detection and classification simultaneously with no need of any *a priori* or *a posteriori* target information. Chapter 15 borrows the concept of the well-known blind source separation technique, independent component analysis (ICA) to establish a linear spectral random mixture analysis (LSRMA) which models potential target sources as unknown random signal sources. Chapter 16 extends the concept of principal components analysis (PCA) and ICA to projection pursuit (PP) with a project index designed to capture interesting structures of image data. When the projection index is the data variance, PP turns out to be PCA. When the project index measures the statistical independency among all projection images, PP becomes the ICA.

One of most challenging problems encountered in unsupervised target detection is how many targets are assumed to be in image data. This issue is addressed in Chapter 17. Instead of using the traditional terminology, intrinsic dimensionality (ID), we introduce a new definition, virtual dimensionality (VD), which we think it more appropriately reflects the nature of the problem. It is defined as the number of spectrally distinct signal sources present in image data, which may include image endmembers, unidentified signal sources, unknown interferers. Three Neyman-Pearson detection theory-based eigen-thresholding methods are proposed in Chapter 17 to estimate the VD.

1.2.4 Hyperspectral Data to Be Used in the Book

Three data sets will be used for experiments throughout this book. Each of these data sets has a particular purpose of usage. The first data set was presented in (Harsanyi and Chang, 1994) and is AVIRIS reflectance data shown in Fig. 1.5, which will be mainly used for computer simulations. There are five field reflectance spectra, blackbrush, creosote leaves, dry grass, red soil and sagebrush with spectral coverage from 0.4 μm to 2.5 μm and 158 bands after the water bands are removed. The second and third data sets to be used for experiments in this book are AVIRIS and HYDICE data. The AVIRIS image shown in Fig. 1.6(a) is the same image data considered in Harsanyi and Chang (1994) and Chang et al. (1998b) which has 10nm spectral resolution and 20m spatial resolution. There are five target signatures of interest, cinders, rhyolite, playa (dry lake), vegetation and shade. It is a subscene of 200×200 pixels extracted from the upper left corner of the Lunar Crater Volcanic Field in Northern Nye County, Nevada delineated by white lines as shown in Figure 1.6(b).

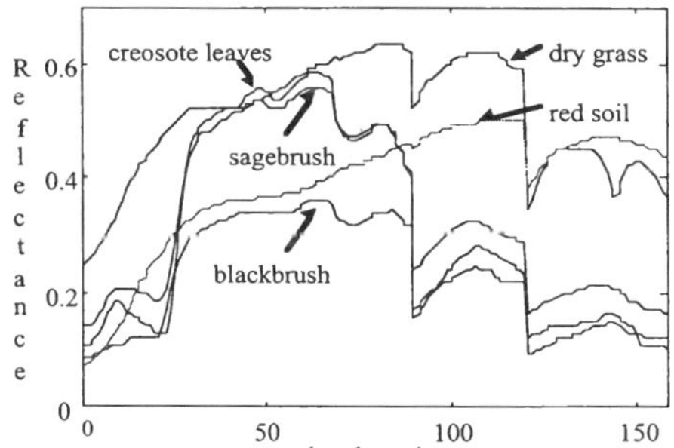

Figure 1.5. Spectra of five AVIRIS reflectances

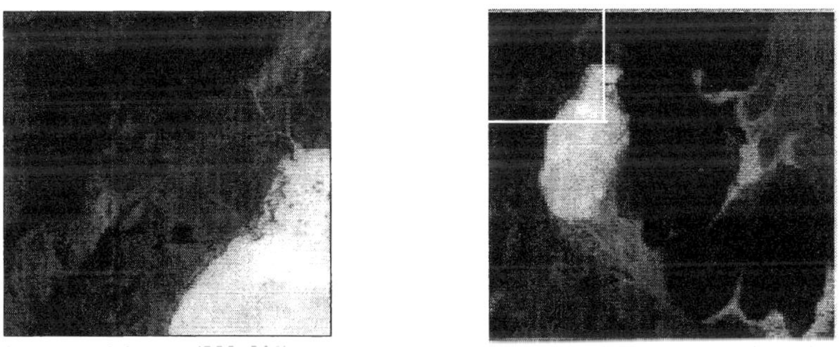
(a) LCVF Subscene (200x200) (b) AVIRIS LCVF scene (512x614)
Figure 1.6. AVIRIS LCVF image

The HYDICE image shown in Fig. 1.7(a) has size of size 64×64 with 15 panels in the scene. Within the scene there has a large grass field background, a forest on the left edge and a barely visible road running one the right edge of the scene. Low signal/high noise bands: bands 1-3 and bands 202-210; and water vapor absorption bands: bands 101-112 and bands 137-153 were removed. The spatial resolution is 1.5m and spectral resolution is 10nm. There are 15 panels located in the center of the grass field and are arranged in a 5×3 matrix as shown in Fig. 1.7(b) which provides the ground truth map of Fig. 1.7(a). Each element in this matrix is a square panel and denoted by p_{ij} with row indexed by $i = 1,\cdots,5$ and column indexed by $j = 1,2,3$. For each row $i = 1,\cdots,5$, the three panels p_{i1}, p_{i2}, p_{i3} were painted by the same material but have three different sizes. $j = 1,2,3$, the five panels p_{1j}, p_{2j}, p_{3j}, p_{4j}, p_{5j} have the same size but were painted by five different materials. It should be noted that the panels in rows 2 and 3 are made by the same material with different paints, so did the panels in rows 4 and 5. Nevertheless, they were still considered as different materials. The sizes of the panels in the first, second and third columns are $3\text{m} \times 3\text{m}$, $2\text{m} \times 2\text{m}$ and $1\text{m} \times 1\text{m}$ respectively. So, the 15 panels have five different materials and three different sizes. Fig. 1.7(b) shows the precise spatial locations of these 15 panels where red pixels (R pixels) are the panel center pixels and the pixels in yellow (Y pixels) are panel pixels mixed with background. The 1.5m-spatial resolution of the image scene suggests that most of the 15 panels are one pixel in size except that p_{21}, p_{31}, p_{41}, p_{51} which are two-pixel panels. Since the size of the panels in the third column is $1\text{m} \times 1\text{m}$, they cannot be seen visually from Fig. 1.7(a) due to the fact that its size is less than the 1.5m pixel resolution.

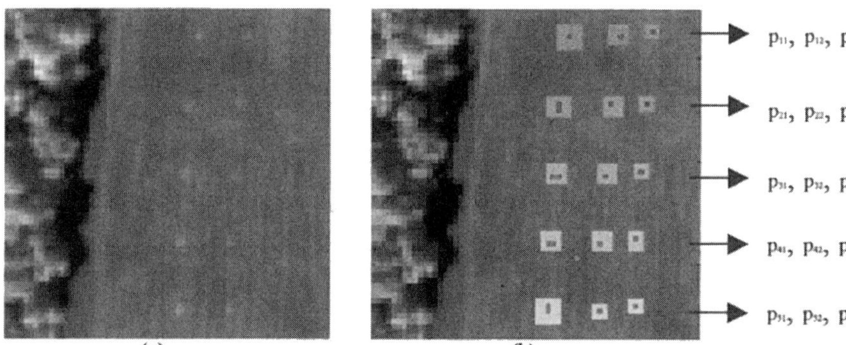

(a) (b)

Figure 1.7. (a) A HYDICE panel scene which contains 15 panels; (b) Ground truth map of spatial locations of the 15 panels

Fig. 1.8 plots the five panel spectral signatures obtained from Fig. 1.7(b), where the i-th panel signature, denoted by Pi was generated by averaging the red panel center pixels in row i. These panel signatures will be used to represent target knowledge of the panels in each row.

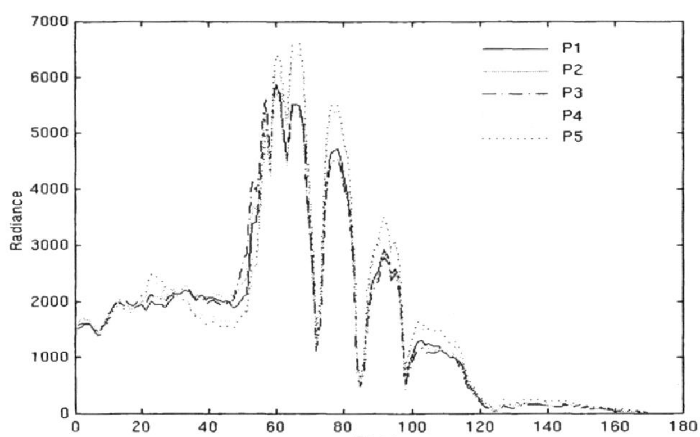

Figure 1.8. Spectra of P1, P2, P3, P4 and P5

1.2.5 Notations to Be Used in the Book

Since this book primarily deals with real hyperspectral data, the image pixels are generally mixed and not necessarily pure, the term "endmember" is not used here;

instead, a general term "signature" or "signature vector" is used. In addition, since we are only interested in target detection and classification, the term "targets" instead of "materials" is also used throughout the book. In order to make a distinction between a target pixel and its spectral signature vector, we use notation $\mathbf{t}$ to represent a target pixel or signature, $\mathbf{r}$ for an image pixel vector and $\mathbf{m}$ for a spectral signature vector. We also use bold upper case for matrices and bold lower case for vectors. The *italic* upper case L will be used for the total number of spectral bands, $\mathbf{K}_{L\times L}$ for the sample spectral covariance matrix and $\mathbf{R}_{L\times L}$ for the sample spectral correlation matrix. Also, $\delta_*(\mathbf{r})$ is used to represent a detector or classifier that operates on an image pixel vector $\mathbf{r}$ where the subscript * in $\delta_*(\mathbf{r})$ specifies what type of a detector or classifier to be used. It should be noted that $\delta_*(\mathbf{r})$ is a real-valued function that takes a form of inner product of a filter vector $\mathbf{w}^*$ with $\mathbf{r}$, that is, $\delta_{\mathbf{w}^*}(\mathbf{r}) = \left(\mathbf{w}^*\right)^T \mathbf{r}$ with the filter vector $\mathbf{w}^*$ specified by a particular detector or classifier. We also use "α" and $\hat{\alpha}$ to represent the abundance vector and its estimate where the notation "hat" over "α" indicates "estimate". Finally, all the acronyms of terminology used in this book are provided in the glossary at the end of the book.

I

HYPERSPECTRAL MEASURES

Spectral angle mapper (SAM) has been widely used as a spectral similarity measure in remote sensing. It measures the angle between two spectral signature vectors, which provides a useful feature for material identification. Euclidean distance (ED) is a distance measure commonly used in spatial-based image processing. It is generally used to measure the closeness between two data samples in terms of distance. When it is applied to spectral pixel vectors, it can be also used to measure the similarity between two spectral pixel vectors. Despite their difference in form, SAM and ED turn out to be a same measure as the angle of two spectral pixels becomes small. A common drawback of SAM and ED is that they are deterministic measures in the sense that they do not provide stochastic description about pixel vectors. As a hyperspectral sensor significantly improves the spectral resolution, its sensitivity to noise and level of information is also increased. Such sensitivity may be better characterized by uncertainty. Using a deterministic measure, such as SAM and ED may not be an effective way to describe random behavior of spectral variability. PART I consists of a single chapter, Chapter 2 that addresses this issue by considering a hyperspectral pixel vector as a random variable with its spectrum as a probability distribution. As a result of this interpretation, a spectral information measure (SIM) can be defined by the concept of self-information borrowed from information theory. With the use of the self-information two information theoretic measures, called spectral information divergence (SID) and hidden Markov model-based information divergence (HMMID) can be derived to measure similarity between hyperspectral image pixel vectors. In the past, many spectral similarity measures developed in pattern classification have been also used for hyperspectral measures. In order to compare one against another, a new definition of relative spectral discriminatory power (RSDPW) is also suggested in Chapter 2. On some occasions we are also interested in identifying a pixel vector of interest using an existing database or a spectral library. For this purpose, a new criterion, called relative spectral discriminatory probability (RSDPB) is also proposed. Finally, an uncertainty measure, called relative spectral discriminatory entropy (RSDE) can be also derived from RSDPB, which can be used to measure the uncertainty of a similarity measure in material identification using a spectral library.

2

HYPERSPECTRAL MEASURES FOR SPECTRAL CHARACTERIZATION

A hyperspectral image can be considered as an image cube where the third dimension is represented by hundreds of contiguous spectral bands. As a result, a hyperspectral pixel is actually a column vector with dimensions equal to the number of spectral bands. Such between-band spectral information is very useful and can be used for spectral characterization. Many measures proposed in signal processing and pattern recognition can be used for this purpose. Nevertheless, most of them are spatial-based measures, and they are not particularly designed to measure spectral properties inherent in a single pixel vector. In this chapter, two new hyperspectral measures, spectral information measure (SIM) (Chang, 2000) and hidden Markov model (HMM)-based spectral measure (Du and Chang, 2001) are presented, both of which are derived from the Kullback-Leibler information distance to capture the spectral variability of a pixel vector. Additionally, spectral information divergence (SID), relative spectral discriminatory probability, relative spectral discriminatory power and relative spectral discriminatory entropy are also introduced to further account for spectral similarity and discrminability among pixel vectors.

2.1 MEASURES OF SPERCTRAL VARIABILITY

A hyperspectral image is generally acquired by hundreds of spectral channels. As a result, a scene pixel vector is usually represented by a column vector, in which each component contains specific spectral information provided by a particular channel. Therefore, a greater number of spectral channels translate more spectral information. This implies that a hyperspectral image pixel vector generally contains more spectral information than does a multispectral image pixel vector. In many situations, such spectral information is valuable and crucial in data analysis. In order to capture and characterize the spectral properties vector provided in a single pixel vector by hundreds bands, two statistical spectral measures are introduced in this section.

2.1.1 Spectral Information Measure (SIM)

The spectral information measure (SIM) is an information theoretic measure that models the spectral band-to-band variability as uncertainty resulting from randomness. It considers each pixel vector as a random variable with the probability distribution obtained by normalizing its spectral histogram to unity. With this interpretation, SIM can measure the spectral variability of a single hyperspectral pixel vector resulting from band-to-band correlation. It not only can describe the randomness of a pixel vector, but also can generate high-order statistics of each pixel vector based on its spectral histogram. So, SIM can be considered as a single pixel vector-based stochastic measure. It is particularly useful for hyperspectral image analysis. This is because the spectral information provided by each hyperspectral image pixel vector can improve material detection, discrimination, classification and identification. However, this advantage is also traded for extraction of many unknown interferers, which can be identified *a priori* (Chang et al., 1998). The effects caused by such interference can be only described by randomness but cannot be characterized deterministically. SIM is designed to meet this need. It can capture the uncertainty created by unknown signal sources in a stochastic manner. Therefore, the higher the data dimensionality, the more the randomness, which enhances the relative effectiveness of SIM. Because SIM is a statistical measure, it can generate higher order statistics that can further be used to characterize spectral variability such as variance (second order statistic used to measure for standard deviation), skewness (third order statistic used to measure for symmetry), kurtosis (fourth order statistic used to measure flatness). This advantage cannot be achieved by any deterministic measure such as Euclidean distance or spectral angle mapper (SAM) (Schowengerdt, 1997).

For a given hyperspectral pixel vector $\mathbf{x} = (x_1, \cdots, x_L)^T$, each component x_l represents a pixel in band image B_l which is acquired by a certain wavelength ω_l in a specific spectral range. Let $\mathbf{s} = (s_1, \cdots, s_L)^T$ be the corresponding spectral signature (i.e., spectrum) of $\mathbf{x}$ where s_l represents its spectral signature of x_l in the form of either radiance or reflectance values. Suppose that $\{\omega_l\}_{l=1}^{L}$ is a set of L wavelengths, each of which corresponds to a spectral band channel. Then $\mathbf{x}$ can be modeled as a random variable by defining an appropriate probability space (Ω, Σ, P) associated with it where Ω is a sample space, Σ is an event space and P is a probability measure. In this case, we let $\Omega = \{\omega_1, \omega_2, \cdots, \omega_{L-1}, \omega_L\}$ be the sample space and Σ be the power set of Ω, i.e., the set of all subsets of Ω and $\mathbf{x}(\omega_l) = s_l$. In order to define a legitimate probability measure P for $\mathbf{x}$, we first assume that all components s_l's associated with $\mathbf{x}$ are nonnegative. This is generally a valid assumption due to the nature of radiance or reflectance. With this assumption, we can normalize x_l's to the range of [0, 1] as follows,

$$p_l = s_l \,/\, \textstyle\sum_{k=1}^{L} s_k \,. \tag{2.1}$$

Using (2.1) we define a probability measure P for $\mathbf{x}$ by

$$P(\{\omega_l\}) = p_l \,. \tag{2.2}$$

The probability vector $\boldsymbol{p} = (p_1, p_2, \cdots, p_L)^T$ is the desired probability distribution of pixel vector $\mathbf{x}$. By means of this probability interpretation, any pixel vector $\mathbf{x} = (x_1, \cdots, x_L)^T$ can be viewed as a single information source with its statistics governed by $\boldsymbol{p} = (p_1, p_2, \cdots, p_L)^T$ defined by (2.1) and (2.2). As a result, these statistics can be used to describe the spectral variability of a pixel vector. For instance, we can define its statistics of different orders, such as mean $\mu(\mathbf{x}) = \sum_{l=1}^{L} p_l s_l$, variance $\sigma^2(\mathbf{x}) = \sum_{l=1}^{L} p_l \left(s_l - \mu(\mathbf{x})\right)^2$, third central moment $\kappa_3(\mathbf{x}) = \sum_{l=1}^{L} p_l \left(s_l - \mu(\mathbf{x})\right)^3$, fourth central moment, $\kappa_4(\mathbf{x}) = \sum_{l=1}^{L} p_l \left(s_l - \mu(\mathbf{x})\right)^4$, etc. Like Taylor's series which is widely used to approximate a deterministic function, we can also use the moment generation equation to fully describe the probabilistic behavior of the spectral signature of each hyperspectral image pixel vector where all moments can be obtained via (2.1) and (2.2).

Since a hyperspectral image pixel vector $\mathbf{x}$ can be considered as an information source described by (2.1) and (2.2), from information theory (Fano, 1960) we can further define its self-information provided by band l by

$$I_l(x) = -\log p_l. \tag{2.3}$$

Using (2.3) the entropy of each hyperspectral image pixel vector $\mathbf{x}$, $H(\mathbf{x})$ can be obtained as

$$H(\mathbf{x}) = -\sum_{l=1}^{L} p_l \log p_l \tag{2.4}$$

which can be used to describe the uncertainty resulting from the pixel vector $\mathbf{x}$.

2.1.2 Hidden Markov Model (HMM)-Based Measure

Another statistical measure makes use of a hidden Markov model (HMM) to capture the unobserved and hidden spectral properties of a hyperspectral image pixel vector. HMM has been widely used in speech recognition (Rabiner and Juang, 1993) to model a speech signal as a doubly stochastic process with a hidden state process that can be only observed through a sequence of observations. Since the temporal variability of a speech signal is similar to the spectral variability of a hyperspectral image pixel vector, a similar idea can be applied to a hyperspectral spectral vector. In this case, a hidden Markov process is used to characterize spectral correlation as well as band-to-band variability with model parameters determined by the spectrum of the pixel vector that forms an observation sequence. In speech processing, the same word spoken in different times generally results in different speech spectra. This is also true for the case that the spectrum of the same material taken at different times varies.

In analogy with SID, we introduce a new HMM-based spectral measure, referred to as HMM information divergence (HMMID) that can be also derived from self-information specified by (2.3) and (2.4) (Du and Chang, 2001).

Let $\mathbf{o} = \left(o_1, o_2, \cdots, o_T\right)$ be an observation process with o_t being the observation taken place at time t and T is the number of observations made in the process. Assume that there are N states denoted by $\{1,2,\cdots,N\}$ and the state at time t is denoted by q_t. Let $\mathbf{A} = \{a_{ij}\}_{1\le i,j\le N}$ be the state transition matrix with a_{ij} given by

$$a_{ij} = P(q_{t+1} = j|q_t = i) \tag{2.5}$$

and $\mathbf{B} = \left[b_j(o_t)\right]_{1\le j\le N, 1\le t\le T}$ be the observation probability density matrix where $b_j(o_t)$ is the probability density of the observation o_t at time t in the j-th state. We further assume that $\pi = (\pi_1, \pi_2, \cdots, \pi_N)$ is the initial state distribution with π_j given by

$$\pi_j = P(q_1 = j). \tag{2.6}$$

So, an HMM can be uniquely defined by a parameter triplet, denoted by $\lambda = (\mathbf{A}, \mathbf{B}, \pi)$ which can be estimated by the Baum-Welch algorithm using the maximum likelihood estimation as follows (Rabiner and Juang, 1993).

Assume that the probability density $b_j(o_t)$ is a Gaussian mixture. It has been shown in Rabiner and Juang (1986) that it is equivalent to a multistage single Gaussian density given by

$$b_j(o_t) = \frac{1}{\sqrt{2\pi\sigma_j^2}} \exp\left[-\frac{(o_t - \mu_j)^2}{2\sigma_j^2}\right] \tag{2.7}$$

where μ_j and σ_j^2 are the mean and variance of the observation o_t in the j-th state respectively.

Now, we define the forward probability $\alpha_j(t)$ as the joint probability of observing the first t observations $o_1, o_2, \cdots, o_t$ in the j-th state at time t. It can be solved by

$$\alpha_j(1) = \pi_j b_j(o_1) \tag{2.8}$$

$$\alpha_j(t) = \left[\sum_{i=1}^{N} \alpha_i(t-1) a_{ij}\right] b_j(o_t) \text{ for } 1 < t \le T. \tag{2.9}$$

Similarly, we define the backward probability $\beta_j(t)$ as the conditional probability of observing the observations $o_{t+1}, o_{t+2}, \cdots, o_T$ given that the state at time t is j. It can be solved by

$$\beta_j(T) = 1 \tag{2.10}$$

$$\beta_j(t) = \sum_{i=1}^{N} a_{ji} b_i(o_{t+1}) \beta_i(t+1) \text{ for } 1 < t \le T. \tag{2.11}$$

Let $\gamma_j(t)$ denote the conditional probability of $b_j(o_t)$ given that the observation $\mathbf{o} = (o_1, o_2, \cdots, o_T)$ and $\xi_{ij}(t)$ denote the conditional probability of a transition from state i to state j at time t+1 given the observation $\mathbf{o} = (o_1, o_2, \cdots, o_T)$. Then they can be calculated as follows

$$\gamma_j(t) = \frac{\alpha_j(t)\beta_j(t)}{\sum_{i=1}^{N} \alpha_i(t)\beta_i(t)} \tag{2.12}$$

$$\xi_{ij}(t) = \frac{\alpha_i(t)a_{ij}b_j(o_{t+1})\beta_j(t+1)}{\sum_{k=1}^{N}\sum_{l=1}^{N} \alpha_k(t)a_{kl}b_l(o_{t+1})\beta_l(t+1)} . \tag{2.13}$$

Using (8) and (9) the μ_j, σ_j^2, a_{ij} and π_j can be estimated by the following equations

$$\hat{\mu}_j(t) = \frac{\sum_{t=1}^{T}\left[\gamma_j(t)o_t\right]}{\sum_{t=1}^{T}\gamma_j(t)} \tag{2.14}$$

$$\hat{\sigma}_j^2(t) = \frac{\sum_{t=1}^{T}\left[\gamma_j(t)\left(o_t - \hat{\mu}_j(t)\right)^2\right]}{\sum_{t=1}^{T}\gamma_j(t)} \tag{2.15}$$

$$\hat{a}_{ij}(t) = \frac{\sum_{t=1}^{T-1}\xi_{ij}(t)}{\sum_{t=1}^{T-1}\gamma_i(t)} \tag{2.16}$$

$$\hat{\pi}_j = \gamma_j(1) . \tag{2.17}$$

As indicated previously, an endmember may be represented by variants of its true spectral signature from pixel vector to pixel vector because of unpredicted mixing occurring in a pixel vector. However, there must have some unobserved properties governed by this particular endmember that can distinguish itself from other endmembers. This spectral characterization is very similar to speech signals where one of their key features is pitches. If we assume the observation sequence $\mathbf{o} = (o_1, o_2, \cdots, o_T)$ is represented by the spectral signature **s** of a pixel vector **x** in a hyperspectral image, we can use the HMM to capture the unobserved and hidden spectral properties of **s**. Let λ_s be the parameter vector used to specify **s** and HMM(λ_s) be the HMM determined by λ_s. In analogy with (2.3) we can define the self-information of **s** provided by HMM(λ_s), denoted by $I_{\mathrm{HMM}(\lambda_s)}(\mathbf{s})$ as follows

$$I_{\mathrm{HMM}(\lambda_s)}(\mathbf{s}) = -1/T\left[\log P(\mathbf{s}|\lambda_s)\right]. \tag{2.18}$$

2.2 SPECTRAL SIMILARITY MEASURES

In the previous section, two new hyperspectral measures were suggested to capture spectral variations of a single pixel vector. In this section, we consider the issue of how to measure the similarity between two pixel vectors. In the past, many pure pixel vector-based similarity measures have been used to evaluate the similarity between two pixel vectors. However, they may not be effective for hyperspectral pixel vectors since they do not take full advantage of band-to-band spectral correlation. In order to make comparison, we describe several commonly used measures which can be used for spectral similarity. First, we assume that there are two pixel vectors $\mathbf{r}_i = (r_{i1}, \cdots, r_{iL})^T$ and $\mathbf{r}_j = (r_{j1}, \cdots, r_{jL})^T$ with their respective spectral signatures given by $\mathbf{s}_i = (s_{i1}, \cdots, s_{iL})^T$ and $\mathbf{s}_j = (s_{j1}, \cdots, s_{jL})^T$.

2.2.1 Commonly Used Measures

Three different types of measures are of interest in pattern classification and can be briefly described as follows.

2.2.1.1 Distance-Based Measures

In multivariate analysis, communications and signal processing, distance-based measures are most commonly used metrics to measure the distance between sample data points. Three metrics that calculate the distance between the spectral signatures of two pixel vectors, $\mathbf{s}_i$ and $\mathbf{s}_j$ can be derived from l_1, l_2, l_∞-norms in real analysis.

- City block distance (CBD) corresponding to l_1-norm

$$\mathrm{CBD}(\mathbf{s}_i, \mathbf{s}_j) = \Sigma_{l=1}^{L} |s_{il} - s_{jl}| \ . \tag{2.19}$$

- Euclidean distance (ED) corresponding to l_2-norm

$$\mathrm{ED}(\mathbf{s}_i, \mathbf{s}_j) = \|\mathbf{s}_i - \mathbf{s}_j\| \equiv \left[\Sigma_{l=1}^{L} (s_{il} - s_{jl})^2\right]^{1/2} \ . \tag{2.20}$$

- Tchebyshev distance (TD) or maximum distance corresponding to l_∞-norm

$$\mathrm{TD}(\mathbf{s}_i, \mathbf{s}_j) = \max_{1 \le l \le L} \left\{ |s_{il} - s_{jl}| \right\} \ . \tag{2.21}$$

2.2.1.2 Orthogonal Projection-Based Measures

Two measures can be derived from orthogonal projection, SAM and orthogonal projection divergence (OPD) (Ren and Chang, 1998).

SAM is a widely used spectral similarity metric in remote sensing. It measures spectral similarity by finding the angle between the spectral signatures of two pixel vectors, $\mathbf{s}_i$ and $\mathbf{s}_j$.

$$\begin{aligned}\mathrm{SAM}(\mathbf{s}_i,\mathbf{s}_j) &\equiv \cos^{-1}\left(\mathbf{s}_i \cdot \mathbf{s}_j \,/\, \|\mathbf{s}_i\|\|\mathbf{s}_j\|\right) \\ &= \cos^{-1}\left(\sum_{l=1}^{L} s_{il}s_{jl} \,/\, \left[\sum_{l=1}^{L} s_{il}^2\right]^{1/2}\left[\sum_{l=1}^{L} s_{jl}^2\right]^{1/2}\right)\end{aligned} \tag{2.22}$$

The concept of the OPD is originated from the orthogonal subspace projection (OSP) (Harsanyi and Chang, 1994). It finds the residuals of orthogonal projections resulting from two pixel vectors, $\mathbf{s}_i$ and $\mathbf{s}_j$ given by

$$\mathrm{OPD}(\mathbf{s}_i,\mathbf{s}_j) = \left(\mathbf{s}_i^T P_{\mathbf{s}_j}^{\perp}\mathbf{s}_i + \mathbf{s}_j^T P_{\mathbf{s}_i}^{\perp}\mathbf{s}_j\right)^{1/2} \tag{2.23}$$

where $P_{\mathbf{s}_k}^{\perp} = \mathbf{I}_{L\times L} - \mathbf{s}_k(\mathbf{s}_k^T\mathbf{s}_k)^{-1}\mathbf{s}_k^T$ for $k = i, j$ and $\mathbf{I}_{L\times L}$ is the $L\times L$ identity matrix.

It is worth noting that if both $\mathbf{s}_i$ and $\mathbf{s}_j$ are normalized to unity, the relationship between $\mathrm{ED}(\mathbf{s}_i,\mathbf{s}_j)$ and $SAM(\mathbf{s}_i,\mathbf{s}_j)$ can be established as follows.

$$\begin{aligned}\mathrm{ED}(\mathbf{s}_i,\mathbf{s}_j) &= \sqrt{2 - 2\langle\mathbf{s}_i,\mathbf{s}_j\rangle} = \sqrt{2\left(1 - \cos\left(\mathrm{SAM}(\mathbf{s}_i,\mathbf{s}_j)\right)\right)} \\ &= 2\sqrt{1 - \cos\left(\mathrm{SAM}(\mathbf{s}_i,\mathbf{s}_j)\right)/2} = 2\sin\left(\mathrm{SAM}(\mathbf{s}_i,\mathbf{s}_j)/2\right)\end{aligned} \tag{2.24}$$

where $\langle\mathbf{s}_i,\mathbf{s}_j\rangle = \sum_{l=1}^{L} s_{il}s_{jl}$.

When $SAM(\mathbf{s}_i,\mathbf{s}_j)$ is small, $2\sin\left(SAM(\mathbf{s}_i,\mathbf{s}_j)/2\right) \approx SAM(\mathbf{s}_i,\mathbf{s}_j)$, in which case $SAM(\mathbf{s}_i,\mathbf{s}_j)$ is nearly the same as $\mathrm{ED}(\mathbf{s}_i,\mathbf{s}_j)$. Fig. 2.1 shows the geometric interpretations of ED and SAM, and their relationship.

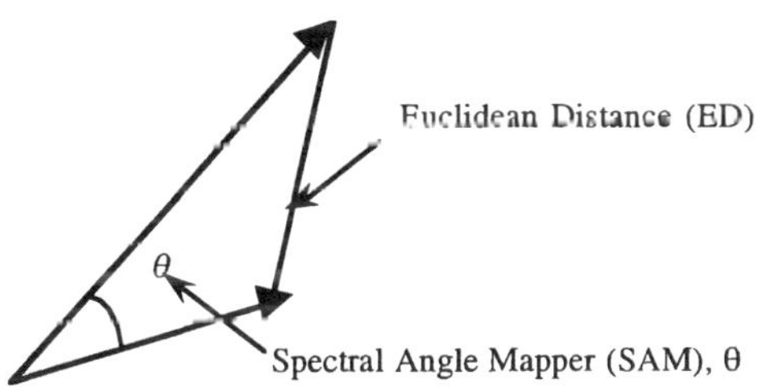

Figure 2.1. Geometric relationship between ED and SAM

2.2.2 Spectral Information Divergence (SID)

A new criterion to measure spectral similarity, called spectral information divergence (SID) is presented in this section (Chang, 2000). It originates from the concept of divergence in information theory and measures the discrepancy of probabilistic behaviors between the spectral signatures of two pixel vectors. In other words, the spectral similarity between two pixel vectors is measured by SID based on the discrepancy between their corresponding spectral signature-derived probability distributions. The idea of using divergence is not new and has been also found in pattern recognition (Tou and Gonzalez, 1974) and band selection (Mausel et al., 1990, Conese and Maselli, 1993, Stearns et al. 1993, Jensen, 1996) such as Jeffries-Matusita measure. But what is new for SID is that it is designed from SIM for spectral similarity. Comparing to SAM and ED that extract geometric features, i.e. angle and spatial distance between two pixel vectors, SID measures the distance between the probability distributions produced by the spectral signatures of two pixel vectors. Accordingly, SID may be more effective than ED and SAM in capturing spectral variability.

Now, following (2.1) we calculate the probability vectors $\boldsymbol{p} = (p_1, p_2, \cdots, p_L)^T$ and $\boldsymbol{q} = (q_1, q_2, \cdots, q_L)^T$ for the spectral signatures of two pixel vectors, $\mathbf{s}_i$ and $\mathbf{s}_j$ where $p_k = s_{ik} / \sum_{l=1}^L s_{il}$ and $q_k = s_{jk} / \sum_{l=1}^L s_{jl}$. So, the self-information provided by $\mathbf{r}_j$ for band l is defined by (2.3) and given by

$$I_l(\mathbf{r}_j) = -\log q_l. \tag{2.25}$$

Using (2.3) and (2.25) we can further define $D_l(\mathbf{r}_i \| \mathbf{r}_j)$(Cover and Thomas, 1993), the discrepancy in the self-information of band l in $\mathbf{r}_j$ relative to the self-information of band l in $\mathbf{r}_i$ by

$$\begin{aligned} D_l(\mathbf{r}_i \| \mathbf{r}_j) &= I_l(\mathbf{r}_j) - I_l(\mathbf{r}_i) = \left(-\log q_l\right) - \left(-\log p_l\right) \\ &= \log\left(p_l / q_l\right) \end{aligned}. \tag{2.26}$$

Averaging $D_l(\mathbf{r}_i \| \mathbf{r}_j)$ in (2.16) over all bands $1 \le l \le L$ with respect to $\mathbf{r}_i$ results in

$$\begin{aligned} D(\mathbf{r}_i \| \mathbf{r}_j) &= \sum_{l=1}^L p_l D_l(\mathbf{r}_i \| \mathbf{r}_j) = \sum_{l=1}^L p_l \left(I_l(\mathbf{r}_j) - I_l(\mathbf{r}_i)\right) \\ &= \sum_{l=1}^L p_l \log\left(p_l / q_l\right) \end{aligned} \tag{2.27}$$

where $D(\mathbf{r}_i \| \mathbf{r}_j)$ is the average discrepancy in the self-information of $\mathbf{r}_j$ relative to the self-information of $\mathbf{r}_i$. In context of information theory $D(\mathbf{r}_i \| \mathbf{r}_j)$ in (2.27) is called the relative entropy of $\mathbf{r}_j$ with respect to $\mathbf{r}_i$ which is also known as Kullback-Leibler information measure, directed divergence or cross entropy (Kullback, 1968). Similarly, we can also define the average discrepancy in the self-information of $\mathbf{r}_i$ relative to the self-information of $\mathbf{r}_j$ by

$$\begin{aligned} D(\mathbf{r}_j \| \mathbf{r}_i) &= \sum_{l=1}^L q_l D_l(\mathbf{r}_j \| \mathbf{r}_i) = \sum_{l=1}^L q_l \left(I_l(\mathbf{r}_i) - I_l(\mathbf{r}_j)\right) \\ &= \sum_{l=1}^L q_l \log\left(q_l / p_l\right) \end{aligned}. \tag{2.28}$$

Summing (2.27) and (2.28) yields spectral information divergence (SID) defined by

$$\mathrm{SID}(\mathbf{r}_i,\mathbf{r}_j) = D(\mathbf{r}_i\|\mathbf{r}_j) + D(\mathbf{r}_j\|\mathbf{r}_i), \tag{2.29}$$

which can be used to measure the spectral similarity between two pixel vectors $\mathbf{r}_i$ and $\mathbf{r}_j$. It should be noted that while $\mathrm{SID}(\mathbf{r}_i,\mathbf{r}_j)$ is symmetric, $D(\mathbf{r}_i\|\mathbf{r}_j)$ is not. This is because $\mathrm{SID}(\mathbf{r}_i,\mathbf{r}_j) = \mathrm{SID}(\mathbf{r}_j,\mathbf{r}_i)$, $D(\mathbf{r}_i\|\mathbf{r}_j) \neq D(\mathbf{r}_j\|\mathbf{r}_i)$. Compared to other popular similarity measures, SID offers a new look of spectral similarity by taking advantage of relative entropy to account for the spectral information provided by each pixel vector.

A measure similar to SID, called Jeffries-Matusita measure that has been used for band selection (Jensen, 1996) can be also defined as Jeffries-Matusita distance (JMD) by

$$\mathrm{JMD}(\mathbf{r}_i,\mathbf{r}_j) = \sqrt{\sum_{l=1}^{L}\left[\sqrt{p_l} - \sqrt{q_l}\right]^2}. \tag{2.30}$$

2.2.3 Hidden Markov Model-Based Information Divergence (HMMID)

Following an approach similar to SID, we can also define an HMM-based information divergence for two pixel vectors. Suppose that a hyperspectral pixel vector $\mathbf{s}_k$ specified by HMM($\lambda_{\mathbf{s}_k}$) can be also modeled by another HMM(λ). In this case, the information discrepancy of $\mathbf{s}_k$ provided by HMM($\lambda_{\mathbf{s}_k}$) and HMM(λ) is the entropy of HMM(λ) relative to HMM($\lambda_{\mathbf{s}_k}$) and is given by

$$\begin{aligned} J_{\mathrm{HMM}}(\lambda_{\mathbf{s}_k};\lambda) &= I_{\mathrm{HMM}(\lambda_{\mathbf{s}_k})}(\mathbf{s}_k) - I_{\mathrm{HMM}(\lambda)}(\mathbf{s}_k) \\ &= 1/T\left[\log P(\mathbf{s}_k|\lambda) - \log P(\mathbf{s}_k|\lambda_{\mathbf{s}_k})\right] \end{aligned}. \tag{2.31}$$

Using (2.31) we can further define an HMM-based information distance between two hyperspectral pixel vectors $\mathbf{r}_i = (r_{i1},\cdots,r_{iL})^T$ and $\mathbf{r}_j = (r_{j1},\cdots,r_{jL})^T$. Let $\mathbf{s}_i = (s_{i1},\cdots,s_{iL})^T$ and $\mathbf{s}_j = (s_{j1},\cdots,s_{jL})^T$ be their respective spectral signatures specified by their associated hidden Markov models, HMM($\lambda_{\mathbf{s}_i}$) and HMM($\lambda_{\mathbf{s}_j}$). An HMM information divergence (HMMID) between $\mathbf{r}_i$ and $\mathbf{r}_j$ is defined by Du and Chang (2001)

$$\mathrm{HMMID}(\mathbf{r}_i;\mathbf{r}_j) = J_{\mathrm{HMM}}(\lambda_{\mathbf{s}_i};\lambda_{\mathbf{s}_j}) + J_{\mathrm{HMM}}(\lambda_{\mathbf{s}_j};\lambda_{\mathbf{s}_i}). \tag{2.32}$$

2.3 MEASURES OF SPECTRAL DISCRIMINABILITY

In Section 2.2, spectral similarity measures between two pixel vectors were developed. When there are more than two pixel vectors, how do we discriminate one from another?

In this section, we study three measures that can be used for spectral discriminability among a set of pixel vectors (Chang, 2000).

2.3.1 Relative Spectral Discriminatory ProBability (RSDPB)

In many remote sensing applications it is often the case that we are required to identify a pixel vector of interest using an existing spectral library or data base. Under this circumstance, we are interested in how much likelihood of a particular spectral signature **t** will be identified by a selective set of spectral signatures, Δ, which serves as a spectral database or spectral library. To meet this need, a new criterion, called relative spectral discriminatory probability (RSDPB) is proposed. It calculates the discriminatory probabilities of all the spectral signatures in a spectral library or database relative to the spectral signature of a pixel vector to be identified. The resulting probabilities show the likelihood of the pixel vector to be identified by the materials in the library or database. Although SAM has been also widely used for this purpose, it is shown in experiments that SID seems to have advantages over SAM in characterizing spectral similarity and variability. As explained previously, this may be due to the fact that SAM only preserves the angle between two pixel vectors rather than their random behaviors in nature. Since remotely sensed image pixel vectors are generally corrupted by noise and other unknown factors during acquisition, using information theoretic-based measures may be a better alternative to traditional measures, which allow each pixel vector to have a certain degree of spectral variations.

Let $\{\mathbf{s}_k\}_{k=1}^{K}$ be K spectral signatures in the set Δ which can be considered as a database and **t** be any specific target spectral signature to be identified using Δ. We define the spectral discriminatory probabilities of all $\mathbf{s}_k$'s in Δ relative to **t** as follows.

$$p_{\mathbf{t},\Delta}(k) = m(\mathbf{t},\mathbf{s}_k) \,/\, \textstyle\sum_{j=1}^{L} m(\mathbf{t},\mathbf{s}_j) \text{ for } k = 1,\cdots,K \tag{2.33}$$

where $\sum_{j=1}^{L} m(\mathbf{t},\mathbf{s}_j)$ is a normalization constant determined by **t** and Δ. The resulting probability vector $\mathbf{p}_{\mathbf{t},\Delta} = (p_{\mathbf{t},\Delta}(1), p_{\mathbf{t},\Delta}(2),\cdots,p_{\mathbf{t},\Delta}(K))^T$ is called relative spectral discriminatory probability (RSDPB) of Δ with respect to **t** or spectral discriminatory probability vector of Δ relative to **t**. Then, using (2.33) we can identify **t** via Δ by selecting the one with the smallest relative spectral discriminability probability. If there is a tie, either one can be used to identify **t**. Fig. 2.2 shows a graphical representation of RSDPB of Δ with respect to **t**.

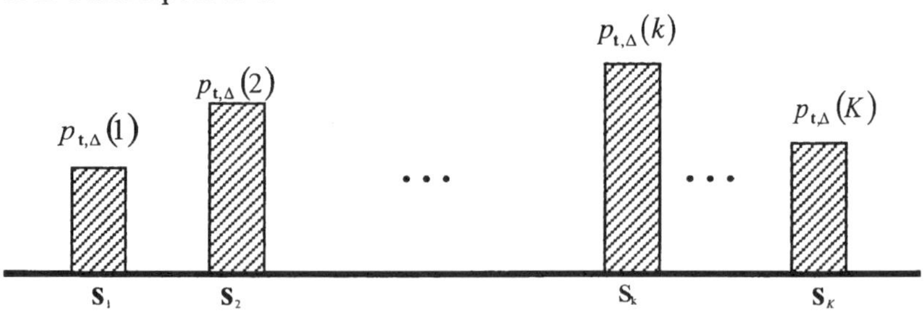

Figure 2.2. Graphical representation of RSDPB, $\mathbf{p}_{\mathbf{t},\Delta}$ of Δ with respect to **t**

2.3.2 Relative Spectral Discriminatory PoWer (RSDPW)

In the previous section, we focused on the similarity between the spectral signatures of two pixel vectors. However, if we are given two spectral similarity measures, how do we evaluate which one is more effective than the other? In order to cope with this issue, a new criterion, called relative spectral discriminatory power (RSDPW) is suggested in this section. It is designed based on the power of discriminating one pixel vector from another relative to a reference pixel vector.

Assume that $m(\cdot,\cdot)$ is any given hyperspectral measure. Let $\mathbf{d}$ be the spectral signature of a reference pixel vector and $\mathbf{s}_i$, $\mathbf{s}_j$ be the spectral signatures of any pair of two pixel vectors. The RSDPW of $m(\cdot,\cdot)$, denoted by $\text{RSDPW}_m(\mathbf{s}_i,\mathbf{s}_j;\mathbf{d})$ is defined by

$$\text{RSDPW}_m(\mathbf{s}_i,\mathbf{s}_j;\mathbf{d}) = \max\left\{m(\mathbf{s}_i,\mathbf{d})\,/\,m(\mathbf{s}_j,\mathbf{d}), m(\mathbf{s}_j,\mathbf{d})\,/\,m(\mathbf{s}_i,\mathbf{d})\right\}. \quad (2.34)$$

More precisely, $\text{RSDPW}_m(\mathbf{s}_i,\mathbf{s}_j;\mathbf{d})$ selects as the discriminatory power of $m(\cdot,\cdot)$ the maximum of two ratios, ratio of $m(\mathbf{s}_i,\mathbf{d})$ to $m(\mathbf{s}_j,\mathbf{d})$ and ratio of $m(\mathbf{s}_j,\mathbf{d})$ to $m(\mathbf{s}_i,\mathbf{d})$. The $\text{RSDPW}_m(\mathbf{s}_i,\mathbf{s}_j;\mathbf{d})$ defined by (2.34) provides a quantitative index of spectral discrimination capability of a specific hyperspectral measure $m(\cdot,\cdot)$ between two spectral signatures $\mathbf{s}_i$, $\mathbf{s}_j$ relative to $\mathbf{d}$. Obviously, the higher the $\text{RSDPW}_m(\mathbf{s}_i,\mathbf{s}_j;\mathbf{d})$ is, the better discriminatory power the $m(\cdot,\cdot)$ is. In addition, $\text{RSDPW}_m(\mathbf{s}_i,\mathbf{s}_j;\mathbf{d})$ is symmetric and bounded below by one, i.e., $\text{RSDPW}_m(\mathbf{s}_i,\mathbf{s}_j;\mathbf{d}) \geq 1$ with equality if and only if $\mathbf{s}_i = \mathbf{s}_j$.

2.3.3 Relative Spectral Discriminatory Entropy (RSDE)

Since $\mathbf{p}_{\mathbf{t},\Delta} = (p_{\mathbf{t},\Delta}(1), p_{\mathbf{t},\Delta}(2), \cdots, p_{\mathbf{t},\Delta}(K))^T$ given by (2.33) is the relative spectral discriminability probability vector of $\mathbf{t}$ using a selective set of spectral signatures, $\Delta = \{\mathbf{s}_k\}_{k=1}^K$, we can further define the relative spectral discriminatory entropy (RSDE) of the spectral signature $\mathbf{t}$ with respect to the set Δ, denoted by $H_{\text{RSDE}}(\mathbf{t};\Delta)$ by

$$H_{\text{RSDE}}(\mathbf{t};\Delta) = -\textstyle\sum_{k=1}^K p_{\mathbf{t},\Delta}(k)\log p_{\mathbf{t},\Delta}(k). \quad (2.35)$$

Equation (2.35) provides an uncertainty measure of identifying $\mathbf{t}$ resulting from using $\Delta = \{\mathbf{s}_k\}_{k=1}^K$. A higher $H_{\text{RSDE}}(\mathbf{t};\Delta)$ may have a less chance to identify $\mathbf{t}$. For example, if $\mathbf{p}_{\mathbf{t},\Delta}$ is uniformly distributed, i.e., $p_{\mathbf{t},\Delta}(k) = 1/K$ for $k = 1,\cdots,K$, then $H_{\text{RSDE}}(\mathbf{t};\Delta)$ achieves its maximum, $\log K$, namely, $H_{\text{RSDE}}(\mathbf{t};\Delta) \leq \log K$. This is the worst scenario of material identification. In other words, when the probability of using any $\mathbf{s}_k$ in Δ to identify $\mathbf{t}$ is equally likely, there is no way for us to know which signature is more likely to be $\mathbf{t}$. In this case, any signature can be used to identify $\mathbf{t}$. However, if $H_{\text{RSDE}}(\mathbf{t};\Delta)$ is small, it only provides information that some candidates in Δ are more likely to be used to identify $\mathbf{t}$, but it does not necessarily mean that their relative discriminability rates are high. As an illustrative example, we consider a hypothetical case that the relative spectral discriminability probabilities for two signatures $\mathbf{u}$, $\mathbf{v}$ in Δ are very small with equal probability 2^{-4}, while others are all zeros except for one signature $\mathbf{s}_k \in \Delta$ with

$p_{\mathbf{t},\Delta}(k) = 14/16$. The RSDE $H_{\text{RSDE}}(\mathbf{t};\Delta) = 1/2 + 14/16\left(4 - \log_2 14\right)$ bits, which is 0.6686, bits less than one bit. So, the chance to identify **t** must be good. However, this is only partly true in the sense that we may conclude that **t** must be either **u** or **v** but there is no way for us to know which one would be since their relative spectral discriminability probabilities are equal to 2^{-4}. Under this circumstance, one of them can be used to identify **t**.

2.4 EXPERIMENTS

Two sets of hyperspectral image data, AVIRIS and HYDICE will be used for experiments. Since the performance of CBD, TD, OPD and JMD is very similar to that of ED and SAM according to experiments conducted based on these two data sets, only four representative spectral measures, ED, SAM, SID and HMMID, one from each category described in Section 2.2 are selected for performance evaluation.

2.4.1 AVIRIS Data

The data set to be used in this section is Airborne Visible/Infrared Imaging Spectrometer (AVIRIS) reflectance data shown in Fig. 1.5. For spectral variability, Tables 2.1 tabulates the statistics generated by SIM for these five signatures up to four moments (not central moments) using (2.1) and (2.2) where μ_i is the i-th moment, and Fig.2.3 plots values in Table 2.1 for graphical interpretation.

Table 2.1. Four moments generated by SIM for the five signatures in Fig. 1.5

	μ_1	μ_3	μ_3	μ_4
blackbrush	0.2670	0.0786	0.0245	0.0079
creosote leaves	0.4091	0.1941	0.0982	0.0512
drygrass	0.5310	0.2905	0.1628	0.0929
redsoil	0.4202	0.1828	0.0812	0.0366
sagebrush	0.3979	0.1774	0.0846	0.0419

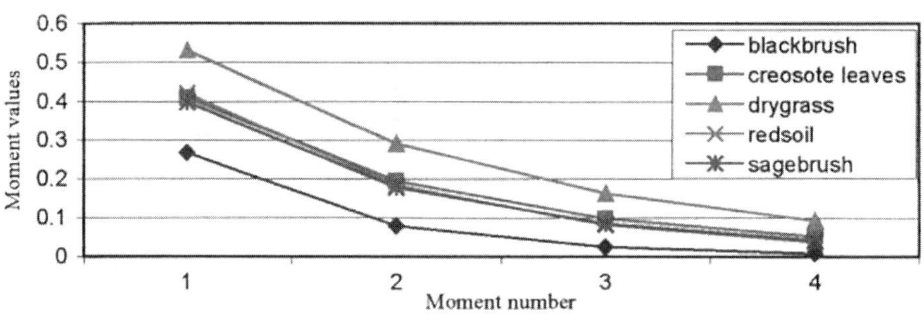

Figure 2.3. Plots of Table 2.1

Table 2.2 also tabulates the self-information of these five signatures using SID specified by (2.3) and HMM specified by (2.18) respectively and Figure 2.4 is their graphical plots.

Table 2.2. Self-information of the five signatures in Fig. 1.5 using SIM and HMM

	blackbrush	creosote leaves	drygrass	redsoil	sagebrush
SID	7.1814	7.0564	7.2740	7.2356	7.1709
HMM	3.5639	3.3771	3.9115	3.6489	3.5959

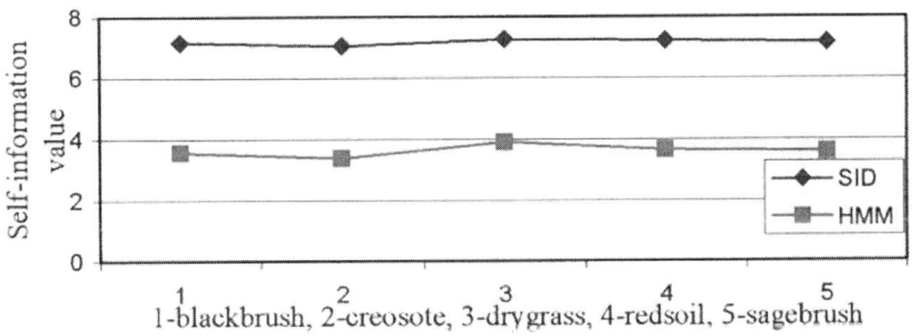

Figure 2.4. Plots of Table 2.2

As we can see in Tables 2.1 and 2.2, the HMM-based measure offers more valuable information than SID does. For example, SID produces very close values of self-information for all the five signatures. On the contrary, HMM categorizes the values of self-information of these five signatures into three classes, one for three signatures: blackbrush, creosote leaves and sagebrush, one for red soil and another for dry grass. From Fig. 1.5 we can see that blackbrush (open circle) creosote leaves (asterisked line) and sagebrush (diamonded line) have very similar spectra while red soil and dray grass are very distinct. This fact is also reflected in Table 2.1 where all their four moments generated by SIM are very close. However, it is very difficult to determine which signature is more close to another by visual inspection from the figure.

Table 2.3. Similarity values produced by ED among the five signatures in Fig. 1.5

	blackbrush	creosote leaves	drygrass	redsoil	sagebrush
blackbrush	0	0.1765	0.2568	0.4031	0.0681
Creosote leaves		0	0.4182	0.5637	0.1288
drygrass			0	0.2175	0.2957
redsoil				0	0.4477
sagebrush					0

Table 2.4. Similarity values produced by SAM among the five signatures in Fig. 1.5

	blackbrush	Creosote leaves	drygrass	redsoil	sagebrush
blackbrush	0	0.1767	0.2575	0.4058	0.0681
Creosote leaves		0	0.4213	0.5714	0.1289
drygrass			0	0.2179	0.2968
redsoil				0	0.4515
sagebrussh					0

Table 2.5. Similarity values produced by SID among the five signatures in Fig. 1.5

	blackbrush	creosote leaves	drygrass	redsoil	sagebrush
blackbrush	0	0.0497	0.0766	0.1861	0.0063
creosote leaves		0	0.2298	0.4154	0.0303
drygrass			0	0.0640	0.0973
redsoil				0	0.2340
sagebrush					0

Table 2.6. Similarity values produced by HMMID among the five signatures in Fig. 1.5

	blackbrush	creosote leaves	drygrass	redsoil	sagebrush
blackbrush	0	1.5390	3.6717	3.9182	0.2263
creosote leaves		0	6.2549	6.7333	1.4102
drygrass			0	1.1302	4.0073
redsoil				0	3.4166
sagebrush					0

Tables 2.3-2.6 are produced for spectral similarity among these five signatures by ED, SAM, SID and HMMID using (2.20), (2.22), (2.29) and (2.32) respectively. According to Tables 2.3-2.6, the smaller value between two signatures is, the more similar the two signatures are. The results produced by ED and SAM were nearly the same because of (2.24). Moreover, as we can see, blackbrush is closest to sagebrush, while creosote leaves is closest to sagebrush. If we examine the last column under sagebrush in Tables 2.3-2.6, it found that sagebrush is more closer to creosote leaves than to blackbrush. The similarity values produced by ED and SAM between blackbrush and sagebrush was about twice between creosote leaves. The similarity values produced by SID between blackbrush and sagebrush was about five times between creosote leaves. The similarity values produced by HMMID between blackbrush and sagebrush was about six times between creosote leaves. For signatures whose spectra are dissimilar, HMMID produced even much greater values than do other three measures. For example, the similarity values between red soil and blackbrush, between red soil and creosote leaves, between red soil and sagebrush were 3.9182, 6.7333 and 3.4166 respectively which are much greater values compared to their counterparts produced by ED, SAM and SID. An assessment based on Tables 2.3-2.6 may be subjective and difficult to compare the discriminatory power of the four measures. In order to better illustrate the results in Tables 2.3-2.6, we also plot their values in Fig. 2.5 where B,C,D,R,S represent blackbrush, creosote leaves, dray grass, red soil and sagebrush respectively where the plots produced by ED and SAM were nearly identical.

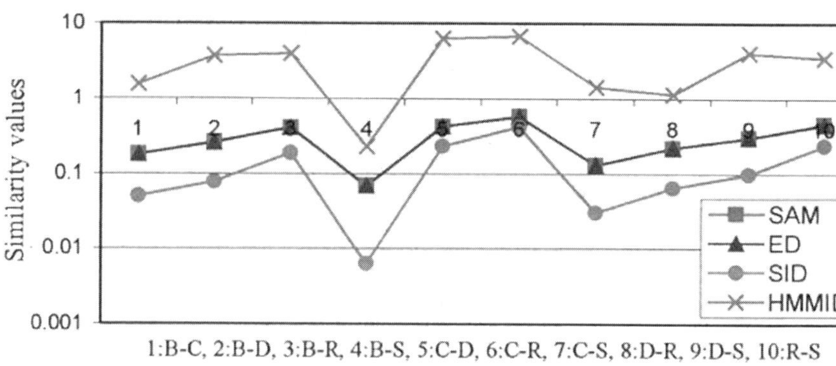

Figure 2.5. Plots of Tables 2.3-2.6

However, it should be noted that the magnitude of the similarity values in Fig. 2.5 does not imply the discriminatory power of each measure. In this case, we calculate RSDPW for ED, SAM, SID and HMMID to evaluate their spectral discriminatory power relative to a reference signature $\mathbf{d}$, which is chosen to be the red soil. As we know, the signature of dry grass is very close to that of red soil. So, the power of HMMID discriminating $\mathbf{s}_i$ = dry grass from the other three signatures $\mathbf{s}_j$ = blackbrush, creosote leaves, sagebrush, were given by 3.4665, 5.9571, 3.0227 respectively. The values of RSDPW produced by HMMID are much greater than those produced by ED, SAM and SID. This implies that the signature of dry grass is very different from those of blackbrush, creosote leaves and sagebrush.

Comparing $\text{RSDPW}_{\text{HMMID}}(\mathbf{s}_i = \text{dry grass}, \mathbf{s}_j = \text{blackbrush}; \mathbf{d} = \text{red soil}) \approx 63$ to $\text{RSDPW}_{\text{HMMID}}(\mathbf{s}_i = \text{dry grass}, \mathbf{s}_j = \text{sagebrush}; \mathbf{d} = \text{red soil}) \approx 69$ suggests that both blackbrush and sagebrush must have very similar signatures, which is indeed the case. This evidence is also provided by finding RSDPWs of blackbrush and sagebrush with respect to the reference signature **d** specified by red soil.

$$\text{RSDPW}_{\text{SAM}}(\mathbf{s}_i = \text{blackbrush}, \mathbf{s}_j = \text{sagebrush}; \mathbf{d} = \text{red soil}) = 1.1126 \quad (2.36)$$

$$\text{RSDPW}_{\text{ED}}(\mathbf{s}_i = \text{blackbrush}, \mathbf{s}_j = \text{sagebrush}; \mathbf{d} = \text{red soil}) = 1.1106 \quad (2.37)$$

$$\text{RSDPW}_{\text{SID}}(\mathbf{s}_i = \text{blackbrush}, \mathbf{s}_j = \text{sagebrush}; \mathbf{d} = \text{red soil}) = 1.2576 \quad (2.38)$$

$$\text{RSDPW}_{\text{HMMID}}(\mathbf{s}_i = \text{blackbrush}, \mathbf{s}_j = \text{sagebrush}; \mathbf{d} = \text{red soil}) = 1.1468 \quad (2.39)$$

which also show how close the signatures of blackbrush and sagebrush are and is smaller than 1.1106, 1.1126 and 1.2576 produced by ED, SAM and SID respectively.

Tables 2.7-2.10 tabulate the RSDPWs of ED, SAM, SID and HMMID respectively with red soil used as a reference signature **d** where Fig. 2.6 plots their corresponding values.

Table 2.7. RSDPW of ED using red soil as a reference signature **d**

	blackbrush	creosote leaves	drygrass	sagebrush
blackbrush	1	1.3984	1.8533	1.1106
creosote leaves		1	2.5917	1.2591
drygrass			1	2.0584
sagebrush				1

Table 2.8. RSDPW of SAM using red soil as a reference signature **d**

	Blackbrush	creosote leaves	drygrass	sagebrush
blackbrush	1	1.4081	1.8623	1.1126
creosote leaves		1	2.6223	1.2656
drygrass			1	2.0721
sagebrus				1

Table 2.9. RSDPW of SID using red soil as a reference signature **d**

	blackbrush	creosote leaves	drygrass	sagebrush
blackbrush	1	2.2321	2.9078	1.2574
creosote leaves		1	6.4906	1.7752
drygrass			1	3.6562
sagebrush				1

Table 2.10. RSDPW of HMMID using red soil as a reference signature **d**

	blackbrush	creosote leaves	drygrass	sagebrush
blackbrush	1	1.7185	3.4665	1.1468
creosote leaves		1	5.9571	1.9708
drygrass			1	3.0227
sagebrush				1

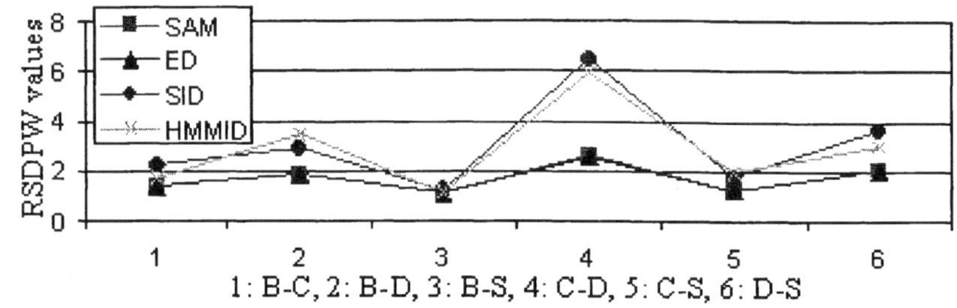

Figure 2.6. Plots of Tables 2.7-2.10

From Fig. 2.6 it is obvious that the stochastic information measures, HMMID and SID were more effective than deterministic measures ED and SAM. In order to evaluate which measure is more effective in terms of spectral discriminability, a mixed spectral signature was randomly generated to be used as a target signature, **t** for identification. It was composed of 0.1055 blackbrush, 0.0292 creosote leaves, 0.0272 dry grass, 0.7588 red soil and 0.0974 sagebrush. It should be noted that the **t** was generated randomly, not intended for a particular preference. From Tables 2.1-2.4, the spectrum of red soil is very similar to that of dry grass. Using (2.33) Table 2.11 tabulates the RSDPB of ED, SAM, SID and HMMID where Fig. 2.7 also plots the values of Table 2.11.

Table 2.11. RSDPB produced by ED, SAM, SID and HMMID with **t** chosen to be a mixture of 0.1055 blackbrush, 0.0292 creosote leaves, 0.0272 dry grass, 0.7588 red soil and 0.0974 sagebrush

	blackbrush	creosote leaves	drygrass	redsoil	sagebrush
ED	0.2215	0.3417	0.1049	0.0773	0.2547
SAM	0.2212	0.3430	0.1044	0.0769	0.2546
SID	0.1897	0.4933	0.0588	0.0112	0.2500
HMMID	0.2395	0.4911	0.0511	0.0036	0.2147

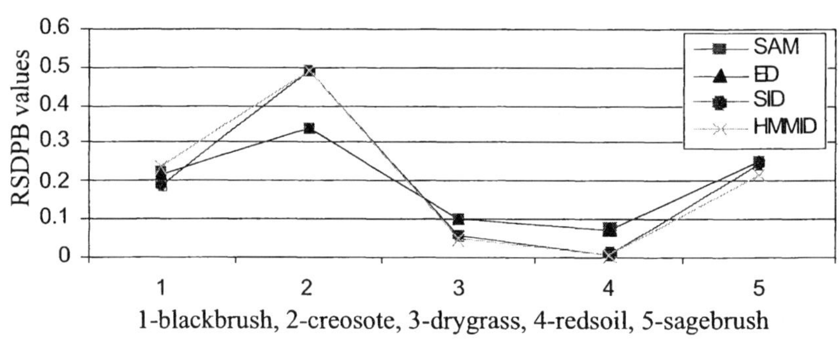

Figure 2.7. Plot of Table 2.11

According to Table 2.11, the ratio of using ED and SAM to identify **t** as red soil to as dry grass was $0.1049{:}0.0773 \approx 0.1044{:}0.0769 \approx 1.36$. Compared to ED and SAM, SID and HMMID yielded $0.0588{:}0.0112 \approx 5.25$ and $0.0511{:}0.0036 \approx 14.19$ respectively. A similar observation from Fig. 2.7 also shows that SID and HMMID were more effective than other three measures in identifying **t** as red soil with nearly zero value of spectral discriminatory probability at red soil. Table 2.12 calculated RSDE of ED, SAM, SID and HMMID using Table 2.11 where HMMID produced the least entropy with Table 2.13 produced by the RSDE results of HMMID using different numbers of states.

Table 2.12. RSDE of Table 2.11 produced by ED, SAM, SID and HMMID

	ED	SAM	SID	HMMID
RSDE	1.4835	1.4822	1.2274	1.1940

Table 2.13. RSDEs resulting from different number of states

# of states	3	4	5	6	7	8	9	10
RSDE	1.1940	1.1939	1.1938	1.1936	1.1932	1.1943	1.1941	1.1941

It was found that RSDE decreased slightly as the number of states increased. However, a high number of states generally increases computational complexity of parameter estimation process significantly. So, according to our experiments, a reasonable number of states ranges from 4 to 6 that allows us to achieve good compromise.

2.4.2 HYDICE Data

Unlike the AVIRIS data studied in the previous section, the HYDICE data used in the following experiments were directly extracted from a HYDICE image scene of size 64×64 shown in Fig. 1.6(a). The spectra of P1, P2, P3, P4 and P5 in Fig. 1.7(c) form a data set $\Delta = \{P1, P2, P3, P4, P5\}$ that would be used for panel identification. Tables 2.14-2.17 tabulate the similarity values resulting from ED, SAM, SID and HMMID.

Table 2.14. Similarity values produced by ED among the five panel signatures in Fig. 2.5

	P1	P2	P3	P4	P5
P1	0	1301.6	2033.3	4107.3	4831.6
P2		0	1340.4	5064.1	5733.0
P3			0	5434.1	5968.7
P4				0	1125.4
P5					0

Table 2.15. Similarity values produced by SAM among the five panel signatures in Fig. 2.5

	P1	P2	P3	P4	P5
P1	0	0.0435	0.0673	0.1144	0.1240
P2		0	0.0430	0.1479	0.1567
P3			0	0.1652	0.1710
P4				0	0.0248
P5					0

Table 2.16. Similarity values produced by SID among the five panel signatures in Fig. 2.5

	P1	P2	P3	P4	P5
P1	0	0.0039	0.0086	0.0233	0.0313
P2		0	0.0033	0.0385	0.0484
P3			0	0.0476	0.0570
P4				0	0.0025
P5					0

Table 2.17. Similarity values produced by HMMID among the five panel signatures in Fig. 2.5

	P1	P2	P3	P4	P5
P1	0	0.0255	0.0291	0.2935	0.4798
P2		0	0.0215	0.2891	0.4483
P3			0	0.3590	0.5483
P4				0	0.0186
P5					0

It is interesting to note that unlike Tables 2.3-2.4 which show that ED and SAM generated very close values, the values generated by ED and SAM in Tables 2.14-2.15 are very different. However, if the five panel signatures were normalized, the results produced by ED and SAM turned out to be very close. In analogy with Fig. 2.5, Fig. 2.8 also plots similarity values of the four measures in Tables 2.14-2.17 for comparison.

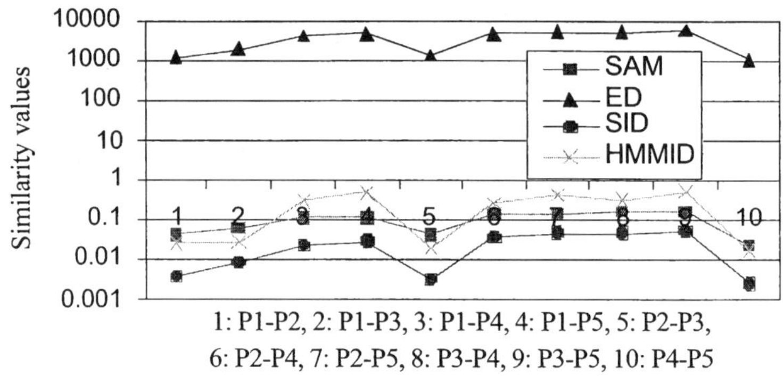

Figure 2.8. Plots of Tables 2.14-2.17

The above HYDICE experiments also showed that ED and SAM performed very similarly because of (2.24). Tables 2.14-2.17 and Fig. 2.8 show that the signatures of P1, P2 and P3 are very close. Similarly, both P4 and P5 have very close signatures, but are very distinct from those of P1, P2 and P3.

Tables 2.18-2.21 tabulate RSDPW values for ED, SAM, SID and HMMID where P2 was used as a reference signature **d**. These tables also demonstrate that SID and HMMID outperformed than ED and SAM in terms of RSDPW and HMMID performed better than SID.

Table 2.18. RSDPW of ED using P2 as a reference signature **d**

	P1	P3	P4	P5
P1	1	1.0298	3.8907	4.4046
P3		1	3.7781	4.2771
P4			1	1.1321
P5				1

Table 2.19. RSDPW of SAM using P2 as a reference signature **d**

	P1	P3	P4	P5
P1	1	1.0116	3.4000	3.6023
P3		1	3.4395	3.6442
P4			1	1.0595
P5				1

Table 2.20. RSDPW of SID using P2 as a reference signature **d**

	P1	P3	P4	P5
P1	1	1.1818	9.8718	12.4103
P3		1	11.6667	14.6667
P4			1	1.2571
P5				1

Table 2.21. RSDPW of HMMID using P2 as a reference signature **d**

	P1	P3	P4	P5
P1	1	1.1860	11.3373	17.5804
P3		1	13.4465	20.8512
P4			1	1.5507
P5				1

From Tables 2.18-2.21 we also plot their RSDPW values in Fig. 2.9. Like the AVIRIS experiments, the stochastic information measures, HMMID and SID performed more effectively than deterministic measures ED and SAM.

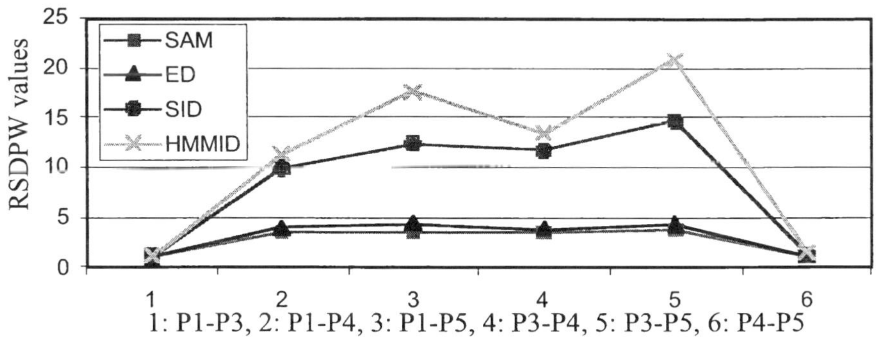

Figure 2.9. Plots of Tables18-2.21

In order to evaluate the RSDPB, a target pixel vector randomly extracted from the white mask of p_{23} was chosen as **t** for identification. It was based on the fact that P2 is very close to both P1 and P3, but more closer to P3. This selection enables us to evaluate the effectiveness of RSDPB. The pixel vector **t** was a panel edge pixel vector mixed with the background grass signature. Table 2.22 tabulates its discriminatory probabilities against five panel signatures using ED, SAM, SID and HMMID where Fig. 2.10 plots their RSDPB values in Table 2.22.

Table 2.22. RSDPB values produced by ED, SAM, SID and HMMID with **t** chosen from a pixel vector in the white mask of p_{23}

	P1	P2	P3	P4	P5
ED	0.1530	0.1339	0.1578	0.2631	0.2922
SAM	0.1544	0.1108	0.1482	0.2837	0.3028
SID	0.1029	0.0520	0.0813	0.3419	0.4218
HMMID	0.0994	0.0419	0.0939	0.3680	0.3968

As we can see from Table 2.22, the RSDPBs among P1, P2 and P3 using ED and SAM were very close, but both SID and HMMID were clearly better than ED and SAM. This fact can be better demonstrated in Fig. 2.10, where SID and HMMID produced lowest discriminatory probability of **t** against P2.

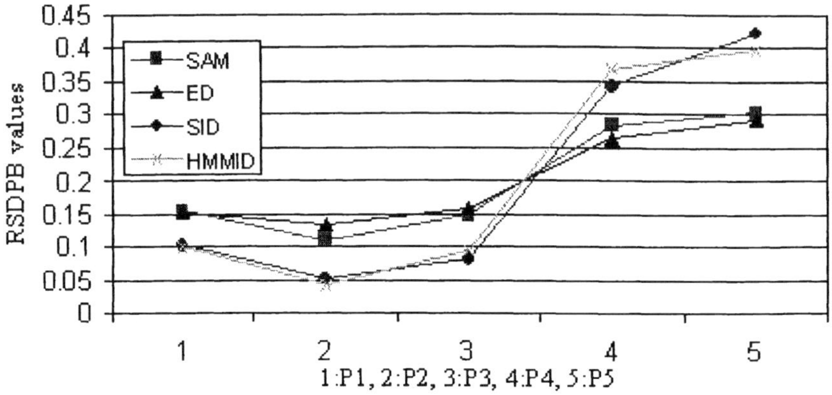

Figure 2.10. Plot of Table 2.22

If we further calculate the ratio of the second smallest RSDPB to the smallest RSDPB for ED, SAM, SID and HMMID respectively,

$$p_{\mathbf{t},\Delta}^{\mathrm{ED}}(\mathrm{P3}):p_{\mathbf{t},\Delta}^{\mathrm{ED}}(\mathrm{P2}) = 0.1530:0.1339 \approx 1.14, \quad (2.32)$$

$$p_{\mathbf{t},\Delta}^{\mathrm{SAM}}(\mathrm{P3}):p_{\mathbf{t},\Delta}^{\mathrm{SAM}}(\mathrm{P2}) = 0.1482:0.1108 \approx 1.34, \quad (2.33)$$

$$p_{\mathbf{t},\Delta}^{\mathrm{SID}}(\mathrm{P3}):p_{\mathbf{t},\Delta}^{\mathrm{SID}}(\mathrm{P2}) = 0.0813:0.0520 \approx 1.56 \quad (2.34)$$

$$p_{\mathbf{t},\Delta}^{\mathrm{HMMID}}(\mathrm{P3}):p_{\mathbf{t},\Delta}^{\mathrm{HMMID}}(\mathrm{P2}) = 0.0939:0.0419 \approx 2.24 \quad (2.35)$$

HMMID was about twice as effective as ED, SAM, and SID to identify **t** as P2.

Table 2.23. RSDE of Table 2.21 produced by ED, SAM, SID and HMMID

ED	SAM	SID	HMMID
1.5586	1.5344	1.3230	1.3190

Table 2.23 tabulates their respective RSDEs. Once again, HMMID produced the least entropy. More experiments can be found in Chang (2000), Du and Chang (2001) and Du (2000).

2.5 CONCLUSIONS

This chapter presents two new information theoretic hyperspectral measures, spectral information measure (SIM) and hidden Markov model (HMM)-based measure for spectral characterization. Both measures use self-information to characterize spectral variability, similarity, and discrimination for hyperspectral image analysis. SIM considers a hyperspectral image pixel vector as a random variable so that the spectral variability of the pixel vector can be more effectively described by randomness in nature. The Hidden Markov Model (HMM) is introduced to model the unobserved and hidden spectral properties of a hyperspectral image pixel vector as a Markov random process. With these interpretations, a SIM-derived Spectral Information Divergence (SID) and an HMM-derived Information Divergence (HMMID) are further developed to measure the spectral similarity between two pixel vectors. In order to evaluate the effectiveness of a hyperspectral measure, a new definition of spectral discriminatory power is suggested for performance analysis. Finally, the concept of spectral discriminatory probability vector is also developed for the purpose of material identification. Experiments have demonstrated that SID and HMMID capture spectral characteristics more effectively than do commonly used spectral similarity measures, Euclidean distance (ED), spectral angle mapper (SAM). This is because both SID and HMMID are statistical measures as opposed to ED, SAM, which are considered to be deterministic measures. In particular, the HMM can be used to specify complicated hidden spectral properties that cannot be observed in the spectrum of a hyperspectral pixel vector. As the number of spectral channels grows, the uncertainty and unpredictability in spectral analysis increase. Under this circumstance, SID and HMMID have better ability than do ED, SAM in characterizing spectral properties. This has been demonstrated in the experiments. The disadvantage of HMMID is the complexity in the implementation of HMM where the parameter vector $\lambda = (\mathbf{A}, \mathbf{B}, \pi)$ used in HMM must be estimated before it is used. As a final note, a discrete HMM-based spectral measure was also developed in (Du and Chang, 1999; Du, 2000) which was derived from the continuous HMM described in Section 2.1.2 by uniformly quantizing the spectrum of each pixel vector into a finite number of discrete values. The quantization used in the discrete HMM can simply analysis, but may lead to loss of information since an unsupervised clustering technique such as C-means (or K-means) method must be used to minimize the information loss and capture the major features of a pixel vector.

II

SUBPIXEL DETECTION

Target detection in remotely sensed images can be conducted spatially, spectrally or both. The difficulty with spatial analysis-based target detection arises from the fact that the ground sampling distance (GSD) is generally larger than the size of targets of interest, in which case targets are embedded in a single pixel and cannot be detected spatially. Under this circumstance target detection must be carried out at subpixel level and spectral analysis offers a valuable alternative. Since a hyperspectral image is generally represented by an image cube, an image pixel **r** is typically an L-dimensional column vector where L is the number of spectral channels used for data acquisition. As a hyperspectral image vector, **r** can be completely characterized by two features, its vector length and vector direction. In the light of this interpretation, the commonly used linear spectral mixture analysis (LSMA) basically deals with the vector length of **r** whose projection along each coordinate corresponds to the abundance fraction of a particular spectral channel. When there are no constraints imposed on the pixel vector **r**, the LSMA is referred to as unconstrained LSMA. Many algorithms have been developed along this line over the past years. If the sum of the projections is constrained to one, LSMA becomes sum-to-one constrained LSMA, which seeks solutions in an L-dimensional surface of a polygon (Settle and Drake, 1993, Ashton and Schaum, 1998). On the other hand, if each of the projections is constrained to being nonnegative, the LSMA is referred to as nonnegativity constrained LSMA, which finds solutions in a nonnegative region in the L-dimensional space (Lawson and Hanson, 1995; Bro and Jong, 1997). If both sum-to-one and nonnegativity constraints imposed on the projections, the resulting LSMA is called fully constrained LSMA. In this case, it searches for solutions in an L-dimensional unit convex hull in a nonnegative region (Shimabukuro and Smith, 1991; Settle and Drake, 1993; Ashton and Schaum, 1998; Heinz and Chang, 1999b; Heinz and Chang, 2001). A similar approach to convex hull, called convex cone analysis was also recently investigated in Ifarraguerri and Chang (1999). It should be noted that in the fully constrained LSMA the vector length of the projections is actually the city block distance specified by (2.19). Interestingly, the vector direction of an image pixel is also as important as its vector length. Finding a vector direction of interest has been studied extensively in passive sensor array processing where the arrival of signals from desired directions is of major interest. It seems that such a concept has not been explored in LSMA.

In PART II, the problem of subpixel target detection in remote sensing images is considered, where the two constrained target detection approaches described above are studied and compared. One is a target abundance-constrained approach, which can be viewed as vector length-constrained technique and another is a target signature-constrained approach, which can be considered as a vector direction-constrained technique. The former uses the target signature vectors used in LSMA as coordinates along which the projections of an image vector **r** are considered as target abundance fractions. It is based on the principle of orthogonality and least-squares error from which several detection algorithms can be derived, unconstrained orthogonal subspace projection (OSP) method, sum-to-one constrained least squares (SCLS) method and nonnegatively constrained least squares (NCLS) method which will be discussed in Chapter 3. The latter is a linearly constrained minimum variance (LCMV) approach, which will be studied in Chapter 4. It makes use of a linear constraint to lock on directions of target pixel vectors of interest while minimizing the energy of signal sources coming from other directions. In this case, two pixel vectors pointing to close directions will be considered as similar. This concept is similar to the spectral angle mapper (SAM) specified by (2.12) in Chapter 2 to measure spectral similarity. Two special versions of LCMV, constrained energy minimization (CEM) method and target-constrained inference-minimized filter (TCIMF) will be also introduced in Chapter 4. By means of a QR-decomposition, both CEM and TCIMF can be processed in real time to achieve computational efficiency. Despite their difference in algorithm design, both target abundance-constrained approach and target signature-constrained approach arrive at a same functional form of a matched filter using different levels of information that are determined by the spectral information used in the LSMA and the sample spectral information used in LCMV. Their relationship will be also explored in this chapter. In order to extend the methods in Chapters 3 and 4 to automatic subpixel detection, two types of automatic subpixel detection are considered, unsupervised subpixel detection in Chapter 5 and anomaly detection in Chapter 6. Three unsupervised learning algorithms are presented for unsupervised subpixel detection, unsupervised vector quantization algorithm, unsupervised target generation process and unsupervised NCLS to generate the spectral information directly from the image data in an unsupervised manner. For anomaly detection, several anomaly detectors are investigated, which include the RX detector (Reed and Yu, 1990), the low probability target detector (Harsanyi, 1993) and their variants. Since anomalous targets are generally small compared to their surroundings, the line-by-line real-time implementation considered in Chapter 4 is further extended to the pixel-by-pixel real time processing. Finally, Chapter 7 conducts an analysis on issues of sensitivity to target knowledge and noise, which have significant impact on subpixel detection performance.

3

TARGET ABUNDANCE-CONSTRAINED SUBPIXEL DETECTION: PARTIALLY CONSTRAINED LEAST-SQUARES METHODS

Subpixel target detection has received considerable interest in remote sensing image processing in recent years (Sabol et al., 1992). Due to significantly improved spectral resolution by recent advances of remote sensing instruments, imaging spectrometers such as AVIRIS and HYDICE sensors can now uncover and extract targets smaller than the pixel spatial resolution, in which case targets are generally embedded in a single pixel and cannot be detected spatially. As a result, traditional spatial-based image processing techniques are not directly applicable. In order to resolve this problem, we must rely on their spectral properties to detect these targets at subpixel level. In this chapter, linear spectral mixture analysis (LSMA) is investigated for subpixel target detection. In particular, we consider the target abundance-constrained subpixel detection that imposes different partial constraints on the abundance fractions of the target signatures used in the LSMA. Two partially abundance-constrained methods, referred to as sum-to-one constrained least-squares (SCLS) and nonnegatively constrained least-squares (NCLS), will be presented in this chapter. SCLS constrains on the abundance fractions of the target signatures summed to one with no constraint on nonnegativity of the abundance fractions. On the other hand, NCLS requires the abundance fractions of the target signatures to be nonnegative, but discards the abundance sum-to-one constraint.

3.1 INTRODUCTION

The LSMA has been widely used for image endmember classification, which will be discussed in Chapter 8 in great detail. It can be briefly described as follows. Suppose that there are p targets, $\mathbf{t}_1, \mathbf{t}_2, \cdots, \mathbf{t}_p$ present in an image scene. Let $\mathbf{m}_1, \mathbf{m}_2, \cdots, \mathbf{m}_p$ denote their respective spectral signatures (i.e., spectra) and $\mathbf{r}$ be a hyperspectral image pixel vector. LSMA assumes that the spectral signature or spectrum of $\mathbf{r}$ can be represented by a linear mixture of $\mathbf{m}_1, \mathbf{m}_2, \cdots, \mathbf{m}_p$ with appropriate abundance fractions specified by $\alpha_1, \alpha_2, \cdots, \alpha_p$. In general, two constraints should be imposed on this model to produce a desired solution. These are (a) abundance sum-to-one constraint, referred to as the ASC, that is, $\sum_{j=1}^{p} \alpha_j = 1$ and (b) abundance nonnegativity constraint, referred to as the ANC, that is, $\alpha_j \geq 0$ for all $1 \leq j \leq p$. An LSMA imposing these two constraints has been studied in the past (Shimabukuro and Smith, 1991; Settle and Drake, 1993; Ashton and Schaum, 1998; Heinz and Chang, 1999b; Heinz and Chang, 2001). It can be used for quantification of materials present in any pixel vector $\mathbf{r}$ where the abundance fractions $\alpha_1, \alpha_2, \cdots, \alpha_p$ can be estimated accurately to reflect their true abundance fractions. Such material quantification cannot be accomplished by many unconstrained or partially constrained LSMA methods. However, from a target detection point of view, whether or not the estimated amount of a target abundance fraction is accurate may not be essential. As long as the estimated abundance fractions of the desired target pixel vectors can distinguish themselves from their surrounding pixel vectors, the targets can be detected effectively even if the abundance fractions of LSMA do not satisfy ASC or ANC. Such target detectability was demonstrated through a family of unconstrained orthogonal subspace projection (OSP)-based methods (Harsanyi and Chang, 1994; Chang et al., 1998b). Therefore, in some cases, an LSMA-based fully constrained least-squares method (FCLS) may not be as effective as unconstrained methods or partially constrained in terms of target detection (Heinz and Chang, 2001). This is because a fully constrained method requires both ASC and ANC to be satisfied where the estimated abundance fractions must be confined to the range of [0,1] which limits its target detection capability.

Two LSMA-based partially constrained least-squares methods have been considered in the past, sum-to-one constrained least-squares method (SCLS) (Settle and Drake, 1993, Ashton and Schaum, 1998) and nonnegativity constrained least-squares method (NCLS) (Lawson and Hanson, 1995; Bro and Jong, 1997). SCLS imposes ASC while ignoring ANC. On the contrary, NCLS implements ANC while discarding ASC. As a result, both methods generally do not estimate target abundance fractions accurately. Nevertheless, their estimated abundance fractions are generally more accurate than those estimated by unconstrained method and can be still used for target detection. Since SCLS-generated abundance fractions must be summed to one, the magnitudes of the SCLS-estimated target abundance fractions are usually spread out in the range of [0,1]. Accordingly, these abundance fractions will be relatively small in order to satisfy to the sum-to-one constraint. It is particularly true when an image scene contains many target signatures. Therefore, the target detectability is substantially reduced in such cases. This situation becomes even worse if the spectra of targets are very similar. On the other hand, NCLS-generated abundance fractions do not have this problem. Being free of ASC, the NCLS can use whatever values it generates for abundance fractions. Despite the fact that its estimated abundance fractions may not reflect accurate abundance fractions, the target detectability resulting from NCLS may be actually benefited from not satisfying ASC. As a consequence, NCLS performs than SCLS in target detection.

3.2 LINEAR SPECTRAL MIXTURE MODEL

Suppose that L is the number of spectral bands and $\mathbf{t}_1, \mathbf{t}_2, \cdots, \mathbf{t}_p$ are p targets present in an image scene. Let $\mathbf{m}_1, \mathbf{m}_2, \cdots, \mathbf{m}_p$ denote their corresponding target signature vectors, which are generally referred to as digital numbers (DN). A linear spectral mixture of $\mathbf{r}$ is a linear combination of $\mathbf{m}_1, \mathbf{m}_2, \cdots, \mathbf{m}_p$ with appropriate abundance fractions specified by $\alpha_1, \alpha_2, \cdots, \alpha_p$. More precisely, let $\mathbf{r}$ and $\mathbf{M}$ be an $L \times 1$ column pixel vector and an $L \times p$ target spectral signature matrix, denoted by $\left[\mathbf{m}_1\ \mathbf{m}_2\ \cdots\ \mathbf{m}_p\right]$ respectively, where $\mathbf{m}_j$ is an $L \times 1$ column vector for $1 \leq j \leq p$. Also let $\alpha = \left(\alpha_1\ \alpha_2\ \cdots\ \alpha_p\right)^T$ be a $p \times 1$ abundance column vector associated with $\mathbf{r}$ with α_j denoting the abundance fraction of the j-th target signature $\mathbf{m}_j$ present in the pixel vector $\mathbf{r}$. We can model the spectral signature of an image pixel $\mathbf{r}$ by a linear regression form

$$\mathbf{r} = \mathbf{M}\alpha + \mathbf{n} \tag{3.1}$$

where $\mathbf{n}$ is noise or can be interpreted as a measurement error or a model error. Here without confusion the $\mathbf{r}$ is used to indicate either an image pixel or its spectral signature vector interchangeably throughout the book. A classical approach is linear unmixing which finds a set of appropriate abundance fractions, $\alpha_1, \alpha_2, \cdots, \alpha_p$ of the p targets, $\mathbf{t}_1, \mathbf{t}_2, \cdots, \mathbf{t}_p$ that are contained in the pixel vector $\mathbf{r}$.

3.3 ORTHOGONAL SUBSPACE PROJECTION (OSP)

In order to solve $\alpha = \left(\alpha_1\ \alpha_2\ \cdots\ \alpha_p\right)^T$, an unconstrained orthogonal subspace projection (OSP) approach was recently proposed by Harsanyi and Chang (1994). It assumed that there was a desired target, say $\mathbf{t}_p$ among the p targets, $\mathbf{t}_1, \mathbf{t}_2, \cdots, \mathbf{t}_p$ and the remaining $\mathbf{t}_1, \mathbf{t}_2, \cdots, \mathbf{t}_{p-1}$ were undesired targets. Since the desired target $\mathbf{t}_p$ was the only target of interest, all other targets will be considered as interferers to $\mathbf{t}_p$. In this case, the OSP approach eliminated the interfering effects caused by these undesired targets $\mathbf{t}_1, \mathbf{t}_2, \cdots, \mathbf{t}_{p-1}$ before the detection of $\mathbf{t}_p$ took place. As a consequence of annihilation of $\mathbf{t}_1, \mathbf{t}_2, \cdots, \mathbf{t}_{p-1}$, the detectability of $\mathbf{t}_p$ was improved. The main idea of the OSP approach can be described mathematically as follows.

It first separates $\mathbf{t}_p$ from $\mathbf{t}_1, \mathbf{t}_2, \cdots, \mathbf{t}_{p-1}$ and rewrites (3.1) as

$$\mathbf{r} = \mathbf{d}\alpha_p + \mathbf{U}\gamma + \mathbf{n} \tag{3.2}$$

where $\mathbf{d} = \mathbf{m}_p$ is denoted as the desired spectral signature of $\mathbf{t}_p$ and $\mathbf{U} = \left[\mathbf{m}_1\ \mathbf{m}_2\ \cdots\ \mathbf{m}_{p-1}\right]$ is the undesired target spectral signature matrix made up of the remaining $p-1$ undesired spectral signatures $\mathbf{t}_1, \mathbf{t}_2, \cdots, \mathbf{t}_{p-1}$. Here, without loss of generality we assume that the desired target consists of a single target $\mathbf{t}_p$. Nevertheless, it can be extended to multiple targets as discussed in the TCIMF approach in Chapter 4.

Separating $\mathbf{U}$ from $\mathbf{d}$ allows us to design an orthogonal subspace projector to annihilate $\mathbf{U}$ from the pixel vector $\mathbf{r}$ before the detection of $\mathbf{t}_p$ takes place. One such desired orthogonal subspace projector was derived in Harsanyi and Chang (1994) and given by

$$P_{\mathbf{U}}^{\perp} = \mathbf{I}_{L\times L} - \mathbf{U}\mathbf{U}^{\#} \tag{3.3}$$

where $\mathbf{U}^{\#} = (\mathbf{U}^T\mathbf{U})^{-1}\mathbf{U}^T$ is the pseudo-inverse of $\mathbf{U}$ and $\mathbf{I}_{L\times L}$ is the L-dimensional identity matrix. The notation "$\perp$" in $P_{\mathbf{U}}^{\perp}$ indicates that the projector $P_{\mathbf{U}}^{\perp}$ maps the image pixel vector $\mathbf{r}$ into $<\mathbf{U}>^{\perp}$, the space orthogonal to $<\mathbf{U}>$.

Applying $P_{\mathbf{U}}^{\perp}$ to (3.2) leads to an undesired target-eliminated spectral mixture model

$$P_{\mathbf{U}}^{\perp}\mathbf{r} = P_{\mathbf{U}}^{\perp}\mathbf{d}\alpha_p + P_{\mathbf{U}}^{\perp}\mathbf{n} \tag{3.4}$$

where the undesired signatures in $\mathbf{U}$ have been eliminated and the original noise $\mathbf{n}$ has been suppressed to $P_{\mathbf{U}}^{\perp}\mathbf{n}$.

Equation (3.4) represents a standard signal detection problem. If we assume that a detector solving (3.4) is a linear filter specified by a filter vector $\mathbf{w}$, the filter output is then given by

$$\mathbf{w}^T P_{\mathbf{U}}^{\perp}\mathbf{r} = \mathbf{w}^T P_{\mathbf{U}}^{\perp}\mathbf{d}\alpha_p + \mathbf{w}^T P_{\mathbf{U}}^{\perp}\mathbf{n} \tag{3.5}$$

A commonly used criterion to measure detection performance specified by (3.5) is the signal-to-noise ratio (SNR), which is defined by the ratio of output signal energy to noise output energy. According to (3.5) the output signal and noise energies can be obtained by the variances of $\mathbf{w}^T P_{\mathbf{U}}^{\perp}\mathbf{d}\alpha_p$ and $\mathbf{w}^T P_{\mathbf{U}}^{\perp}\mathbf{n}$ respectively. So, the SNR resulting from (3.5) is given by

$$\begin{aligned}\mathrm{SNR}(\mathbf{w}) &= \frac{\left[\mathbf{w}^T P_{\mathbf{U}}^{\perp}\mathbf{d}\right]\alpha_p^2\left[\mathbf{d}^T P_{\mathbf{U}}^{\perp}\mathbf{w}\right]}{\mathbf{w}^T P_{\mathbf{U}}^{\perp} E[\mathbf{n}\mathbf{n}^T] P_{\mathbf{U}}^{\perp}\mathbf{w}} = \left(\alpha_p / \sigma\right)^2 \frac{\left[\mathbf{w}^T \left(P_{\mathbf{U}}^{\perp}\mathbf{d}\right)\left(P_{\mathbf{U}}^{\perp}\mathbf{d}\right)^T \mathbf{w}\right]}{\mathbf{w}^T P_{\mathbf{U}}^{\perp}\mathbf{w}} \\ &= \left(\alpha_p / \sigma\right)^2 \frac{\left[\mathbf{w}^T\left(P_{\mathbf{U}}^{\perp}\mathbf{d}\right)\right]\left[\mathbf{w}^T\left(P_{\mathbf{U}}^{\perp}\mathbf{d}\right)\right]^T}{\mathbf{w}^T P_{\mathbf{U}}^{\perp}\mathbf{w}} = \left(\alpha_p / \sigma\right)^2 \frac{\left[\mathbf{w}^T\left(P_{\mathbf{U}}^{\perp}\mathbf{d}\right)\right]^2}{\mathbf{w}^T P_{\mathbf{U}}^{\perp}\mathbf{w}}\end{aligned} \tag{3.6}$$

where the noise is assumed to white with variance given by σ^2 and $P_{\mathbf{U}}^{\perp}$ is idempotent, i.e., $\left(P_{\mathbf{U}}^{\perp}\right)^2 = P_{\mathbf{U}}^{\perp}$. An optimal detector for (3.4) can be obtained by maximizing the SNR specified by (3.6) over the filter vector $\mathbf{w}$. By virtue of Schwarz's inequality, $\left|\mathbf{w}^T\left(P_{\mathbf{U}}^{\perp}\mathbf{d}\right)\right| \le \|\mathbf{w}\|\left\|P_{\mathbf{U}}^{\perp}\mathbf{d}\right\|$ with equality if and only if $\mathbf{w} = \kappa P_{\mathbf{U}}^{\perp}\mathbf{d}$ for some constant κ where $\|\cdot\|$ is the Euclidean distance defined in (2.20). So, the desired optimal filter vector $\mathbf{w}^*$ can be found by $\mathbf{w}^* = \kappa P_{\mathbf{U}}^{\perp}\mathbf{d}$. Additionally, it is known that the maximum SNR, that is, $\max_{\mathbf{h}} \mathrm{SNR}(\mathbf{h})$ in (3.6) is the largest eigenvalue, $\lambda_{\max}$ of the matrix $\left(P_{\mathbf{U}}^{\perp}\mathbf{d}\right)\left(P_{\mathbf{U}}^{\perp}\mathbf{d}\right)^T$ in the second equality of (3.6) (Stark and Woods, 2002). Since $P_{\mathbf{U}}^{\perp}\mathbf{d}$ is a vector, the rank of the matrix, $\left(P_{\mathbf{U}}^{\perp}\mathbf{d}\right)\left(P_{\mathbf{U}}^{\perp}\mathbf{d}\right)^T$ is one. Therefore, there is only one nonzero eigenvalue, which is λ_{max}. As also shown by Schwarz's inequality, $\mathbf{w}^* = \kappa P_U^\perp \mathbf{d}$ is an optimal solution to (3.6), in which case the eigenvector corresponding to λ_{max} must be of the form $\mathbf{w}^* = \kappa P_U^\perp \mathbf{d}$. If we further define a matched filter $M_\mathbf{w}$ using the filter vector $\mathbf{w}$ as the matched signal by

$$M_\mathbf{w}(\mathbf{x}) = \kappa \mathbf{w}^T \mathbf{x} \text{ for some constant } \kappa, \tag{3.7}$$

the optimal detector solving (3.4) turns out to be $M_{\mathbf{w}^*}$ with matched signal specified by $\mathbf{w}^* = \kappa P_U^\perp \mathbf{d}$. The maximum of SNR($\mathbf{w}$) is then given by and is equal to λ_{max}, that is,

$$\max_\mathbf{w} \mathrm{SNR}(\mathbf{w}) = \mathrm{SNR}(\mathbf{w}^*) = \lambda_{max} = \left(\alpha_p / \sigma\right)^2 \mathbf{d}^T P_U^\perp \mathbf{d} . \tag{3.8}$$

Based on the above approach with the SNR criterion defined by (3.6), an OSP detector denoted by $\delta_{OSP}(\mathbf{r})$, can be derived by a linear matched filter

$$\delta_{OSP}(\mathbf{r}) = \left(\mathbf{w}^{OSP}\right)^T \mathbf{r} = \mathbf{d}^T P_U^\perp \mathbf{r} \tag{3.9}$$

which operates $\mathbf{r}$ with the matched signal given by $\mathbf{w}^{OSP} = P_U^\perp \mathbf{d}$ with $\kappa = 1$ in (3.7).

An alternative approach proposed in Harsanyi and Chang (1994) and Harsanyi (1993) is to operate on (3.4) with $\mathbf{x}$ to maximize SNR. The resulting SNR is given by

$$\mathrm{SNR}(\mathbf{x}) = \frac{\left[\mathbf{x}^T P_U^\perp \mathbf{d}\right] \alpha_p^2 \left[\mathbf{d}^T P_U^\perp \mathbf{x}\right]}{\mathbf{x}^T P_U^\perp E[\mathbf{n}\mathbf{n}^T] P_U^\perp \mathbf{x}} . \tag{3.10}$$

Now maximizing (3.10) is equivalent to solving the following generalized eigenvalue problem for λ (see Theorem 5.5-1 in Stark and Woods (2002)),

$$\left[\sigma^2 P_U^\perp\right]^{-1} \left[P_U^\perp \mathbf{d} \alpha_p^2 \mathbf{d}^T P_U^\perp\right] \mathbf{x} = \lambda \mathbf{x}. \tag{3.11}$$

Since the detection problem specified by (3.4) represents a two-class classification problem, the rank of the matrix, $\left[\sigma^2 P_U^\perp\right]^{-1} \left[P_U^\perp \mathbf{d} \alpha_p^2 \mathbf{d}^T P_U^\perp\right]$ on the left of (3.11) is one. This implies that the only nonzero eigenvalue is the maximum eigenvalue λ_{max}. Interestingly, $\lambda = \left(\alpha_p / \sigma\right)^2 \mathbf{d}^T P_U^\perp \mathbf{d} \neq 0$ and $\mathbf{x} = \kappa \mathbf{d}$ are a solution that satisfies (3.11). Substituting $\left(\alpha_p / \sigma\right)^2 \mathbf{d}^T P_U^\perp \mathbf{d}$ for λ and κ**d** for x in (3.11) results in an equation identical to (3.8). This shows that the solution to (3.4) using (3.10) as a criterion is also a matched filter, $M_\mathbf{w}$ with the matched signal $\mathbf{w}$ which turns out to be the desired signature vector $\mathbf{d}$ and is specified by the eigenvector corresponding to $\lambda_{max} = \left(\alpha_p / \sigma\right)^2 \mathbf{d}^T P_U^\perp \mathbf{d}$. As a result, the linear optimal filter that maximizes (3.10) for (3.4) is exactly the OSP detector, $\delta_{OSP}(\mathbf{r})$ specified by (3.9).

The $\delta_{\text{OSP}}(\mathbf{r})$ defined by (3.9) is implemented by an undesired signature annihilator $P_{\mathbf{U}}^{\perp}$ followed by a matched filter, $\text{M}_{\mathbf{d}}$. More precisely, if we want to detect a target signature **d** in a mixed pixel vector **r**, we first apply $P_{\mathbf{U}}^{\perp}$ to (3.2) to eliminate **U**, then use the matched filter $\text{M}_{\mathbf{d}}$ to extract the **d** from (3.4).

It should be noted that Harsanyi and Chang's OSP detector $\delta_{\text{OSP}}(\mathbf{r})$ did not estimate the abundance vector α in (3.1). In reality, α is generally not known and needs to be estimated. In order to estimate $\alpha = \left(\alpha_1\ \alpha_2\ \cdots\ \alpha_p\right)^T$ directly from the image data, several least-squares based techniques were developed (Tu et al., 1998; Chang et al., 1998b). It is known that the least-squares estimate of $\alpha = \left(\alpha_1\ \alpha_2\ \cdots\ \alpha_p\right)^T$ for (3.1), denoted by $\hat{\alpha}_{LS} = \left(\hat{\alpha}_1, \hat{\alpha}_2, \cdots, \hat{\alpha}_p\right)^T$ can be obtained by (Scharf, 1992)

$$\hat{\alpha}_{LS} = (\mathbf{M}^T\mathbf{M})^{-1}\mathbf{M}^T\mathbf{r}. \tag{3.12}$$

It should be noted that $\hat{\alpha}_{LS}$ is a function of **r**. To simplify our notation **r** is not included in $\hat{\alpha}_{LS}$ to indicate the dependency of $\hat{\alpha}_{LS}$ on **r**. So, an LS detector, denoted by $\delta_{LS}(\mathbf{r})$, can be derived from (3.9) and (3.12) as

$$\delta_{LS}(\mathbf{r}) = \left(\mathbf{d}^T P_{\mathbf{U}}^{\perp}\mathbf{d}\right)^{-1}\delta_{\text{OSP}}(\mathbf{r}). \tag{3.13}$$

The detection of the desired target $\mathbf{t}_p$ specified by the spectral signature **d** can be accomplished via $\delta_{LS}(\mathbf{r})$ by finding the estimated abundance fraction $\hat{\alpha}_p$ contained in $\hat{\alpha}_{LS}$. The LS detector $\delta_{LS}(\mathbf{r})$ given by (3.13) was referred to as *a posteriori* OSP in Chang et al. (1998a) and LSOSP in Tu et al. (1997), whereas Harsanyi and Chang's OSP was referred to as *a priori* OSP. More discussions on least-squares estimation will be given in Chapter 8.

3.4 SUM-TO-ONE CONSTRAINED LEAST-SQUARES METHOD (SCLS)

Equation (3.1) is a general linear mixture model with no constraints imposed on the abundance vector $\alpha = \left(\alpha_1\ \alpha_2\ \cdots\ \alpha_p\right)^T$. So, it does not provide accurate estimates of abundance fractions and only offers an unconstrained solution. In this section, we consider a partially target abundance-constrained approach by imposing the ASC on (3.1) that results in the following SCLS linear mixing problem:

$$\min_{\alpha\in\Delta}\left\{(\mathbf{r}-\mathbf{M}\alpha)^T(\mathbf{r}-\mathbf{M}\alpha)\right\} \text{ subject to } \Delta = \left\{\alpha \mid \textstyle\sum_{j=1}^{p}\alpha_j = 1\right\}. \tag{3.14}$$

The solution, $\hat{\alpha}_{\text{SCLS}}$ to (3.14) can be derived as

$$\hat{\alpha}_{\text{SCLS}} = P^{\perp}_{\mathbf{M},\mathbf{1}}\hat{\alpha}_{LS}(\mathbf{r}) + \left(\mathbf{M}^T\mathbf{M}\right)^{-1}\mathbf{1}\left[\mathbf{1}^T\left(\mathbf{M}^T\mathbf{M}\right)^{-1}\mathbf{1}\right]^{-1} \quad (3.15)$$

where $\hat{\alpha}_{LS}$ is given by (3.12) and

$$P^{\perp}_{\mathbf{M},\mathbf{1}} = \mathbf{I}_{L\times L} - \left(\mathbf{M}^T\mathbf{M}\right)^{-1}\mathbf{1}\left[\mathbf{1}^T\left(\mathbf{M}^T\mathbf{M}\right)^{-1}\mathbf{1}\right]^{-1}\mathbf{1}^T \quad (3.16)$$

with $\mathbf{1} = (\underbrace{11\cdots1}_{p})^T$ being a p-dimensional column vector. Using (3.15), an SCLS detector, denoted by $\delta_{\text{SCLS}}(\mathbf{r})$ can be obtained by

$$\delta_{\text{SCLS}}(\mathbf{r}) = \hat{\alpha}_{\text{SCLS}} \, . \quad (3.17)$$

Note that (3.15) was also derived in Settle and Drake (1997), Ashton and Schaum (1998). Like $\delta_{LS}(\mathbf{r})$, the detection of the desired target $\mathbf{t}_p$ specified by the spectral signature $\mathbf{d}$ can be achieved via (3.17) by finding the estimated abundance fraction $\hat{\alpha}_p$ contained in $\hat{\alpha}_{\text{SCLS}}$.

3.5 NONNEGATIVELY CONSTRAINED LEAST-SQUARES METHOD (NCLS)

Because the SCLS described above is only based on the constraint $\sum_{j=1}^{p}\alpha_j = 1$, its solution $\hat{\alpha}_{\text{SCLS}}$ does not guarantee that the estimated abundance fractions are nonnegative, i.e., $\alpha_j \geq 0$ for all $1 \leq j \leq p$. As an alternative, we consider to impose ANC ($\alpha_j \geq 0$ for each $1 \leq j \leq p$) rather than ASC on (3.1). The resulting problem is referred to an NCLS problem. Due to the fact that the nonnegativity constraint consists a set of inequalities, no analytical solution can be derived for a closed form. Furthermore, since NCLS does not satisfy ASC, its estimated abundance fractions are generally not accurate. So, why is it important to consider NCLS? The truth is that NCLS may not be as good as a fully constrained method considered in Heinz and Chang (1999b, 2001) for material quantification. But, as a target detector, an NCLS-based detector without satisfying the ASC may be more effective than a fully constrained quantifier in target detection. This disadvantage turns out to be an advantage for NCLS in enhancement of target detectability.

In general, an NCLS approach solves the following optimization problem

$$\text{Minimize LSE} = \left(\mathbf{M}\alpha - \mathbf{r}\right)^T\left(\mathbf{M}\alpha - \mathbf{r}\right) \text{ subject to } \alpha \geq 0 \quad (3.18)$$

where LSE is the least-squares error used as a criterion for optimality and $\alpha \geq 0$ represents the nonnegativity constraint: $\alpha_j \geq 0$ for all $1 \leq j \leq p$. As noted, $\alpha \geq 0$ is a set of inequalities. The Lagrange multiplier method is not applicable to solving optimal solutions. In order to mitigate this dilemma, we introduce a p-dimensional unknown positive constraint vector $\mathbf{c} = [c_1 c_2 \cdots c_p]^T > 0$ with $c_j > 0$ for $1 \le j \le p$ to take care of ANC. By means of **c** we can form a Lagrangian $J(\alpha)$ as follows:

$$J(\alpha) = (1/2)(\mathbf{M}\alpha - \mathbf{r})^T(\mathbf{M}\alpha - \mathbf{r}) + \lambda(\alpha - \mathbf{c}) \tag{3.19}$$

subject to the constraint, $\alpha = \mathbf{c}$. Differentiating $J(\alpha)$ with respect to α yields

$$\left.\frac{\partial J(\alpha)}{\partial \alpha}\right|_{\hat{\alpha}_{\text{NCLS}}} = 0 \Rightarrow \mathbf{M}^T\mathbf{M}\hat{\alpha}_{\text{NCLS}} - \mathbf{M}^T\mathbf{r} + \lambda = 0 \tag{3.20}$$

which results in the following two iterative equations given by

$$\hat{\alpha}_{\text{NCLS}} = \left(\mathbf{M}^T\mathbf{M}\right)^{-1}\mathbf{M}^T\mathbf{r} - \left(\mathbf{M}^T\mathbf{M}\right)^{-1}\lambda = \hat{\alpha}_{\text{LS}} - \left(\mathbf{M}^T\mathbf{M}\right)^{-1}\lambda \tag{3.21}$$

and

$$\lambda = \mathbf{M}^T\left(\mathbf{r} - \mathbf{M}\hat{\alpha}_{\text{NCLS}}\right). \tag{3.22}$$

Equations (3.21) and (3.22) can be used iteratively to solve the optimal solution $\hat{\alpha}_{\text{NCLS}}$ via the Lagrange multiplier vector $\lambda = (\lambda_1, \lambda_2, \cdots, \lambda_p)^T$. An NCLS detector, denoted by $\delta_{\text{NCLS}}(\mathbf{r})$, is then derived by

$$\delta_{\text{NCLS}}(\mathbf{r}) = \hat{\alpha}_{\text{NCLS}} \tag{3.23}$$

where the detection of the desired target $\mathbf{t}_p$ specified by the spectral signature **d** is achieved by finding the estimated abundance fraction $\hat{\alpha}_p(\mathbf{r})$ contained in $\hat{\alpha}_{\text{NCLS}}$.

The nonnegativity constraint optimization problem given by (3.18) was previously explored by Lawson and Hanson (1995) and called nonnegative least-squares method (NNLS). Based on Lawson and Hanson's NNLS two Fast NNLS algorithms, referred to as FNNLS and FNNLSb were further developed by Bro and Jong (1996). Their idea was to first decompose the components of the estimate $\hat{\alpha}_{LS}$ into two index sets, called active set, R, and passive set, P where R consists of all indices corresponding to negative (or zero) components in the estimate $\hat{\alpha}_{LS}$ and P contains all indices corresponding to positive components in the estimate $\hat{\alpha}_{LS}$. The NNLS and the FNNLS started off with an empty passive set $P = \varnothing$ and assumed the active set containing all components of $\hat{\alpha}_{LS}$, i.e., $R = \{1, 2, \cdots, p\}$. They then adjusted both sets P and R via iterations using (3.21)-(3.22). It has been shown in Lawson and Hanson (1995) that when an optimal solution was reached, the Lagrange multiplier vector λ must satisfy

$$\lambda_j = 0 \text{ for } j \in P; \ \lambda_j < 0 \text{ for } j \in R. \tag{3.24}$$

The final generated passive set identified which components were legitimate to be used in the abundance estimation, $\hat{\alpha}_{LS}$.

Following a similar idea to Bro and Jong's approach, the NCLS algorithm can be implemented as follows.

NCLS Algorithm

1. Initialization:

   Set the passive set $P^{(k)} = \{1,2,\cdots,p\}$ and active set $R^{(0)} = \varnothing$. Set $k = 0$.
2. Compute $\hat{\alpha}_{LS}$ using (3.12). Let $\hat{\alpha}_{\text{NCLS}}^{(k)} = \hat{\alpha}_{LS}$
3. At the k-th iteration, if all components in $\hat{\alpha}_{\text{NCLS}}^{(k)}$ are positive, the algorithm is terminated; otherwise, continue.
4. Let $k \leftarrow k+1$.
5. Move all indices in $P^{(k-1)}$ that correspond to negative components of $\hat{\alpha}_{\text{NCLS}}^{(k-1)}$ to $R^{(k-1)}$. Let the resulting index sets be denoted by $P^{(k)}$ and $R^{(k)}$ respectively. Create a new index set $S^{(k)}$ and set it equal to $R^{(k)}$.
6. Let $\hat{\alpha}_{R^{(k)}}$ denote the vector consisting of all components $\hat{\alpha}_{LS}$ in $R^{(k)}$.
7. Form a steering matrix $\Phi_{\alpha}^{(k)}$ by deleting all rows and columns in the matrix $\hat{\alpha}_{R^{(k)}}$ specified by $P^{(k)}$.
8. Calculate $\lambda^{(k)} = \left(\Phi_{\alpha}^{(k)}\right)^{-1}\hat{\alpha}_{R^{(k)}}$. If all components in $\lambda_{\max}^{(k)}$ are negative, go to step 13; otherwise, continue.
9. Calculate $\lambda_{\max}^{(k)} = \max_j \lambda_j^{(k)}$ and move its index in $R^{(k)}$ to $P^{(k)}$
10. Form another matrix $\Psi_{\lambda}^{(k)}$ by deleting every column of $\left(\mathbf{M}^T\mathbf{M}\right)^{-1}$ specified by $P^{(k)}$.
11. Set $\hat{\alpha}_{S^{(k)}} = \hat{\alpha}_{LS} - \Psi_{\lambda}^{(k)}\lambda^{(k)}$.
12. If any component of $\hat{\alpha}_{S^{(k)}}$ in $S^{(k)}$ is negative, then move it from $P^{(k)}$ to $R^{(k)}$ and go to step 6.
13. Form another matrix $\Psi_{\lambda}^{(k)}$ by deleting every column of $\left(\mathbf{M}^T\mathbf{M}\right)^{-1}$ specified by $P^{(k)}$.
14. Set $\hat{\alpha}_{\text{NCLS}}^{(k)} = \hat{\alpha}_{\text{LS}} - \Psi_{\lambda}^{(k)}\lambda^{(k)}$. Go to step 3.

To summarize the above procedure, we assume that the current iteration is k. The NCLS algorithm begins by calculating the unconstrained least-squares solution, $\hat{\alpha}_{LS}$. If all components in $\hat{\alpha}_{LS}$ are positive, the algorithm is terminated. Otherwise, all negative components are identified and their corresponding indices are moved to the active set $R^{(k)}$. In the mean time, a duplicate set of $R^{(k)}$, referred to as $S^{(k)}$ is introduced for the purpose of keeping track of the current negative components of $\hat{\alpha}_{\text{NCLS}}^{(k-1)}$ during the k-th iteration. The steering matrix $\Phi_{\alpha}^{(k)}$ is then formed and the Lagrange multiplier vector $\lambda^{(k)}$ that will be used to steer each negative component of $\hat{\alpha}_{\text{NCLS}}^{(k-1)}$ back to zero is calculated.

component of $\lambda^{(k)}$ is shuffled from $R^{(k)}$ to $P^{(k)}$. Since the loop from step 6 to step 12 may be repeated over and over again during a single iteration, $S^{(k)}$ is used to check if all previously identified indices of maximum components of $\lambda^{(k)}$ should be retained in $P^{(k)}$ or moved back to $R^{(k)}$. Once all the values of $\lambda^{(k)}$ are negative, $\hat{\alpha}^{(k)}_{\text{NCLS}}$ is recalculated. It should be noted that the two iterative equations, specified by (3.21) and (3.22) are carried out by step 14 and step 8 respectively.

Two comments on NCLS are noteworthy. The application of NCLS in subpixel target detection has been overlooked in the past because it requires the complete prior knowledge of targets present in an image and NCLS has been primarily used for unmixing abundance fractions of all targets rather than detecting a particular target in a linear mixture. As for quantification of abundance fractions, NCLS cannot compete against the FCLS in Heinz and Chang (1999b) and Heinz and Chang (2001). Another is that the NCLS algorithm has been shown to converge in all the experiments. However, on some rare occasions the NCLS algorithm could oscillate between two passive sets because it steers back to zero a passive set which usually contains more than one negative component during iteration. Should it occur, the NCLS algorithm would adjust only one component at a time at each iteration. More details about the analysis of the NCLS algorithm can be found in Heinz and Chang (1999a).

3.6 HYPERSPECTRAL IMAGE EXPERIMENTS

In this section, experiments using two image scenes in Figs. 1.6(a) and 1.7(a) are conducted to evaluate OSP, SCLS and NCLS. A thorough comparative analysis including various scenarios of computer simulations will be postponed to Chapter 7 where all the subpixel detection methods introduced in Chapters 3-6 will be studied in great detail. Figs. 3.1-3.3 show the results of applying OSP, SCLS and NCLS to the AVIRIS image in Fig. 1.6(a) where figures labeled by (a), (b), (c) and (d) show cinders, playa, rhyolite and vegetation as desired targets respectively and figures labeled by (e) are results of the shade.

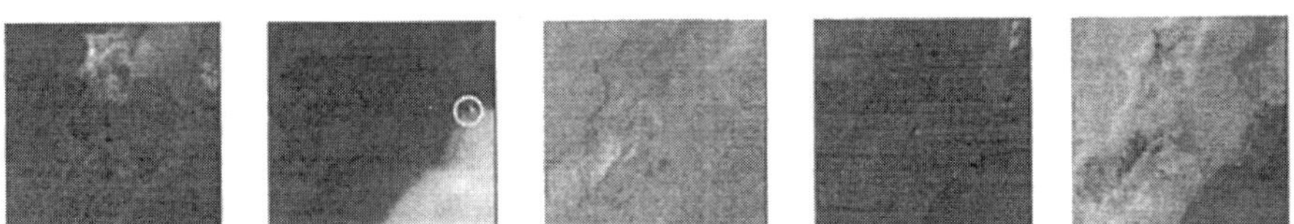

(a) cinders (b) playa (c) rhyolite (d) vegetation (e) shade

Figure 3.1. Results of OSP

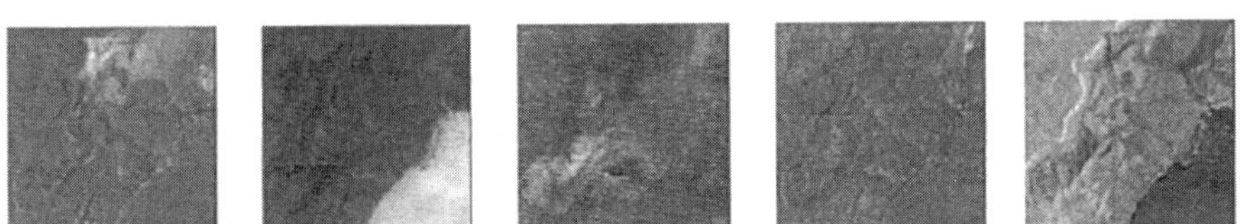

(a) cinders (b) playa (c) rhyolite (d) vegetation (e) shade

Figure 3.2. Results of SCLS

(a) cinders (b) playa (c) rhyolite (d) vegetation (e) shade

Figure 3.3. Results of NCLS

The prior knowledge of the five signatures used in Figs. 3.1-3.3 is obtained directly from the scene as was done in Harsanyi and Chang (1994). As we can see from these figures, NCLS outperformed SCLS and OSP while both OSP and SCLS performed similarly. It is interesting to note that OSP picked up an anomaly marked by a white circle in Fig. 3.1(b), which was not estimated by SCLS and NCLS. However, this anomaly was detected when NCLS was implemented in an unsupervised manner (Chang and Heinz, 2000a; Chang and Heinz, 2000b). This anomalous target will be discussed in Chapter 6.

A second scene used for experiments is the 15 panel HYDICE scene shown in Fig. 1.7. Figs. 3.4-3.6 show the results of OSP, SCLS and NCLS in detection of the 15 panels where the signature of panels used for target information in **M** were given by P1. P2, P3, P4 and P5 plotted in Fig. 1.8.

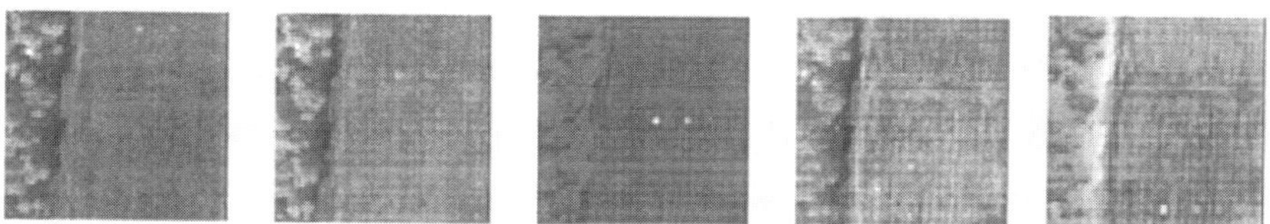

P1 P2 P3 P4 P5

Figure 3.4. Results of OSP

P1 P2 P3 P4 P5

Figure 3.5. Results of SCLS

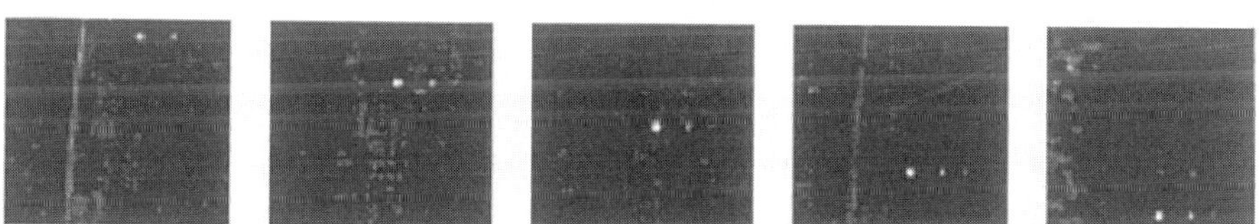

P1 P2 P3 P4 P5

Figure 3.6. Results of NCLS

Like the AVIRIS experiments, NCLS once again performed significantly better than OSP and SCLS where the panels in the first and second columns were detected with small estimated amounts of interferers. In all cases, none of OSP, SCLS and NCLS detected the panels with size $1\text{m} \times 1\text{m}$ in column 3. Compared to SCLS, OSP performed slightly better than SCLS in the sense that it detected some panels in row 2 and less amounts of interferers. This may be due to the fact that in order to accommodate ASC, the abundance fractions estimated by SCLS must be constrained in the range of [0,1] in which case the amounts of the estimated abundance fractions were very small and were not good for detection. Such a situation will become even worse when the image background is complicated and many target signatures are present in the image.

3.7 CONCLUSIONS

In this chapter, least-squares based approaches to subpixel detection are studied. It first introduces an orthogonal subspace projection (OSP) which does not impose any constraint on target abundance fractions. It makes use of an orthogonal subspace projector to annihilate undesired target signatures prior to detection. Despite the fact that the used criteria of signal-to-noise ratio (SNR) in (3.6) and (3.10) assume that the noise is additive (i.e., uncorrelated) white, this assumption can be relaxed by a whitening process described in Section 12.3 of Chapter 12. Also, it is worth noting that unlike maximum likelihood approach (Settle, 1996) OSP does not make an assumption that noise must be Gaussian. As shown in Du and Ren (2002) that OSP can be slightly improved if noise was appropriately whitened. However, how to reliably estimate noise correlation matrix is generally a challenging problem, which will be discussed in Section 17.3 of Chapter 17. So, a common practice is to use OSP without noise whitening, since hyperspectral imagery generally has high signal-to-noise ratio (SNR) and the interference is more dominant than noise. Additionally, due to high spectral resolution of hyperspectral imagery the noise correlation is considered to be much less than it is in multispectral imagery. Therefore, such compromise does not have much impact on OSP performance.

Although OSP is shown to perform well in general, its detection performance can be further improved by imposing constraints on target abundance fractions. Two partially constrained least-squares methods, SCLS and NCLS are investigated in this chapter. With the abundance sum-to-one constraint, SCLS subpixel detector can be solved analytically and expressed by an unconstrained least-squares solution plus a correction term. Since no closed-form solutions can be derived for NCLS due to the nonnegativity abundance constraint, a fast efficient numerical algorithm is developed to generate a desired optimal subpixel detector. The reason that only partial constraints are considered is because we are interested in subpixel target detection rather than quantification of target abundance fractions. To implement full constraints (i.e., ASC and ANC), the target abundance fractions must be constrained to the range of [0,1]. On many occasions such as complicated background where many targets may be present in an image, the estimated abundance fractions among targets may not differ very much for detection. This may result in poor target detection. On the other hand, if there is no constraint imposed on target abundance, the estimated abundance fractions may vary to a very large degree of deviation in a wide range. This may result in poor performance (Chang and Heinz, 2000a; Chang and Heinz, 2000b). In either case, constrained subpixel detection may be useful. As demonstrated in experiments, partially constrained least-squares based subpixel detectors generally perform significantly better than the unconstrained subpixel detectors. However, as indicated previously, in some cases where many target signatures are present in an image scene, SCLS may not perform as well as unconstrained subpixel detectors in detection because of ASC. Under this circumstance, NCLS is always preferred to SCLS.

4

TARGET SIGNATURE-CONSTRAINED SUBPIXEL DETECTION: LINEARLY CONSTRAINED MINIMUM VARIANCE (LCMV)

In Chapter 3, we studied partially constrained least squares approaches to target abundance-constrained subpixel detection. One of their practical limitations is the requirement of prior knowledge about the target signature matrix **M**. In reality, such information is very difficult to obtain, if not impossible. It is particularly true for hyperspectral imagery, which may contain many unknown signal sources extracted by high spectral resolution sensors such as HYDICE. These unidentified sources may include interferers, nonstationary background signatures and natural signatures that cannot be visually inspected from an image scene. Under such circumstance, finding a well-represented target signature matrix **M** may not be realistic. In order to resolve this problem, this chapter develops an alternative approach, referred to as a linearly constrained minimum variance (LCMV) approach, which only constrains targets of interest while minimizing the energy of unknown signal sources. Two special versions can be derived from LCMV, which are constrained energy minimization (CEM) filter and target-constrained interference-minimized Filter (TCIMF) where TCIMF can be viewed as an extension of the CEM filter. Interestingly, there is a close relationship among CEM, TCIMF and OSP where all the three operate a same functional form as a matched filter using different levels of target knowledge.

4.1 INTRODUCTION

Hyperspectral image analysis has become increasingly important in remote sensing image processing because hyperspectral image sensors are now capable of extracting many subtle material substances which usually cannot be resolved by multispectral sensors. However, because of their very high spectral resolution, many unknown and unidentified signal sources (will be referred to as interferers here) may also be extracted as well. These unexpected signal sources generally introduce additional interfering effects on target detection. Such phenomena have been demonstrated in Chang et al. (1998a). Since the interference is generally unknown in nature and cannot be identified from an image scene by visual inspection, a challenging problem associated with it is how to minimize its interfering effect without actually identifying interferers.

Over the past years many algorithms have been developed for hyperspectral image analysis. In particular, linear spectral mixture analysis discussed in Chapter 3 has been widely used for multispectral and hyperspectral image classification as well as for subpixel target detection. Unfortunately, it requires the complete knowledge of targets present in an image scene. In many practical applications, such knowledge including interferers is not known *a priori* and must be obtained directly from the scene. Three unsupervised methods were recently proposed in Chang et al. (2001b), which can be used to generate necessary interference information directly from an image scene. They will be discussed in depth in Chapter 5. One drawback of these approaches is that it needs to know how many interferers are needed to generate.

In order to mitigate this dilemma, an alternative approach, called constrained energy minimization (CEM) was first proposed in Harsanyi (1993) and later published in Harsanyi et al. (1994). It did not require the knowledge of interferers. Instead, the only required knowledge was the desired target to be detected. Using a specific constraint, CEM designed an finite impulse response (FIR) filter to pass through the desired target while minimizing its output energy resulting from target sources other than the desired target. The success of the CEM filter in AVIRIS and HYDICE data exploitation was further demonstrated in Farrand and Harsanyi (1997) and Resmini et al. (1997).

One disadvantage of the CEM filter is its single-target detection. If we had known that there were targets in an image scene which were not of our interest, e.g. natural and background signatures, these targets could have been eliminated prior to detection rather than being minimized as was done by the CEM filter. In this case, the CEM filter still considered the known undesired targets as unknown interferers. Thus, instead of eliminating these undesired targets, it minimized their energies. Therefore, CEM does not take advantage of preprocessing the undesired targets to enhance its detection performance. This can be actually accomplished by a target-constrained interference-minimized Filter (TCIMF) recently developed by Ren and Chang (2001). Coincidentally, the relationship between OSP and CEM can be further explored by TCIMF.

Another disadvantage of CEM is that the CEM filter can only detect one target at a time and very sensitive to the target knowledge that is used in the constraint. If there are targets of same type with slight spectral variability, the CEM filter may miss these targets. This drawback can be remedied by a generalization of the CEM approach, referred to as linearly constrained minimum variance (LCMV) approach, which includes both CEM and TCIMF as special cases.

The LCMV idea can be traced back to Frost's work (Frost III, 1972) in adaptive beamforming. In passive sensor array processing, what we are interested is the desired direction of signal arrival from an array of sensors. If there is one direction of interest, we can lock on that direction by designing a constrained FIR filter with a specific filter gain to look for signals coming from this particular direction. When the gain is constrained to a single direction with unity, the resulting FIR filter is called the minimum variance distortionless response (MVDR) beamformer (Van Veen and Buckley, 1988; Haykin 1994). Now, if we interpret an array of sensors as a bank of spectral channels used in remote sensing instruments and the desired direction of signal arrival as the vector direction of the desired target signature, the MVDR beamformer becomes the CEM filter. An early idea of CEM was explored in chemical remote sensing (Althouse and Chang, 1991). More recently, an extension of the CEM filter to detection of multiple targets was also studied in Chang and Ren (1999) and Chang et al. (2001c). It used Frost's idea to linearly constrain a set of multiple targets so that these targets can be passed through a designed FIR filter constrained by a set of filter gains. In the mean time, the filter output energy resulting from target sources other than the desired directions is minimized in the sense of variance. This can be achieved by augmenting the scalar unity used to constrain the CEM filter to a constraint vector, of which each component was used to constrain a specific target.

In some applications, there is always partial knowledge available about the image data, in which case we should take advantage of rather than discard this information. Specifically, we can categorize signal sources in an image scene into three signal classes. One is a desired target class, which consists of targets of interest. A second signal class, called an undesired target class comprises of all targets that are known, but either unwanted or uninterested targets. A third signal class is an interference class made up of all unknown or unidentified signal sources resident in the image. An effective detector should best utilize its available target knowledge to achieve its optimal performance. In other words, it should detect the desired targets, eliminate the undesired targets, and minimize the effects caused by all unknown interferers simultaneously. The TCIMF developed in Ren and Chang (2001) accomplished these three tasks in one single operation where TCIMF divided known targets into desired targets and undesired targets to detect the desired targets and eliminate the undesired targets at the same time. This advantage cannot be achieved by CEM. Additionally, TCIMF also preserves the same strength of CEM, which does not need to identify interferers but still can minimize interference as the way carried out by CEM. In a way, TCIMF combines Frost's idea with the oblique subspace projection (Scharf, 1991; Chang et al. 1998b) in such a manner that the desired targets are projected into the range space while undesired targets being annihilated by mapping them into the null space. Specially, TCIMF accomplishes detection of the desired targets, annihilation of the undesired targets and minimization of interfering effects at the same time as opposed to CEM that only achieves detection of a desired target and minimization of interfering effects with no elimination of undesired targets. As a consequence, TCIMF generally performs better than CEM. From this point of view, CEM can be viewed as a special version of TCIMF.

Since both CEM and TCIMF deal with an over-determined system where the number of pixel samples is generally greater than that of spectral bands, the rank of the sample correlation matrix that is used to design optimal weights for the filters will become crucial. This issue is determined by noise sensitivity and will be investigated in Chapter 7 in detail. Furthermore, it is also very closely related to the intrinsic dimensionality of the image data that will be discussed in Chapter 17.

4.2 LCMV TARGET DETECTOR

Assume that $\{\mathbf{r}_1, \mathbf{r}_2, \cdots, \mathbf{r}_N\}$ is a collection of N image pixels in a remotely sensed image where $\mathbf{r}_i = (r_{i1}, r_{i2}, \cdots, r_{iL})^T$ for $1 \le i \le N$ is an L-dimensional pixel vector. As defined in Chapter 3, $\mathbf{m}_1, \mathbf{m}_2, \cdots, \mathbf{m}_p$ are the spectral signatures of p targets $\mathbf{t}_1, \mathbf{t}_2, \cdots, \mathbf{t}_p$ present in the image and $\mathbf{M}$ is a target signature matrix formed by $[\mathbf{m}_1\ \mathbf{m}_2 \cdots \mathbf{m}_p]$. The goal is to design a constrained FIR linear filter with by an L-dimensional weight vector $\mathbf{w} = (w_1, w_2, \cdots, w_L)^T$ specified by a set of L filter coefficients $\{w_1, w_2, \cdots, w_L\}$ that minimizes the filter output energy subject to the following constraint

$$\mathbf{M}^T\mathbf{w} = \mathbf{c} \text{ where } \mathbf{m}_j^T\mathbf{w} = \textstyle\sum_{l=1}^{L} m_{jl}w_l = c_j \text{ for } 1 \le j \le p \tag{4.1}$$

where $\mathbf{c} = \left(c_1, c_2, \cdots, c_p\right)^T$ is a constraint vector.

Now, let y_i denote the output of the designed FIR filter resulting from the input $\mathbf{r}_i$. Then y_i can be expressed by

$$y_i = \sum_{l=1}^{L} w_l r_{il} = \mathbf{w}^T \mathbf{r}_i = \mathbf{r}_i^T \mathbf{w}. \tag{4.2}$$

Using (4.1)-(4.2) Frost's LCMV beamformer becomes an LCMV-based target detector, which minimizes the average energy of the total filter outputs

$$\begin{aligned}(1/N)\left[\sum_{i=1}^{N} y_i^2\right] &= (1/N)\left[\sum_{i=1}^{N}\left(\mathbf{r}_i^T\mathbf{w}\right)^T \mathbf{r}_i^T\mathbf{w}\right] \\ &= \mathbf{w}^T\left(1/N\left[\sum_{i=1}^{N}\mathbf{r}_i\mathbf{r}_i^T\right]\right)\mathbf{w} = \mathbf{w}^T\mathbf{R}_{L\times L}\,\mathbf{w}\end{aligned} \tag{4.3}$$

where $\mathbf{R}_{L\times L} = (1/N)\left[\sum_{i=1}^{N}\mathbf{r}_i\mathbf{r}_i^T\right]$ is the sample spectral matrix of the image. If we let $\mathbf{X} = \left[\mathbf{r}_1\,\mathbf{r}_2\cdots\mathbf{r}_N\right]$ be the data matrix formed by $\left\{\mathbf{r}_1, \mathbf{r}_2, \cdots, \mathbf{r}_N\right\}$, then $\mathbf{R}_{L\times L}$ is also equal to $\mathbf{R}_{L\times L} = (1/N)\left[\mathbf{X}\mathbf{X}^T\right]$. Combining (4.3) with the constraint equation (4.1) results in the following linearly constrained optimization problem

$$\min_{\mathbf{w}}\left\{\mathbf{w}^T\mathbf{R}_{L\times L}\,\mathbf{w}\right\} \text{ subject to } \mathbf{M}^T\mathbf{w} = \mathbf{c}. \tag{4.4}$$

The solution to (4.4) can be obtained by (Chang et al. 1999; Chang et al. 2001c)

$$\mathbf{w}^{\text{LCMV}} = \mathbf{R}_{L\times L}^{-1}\mathbf{M}\left(\mathbf{M}^T\mathbf{R}_{L\times L}^{-1}\mathbf{M}\right)^{-1}\mathbf{c}\,. \tag{4.5}$$

So, an LCMV detector, $\delta_{\text{LCMV}}(\mathbf{r})$ can be derived by substituting $\mathbf{w}^{\text{LCMV}}$ for $\mathbf{w}$ in (4.2) and it is given by

$$\delta_{\text{LCMV}}(\mathbf{r}) = (\mathbf{w}^{\text{LCMV}})^T\mathbf{r} = \left(\mathbf{R}_{L\times L}^{-1}\mathbf{M}\left(\mathbf{M}^T\mathbf{R}_{L\times L}^{-1}\mathbf{M}\right)^{-1}\mathbf{c}\right)^T\mathbf{r}. \tag{4.6}$$

It should be noted that the optimal weight $\mathbf{w}^{\text{LCMV}}$ in (4.6) in the designed LCMV-based target detector uses the sample spectral correlation matrix $\mathbf{R}_{L\times L}$ to capture the statistics of the first two orders among the image pixel vectors $\mathbf{r}_1, \mathbf{r}_2, \cdots, \mathbf{r}_N$. Two special cases of (4.1) are of particular interest.

4.2.1 Constrained Energy Minimization (CEM)

If we are only interested in a single target signature **d**, that is, $\mathbf{M} = \mathbf{d}$, the constraint vector **c** in (4.1) is then substituted by $\mathbf{d}^T\mathbf{w} = \sum_{l=1}^{L} d_l w_l = 1$ and becomes a constraint unity scalar. In this particular case, (4.4) is reduced to

$$\min_{\mathbf{w}}\left\{\mathbf{w}^T\mathbf{R}_{L\times L}\,\mathbf{w}\right\} \text{ subject to } \mathbf{d}^T\mathbf{w}=1 \tag{4.7}$$

with the optimal solution $\mathbf{w}^{\text{CEM}}$ given by

$$\mathbf{w}^{\text{CEM}}=\frac{\mathbf{R}_{L\times L}^{-1}\mathbf{d}}{\mathbf{d}^T\mathbf{R}_{L\times L}^{-1}\mathbf{d}}. \tag{4.8}$$

Substituting the optimal weight $\mathbf{w}^{\text{CEM}}$ for $\mathbf{w}$ in (4.2) results in the CEM filter, which implements a detector, $\delta_{\text{CEM}}(\mathbf{r})$ given by

$$\delta_{\text{CEM}}(\mathbf{r})=(\mathbf{w}^{\text{CEM}})^T\mathbf{r}=\left(\mathbf{d}^T\mathbf{R}_{L\times L}^{-1}\mathbf{d}\right)^{-1}\left(\mathbf{R}_{L\times L}^{-1}\mathbf{d}\right)^T\mathbf{r}. \tag{4.9}$$

The approach to solving (4.7) via (4.8) is referred to as CEM in Harsanyi's dissertation (1993).

4.2.2 Target-Constrained Interference-Minimized Filter (TCIMF)

It has been demonstrated that interference played a significant role in hyperspectral image analysis. This is primarily due to the fact that many subtle materials can be revealed by hyperspectral image sensors with very fine spatial and spectral resolution. Unfortunately, this also results in extraction of additional unknown signal sources. A recently developed target-constrained interference-minimized filter (TCIMF) assumed that an image scene was made up of three separate signal sources, **D** (desired targets), **U** (undesired targets) and **I** (interference). The idea to separate interference from a signal model as an independent and separate source was previously explored in Chang et al. (1998a) and Chang and Du (1999). The CEM filter takes care of interference problem using (4.7) by constraining the desired target signature **d** while minimizing the output energy resulting from other target sources. However, in case that there is information available about the undesired target signatures in **U**, simply discarding this information may not be the best way to achieve optimal detection performance. TCIMF resolves this problem. It implements a vector that simultaneously constrains **D** and **U** in such a way that the desired target signatures in **D** can be detected, while the undesired target signatures in **U** can be eliminated at the same time. A similar LCMV-based filter was also proposed to extend CEM to detect multiple desired target signatures through a constraint vector (Chang and Ren, 1999). But, the used constraint vector was designed only for the purpose of multiple-target detection, not for annihilation of undesired target signatures in **U** as was done in TCIMF to enhance the target detectability. The idea of TCIMF can be briefly described as follows.

Let $\mathbf{D}=\left[\mathbf{d}_1\mathbf{d}_2\cdots\mathbf{d}_p\right]$ and $\mathbf{U}=\left[\mathbf{u}_1\mathbf{u}_2\cdots\mathbf{u}_q\right]$ denote the desired target signature matrix and the undesired target signature matrix respectively. A constraint vector can be derived from (4.1) by replacing the target signature matrix **M** and the constraint vector **c** with the desired-undesired target signature matrix $\left[\mathbf{D}\,\mathbf{U}\right]$ and the desired-undesired target signature constraint vector $\mathbf{c}=\left(\mathbf{1}_{p\times 1}^T,\mathbf{0}_{q\times 1}^T\right)$ as follows,

$$[\mathbf{D}\,\mathbf{U}]^T \mathbf{w} = \begin{bmatrix} \mathbf{1}_{p\times 1} \\ \mathbf{0}_{q\times 1} \end{bmatrix} \tag{4.10}$$

where $\mathbf{1}_{p\times 1}$ is a $p \times 1$ column vector with ones in all components and $\mathbf{0}_{q\times 1}$ is a $q \times 1$ column constraint vector with all zeros in its components. It is worth noting that in (4.10), $\mathbf{1}_{p\times 1}$ is used to constrain the desired target signatures in $\mathbf{D}$ performed by the CEM filter in (4.7), whereas $\mathbf{0}_{q\times 1}$ is included to annihilate the undesired target signatures, $\mathbf{u}_1, \mathbf{u}_2, \cdots, \mathbf{u}_q$ in a similar fashion done by the oblique subspace projection (Scharf, 1991, Chang et al., 1998a-1998b). By taking advantage of (4.10) CEM can be extended to the following linearly constrained optimization problem

$$\min_{\mathbf{w}} \left\{ \mathbf{w}^T \mathbf{R}_{L\times L}\, \mathbf{w} \right\} \text{ subject to } [\mathbf{D}\,\mathbf{U}]^T \mathbf{w} = \begin{bmatrix} \mathbf{1}_{p\times 1} \\ \mathbf{0}_{q\times 1} \end{bmatrix}. \tag{4.11}$$

where the optimal weight vector $\mathbf{w}^{\mathrm{TCIMF}}$ can be solved by

$$\mathbf{w}^{\mathrm{TCIMF}} = \mathbf{R}_{L\times L}^{-1} [\mathbf{D}\,\mathbf{U}] \left([\mathbf{D}\,\mathbf{U}]^T \mathbf{R}_{L\times L}^{-1} [\mathbf{D}\,\mathbf{U}] \right)^{-1} \begin{bmatrix} \mathbf{1}_{p\times 1} \\ \mathbf{0}_{q\times 1} \end{bmatrix}. \tag{4.12}$$

The filter specified by the weight vector given by $\mathbf{w}^{\mathrm{TCIMF}}$ in (4.12) is called TCIMF and can be implemented as a detector, $\delta_{\mathrm{TCIMF}}(\mathbf{r})$ by

$$\delta_{\mathrm{TCIMF}}(\mathbf{r}) = \left(\mathbf{w}^{\mathrm{TCIMF}} \right)^T \mathbf{r}. \tag{4.13}$$

4.3 RELATIONSHIP AMONG OSP, CEM AND TCIMF

In this section, we present an interesting finding (Chang, 2002a). If we compare $\delta_{\mathrm{OSP}}(\mathbf{r}) = \mathbf{d}^T P_{\mathbf{U}}^{\perp} \mathbf{r}$ specified by (3.9) against $\delta_{\mathrm{CEM}}(\mathbf{r}) = \left(\mathbf{d}^T \mathbf{R}_{L\times L}^{-1} \mathbf{d} \right)^{-1} \mathbf{d}^T \mathbf{R}_{L\times L}^{-1} \mathbf{r}$ specified by (4.9), we will discover that there is a close relationship between $P_{\mathbf{U}}^{\perp}$ used in $\delta_{\mathrm{OSP}}(\mathbf{r})$ and $\mathbf{R}_{L\times L}^{-1}$ used in $\delta_{\mathrm{CEM}}(\mathbf{r})$. They both operate a functional form as a matched filter specified by (3.7) with the same matched signal $\mathbf{d}$ but different scale constant κ. Interestingly, it has been an oversight of significance in the constant κ. In the matched filter derivation of (3.7), the constant κ resulting from Schwarz's inequality generally plays no role in signal detection and is usually assumed to be 1 without loss of signal detectability, e.g. $\delta_{\mathrm{OSP}}(\mathbf{r})$. This is not the case for abundance estimation. As noted in Settle (1996), the constant κ determines the estimation error of the abundance vector α. So, the relationship between $\delta_{\mathrm{OSP}}(\mathbf{r})$ and $\delta_{\mathrm{CEM}}(\mathbf{r})$ can be described from two viewpoints, signal detection and signal estimation.

In signal detection, the knowledge of $P_{\mathbf{U}}^{\perp}$ which is assumed to be known in $\delta_{\text{OSP}}(\mathbf{r})$ is not available in $\delta_{\text{CEM}}(\mathbf{r})$. The $\delta_{\text{CEM}}(\mathbf{r})$ must estimate $P_{\mathbf{U}}^{\perp}$ directly from the image data. One way of doing so is to approximate the " $P_{\mathbf{U}}^{\perp}$ " in the sense of minimum least squares error by the sample spectral information, $\mathbf{R}_{L\times L}^{-1}$ that can be obtained by the inverse of the sample correlation matrix from the image data. More specifically, $\delta_{\text{CEM}}(\mathbf{r})$ makes use of the *a posteriori* information, $\mathbf{R}_{L\times L}^{-1}$ to approximate the *a priori* information $P_{\mathbf{U}}^{\perp}$ to accomplish what $\delta_{\text{OSP}}(\mathbf{r})$ does. Since $\delta_{\text{OSP}}(\mathbf{r})$ assumes the prior knowledge of the abundance vector α, there is no need of abundance estimation, in which case, $\kappa = 1$.

Mathematically, we can further demonstrate that both $\delta_{\text{OSP}}(\mathbf{r})$ and $\delta_{\text{CEM}}(\mathbf{r})$ are special versions of $\delta_{\text{TCIMF}}(\mathbf{r})$ where $\delta_{\text{OSP}}(\mathbf{r})$ can be shown to be related to $\delta_{\text{CEM}}(\mathbf{r})$ through $\delta_{\text{TCIMF}}(\mathbf{r})$. On one hand, if the knowledge of $\mathbf{U}$ is absent in the $\mathbf{w}^{\text{TCIMF}}$ in (4.12), $\mathbf{U} = \varnothing$, that is the empty set, then $\delta_{\text{TCIMF}}(\mathbf{r})$ becomes $\delta_{\text{CEM}}(\mathbf{r})$. On the other hand, if we assume that $\delta_{\text{TCIMF}}(\mathbf{r})$ does not use spectral correlation provided by data samples, $\mathbf{R}_{L\times L} = \mathbf{I}_{L\times L}$, in which case, the $\mathbf{w}^{\text{TCIMF}}$ is denoted by $\mathbf{w}^{\text{TCIMF}}_{\mathbf{R}_{L\times L}=\mathbf{I}_{L\times L}}$. In order to relate to $\delta_{\text{OSP}}(\mathbf{r})$, let $\mathbf{D} = \mathbf{d}$ and $\mathbf{U} = \left[\mathbf{m}_1\ \mathbf{m}_2 \cdots \mathbf{m}_{p-1}\right]$. Then $n_{\mathbf{D}} = 1$ and $n_{\mathbf{U}} = p-1$ and $\mathbf{c} = \left(\mathbf{1}_{p\times 1}^T, \mathbf{0}_{q\times 1}^T\right)$ in (4.10) becomes $\mathbf{c} = \left(1, \mathbf{0}_{(p-1)\times 1}^T\right)$. Using a matrix inverse formula such as the equalities in Settle (1996), we can prove that $\delta_{\text{TCIMF}}(\mathbf{r})$ turns out to be $\delta_{\text{OSP}}(\mathbf{r})$.

$$
\begin{aligned}
\mathbf{w}^{\text{TCIMF}}_{\mathbf{R}_{L\times L}=\mathbf{I}_{L\times L}} &= \mathbf{I}_{L\times L}^{-1}[\mathbf{d}\ \mathbf{U}]\left([\mathbf{d}\ \mathbf{U}]^T \mathbf{I}_{L\times L}^{-1}[\mathbf{d}\ \mathbf{U}]\right)^{-1}\begin{bmatrix}1\\ \mathbf{0}_{(p-1)\times 1}\end{bmatrix} \\
&= [\mathbf{d}\ \mathbf{U}]\left(\begin{bmatrix}\mathbf{d}^T\mathbf{d} & \mathbf{d}^T\mathbf{U}\\ \mathbf{U}^T\mathbf{d} & \mathbf{U}^T\mathbf{U}\end{bmatrix}\right)^{-1}\begin{bmatrix}1\\ \mathbf{0}_{(p-1)\times 1}\end{bmatrix} \\
&= [\mathbf{d}\ \mathbf{U}]\begin{bmatrix}\kappa & \kappa\,\mathbf{d}^T\left(\mathbf{U}^{\#}\right)^T\\ -\kappa\mathbf{U}^{\#}\mathbf{d} & \left(\mathbf{U}^T\mathbf{U}\right)^{-1} + \kappa\,\mathbf{U}^{\#}\mathbf{d}\mathbf{d}^T\left(\mathbf{U}^{\#}\right)^T\end{bmatrix}\begin{bmatrix}1\\ \mathbf{0}_{(p-1)\times 1}\end{bmatrix} \\
&= [\mathbf{d}\ \mathbf{U}]\begin{bmatrix}\kappa\\ -\kappa\,\mathbf{U}^{\#}\mathbf{d}\end{bmatrix} = \kappa\,\mathbf{d} - \kappa\,\mathbf{U}\mathbf{U}^{\#}\mathbf{d} = \kappa\left(\mathbf{I} - \mathbf{U}\mathbf{U}^{\#}\right)\mathbf{d} = \kappa\,P_{\mathbf{U}}^{\perp}\mathbf{d}
\end{aligned}
\tag{4.14}
$$

where $\kappa = \left(\mathbf{d}^T\mathbf{R}_{L\times L}^{-1}\mathbf{d}\right)^{-1}$ and $\mathbf{U}^{\#} = (\mathbf{U}^T\mathbf{U})^{-1}\mathbf{U}^T$ is the pseudo-inverse of $\mathbf{U}$ defined in (3.3), and

$$
\left(\mathbf{w}^{\text{TCIMF}}_{\mathbf{R}_{L\times L}=\mathbf{I}_{L\times L}}\right)^T\mathbf{r} = \kappa\delta_{\text{OSP}}(\mathbf{r}). \tag{4.15}
$$

Equation (4.15) implies that when no sample spectral correlation is considered, $\delta_{\text{TCIMF}}(\mathbf{r})$ is actually *a posteriori* OSP, which is identical to $\delta_{\text{OBSP}}(\mathbf{r})$ or $\delta_{\text{GML}}(\mathbf{r})$ specified by (8.24) and (8.30) discussed in Chapter 8.

On the other hand, (4.14) suggests that $\delta_{\mathrm{OSP}}(\mathbf{r})$ can be considered as a whitened $\delta_{\mathrm{TCIMF}}(\mathbf{r})$ with the image data zero-mean whitened, i.e. $\mathbf{R}_{L\times L} = \mathbf{I}_{L\times L}$. However, it should bear in mind that there is a significant difference in filter design between $\delta_{\mathrm{OSP}}(\mathbf{r})$ and $\delta_{\mathrm{TCIMF}}(\mathbf{r})$. The $\delta_{\mathrm{OSP}}(\mathbf{r})$ is derived by a linear mixture signal model specified by (3.2) where the noise is assumed to be additive, but not necessarily Gaussian assumed in the maximum likelihood estimation. By contrast, $\delta_{\mathrm{TCIMF}}(\mathbf{r})$ minimizes the filter output energy resulting from signals other than **d** and **U** without making an assumption of signal detection model, in which case $\delta_{\mathrm{TCIMF}}(\mathbf{r})$ suppresses noise and all signal sources except **d** and **U**. This explains why there is an additional constant κ in (4.15), which is a consequence of interference and noise suppression resulting from $\delta_{\mathrm{TCIMF}}(\mathbf{r})$.

If we assume that a detector is a linear filter specified by a filter vector **w**, a detector can be expressed by an inner product of a filter vector **w** with an image pixel vector **r**. Namely, $\delta_{\mathbf{w}}(\mathbf{r}) = \mathbf{w}^T\mathbf{r}$ where the filter vector **w** is specified by a particular detector or classifier. Table 4.1 summarizes the relationship among $\delta_{\mathrm{OSP}}(\mathbf{r})$, $\delta_{\mathrm{CEM}}(\mathbf{r})$ and $\delta_{\mathrm{TCIMF}}(\mathbf{r})$.

Table 4.1. Relationship among $\delta_{\mathrm{OSP}}(\mathbf{r})$, $\delta_{\mathrm{CEM}}(\mathbf{r})$ and $\delta_{\mathrm{TCIMF}}(\mathbf{r})$

	κ	matched signatures	annihilated signatures	filter vector **w**	spectral correlation
$\delta_{\mathrm{OSP}}(\mathbf{r})$	1	**d**	**U**	$\mathbf{d}^T P_{\mathbf{U}}^{\perp}$ (3.9)	$\mathbf{I}_{LxL}$
$\delta_{\mathrm{CEM}}(\mathbf{r})$	$\left(\mathbf{d}^T\mathbf{R}_{L\times L}^{-1}\mathbf{d}\right)^{-1}$	**d**	none	$\mathbf{d}^T\mathbf{R}_{L\times L}^{-1}$ (4.8)	$\mathbf{R}_{LxL}$
$\delta_{\mathrm{TCIMF}}(\mathbf{r})$	1	**D**	**U**	$\mathbf{R}_{L\times L}^{-1}[\mathbf{D}\,\mathbf{U}]\left([\mathbf{D}\,\mathbf{U}]^T\mathbf{R}_{L\times L}^{-1}[\mathbf{D}\,\mathbf{U}]\right)^{-1}\left[\mathbf{1}_{p\times 1}^T\ \mathbf{0}_{q\times 1}^T\right]$ (4.12)	$\mathbf{R}_{LxL}$

4.4 A COMPARATIVE ANALYSIS BETWEEN CEM AND TCIMF

This section conducts a comparative analysis between CEM and TCIMF experiments to demonstrate their relative performance using a series of computer simulations and real hyperspectral data.

4.4.1 Computer Simulations

The AVIRIS laboratory data set shown in Fig. 1.5 and considered in Chapter 2 will be used for performance evaluation. Three signatures, blackbrush, creosote leaves and sagebrush were used as targets of interest, red soil as a background signature and dry grass as an interferer. A set of 300 mixed pixels were simulated. Each simulated pixel contained one background signature which is red soil with abundance fixed at 5% and one interferer which is dry grass with abundance fixed at 5%. In addition, each pixel also contained two undesired target signatures creosote leaves and sagebrush with even split abundance, 45%. Now at pixels 50, 100, 150, 200, 250, 300 we added the desired target signature, blackbrush with abundance 5%, 10%, 20%, 40%, 60%, 80% respectively while evenly reducing the abundance fractions of the two undesired target signatures, creosote leaves and sagebrush. For example, at pixel 50, it contained 5% blackbrush as the desired target signature, 42.5% creosote leaves and 42.5% sagebrush as undesired target signatures, 5% dry grass as an interferer and 5% red soil as a background signature. The abundance fractions of five signatures in these 300 simulated mixed pixels were shown in Fig. 4.1.

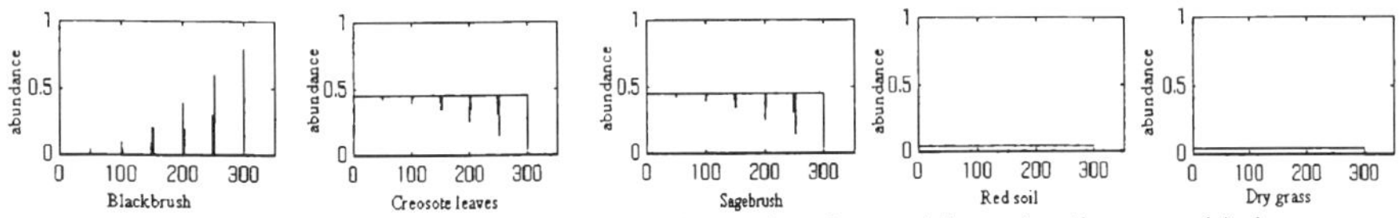

(a) blackbrush (b) creosote leaves (c) sagebrush (d) red soil (e) dry grass

Figure 4.1. 300 simulated mixed pixels

In addition, a white Gaussian noise was added to each pixel to achieve 30:1 signal-to-noise ratio (SNR) which was defined as 50% reflectance divided by the standard deviation of the noise (Harsanyi and Chang, 1994). Using blackbrush as the desired signature **d**, three variants of TCIMF were implemented depending upon the how the undesired signature matrix **U** was selected. One used the two undesired signatures, creosote leaves and sagebrush to make up the undesired signature matrix **U** with the constraint vector chosen to be $(1,0,0)^T$ while the other two used only one undesired signature to form **U** with the constraint vector given by $(1,0)^T$. Fig. 4.2(a-c) show their respective detection results and Fig. 4.2(d) is the detection results of CEM filter with blackbrush as the desired signature **d** while discarding creosote leaves and sagebrush. As we can see from Fig. 4.2(a), TCIMF detected target pixels 150 (barely detected), 200, 250, 300 but missed the target pixels 50, 100. The performance of TCIMF was slightly degraded shown in Figs. 4.2(b) and 4.2(c) when only one signature used for **U**. The performance of CEM in Fig. 4.2(d) was the worst because it missed all of the six target pixels. This experiment demonstrated the advantage of constraining the undesired target signature matrix **U** to zero over minimizing the energy contributed from **U**.

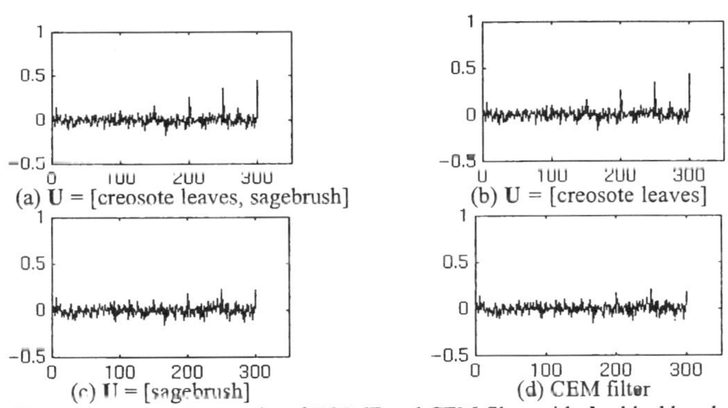

(a) **U** = [creosote leaves, sagebrush] (b) **U** = [creosote leaves]

(c) **U** = [sagebrush] (d) CEM filter

Figure 4.2. Detection results of TCIMF and CEM filter with **d** = blackbrush

Similar computer simulations to Fig. 4.1 were also conducted with using creosote leaves as the desired target signature and **U** = [blackbrush, sagebrush] as the undesired targets. The detection results are shown in Fig. 4.3.

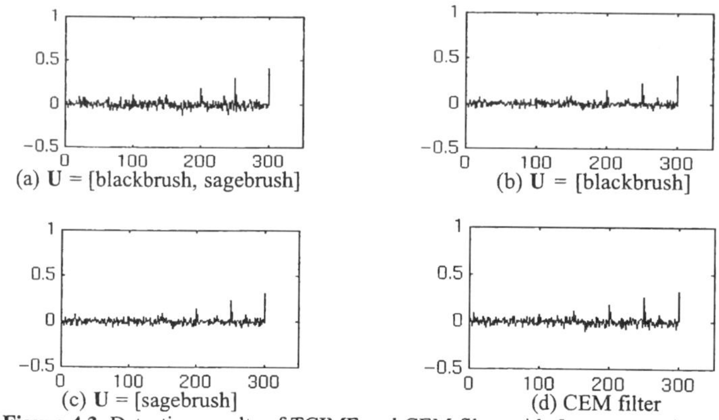

(a) **U** = [blackbrush, sagebrush] (b) **U** = [blackbrush]

(c) **U** = [sagebrush] (d) CEM filter

Figure 4.3. Detection results of TCIMF and CEM filter with **d** = creosote leaves

Interestingly, in this case, the detection performance of TCIMF was only slightly better than CEM and TCIMF with **U** using only one undesired target signature whose detection performance was nearly the same as shown in Fig. 4.3(b-d) where CEM detected creosote leaves at pixels 200, 250 and 300. However, when the same simulations were conducted using sagebrush as the desired target signature and **U** = [blackbrush, creosote leaves] as the undesired target signatures, the detection results shown in Fig. 4.4 are very different from those in Figs. 4.2 and 4.3. In this case, the TCIMF with **U** = [blackbrush, creosote leaves] and the TCIMF using **U** = [creosote leaves] performed nearly the same by extracting sagebrush at pixels 200, 250 and 300. They both performed significantly better than CEM and the TCIMF using only blackbrush as the undesired target signature whose detection performance missed all six targets except for the case of Fig. 4.4(b) where sagebrush was barely detected at pixel 300 by the TCIMF with **U** = [blackbrush].

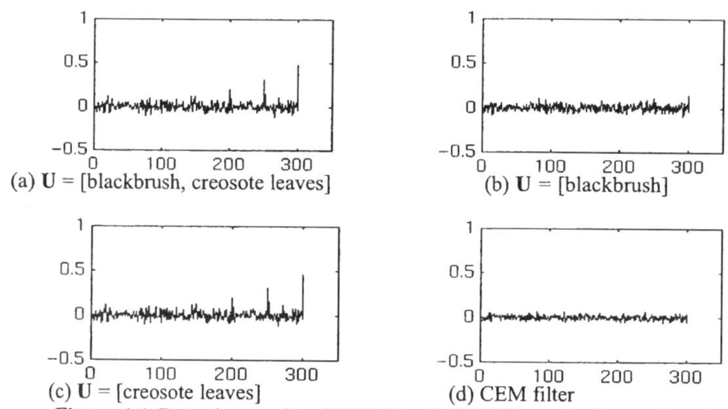

(a) **U** = [blackbrush, creosote leaves] (b) **U** = [blackbrush]

(c) **U** = [creosote leaves] (d) CEM filter

Figure 4.4. Detection results of TCIMF and CEM filter with **d** = sagebrush

All these phenomena can be explained by their spectral similarity values of the five targets in Fig. 1.5 tabulated in Tables 2.3-2.6. For example, Table 2.5 (Table 2.6) shows that blackbrush and sagebrush had most similar spectral signatures with its SID-measured spectral similarity value as small as 0.0063 (0.2263 measured by HMMID). Compared to the SID-measured spectral similarity values, 0.4970 (1.5390 measured by HMMID) between blackbrush and creosote leaves, and 0.0303 (1.4102 measured by HMMID) between creosote leaves and sagebrush, the SID-measured spectral similarity value between blackbrush and sagebrush is so small that it was very difficult to differentiate one from another. Therefore, if one was used as the desired target signature while the other as the undesired target signature, most spectral abundance of the desired target signature was eliminated by TCIMF or minimized by CEM. As a result, the detection performance was significantly reduced as shown in Fig. 4.2(c-d) as well as Fig. 4.4(b,d).

4.4.2 Hyperspectral Image Experiments

In this section, the 15-panel HYDICE scene in Fig. 1.7 was used for experiments where the five panel signatures in Fig. 1.8 were used for target information. When CEM was implemented, only the panels of interest was designated as the desired target while discarding all the information of other target panels. By contrast, TCIMF used the target knowledge to pass the desired target panels but also eliminate undesired target panels. The detection results of TCIMF and CEM are shown in Figs. 4.5 and 4.6 respectively. According to Tables 2.13-2.16 the spectral signatures of panels in rows 4 and 5 are very similar where their spectral similarity value measured by SID was 0.0017. As expected, CEM detected panels in one row also picked up a few panel pixels in another row. A similar phenomenon was also observed in the detection of panels in rows 2 and 3 because the spectral signatures of panels in rows 2 and 3 are very similar with their SID-measured spectral similarity value, 0.0023. By contrast, TCIMF performed better than CEM by nulling undesired panel pixels. Despite that the SID-measured similarity value between P1 and P2, 0.0027 is also small compared to 0.0023 between P2 and P3, the SID-measured similarity value between P1 and P3, 0.0060 is almost three times the SID-measured similarity value between P1 and P2. In addition, the spectral shape of P1 is not as close as the spectral shapes among P1 and P2. As a result, both TCIMF and CEM performed equally well in detecting panel pixels in row 1.

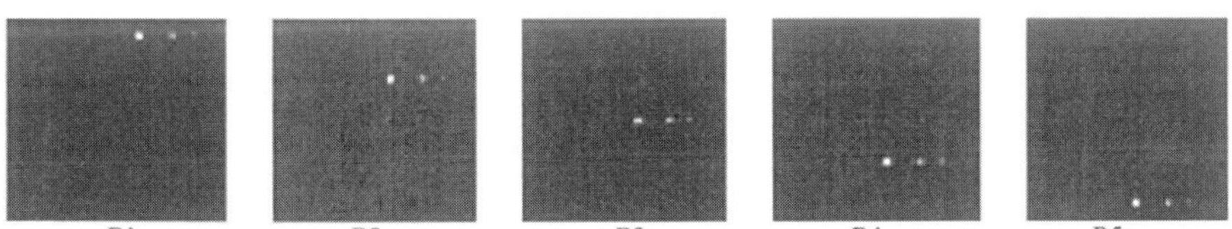

P1 P2 P3 P4 P5

Figure 4.5. Detection results of TCIMF

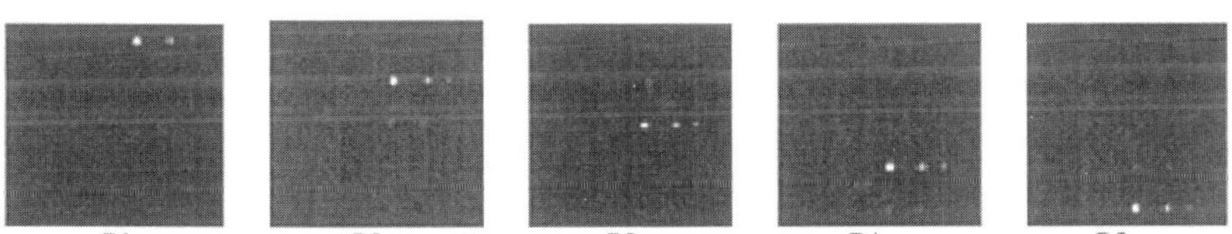

P1 P2 P3 P4 P5

Figure 4.6. Detection results of CEM

This experiment demonstrates that both SID-similarity values and geometric shapes of spectral signatures have significant impacts on target detection performance.

Since Fig. 1.7(b) provides the ground truth map of the 15 panels, we can actually compare the performance of CEM and TCIMF to that of OSP and NCLS detectors in Chapter 3 by tallying how many of red panel pixels are detected correctly. Here we did not include SCLS due to the fact that it generally did not perform well compared to NCLS as shown in Chapter 3. It should be noted that the images generated by all these four detectors are abundance fractional images, which are gray scale. In this case, we need a thresholding method to determine if a pixel is a panel pixel. The abundance percentage cut-off method proposed in Ren (1998) was used to threshold abundance fractional images into binary images. It will be defined in (9.3) and discussed in detail in Chapter 9. If the abundance fraction of the desired target signature present in a pixel is above a certain percentage, the pixel will be declared to be a target pixel and assigned to 1. Otherwise, the pixel will be considered to be a background pixel and assigned to 0. Tables 4.2-4.5 are detection results obtained by using 50%, 25%, 20% and 10% abundance as cut-off threshold values respectively. In these tables, N is total number of panel pixels, N_R is the number of red panel pixels, N_{RD} is the number of red panel pixels detected and classified correctly, N_{FA} is the number of false alarm pixels and R_D is detection rate defined by $R_D = N_{RD} / N_R$

Table 4.2. Detection results of OSP, NCLS, CEM, and TCIMF using 50% cut-off abundance

	N_R	OSP			NCLS			CEM			TCIMF		
		N_{RD}	R_D	N_{FA}	N_{RD}	R_D	N_{FA}	N_{RD}	R_D	N_{FA}	N_{RD}	R_D	N_{FA}
P1	3	1	0.3333	6	1	0.3333	2	2	0.6667	0	2	0.6667	0
P2	4	2	0.5000	338	3	0.7500	2	3	0.7500	0	3	0.7500	0
P3	4	3	0.7500	227	3	0.7500	3	3	0.7500	0	3	0.7500	0
P4	4	0	0.0000	443	3	0.7500	2	3	0.7500	0	3	0.7500	0
P5	4	3	0.7500	0	3	0.7500	2	3	0.7500	0	3	0.7500	0
Total	19	9	0.4737	1014	13	0.6842	11	14	0.7368	0	14	0.7368	0

Table 4.3. Detection results of OSP, NCLS, CEM, and TCIMF using 25% cut-off abundance

	N_R	OSP			NCLS			CEM			TCIMF		
		N_{RD}	R_D	N_{FA}	N_{RD}	R_D	N_{FA}	N_{RD}	R_D	N_{FA}	N_{RD}	R_D	N_{FA}
P1	3	2	0.6667	35	2	0.6667	6	2	0.6667	0	2	0.6667	0
P2	4	2	0.5000	1572	3	0.7500	3	4	1.0000	1	4	1.0000	0
P3	4	4	1.0000	712	4	1.0000	4	4	1.0000	0	4	1.0000	0
P4	4	4	1.0000	2046	3	0.7500	5	4	1.0000	0	4	1.0000	0
P5	4	3	0.7500	13	3	0.7500	5	3	0.7500	0	3	0.7500	0
Total	19	17	0.8947	4378	15	0.7895	23	17	0.8947	1	17	0.8947	0

Table 4.4. Detection results of OSP, NCLS, CEM, and TCIMF using 20% cut-off abundance

	N_B	OSP			NCLS			CEM			TCIMF		
		N_{RD}	R_D	N_{FA}	N_{RD}	R_D	N_{FA}	N_{RD}	R_D	N_{FA}	N_{RD}	R_D	N_{FA}
P1	3	3	0.6667	51	2	0.6667	7	3	1.0000	0	3	1.0000	1
P2	4	4	1.0000	1879	4	1.0000	3	4	1.0000	2	4	1.0000	0
P3	4	4	1.0000	774	4	1.0000	5	4	1.0000	0	4	1.0000	0
P4	4	4	1.0000	2468	4	1.0000	5	4	1.0000	0	4	1.0000	0
P5	4	3	0.7500	21	3	0.7500	6	4	1.0000	0	4	1.0000	0
Total	19	18	0.8947	5193	17	0.8947	26	19	1.0000	2	19	1.0000	1

Table 4.5. Detection results of OSP, NCLS, CEM, and TCIMF using 10% cut-off abundance

	N_B	OSP			NCLS			CEM			TCIMF		
		N_{RD}	R_D	N_{FA}	N_{RD}	R_D	N_{FA}	N_{RD}	R_D	N_{FA}	N_{RD}	R_D	N_{FA}
P1	3	3	1.0000	133	3	1.0000	11	3	1.0000	0	3	1.0000	0
P2	4	4	1.0000	2524	4	1.0000	14	4	1.0000	5	4	1.0000	3
P3	4	4	1.0000	944	4	1.0000	12	4	1.0000	2	4	1.0000	0
P4	4	4	1.0000	3393	4	1.0000	9	4	1.0000	3	4	1.0000	0
P5	4	3	0.7500	59	4	1.0000	14	4	1.0000	3	4	1.0000	0
Total	19	18	0.9474	7053	19	1.0000	60	19	1.0000	13	19	1.0000	3

As shown in these tables, CEM and TCIMF performed significantly better than did OSP and NCLS. As the cut-off abundance percentage was decreased from 50% to 10%, the detection rates of all the four detectors were improved at the expense of an increase of false alarm pixels. But the TCIMF was the best in the sense that it produced the smallest number of false alarmed pixels. In particular, when the 20% abundance was used as a cut-off threshold value, both CEM and the TCIMF already reached 100% detection rate with only one or two false alarm pixels, while OSP and NCLS achieved only the detection rate 0.8947 with many false alarmed pixels.

In order to see how TCIMF performed with using different panels as undesired target signatures, Fig. 4.7(a-c) shows the results of detecting panels in row 5 with **U** made up of panels in row 4, panels in rows 3 and 4, and panels in rows 2-4 rows respectively.

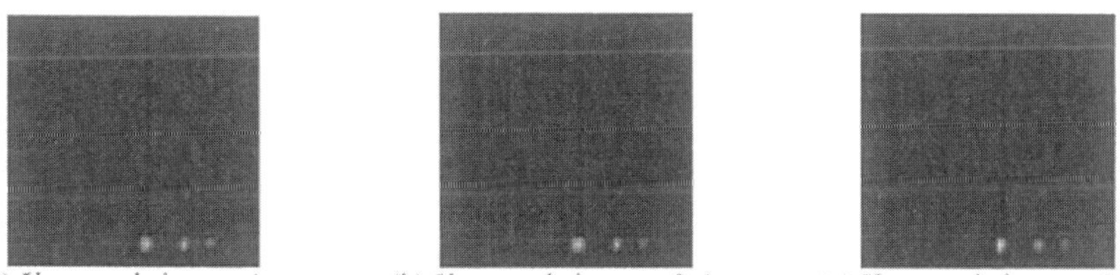

(a) **U** = panels in row 4 (b) **U** = panels in rows 3-4 (c) **U** = panels in rows 2-4

Figure 4.7. Results of detecting panels in row 5

The results shown in Figs. 4.7(a-c) were very close. The reason for that is because all the used **U** contained the panels in row 4 which removed their interfering effects to enhance the detection of the panels in row 5. Whether or not the panels in other rows were included in the **U** had very little impact on the detection performance since their spectral signatures were different from those of panels in row 5.

However, if we removed the panels in row 4 from **U** in the experiment of Fig. 4.7, the detection results of the panels in row 5 are shown in Fig. 4.8(a-c), where Fig. 4.8(a) was precisely the CEM result in shown in Fig. 4.6. As we can see, the panels in row 4 were still barely visible, but could not be nulled out as was done in Fig. 4.5 and Figs. 4.7(a-c) by TCIMF that included the panels in row 4 in the **U**.

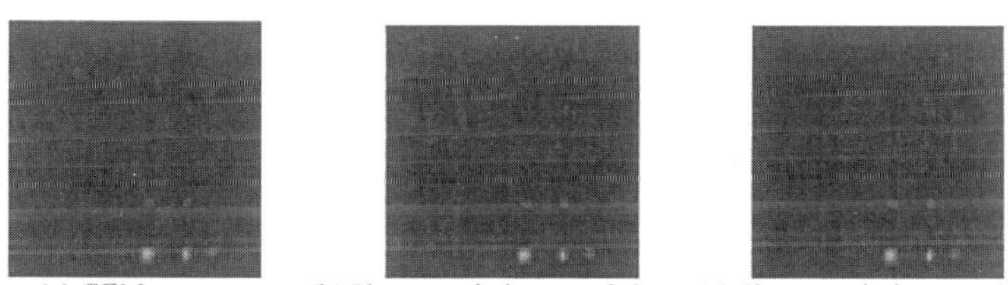

(a) CEM (b) **U** = panels in rows 2-3 (c) **U** = panels in rows 1-3

Figure 4.8. Results of detecting panels in row 5

4.5 SENSITIVITY OF CEM AND TCIMF TO LEVEL OF TARGET INFORMATION

As noted in Section 4.3, $\delta_{\text{CEM}}(\mathbf{r})$ approximates $P_{\mathbf{U}}^{\perp}$ by $\mathbf{R}_{L\times L}^{-1}$. However, it may not be a best approximation in terms of information approximation. This is because the sample correlation matrix $\mathbf{R}_{L\times L}$. includes the desired signature $\mathbf{d}$, which is not included in $P_U^\perp$. So, a better approximation can be achieved by replacing $\mathbf{R}_{L\times L}^{-1}$ with $\tilde{\mathbf{R}}_{L\times L}^{-1}$, which excludes all the pixels specified by the desired target signature $\mathbf{d}$. More specifically, let $\Delta(\mathbf{d})$ denote the set of pixels in the image data that are specified by $\mathbf{d}$. Then $\tilde{\mathbf{R}}_{L\times L} = (N - |\Delta(\mathbf{d})|)^{-1}\left[\sum_{i=1}^{N}\mathbf{r}_i\mathbf{r}_i^T - \sum_{\mathbf{r}_k\in\Delta(\mathbf{d})}\mathbf{r}_k\mathbf{r}_k^T\right]$ where $|\Delta(\mathbf{d})|$ is the number of target pixels in $\Delta(\mathbf{d})$. As will be demonstrated in computer simulations, using $\tilde{\mathbf{R}}_{L\times L}$ in place of $\mathbf{R}_{L\times L}$ can improve the performance of $\delta_{\text{CEM}}(\mathbf{r})$. It is also true for the case that removing target pixels from $\mathbf{R}_{L\times L}$ whose signatures are close and similar to the desired target signatures can further enhance the performance as well. This was also demonstrated in Thai and Healey (2002) where the background subspace was generated by removing those signatures, which are close to target signatures so that background subspace can be effectively annihilated by an orthogonal projector. Similar conclusions can be also drawn for the LCMV-based target detector and TCIMF. In many applications, identifying target pixels $\mathbf{d}$ may be difficult. Under such circumstance, using sample correlation matrix $\mathbf{R}_{L\times L}$ instead of $\tilde{\mathbf{R}}_{L\times L}$ may be a simple way to do. Nevertheless, we will show in the following computer simulations that $\tilde{\mathbf{R}}_{L\times L}$ can indeed improve detection performance of CEM and TCIMF.

4.5.1 Computer Simulations

The used data were three field reflectance spectra, creosote leaves, dry grass and red soil are shown in Fig. 1.5. We simulate 400 mixed pixel vectors, $\{\mathbf{r}_i\}_{i=1}^{400}$, as follows.

We start the first pixel vector with 100% red soil and 0% dry grass, then began to increase 0.25% dry grass and decrease 0.25% red soil every pixel vector until the 400th pixel vector which contained 100% dry grass. We then added creosote leaves to pixel vector numbers 198-202 at abundance fractions 10% while reducing the abundance of red soil and dry grass evenly. For example, after addition of creosote leaves, the resulting pixel vector 200 contained 10% creosote leaves, 45% red soil and 45% dry grass. White Gaussian noise is also added to each pixel vector to achieve a 30:1 signal-to-noise ratio as was defined as 50% reflectance divided by the standard deviation of the noise (Harsanyi and Chang, 1994). Fig. 4.9 shows these 400 simulated mixed pixel vectors.

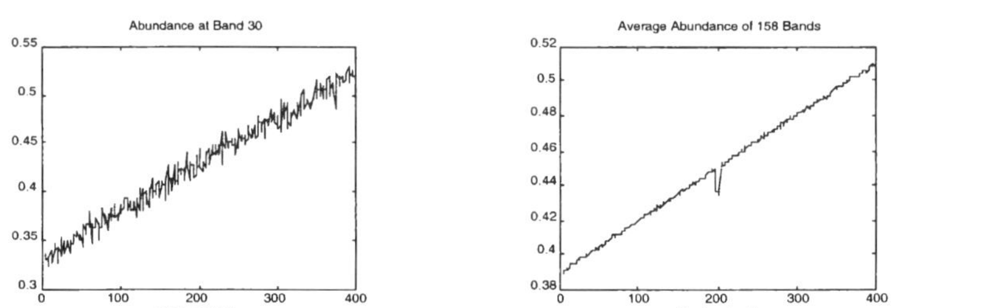

(a) Abundance of each pixel at band 30 (b) Abundance of each pixel averaged over 158 bands

Figure 4.9. 400 simulated mixed pixel vectors

Using these 400 simulated pixel vectors in Fig. 4.9 as data we can closely examine the performance of $\delta_{\text{CEM}}(\mathbf{r})$ using $\mathbf{R}_{L\times L}$ and $\tilde{\mathbf{R}}_{L\times L}$ respectively. Fig. 4.10(a-b) shows the detection results of creosote leaves where the former detected 0.07 of abundance fraction compared to 0.1 of abundance fraction detected by the latter.

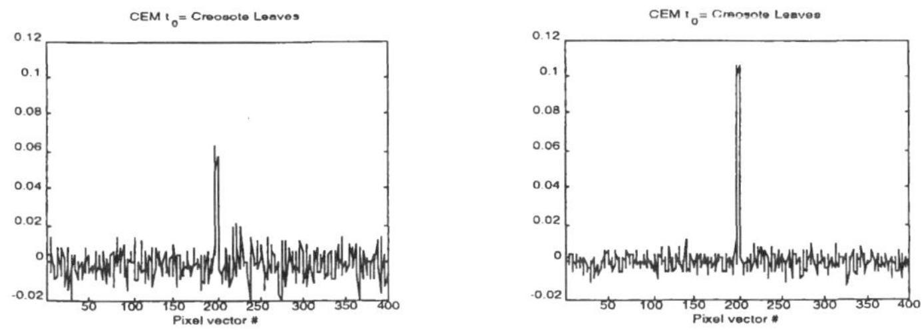

(a) $\delta_{\text{CEM}}(\mathbf{r})$ using $\mathbf{R}_{L\times L}$ (b) $\delta_{\text{CEM}}(\mathbf{r})$ using $\tilde{\mathbf{R}}_{L\times L}$

Figure 4.10. Results of $\delta_{\text{CEM}}(\mathbf{r})$ using $\mathbf{R}_{L\times L}$ and $\tilde{\mathbf{R}}_{L\times L}$ respectively in detection of creosote leaves

Apparently, $\delta_{\text{CEM}}(\mathbf{r})$ using $\tilde{\mathbf{R}}_{L\times L}$ performed better than $\delta_{\text{CEM}}(\mathbf{r})$ using $\mathbf{R}_{L\times L}$ in terms of abundance fraction detection where $\tilde{\mathbf{R}}_{L\times L}$ did not include the desired target pixels from 198-202, $\{\mathbf{r}_i\}_{i=198}^{202}$. Similar results shown in Fig. 4.11 were also obtained by the $\delta_{\text{TCIMF}}(\mathbf{r})$ using $\tilde{\mathbf{R}}_{L\times L}$ in place of $\mathbf{R}_{L\times L}$ where $\delta_{\text{TCIMF}}(\mathbf{r})$ was implemented by letting $\mathbf{d}$ = creosote leaves and $\mathbf{U}$ = [grass, soil]. As we can see, $\delta_{\text{CEM}}(\mathbf{r})$ and $\delta_{\text{TCIMF}}(\mathbf{r})$ using $\tilde{\mathbf{R}}_{L\times L}$ performed better than their counterparts using $\mathbf{R}_{L\times L}$. The advantage of using $\tilde{\mathbf{R}}_{L\times L}$ becomes more evident if the number of target pixels was increased and expanded from 5 pixels to 15 pixels with $\{\mathbf{r}_i\}_{i=193}^{207}$ starting pixel number 193 to pixel number 207 in Figs. 4.12-4.13 and 25 pixels with $\{\mathbf{r}_i\}_{i=188}^{212}$ starting pixel number 188 to pixel number 212 in Figs. 4.14-4.15.

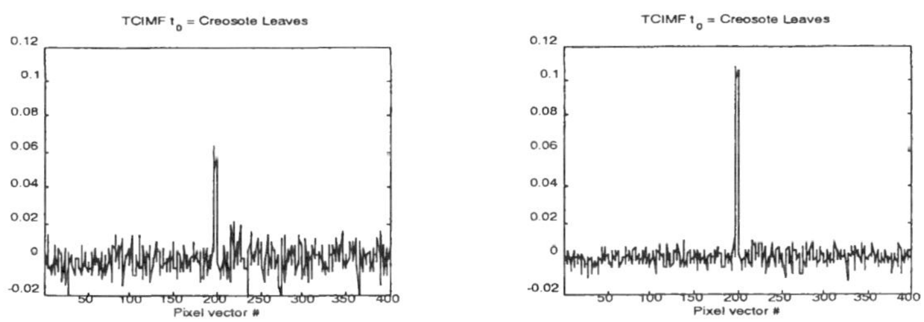

(a) $\delta_{\text{TCIMF}}(\mathbf{r})$ using $\mathbf{R}_{L\times L}$ (b) $\delta_{\text{TCIMF}}(\mathbf{r})$ using $\tilde{\mathbf{R}}_{L\times L}$

Figure 4.11. Results of $\delta_{\text{TCIMF}}(\mathbf{r})$ using $\mathbf{R}_{L\times L}$ and $\tilde{\mathbf{R}}_{L\times L}$ respectively in detection of creosote leaves

Figs. 4.12-4.13 show the results of $\delta_{\text{CEM}}(\mathbf{r})$ and $\delta_{\text{TCIMF}}(\mathbf{r})$ using $\tilde{\mathbf{R}}_{L\times L}$ over their counterparts using $\mathbf{R}_{L\times L}$ was more visible. The detected abundance fractions of the creosote leaves using $\tilde{\mathbf{R}}_{L\times L}$ were more accurate that those detected by using $\mathbf{R}_{L\times L}$. Additionally, and $\delta_{\text{TCIMF}}(\mathbf{r})$ using $\tilde{\mathbf{R}}_{L\times L}$ also suppressed the effects caused by background signatures (dry grass and soil) more effectively than did $\delta_{\text{CEM}}(\mathbf{r})$ and $\delta_{\text{TCIMF}}(\mathbf{r})$ using $\mathbf{R}_{L\times L}$.

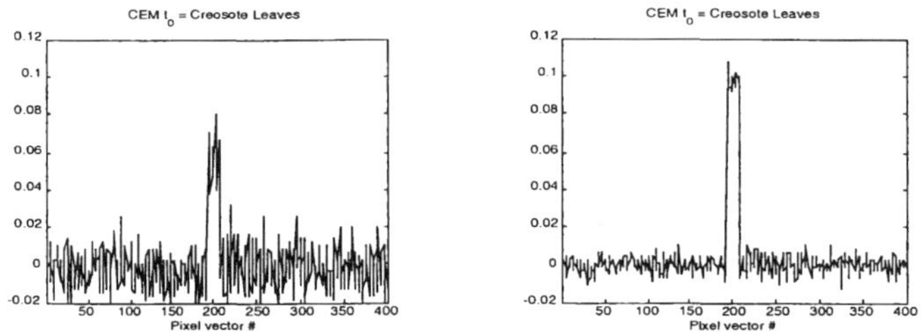

(a) $\delta_{\text{CEM}}(\mathbf{r})$ using $\mathbf{R}_{L\times L}$ (b) $\delta_{\text{CEM}}(\mathbf{r})$ using $\tilde{\mathbf{R}}_{L\times L}$

Figure 4.12. Detection results of 15 creosote leaves pixels, $\{\mathbf{r}_i\}_{i=193}^{207}$ starting pixel number 193 to pixel number 207. (a) $\delta_{\text{CEM}}(\mathbf{r})$ using $\mathbf{R}_{L\times L}$; (b) $\delta_{\text{CEM}}(\mathbf{r})$ using $\tilde{\mathbf{R}}_{L\times L}$

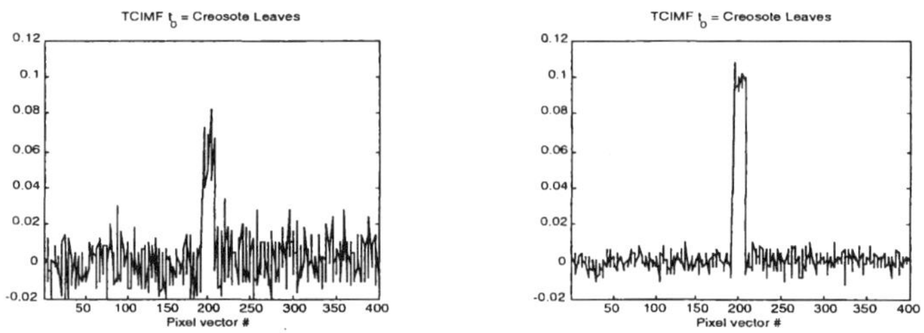

(a) $\delta_{\text{TCIMF}}(\mathbf{r})$ using $\mathbf{R}_{L\times L}$ (b) $\delta_{\text{TCIMF}}(\mathbf{r})$ using $\tilde{\mathbf{R}}_{L\times L}$

Figure 4.13. Detection results of 15 creosote leaves pixels, $\{\mathbf{r}_i\}_{i=193}^{207}$ starting pixel number 193 to pixel number 207. (a) $\delta_{\text{TCIMF}}(\mathbf{r})$ using $\mathbf{R}_{L\times L}$; (b) $\delta_{\text{TCIMF}}(\mathbf{r})$ using $\tilde{\mathbf{R}}_{L\times L}$

Similar experiments to Figs. 4.12-4.13 were also conducted for the case that the number of target pixels was increased from 15 to 25. The results are shown in Figs. 4.14-4.15. As we can see from these figures, the performance of $\delta_{\text{CEM}}(\mathbf{r})$ and $\delta_{\text{TCIMF}}(\mathbf{r})$ using $\mathbf{R}_{L\times L}$ was degraded due to more interfering effects caused by additional target pixels, while the performance of $\delta_{\text{CEM}}(\mathbf{r})$ and $\delta_{\text{TCIMF}}(\mathbf{r})$ using $\tilde{\mathbf{R}}_{L\times L}$ remained nearly the same.

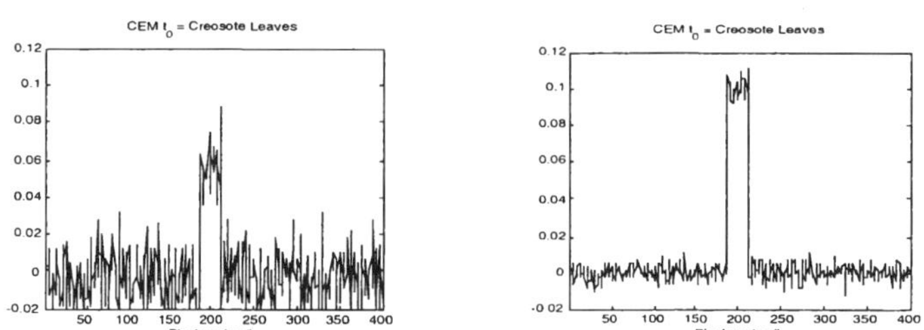

(a) $\delta_{\text{CEM}}(\mathbf{r})$ using $\mathbf{R}_{L\times L}$ (b) $\delta_{\text{CEM}}(\mathbf{r})$ using $\tilde{\mathbf{R}}_{L\times L}$

Figure 4.14. Detection results of 25 creosote leaves pixels, $\{\mathbf{r}_i\}_{i=188}^{212}$ starting pixel number 188 to pixel number 212. (a) $\delta_{\text{CEM}}(\mathbf{r})$ using $\mathbf{R}_{L\times L}$; (b) $\delta_{\text{CEM}}(\mathbf{r})$ using $\tilde{\mathbf{R}}_{L\times L}$

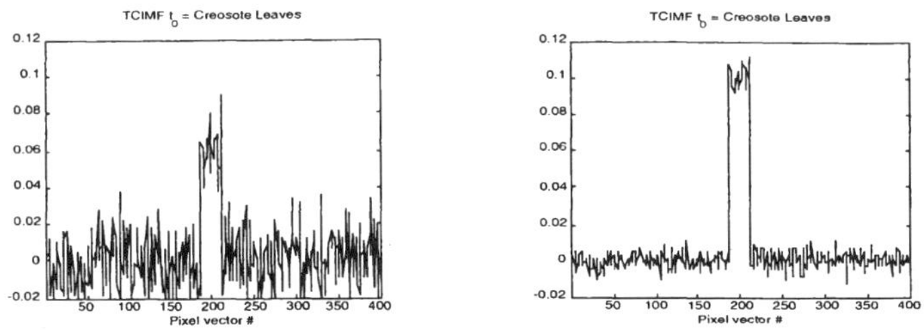

(a) $\delta_{\text{TCIMF}}(\mathbf{r})$ using $\mathbf{R}_{L\times L}$ (b) $\delta_{\text{TCIMF}}(\mathbf{r})$ using $\tilde{\mathbf{R}}_{L\times L}$

Figure 4.15. Detection results of 25 creosote leaves pixels, $\{\mathbf{r}_i\}_{i=188}^{212}$ starting pixel number 188 to pixel number 212. (a) $\delta_{\text{TCIMF}}(\mathbf{r})$ using $\mathbf{R}_{L\times L}$; (b) $\delta_{\text{TCIMF}}(\mathbf{r})$ using $\tilde{\mathbf{R}}_{L\times L}$

4.5.2 Hyperspectral Image Experiments

In Section 4.4, we have seen that CEM and TCIMF worked very effectively for the 15-panel HYDYCE scene. This was because the HYDICE image has very fine spatial resolution (1.5 m) and the used target knowledge was accurate to represent panel signatures. However, if the target information does not well represent the targets of interest, CEM and TCIMF may fail to detect those target pixels, which also belong to the same target class. In other words, CEM and TCIMF are very sensitive to the level of target information that is used in their filter design. Additionally, they are also very sensitive to noise, which is another issue that will be addressed in Chapter 7.

In order to illustrate the sensitivity of CEM to target knowledge, two experiments were conducted using the AVIRIS image in Fig. 1.6. Since the AVIRIS image has 20-m spatial resolution and covers a very large area of LCVF, most of image pixels are likely mixed. In this case, the target information directly obtained from a few pixels in the image scene may not well represent the targets of interest. So, the first experiments used a single target pixel as the target information for target detection marked by white circles in Fig. 4.16. Fig. 4.16 shows the detection of five targets, cinders, playa, rhyolite, shade and vegetation. As we can from this experiment, the CEM could only detect the brightest pixel that was used for target information.

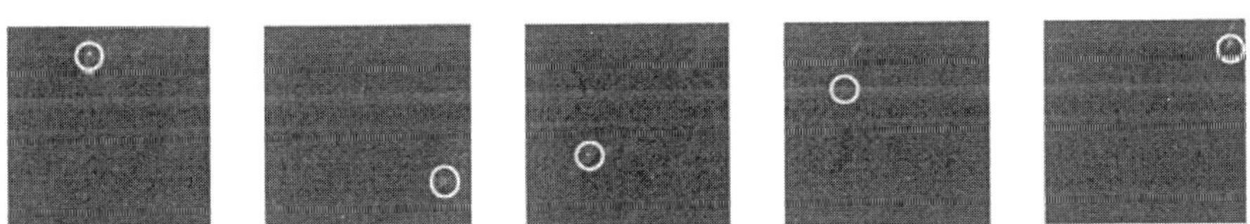

(a) cinders (b) playa (c) rhyolite (d) shade (e) vegetation

Figure 4.16. Detection of CEM using single pixel target information

A second experiment used the full-region target information to detect all five targets. The detection results are shown in Fig. 4.17. Compared to Fig. 10.16, the detection in Fig. 4.17 was significantly improved. In particular, the whole region of the dry lake was extracted and the anomaly marked by a white circle in Fig. 3.1(c) was also shown by a white circle. If we further compare the results in Fig. 4.17 to that in Fig. 3.1 obtained by OSP, the detection results produced by Fig. 4.17 was significantly improved.

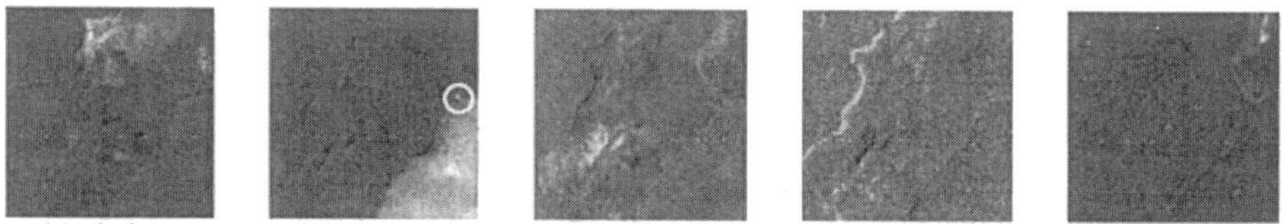

(a) cinders (b) playa (c) rhyolite (d) shade (e) vegetation

Figure 4.17. Detection of CEM using full-region target information

For TCIMF, similar sensitivity phenomena were also observed since the imprecise knowledge of undesired signatures did not effectively eliminate their interfering effects.

4.6 REAL-TIME PROCESSING

One of advantages of CEM and TCIMF is that the data matrix $\mathbf{R}_{L\times L}$ in the optimal weights specified by (4.7) and (4.111) can be decomposed into a product of a unitary matrix $\mathbf{Q}$ and an upper triangular matrix $\mathbf{R}$ by either the Givens rotations or the Householder transform. Such a decomposition is called QR-decomposition (Golub and Van Loan, 1989; Haykin, 1986). By means of QR-decomposition there is no need of computing the inverse of $\mathbf{R}_{L\times L}$, $\mathbf{R}_{L\times L}^{-1}$ directly.

Suppose that data processing is carried out line-by-line from left to right and top to bottom. Let the image to be processed be denoted by a data matrix $\left[\mathbf{r}_{ti}\right]_{t=1,i=1}^{M,N}$ of size $M \times N$ where M is the number of lines (rows) and N is the total number of pixel vectors in one line (i.e., the number of columns in the image). For each line t, we form a data matrix $\mathbf{X}_t = \left[\mathbf{r}_{t1}\,\mathbf{r}_{t2}\cdots\mathbf{r}_{tN}\right]$ where $\mathbf{r}_{ti}$ is the pixel vector being visited at time instance i in the t-th line. In this case, the $\mathbf{R}_{L\times L}$ in (4.3) is replaced by the data correlation matrix of line t in the image, denoted by Σ_t

$$\Sigma_t = (1/N)\left[\Sigma_{i=1}^{N}\mathbf{r}_{ti}\mathbf{r}_{ti}^T\right] = (1/N)\left[\mathbf{X}_t\mathbf{X}_t^T\right]. \tag{4.16}$$

Using a QR-decomposition the matrix, $\mathbf{X}_t$ can be expressed by

$$\mathbf{X}_t = \mathbf{Q}_t\mathbf{R}_t. \tag{4.17}$$

Here $\mathbf{Q}_t$ is a unitary matrix with $\mathbf{Q}_t^{-1} = \mathbf{Q}_t^T$ and $\mathbf{R}_t = \left[\mathbf{R}_t^{upper}\ \mathbf{0}\right]^T$ is not necessarily of full rank where $\mathbf{0}$ is a zero vector. The $\mathbf{R}_t^{upper} = \begin{bmatrix} * & * & \cdots & * \\ 0 & * & \cdots & * \\ \vdots & \ddots & \ddots & \vdots \\ 0 & \cdots & 0 & * \end{bmatrix}$ is an upper triangular matrix and * in $\mathbf{R}_t^{upper}$ is a non-zero element. From (4.12) the inverse of Σ_t can be computed as

$$\Sigma_t^{-1} = N\left(\mathbf{X}_t\mathbf{X}_t^T\right)^{-1} = N\left\{\left(\mathbf{R}_t^{upper}\right)^{-1}\left[\left(\mathbf{R}_t^{upper}\right)^T\right]^{-1}\right\} \tag{4.18}$$

where the unitary matrix $\mathbf{Q}_t$ being canceled out in (4.18) because $\mathbf{Q}_t^{-1} = \mathbf{Q}_t^T$. Substituting Σ_t^{-1} in (4.10) for $\mathbf{R}_{L\times L}^{-1}$ in (4.10) yields

$$\mathbf{w}^* = \left\{\left(\mathbf{R}_t^{upper}\right)^{-1}\left[\left(\mathbf{R}_t^{upper}\right)^T\right]^{-1}\right\}\mathbf{T}\left(\mathbf{T}^T\left\{\left(\mathbf{R}_t^{upper}\right)^{-1}\left[\left(\mathbf{R}_t^{upper}\right)^T\right]^{-1}\right\}\mathbf{T}\right)^{-1}\mathbf{c}\,. \tag{4.19}$$

Since $\mathbf{R}_t^{upper}$ is an upper triangular matrix, so is $\left(\mathbf{R}_t^{upper}\right)^{-1}$. Therefore, (4.19) does not require computing $\mathbf{R}_{L\times L}^{-1}$. As a result, it can be realized by systolic array implementation (Haykin, 1986) and implemented in real-time processing. For details of the systolic array implementation we refer to references Haykin (1986), Chang and Althouse (1991), Ward et al. (1986).

Since CEM and TCIMF share the same detector structure, only CEM will be used to demonstrate its capability of real-time implementation. The HYDICE image scene in Fig. 1.7 was used for experiments where the five panel signatures, P1-P5 in Fig. 1.8 was used for target information in CEM. Figs. 4.8-4.12 show the line-by-line real-time processing of CEM in detecting panels in five different rows respectively where the line number underneath each figure is the line currently being examined. As data processing began, no desired panel pixel detected. The image looked like random noise as shown in the images labeled by (a) in Figs. 4.18-4.22. However, as soon as desired panel pixels were detected, the image background suddenly turned to darkness and only pixels detected by CEM are highlighted by brightness as shown in images labeled by (b-d) in Figs 4.18-4.22. As the process proceeded, the image background remained dark while CEM continuing to look for pixels that matched the desired panel signature as shown in images labeled by (e) in Figs. 4.18-4.22. Finally, the whole process completed when the last pixel in the image is examined as shown in images labeled by (f) in Figs. 4.18-4.22.

(a) line 3 (b) line 4 (c) line 5 (d) line 6 (e) line 7 (f) line 64

Figure 4.18. (a) no panel pixel detected; (b-d) the panel pixels in row 1 detected; (e) completion of detecting panel pixels in row 1, (f) completion of the process

(a) line 17 (b) line 18 (c) line 19 (d) line 20 (e) line 21 (f) line 64

Figure 4.19. (a) no panel pixel detected; (b-d) the panel pixels in row 1 detected; (e) completion of detecting panel pixels in row 2; (f) completion of the process

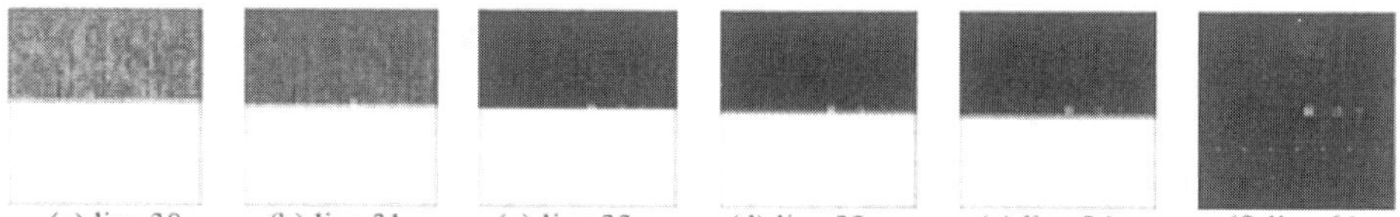

(a) line 30 (b) line 31 (c) line 32 (d) line 33 (e) line 34 (f) line 64

Figure 4.20. (a) no panel pixel detected; (b-d) the panel pixels in row 1 detected; (e) completion of detecting panel pixels in row 3; (f) completion of the process

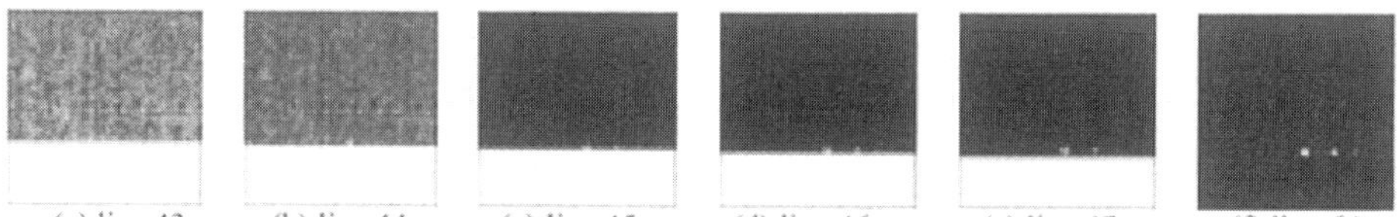

(a) line 43 (b) line 44 (c) line 45 (d) line 46 (e) line 47 (f) line 64

Figure 4.21. (a) no panel pixel detected; (b-d) the panel pixels in row 4 detected; (e) completion of detecting panel pixels in row 1; (f) completion of the process

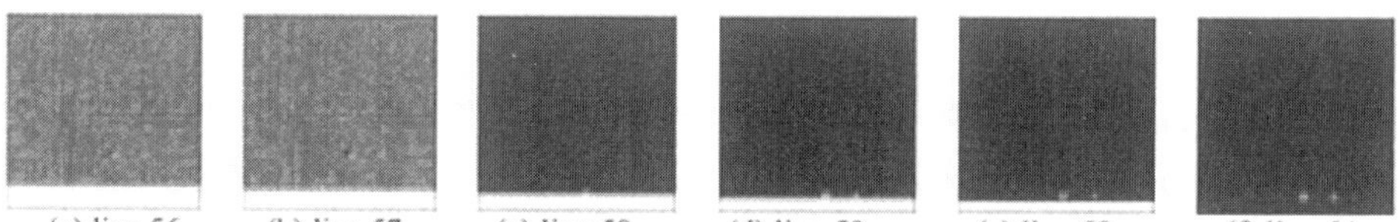

(a) line 56 (b) line 57 (c) line 58 (d) line 59 (e) line 60 (f) line 64

Figure 4.22. (a) no panel pixel detected; (b-d) the panel pixels in row 1 detected; (e) completion of detecting panel pixels in row 5; (f) completion of the process

Several comments are noteworthy.

1) When the sample correlation matrix, $\mathbf{R}_{L\times L}$ is formed by the entire image, it can only capture the global properties within the image. By contrast, if $\mathbf{R}_{L\times L}$ is formed by lines, the information is updated line-by-line. As a result, it can characterize local properties. So, its performance is generally better than the one using global properties.
2) The CEM implemented by the above line-by-line real-time processing only uses the information up to the line currently being processed. There is another real-time implementation, referred to as pixel-by-pixel real-time processing, which utilizes the information up to the pixel currently being examined. So, in both cases, they can be viewed as a causal processing. Since there is no significant difference between these two causal implementations for CEM, only the line-by-line real-time processing was demonstrated in this section. However, we will show in Section 6.4 that there is a noticeable difference in anomaly detection between the line-by-line real-time processing and pixel-by-pixel real-time processing.
3) When the entire image is used for processing, N is the total number of pixel vectors in the image. If the data is processed line-by-line, N is the total number of pixel vectors up to the line currently being visited by CEM. In our HYDICE experiments, the scene has size of 64×64 and the number of pixels in each line is $64 < 210$. In order to avoid singularity problem, CEM used the first four lines to start its real-time processing.

4.7 CONCLUSIONS

A linearly constrained minimum variance-based subpixel approach is presented in this chapter, which takes advantage of the strengths of CEM while improving its weaknesses. It divides targets of interest into a class of desired targets and another class of undesired targets while interference is treated as a separate third class of signal source. With this 3-component signal source model, TCIMF is designed to achieve detection of desired targets, annihilation of undesired targets and minimization of interfering effects all together in one single operation. It improves CEM in the sense that the effects of undesired targets are minimized rather than being eliminated. Since CEM can detect one target at a time, only one single target signature is used for experiments to have a fair comparative analysis between TCIMF and CEM. However, in order that CEM be used as a classifier, CEM must be implemented multiple times to classify different targets. This issue was investigated in Chang (2002b) and will be discussed in Chapter 11. Compared to CEM, TCIMF requires only one implementation to detect and classify multiple targets in the same manner that is carried out by LCMV (Chang and Ren, 1999). As seen in Fig. 1.7(a), without ground truth obtaining the information of targets of interest is extremely difficult. To cope with this problem, developing an unsupervised CEM and TCIMF without appealing for ground truth is highly desirable. Several unsupervised learning methods will be discussed in Chapter 5 (Chang et al., 1998a; Brumbley and Chang, 1999; Chang and Du, 1999; Chang and Ren, 2000; Chang and Heinz, 2000).

5

AUTOMATIC SUBPIXEL DETECTION: UNSUPERVISED SUBPIXEL DETECTION

The subpixel detection studied in Chapters 3 and 4 requires either full or partial target knowledge. In the target abundance-constrained subpixel detection, a linear mixture model is required where the complete knowledge of the target signature matrix **M** must be given *a priori*. In the target signature-constrained subpixel detection, there is no need of a linear mixture model, but the desired target signature **d** must be provided in advance. In many practical applications, such prior knowledge may not be available. One way to remedy this situation is to obtain target information directly from the image data in an unsupervised manner. This chapter considers unsupervised subpixel detection where three unsupervised learning algorithms, unsupervised vector quantization (UVQ) algorithm, unsupervised target generation process (UTGP) and unsupervised NCLS (UNCLS) algorithm are presented. These algorithms extract necessary target information directly from the image data for unsupervised subpixel detection when no prior target information is available. Such generated unsupervised target information is referred to as *a posteriori* target information and can be also used to perform target classification as will be discussed in Chapter 13. One dilemma associated with these unsupervised algorithms is how much target information is sufficient for detection and classification. This is a challenging problem. On many occasions, it is closely related to the virtual dimensionailty, which will be discussed in Chapter 17.

5.1 INTRODUCTION

The need of unsupervised subpixel detection arises from the fact that many unknown signal sources that may affect subpixel detection performance cannot be identified *a priori*. In particular, some of these unknown sources may be considered as interference to detection of the targets in which we are interested. In this case, eliminating these intereferers will certainly improve our desired target detection performance. In order to do so, we must find these interferers, which can be included as the undesired target signatures in the target signature matrix **M**. More precisely, a well-represented target signature matrix **M** must include all image endmembers that completely characterize the image data, which may include image background signatures. Apparently, finding such a well representative set of image endmembers for **M** is rather difficult, specifically for hyperspectral imagery, where many subtle material substances may be present and cannot be known in advance. This chapter addresses this issue. It presents three unsupervised algorithms that can be used to generate necessary target information for **M**. The first algorithm is an unsupervised vector quantization (UVQ) algorithm that uses the well-known Linde-Buzo-Gray (LBG) algorithm (Linde et al., 1980) to produce a codebook with each codeword considered to be a target signature. Then a desired target signature matrix **M** can be formed by the codewords in the UVQ-generated codebook. A second algorithm is an unsupervised target generation process (UTGP) which uses a series of orthogonal projections to generate potential targets from the image data. These orthogonal projection-generated targets are then used to form a desired target signature matrix **M**. A third algorithm is an unsupervised least-squares error based algorithm which takes advantage of the NCLS algorithm developed in Chapter 3 to generate a set of target signatures to form a desired target signature matrix **M**. These three algorithms can be implemented in conjunction with supervised subpixel detection methods such as those discussed in Chapter 3 to achieve unsupervised subpixel detection.

5.2 UNSUPERVISED VECTOR QUANTIZATION (UVQ)-BASED ALGORITHM

One of most widely used unsupervised methods is the nearest neighbor rule (NNR)-based clustering. It has been shown that with an unlimited number of samples, the error rate generated by NNR is never worse than twice the error rate generated by supervised Bayesian rule (Duda and Hart, 1973, pp. 98-103). In this section, we describe an NNR-based unsupervised approach that is derived from LBG vector quantization algorithm (Linde et al., 1980).

Suppose that the number of clusters is p and $\{\mathbf{x}_n\}_{n=1}^{N}$ is a set of data samples. The LBG algorithm, also known as the generalized Lloyd algorithm (Lloyd, 1982), is designed to find an optimal set of p clusters based on the k-means NNR (Duda and Hart, 1973, pp. 103-104). It is an iterative algorithm and can be briefly summarized as follows.

The LBG algorithm starts with an initial codebook, $\text{code}^{(0)} = \{\mu_j^{(0)}\}_{j=1}^{p}$ which can be generated by an algorithm developed by Kasvounides et al. (1994). Using the initial code $\text{code}^{(0)} = \{\mu_j^{(0)}\}_{j=1}^{p}$ we partition the data set $\{\mathbf{x}_n\}_{n=1}^{N}$ into a set of p clusters according to the k-means NNR. Let $R_j^{(0)}$ denote the resulting j-th cluster. If a tie occurs, the data sample is assigned to the cluster with the lowest index. Then we calculate the center of each of new clusters by

$$\mu_j^{(1)} = E\left[\mathbf{X} \mid \mathbf{X} \in R_j^{(0)}\right] \quad \text{for} \quad 1 \le j \le p \tag{5.1}$$

where $\mathbf{X}$ is a random vector to represent the data sample set $\{\mathbf{x}_n\}_{n=1}^{N}$. These new generated cluster centers $\{\mu_j^{(1)}\}_{j=1}^{p}$ will be again used to repartition the data set $\{\mathbf{x}_n\}_{n=1}^{N}$ by the k-means NNR into a new set of p clusters, denoted by $\{R_j^{(1)}\}_{j=1}^{p}$. Using (5.1) again resulting in another new set of cluster centers, $\{\mu_j^{(2)}\}_{j=1}^p$. The entire procedure will be repeated until a stopping rule is satisfied. Here, the minimum squared Euclidean distance between a sample and its nearest cluster center will be used as a measure to stop the procedure. Details of implementing the above procedure are given below.

Unsupervised Vector Quantization (UVQ) Algorithm

1. Initialization: $\text{code}^{(0)} = \{\mu_j^{(0)}\}_{j=1}^p$ is generated by Kasvounides et al.'s algorithm (Kasvounides et al., 1994). Let $\{R_j^{(0)}\}_{j=1}^p$ denote the initial clusters resulting from $\{\mu_j^{(0)}\}_{j=1}^p$ using the *k*-means NNR.
2. At the iteration *k*, calculate a new set of cluster centers from the previous clusters, $\{R_j^{(k-1)}\}_{j=1}^p$ using the following equation,

$$\mu_j^{(k)} = E\left[\mathbf{X} \mid \mathbf{X} \in R_j^{(k-1)}\right] \tag{5.2}$$

3. The new set of cluster centers $\{\mu_j^{(k)}\}_{j=1}^p$ will be used to repartition $\{\mathbf{x}_n\}_{n=1}^N$ to form new clusters $\{R_j^{(k)}\}_{j=1}^p$ using the *k*-means NNR. In other words, we compute the minimum squared Euclidean distance between the data samples and their nearest cluster centers.
4. The reclustering process will be terminated when the minimum squared Euclidean distance resulting from two consecutive iterations is not further reduced or below a prescribed error threshold. In this case, no more data samples are shuffled from one cluster to another.

Note that the above UVQ algorithm was successfully applied to unsupervised target detection for hyperspectral images (Brumbley and Chang, 1999) where the UVQ algorithm was used to find unknown interferers in an image scene so that they can be eliminated prior to target detection.

5.3 UNSUPERVISED TARGET GENERATION PROCESS (UTGP)

As indicated in the introduction, identifying unknown target signatures is almost impossible and prohibitive in practice. Recently, an unsupervised OSP was developed by Ren and Chang (2000) for this purpose. The algorithm, referred to as unsupervised target generation process (UTGP) was designed to extract unidentified targets directly from an unknown image scene and can be briefly described as follows.

There are two ways to implement UTGP depending upon how an initial target signature, denoted by $\mathbf{t}_0$ is chosen. If there is a target, such as man-made objects (vehicles) in an image scene provided by partial knowledge or visual inspection, this target can be chosen for the initial target signature $\mathbf{t}_0$. If there is no prior information available about the image scene, we select a pixel vector with the maximum length (i.e., a pixel with the highest intensity or the brightest pixel) as an initial target signature $\mathbf{t}_0$. In this case, $\mathbf{t}_0$ may not necessarily correspond to a target of interest. It could be an interfering signature, such as rocks, a scratch resulting from a sensor, etc. But the purpose of selecting $\mathbf{t}_0$ is only to initialize UTGP.

After the initial target signature is chosen, we then employ an orthogonal subspace projector $P_{\mathbf{t}_0}^{\perp}$ via (3.3) to project all image pixel vectors into, $<\mathbf{t}_0>^{\perp}$ the orthogonal complement space of $<\mathbf{t}_0>$. Then the maximum length of a pixel vector in $<\mathbf{t}_0>^{\perp}$ will be selected to be a first target signature denoted by $\mathbf{t}_1$. The $\mathbf{t}_0$ differs from $\mathbf{t}_1$ in the sense that $\mathbf{t}_0$ was not generated from OSP, but chosen by partially knowledge or intelligently, such as a pixel with the highest intensity. To avoid selecting similar targets, both $\mathbf{t}_0$ and $\mathbf{t}_1$ must be as distinct as possible. In doing so, we select $\mathbf{t}_1$ as the one with maximum length in the space $<\mathbf{t}_0>^{\perp}$. In this case, $\mathbf{t}_1$ should have the most distinct features from $\mathbf{t}_0$ in terms of orthogonal projection. Next, we need a stopping rule to determine if we would like to terminate UTGP. One criterion is to consider $\mathbf{t}_0$ as the desired signature and $\mathbf{t}_1$ as an undesired signature, then calculate the orthogonal projection correlation index (OPCI) defined by

$$\text{OPCI}(\mathbf{t}_0, \mathbf{U}_1) = \eta_1 = \mathbf{t}_0 P_{\mathbf{U}_1}^{\perp} \mathbf{t}_0 \tag{5.3}$$

where $\mathbf{U}_1 = \mathbf{t}_1$. The OPCI is used to measure the orthogonal projection residual resulting from the projection of $\mathbf{t}_1$ in the direction of $\mathbf{t}_0$. If $\text{OPCI}(\mathbf{t}_0, \mathbf{t}_1)$ is sufficiently small, the process of generating additional target signatures will be terminated. At this stage, the task of UTGP is completed. If $\text{OPCI}(\mathbf{t}_0, \mathbf{t}_1)$ is greater than a prescribed threshold ε, we continue the above process to generate a second target by applying $P_{[\mathbf{t}_0 \mathbf{t}_1]}^{\perp}$ again to the original image. As a result, all image pixel vectors are projected into a space that is orthogonal to the linear space spanned by the target signatures $\mathbf{t}_0$ and $\mathbf{t}_1$. As before, a pixel vector with the maximum length in the space $<\mathbf{t}_0, \mathbf{t}_1>^{\perp}$ will be selected as a second target signature denoted by $\mathbf{t}_2$. Once again, we calculate the OPCI between $\mathbf{t}_0$ and $\mathbf{U}_2 = [\mathbf{t}_1 \mathbf{t}_2]$,

$$\eta_2 = \text{OPCI}(\mathbf{t}_0, \mathbf{U}_2) = \mathbf{t}_0 P_{\mathbf{U}_2}^{\perp} \mathbf{t}_0 \tag{5.4}$$

where $\mathbf{U}_2 = [\mathbf{t}_1 \mathbf{t}_2]$. If η_2 is less than ε, UTGP is terminated. Otherwise, the same procedure is repeated to find a third target signature, a fourth target signature, etc., until at the k-th step, $\text{OPCI}(\mathbf{t}_0, \mathbf{U}_k)$ is small enough and less than ε.

More precisely, let $\{\mathbf{t}_1, \mathbf{t}_2, \cdots, \mathbf{t}_k\}$ be the k-th target set generated by UTGP at the k-th step and the matrix $\mathbf{U}_k = [\mathbf{t}_1 \mathbf{t}_2 \cdots \mathbf{t}_k]$ be a target signature matrix formed by the k target signatures with $\mathbf{t}_1$ as its first column vector, $\mathbf{t}_2$ as its second column vector, $\cdots$, $\mathbf{t}_k$ as its the last column vector. The OPCI between $\mathbf{t}_0$ and $\mathbf{U}_k = [\mathbf{t}_1 \mathbf{t}_2 \cdots \mathbf{t}_k]$, η_k is defined by

$$\eta_k = \text{OPCI}(\mathbf{t}_0, \mathbf{U}_k) = \mathbf{t}_0^T P_{\mathbf{U}_k}^{\perp} \mathbf{t}_0. \tag{5.5}$$

where $P_{\mathbf{U}_k}^{\perp} = \mathbf{I} - \mathbf{U}_k \mathbf{U}_k^{\#}$ and $\mathbf{U}_k^{\#} = (\mathbf{U}_k^T \mathbf{U}_k)^{-1} \mathbf{U}_k^T$. Using OPCI given by (5.5) as a stopping rule, UTGP can be summarized below.

Unsupervised Target Generation Process (UTGP)

1. Initial condition: select a pixel vector with the maximum length as an initial target signature denoted by $\mathbf{t}_0$, i.e.,

$$\mathbf{t}_0 = \arg\{\max_{\mathbf{r}}[\mathbf{r}^T\mathbf{r}]\}. \tag{5.6}$$

   Set $k \leftarrow 1$ and $\mathbf{U}_0 = \phi$.

2. Find the orthogonal projections of all image pixel vectors with respect to $\mathbf{t}_0$ by using $P_{\mathbf{t}_0}^{\perp} = (\mathbf{I} - \mathbf{t}_0 \mathbf{t}_0^{\#})$ where $\mathbf{t}_0^{\#} = (\mathbf{t}_0^T \mathbf{t}_0)^{-1} \mathbf{t}_0^T$ is the pseudo-inverse of $\mathbf{t}_0$.
3. Select a first target signature in the space $< \mathbf{t}_0 >^{\perp}$, denoted by $\mathbf{t}_1$ by finding

$$\mathbf{t}_1 = \arg\left\{\max_{\mathbf{r}}\left[\left(P_{\mathbf{t}_0}^{\perp}\mathbf{r}\right)^T \left(P_{\mathbf{t}_0}^{\perp}\mathbf{r}\right)\right]\right\}. \tag{5.7}$$

4. Use (5.3) to calculate $\eta_1 = \mathbf{t}_0^T P_{\mathbf{U}_1}^{\perp} \mathbf{t}_0$ with $\mathbf{U}_1 = \mathbf{t}_1$ to see if $\eta_1 < \varepsilon$, go to step 7. Otherwise, let $k \leftarrow k+1$ and continue.
5. Find the k-th target signature $\mathbf{t}_k$ generated at the k-th step in the space, $\langle \mathbf{t}_0, \mathbf{U}_{k-1} \rangle^{\perp}$ with $\mathbf{U}_{k-1} = [\mathbf{t}_1 \mathbf{t}_2 \cdots \mathbf{t}_{k-1}]$ which is orthogonal to the space linearly spanned by the initial target signature $\mathbf{t}_0$ and the $k-1$st target signature set $\{\mathbf{t}_1, \mathbf{t}_2, \cdots, \mathbf{t}_{k-1}\}$.

$$\mathbf{t}_k = \arg\left\{\max_{\mathbf{r}}\left[\left(P_{[\mathbf{t}_0 \mathbf{U}_{k-1}]}^{\perp}\mathbf{r}\right)^T \left(P_{[\mathbf{t}_0 \mathbf{U}_{k-1}]}^{\perp}\mathbf{r}\right)\right]\right\} \tag{5.8}$$

   where $P_{[\mathbf{t}_0 \mathbf{U}_{k-1}]}^{\perp}$ is defined by (3.3) and $[\mathbf{t}_0 \mathbf{U}_{k-1}]$ is a matrix made up of $\mathbf{t}_0$ and $\mathbf{U}_{k-1} = [\mathbf{t}_1 \mathbf{t}_2 \cdots \mathbf{t}_{k-1}]$. Let $\mathbf{U}_k = [\mathbf{t}_1 \mathbf{t}_2 \cdots \mathbf{t}_k]$ be the target signature matrix generated at the k-th step.

6. Stopping rule: calculate $\eta_k = \mathbf{t}_0^T P_{\mathbf{U}_k}^{\perp} \mathbf{t}_0$ specified by (5.5) and compare it to the prescribed threshold ε. If $\eta_k > \varepsilon$, go to step 5. Otherwise, continue.
7. At this point, UTGP will be terminated and $\{\mathbf{t}_0, \mathbf{t}_1, \cdots, \mathbf{t}_k\}$ will be the final target signature set for detection.

It should be noted that at each iteration from step 5 to step 6 UTGP generates and detects one target at a time. The detected targets may not be significant and could be something else such as natural background signatures, clutter or interfering signatures. This is a consequence of the unsupervised nature in the UTGP.

One comment on OPCI is useful for implementation of UTGP. The OPCI only provides a guide to terminate UTGP. There is no optimal number of targets required for UTGP to generate. How many targets are needed to be generated by UTGP is determined by the prescribed error threshold ε set for OPCI in step 6. It can be determined empirically. Another way to terminate UTGP is to preset the number of targets to generate. For example, we can estimate the virtual dimensionality (VD) using the techniques developed in Chapter 17. The value of VD can be used to terminate the algorithm. In this case, there is no need of calculating OPCI as a stopping criterion in step 6.

5.4 UNSUPERVISED NCLS (UNCLS) ALGORITHM

The unsupervised method presented in this section is based on the least squares error (LSE) criterion that minimizes the goodness of fit between a linear mixture model and estimated measurements by the NCLS algorithm described in Chapter 3. The resultant approach is referred to as the unsupervised NCLS (UNCLS) algorithm.

Initially, we can select any arbitrary pixel vector as an initial target signature denoted by $\mathbf{t}_0$. However, like UTGP, a good choice may be a pixel vector with the maximum length that corresponds to the brightest pixel in the image scene. Then the NCLS algorithm is used to estimate abundance fraction of $\mathbf{t}_0$, denoted by $\hat{\alpha}_0^{(1)}(\mathbf{r})$ for each pixel vector $\mathbf{r}$ in the scene and the LSE is further calculated between the image pixel vector $\mathbf{r}$ and its estimate $\hat{\alpha}_0^{(1)}(\mathbf{r})\mathbf{t}_0$, i.e.

$$\mathrm{LSE}^{(0)}(\mathbf{r}) = \left[(\mathbf{r} - \hat{\alpha}_0^{(1)}(\mathbf{r})\mathbf{t}_0)^T(\mathbf{r} - \hat{\alpha}_0^{(1)}(\mathbf{r})\mathbf{t}_0)\right]. \tag{5.9}$$

Here, $\mathbf{r}$ is included in the abundance fraction estimate $\hat{\alpha}_0^{(1)}(\mathbf{r})$ to emphasize that the estimated abundance fraction $\hat{\alpha}_0$ is a function of the pixel vector $\mathbf{r}$ and varies with $\mathbf{r}$. The pixel that yields the maximum LSE is then selected as a next target signature, denote by $\mathbf{t}_1$, obtained by

$$\max_{\mathbf{r}} \mathrm{LSE}^{(0)}(\mathbf{r}) = \mathrm{LSE}^{(0)}(\mathbf{t}_1) = \left[(\hat{\alpha}_0^{(1)}(\mathbf{t}_1)\mathbf{t}_0 - \mathbf{t}_1)^T(\hat{\alpha}_0^{(1)}(\mathbf{t}_1)\mathbf{t}_0 - \mathbf{t}_1)\right]. \tag{5.10}$$

Because the LSE between $\hat{\alpha}_0^{(1)}(\mathbf{t}_1)\mathbf{t}_0$ and $\mathbf{t}_1$ is the maximum, it can be expected that $\mathbf{t}_1$ is most dissimilar to $\mathbf{t}_0$. In order to find a second target signature, the UNCLS algorithm estimates the abundance fractions of $\mathbf{t}_0$ and $\mathbf{t}_1$ contained in each pixel vector $\mathbf{r}$ in the image scene, denoted by $\hat{\alpha}_0^{(2)}(\mathbf{r})$ and $\hat{\alpha}_1^{(2)}(\mathbf{r})$. Then the maximum LSE between all image pixel vectors $\mathbf{r}$ and the least squares linear mixture $\hat{\alpha}_0^{(2)}(\mathbf{r})\mathbf{t}_0 + \hat{\alpha}_1^{(2)}(\mathbf{r})\mathbf{t}_1$ estimated by the NCLS algorithm. Once again the pixel vector that yields the maximum

LSE is selected as a second target signature denoted by $\mathbf{t}_2$. The same procedure of using the NCLS algorithm with $\mathbf{M} = [\mathbf{t}_0\, \mathbf{t}_1\, \mathbf{t}_2]$ is repeated until the resulting LSE is small enough and less than a prescribed error threshold. As noted in UTGP, if there is partial knowledge available *a priori*, it can be incorporated in the above process. For example, if we know nothing but the desired target signature $\mathbf{d}$, the initial target pixel signature $\mathbf{t}_0$ can be replaced by this $\mathbf{d}$. If there is more than one known target in the image scene, we can select these target signatures as an initial target signature set and then follow the same procedure described above until the LSE meets a stopping criterion. The procedure outlined as above is called Unsupervised NCLS (UNCLS) algorithm which can be summarized as follows.

Unsupervised NCLS (UNCLS) Algorithm

1. Initial condition:

   Select ε to be a prescribed error threshold and let $\mathbf{t}_0 = \arg\{\max_{\mathbf{r}}[\mathbf{r}^T\mathbf{r}]\}$ where $\mathbf{r}$ is run over all image pixel vectors. Let $k = 0$.
2. Let $k \leftarrow k+1$ and apply the NCLS algorithm with the signature matrix $\mathbf{M} = [\mathbf{t}_0\, \mathbf{t}_1 \cdots \mathbf{t}_{k-1}]$ to estimate the abundance fractions of $\mathbf{t}_0$, $\mathbf{t}_1$, $\cdots, \mathbf{t}_{k-1}$, $\hat{\alpha}_0^{(k)}(\mathbf{r})$, $\hat{\alpha}_1^{(k)}(\mathbf{r})$, $\cdots, \hat{\alpha}_{k-1}^{(k)}(\mathbf{r})$.
3. Find the maximum least squares error defined by

$$\max_{\mathbf{r}} \mathrm{LSE}^{(k-1)}(\mathbf{r}) = \max_{\mathbf{r}}\left\{\left(\mathbf{r} - \left[\Sigma_{j=1}^{k-1}\hat{\alpha}_j^{(k)}(\mathbf{r})\mathbf{t}_j\right]\right)^T\left(\mathbf{r} - \left[\Sigma_{j=1}^{k-1}\hat{\alpha}_j^{(k)}(\mathbf{r})\mathbf{t}_j\right]\right)\right\}. \quad (5.11)$$

   If $\mathrm{LSE}^{(k-1)}(\mathbf{r}) < \varepsilon$ for all $\mathbf{r}$, the algorithm stops; otherwise continue.
4. Find $\mathbf{t}_k = \arg\{\max_{\mathbf{r}} \mathrm{LSE}^{(k-1)}(\mathbf{r})\}$. Go to step 2.

It is worth noting that the superscript (k) in $\hat{\alpha}_j^{(k)}(\mathbf{r})$ is a counter to indicate the number of iterations. It starts with $k=1$. The subscript j in $\hat{\alpha}_j^{(k)}(\mathbf{r})$ starts with $j=1$ and is the index of the j th target signature $\mathbf{t}_j$ generated by UNCLS algorithm. The initial target is represented by $\mathbf{t}_0$ with $j=0$. For example, $\hat{\alpha}_0^{(1)}(\mathbf{r})$ is the abundance estimate of $\mathbf{t}_0$ in the first iteration given by (5.10). It should be also noted that as will be demonstrated in the experiments in Chapter 7, step 4 implemented in the UNCLS algorithm tries to locate pure pixel vectors first. If there is no such pixel vector, it then looks for a mixed pixel vector with the largest possible abundance fraction of any substance in the pixel vector. This implies that a mixed pixel vector with uniform mixture is less likely to be selected by UNCLS as a target signature.

Another comment is worthwhile. The UNCLS algorithm is primarily designed for unsupervised subpixel detection. It can be modified by replacing the NCLS algorithm in step 2 with the fully constrained least squares (FCLS) algorithm in Chapter 10 to make it an unsupervised FCLS (UFCLS) algorithm which can be used for unsupervised subpixel quantification.

5.5 EXPERIMENTS

The UVQ algorithm is derived from a source coding perspective. It makes use of NNR to group image pixels into a set of p clusters that will be used to form a p-codeword codebook with each codeword represented by a cluster center. UTGP and UNCLS algorithm are developed from concepts arising in statistical signal processing, orthogonal projection and least squares error respectively. In order to demonstrate the performance of these three algorithms, the AVIRIS image in Fig. 1.6(a) and the 15-panel HYDICE scene in Fig. 1.7(a) were used for evaluation. The only knowledge required for the unsupervised algorithms is how many target signatures, p, should be generated before the algorithms are terminated. This is closely related to the virtual dimensionality (VD) which will be discussed in Chapter 17. In the following experiments, we used the values of VD in Table 17.2 for p, which are 4, 5, 8 for the AVIRIS image and 13, 15 and 20 for the HYDICE image. After p target signatures were generated, their information was then used to perform OSP specified by (3.9). The UVQ, UTGP and UNCLS algorithm implemented in conjunction with OSP are referred to as UVQ-OSP, UTGP-OSP and UNCLS-OSP respectively.

Example 5.5.1 (AVIRIS Experiments)

Figs. 5.1-5.3 show that the detection results obtained by OSP for the AVRIS image with $p = 4$ using UVQ algorithm, UTGP and UNCLS algorithm to generate 4 target signatures.

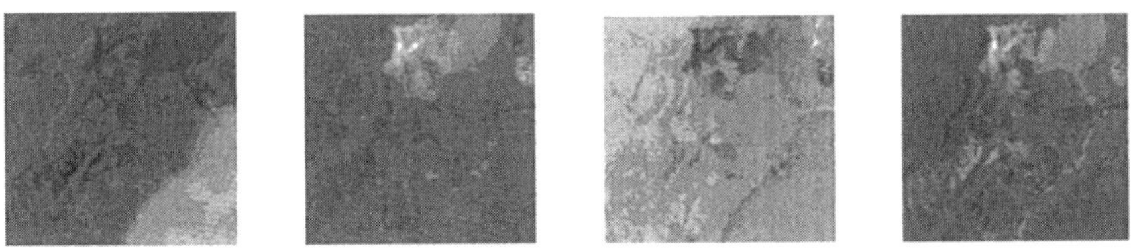

(a) playa (b) cinders (c) vegetation (d) cinders

Figure 5.1. Detection results produced by UVQ-OSP with $p = 4$

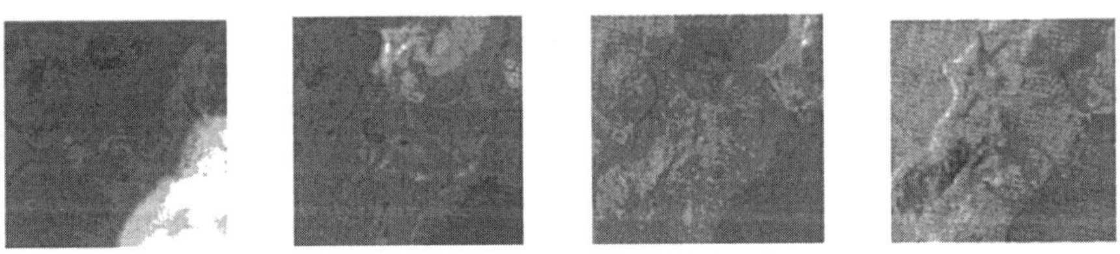

(a) playa (b) cinders (c) vegetation (d) shade

Figure 5.2. Detection results produced by UTGP-OSP with $p = 4$

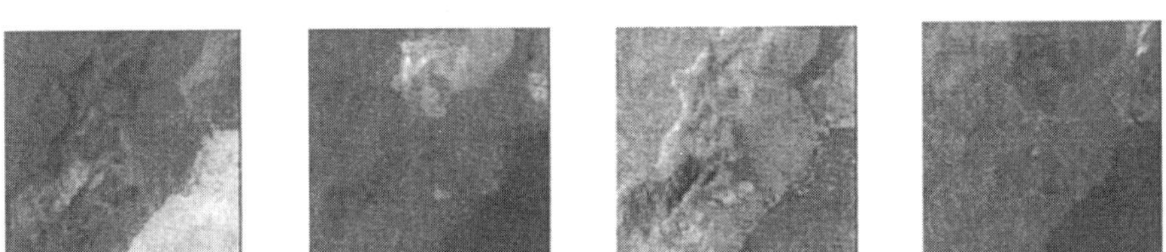

(a) playa,rhyolite (b) cinders (c) shade (d) vegetation

Figure 5.3. Detection results produced by UNCLS-OSP with $p = 4$

As we can see from these figures, the target information produced by both UTGP and UNCLS algorithm was better representative than that produced by UVQ algorithm in terms of target detection. Nevertheless, none of the three algorithms could detect rhyolite.

Similar experiments were also conducted for p = 5 and 8 and their detection results are shown in Figs. 5.4-5.9.

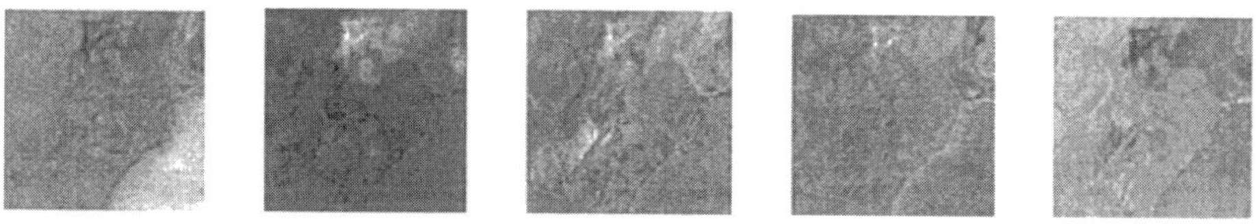

(a) playa (b) cinders (c) cinders,rhyolite,vegetation (d) cinders (e) shade

Figure 5.4. Detection results produced by UVQ-OSP with $p = 5$

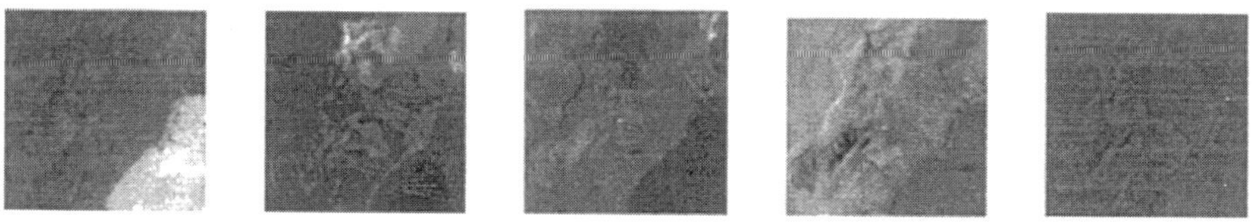

(a) playa (b) cinders (c) vegetation (d) shade (e) anomaly

Figure 5.5. Detection results produced by UTGP-OSP with $p = 5$

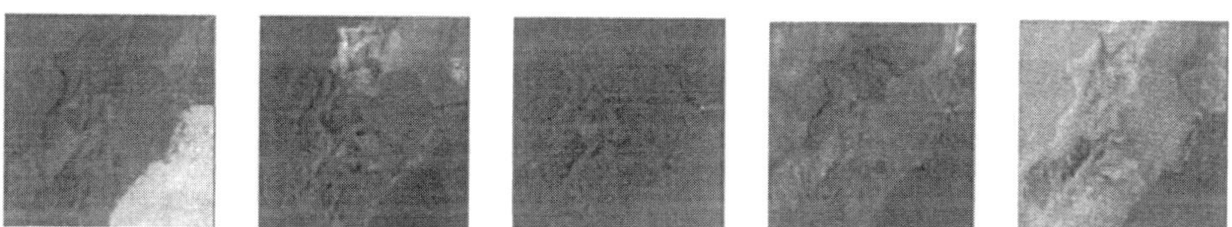

(a) playa (b) cinders (c) anomaly (d) vegetation (e) shade

Figure 5.6. Detection results produced by UNCLS-OSP with $p = 5$

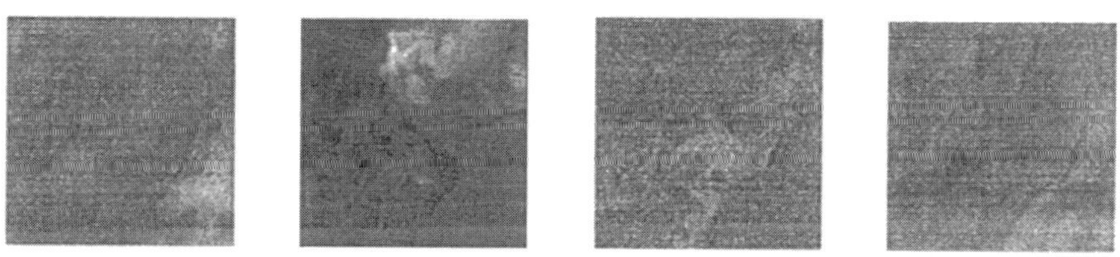

(a) playa (b) cinders (c) (d)

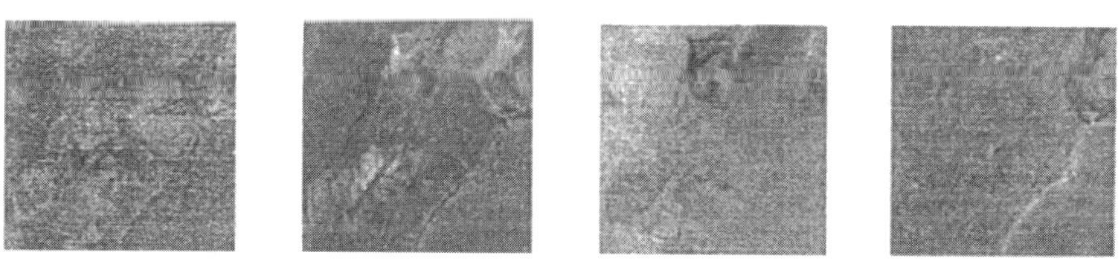

(e) (f) cinders,rhyolite (g) vegetation (h)

Figure 5.7. Detection results produced by UVQ-OSP with $p = 8$

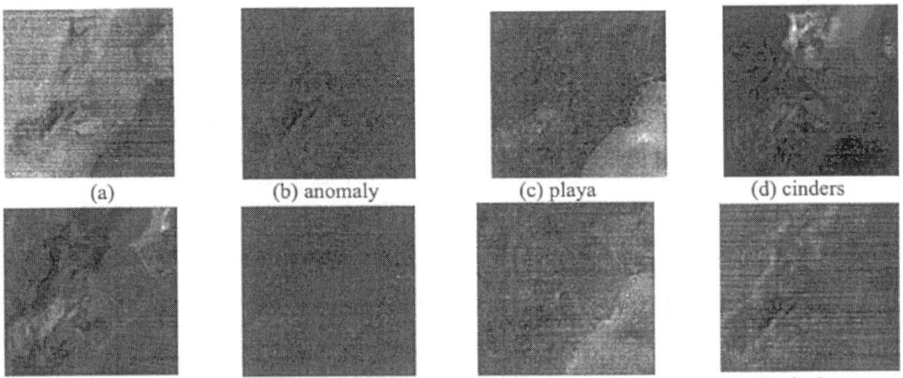

(a) (b) anomaly (c) playa (d) cinders

(e) rhyolite,vegetation (f) anomaly (g) (h) shade

Figure 5.8. Detection results produced by UTGP-OSP with $p = 8$

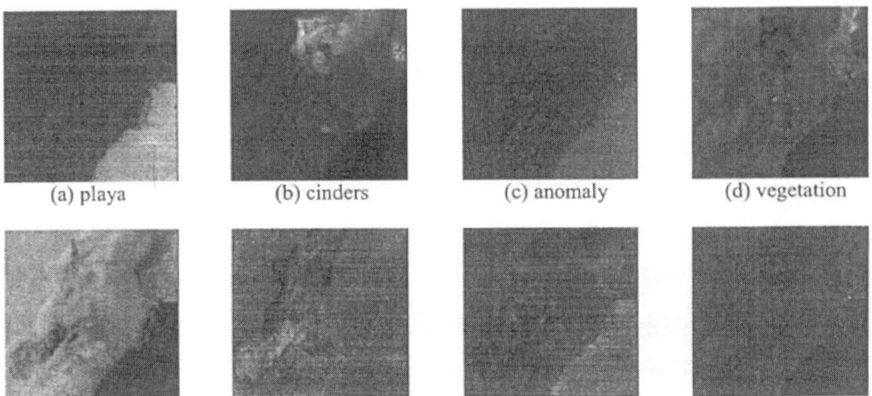

(a) playa (b) cinders (c) anomaly (d) vegetation

(e) shade (f) rhyolite (g) (h) anomaly

Figure 5.9. Detection results produced by UNCLS-OSP with $p = 8$

A slight improvement was observed in Fig. 5.4 for UVQ-OSP where rhyolite was detected along with cinders. Interestingly, UTGP-OSP and UNCLS-OSP detected the anomaly in Figs. 5.5-5.6 and 5.8-5.9 which was missed in Figs. 5.2-5.3 with $p = 4$. Unfortunately, both figures also missed the detection of rhyolite shown in Figs. 5.5-5.6 with $p = 5$. However, as p increased to 8 all the three detectors extracted the rhyolite, but only UNCLS-OSP detected it in a separate image. On the whole, UNCLS-OSP performed better than did the other two. This is consistent with the results obtained in Chapter 3. When p is too small such as $p = 4$, the generated target information may not be sufficient to represent the image scene. In this case, the target detection generally did not perform well regardless of which unsupervised algorithm is used to produce target information. On the other hand, if p is too large, the targets of same type that contain different amounts of abundance will be forced to split into more than one class. For example, the anomaly detected in Figs. 5.5-5.6 has size of two pixels. It was detected in Fig. 5.8(b,f) and Fig. 5.9(c,h) where these two pixels were detected in two separate images. According to our experiments, $p = 8$ seems to be an appropriate number which will be further justified by experiments in Chapter 15.

Example 5.5.2 (HYDICE Experiments)

Same experiments in Example 5.5.1 were conducted for p = 13, 15, 20 using the 15-panel HYDICE scene. As shown in Figs. 5.10-5.15 with p = 13, all the three detectors had difficulty with separating the panels in row 2 from the panels in row 3 as well as separating the panels in row 4 from the panels in row 5. This is because the panels in rows 2 and 3 were made of the same materials with different paints. So are the panels in rows 4 and 5.

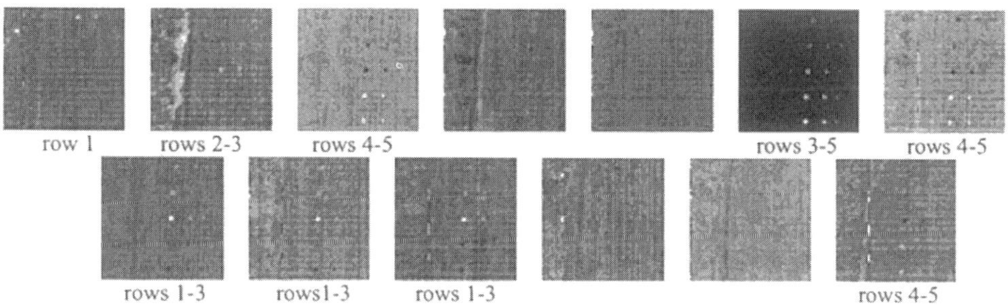

Figure 5.10. Detection results produced by UVQ-OSP with p = 13

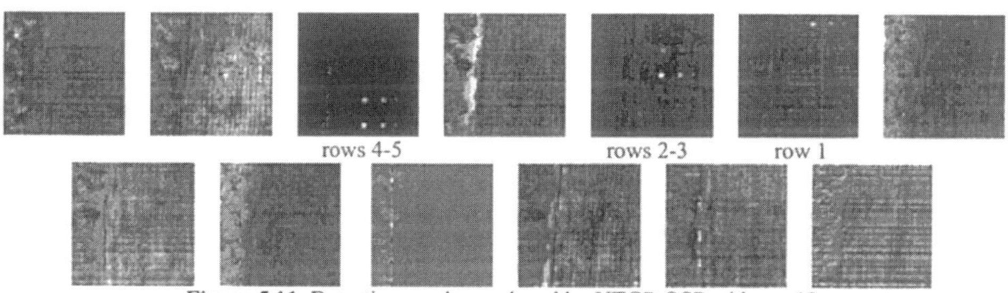

Figure 5.11. Detection results produced by UTGP-OSP with p = 13

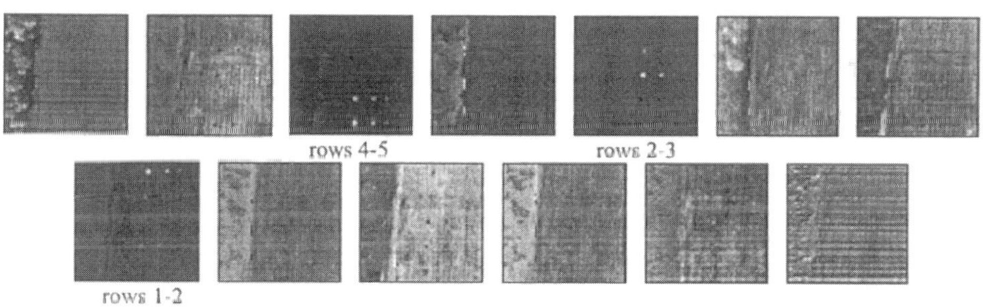

Figure 5.12. Detection results produced by UNCLS-OSP with p = 13

For p = 15, all the three detectors, UVQ-OSP, UTGP-OSP and UNCLS-OSP also had problem with differentiating panels in rows 2 and 3, panels in rows 4 and 5 shown in Figs. 5.13-5.15.

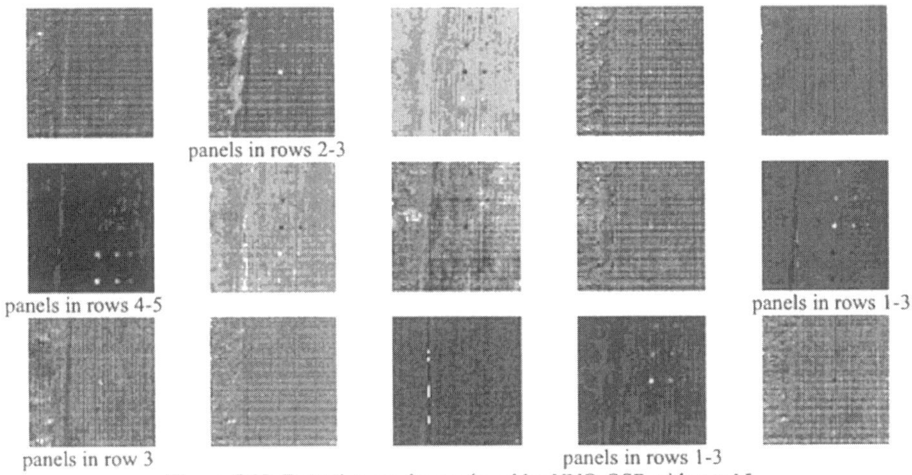

Figure 5.13. Detection results produced by UVQ-OSP with $p = 15$

Figure 5.14. Detection results produced by UTGP-OSP with $p = 15$

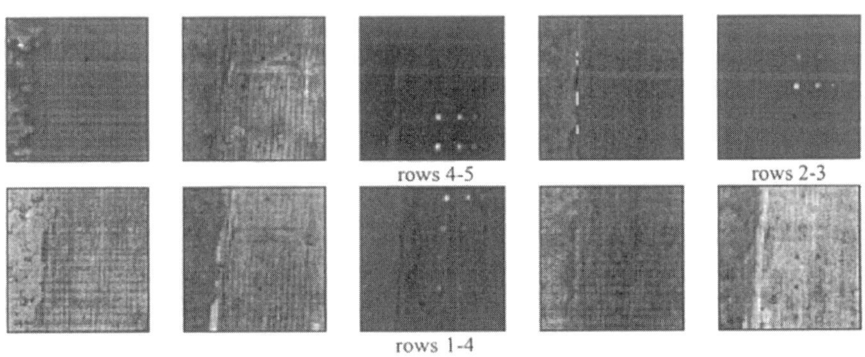

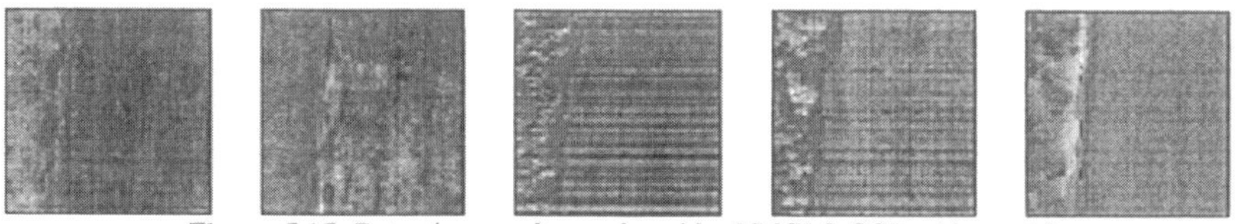

Figure 5.15. Detection results produced by UNCLS-OSP with $p = 15$

As p was increased to 20, only UTGP-OSP was able to detect and discriminate all the 15 panels into five separate images in Figs. 5.17. Both UVQ-OSP and UNCLS-OSP still had the same difficulty with differentiating the panels in rows 1-3 and the panels in rows 3-5 as shown in Figs. 5.16 and 5.18.

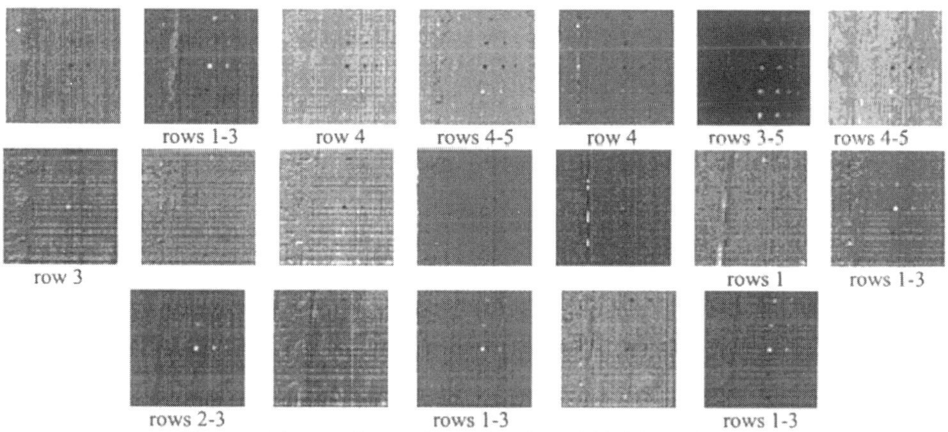

Figure 5.16. Detection results produced by UVQ-OSP with $p = 20$

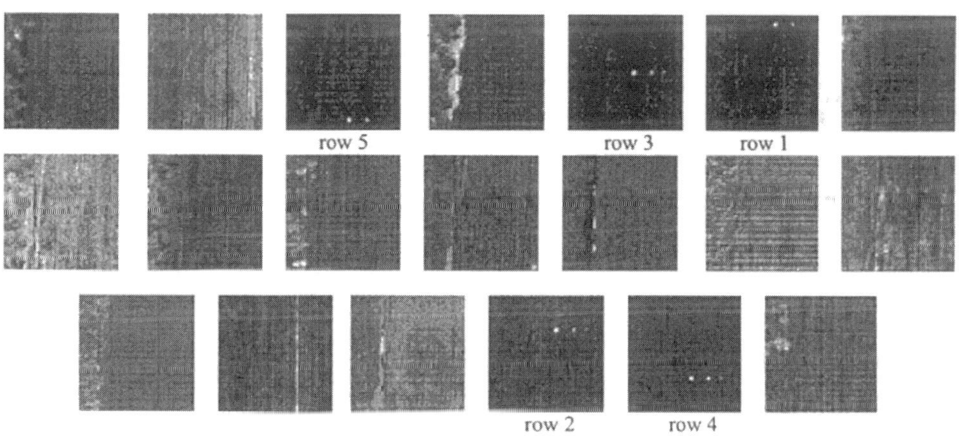

Figure 5.17. Detection results produced by UTGP-OSP with $p = 20$

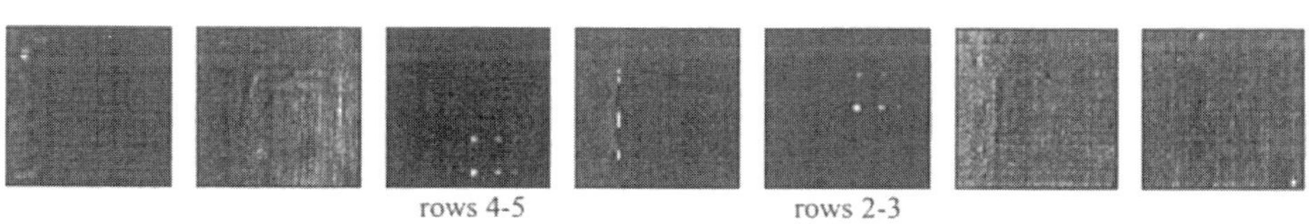

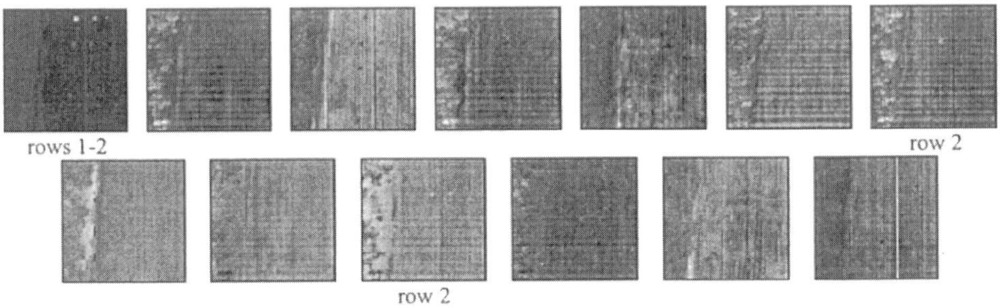

Figure 5.18. Detection results produced by UNCLS-OSP with $p = 20$

In order for UVQ-OSP to effectively detect the 15 panels, 40 target signatures were required to generate necessary target information as shown in Fig. 5.19. Even in this case, no single separate image can discriminate the panels in row 2 from the panels in other rows. However, the detection of the 15 panels can be accomplished by combining these separate images. Interestingly, these results were also consistent with the results obtained by Brumbley and Chang (1999) where the proposed unsupervised vector quantization-based target subspace projection approach is essentially UVQ-OSP, even though a different HYDICE scene was used.

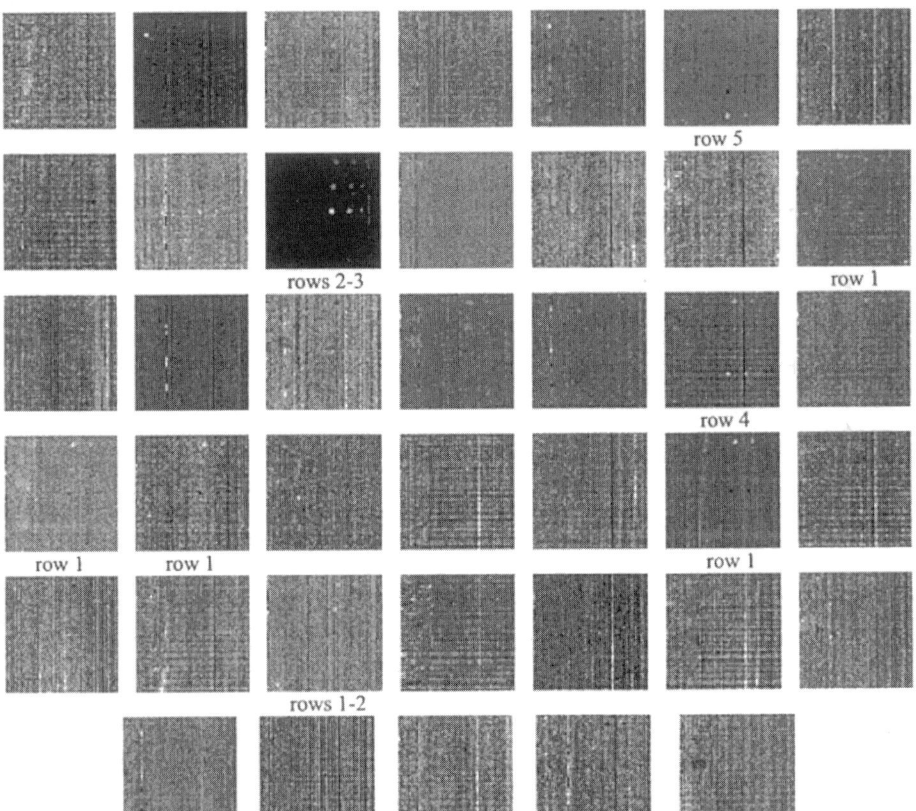

Figure 5.19. Detection results produced by UVQ-OSP with $p = 40$

Similarly, it required 34 target signatures for UNCLS-OSP to effectively separate the 15 panels as shown in Fig. 5.20.

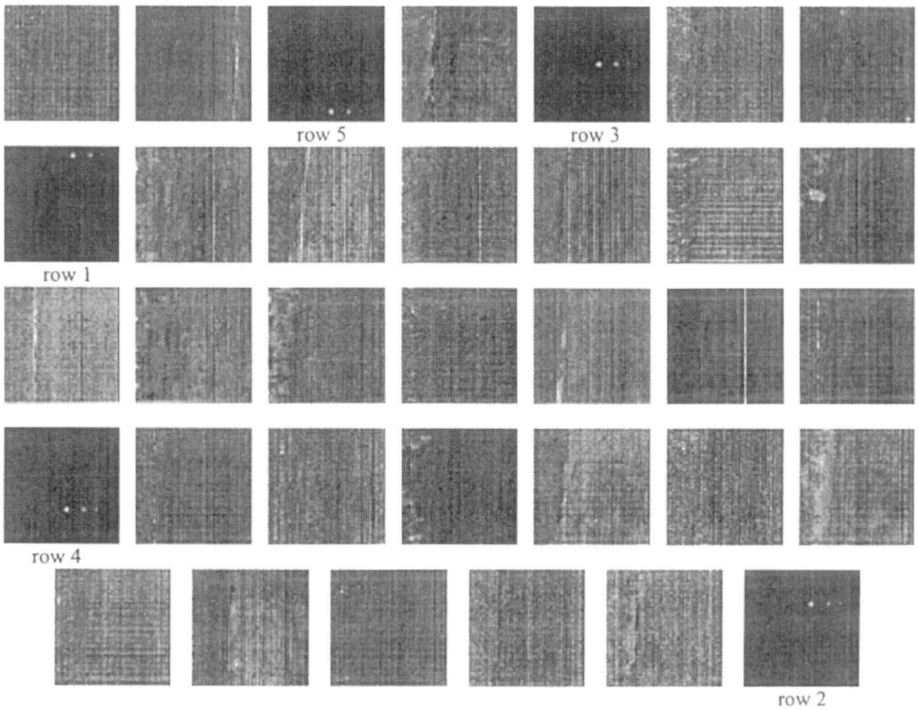

Figure 5.20. Detection results produced by UNCLS-OSP with $p = 34$

The detection results in Fig. 5.20 were significantly better than those obtained by UVQ-OSP and also very close to those obtained by UTGP-OSP using 20 target signatures in Fig. 5.17. In Example 7.9 of Chapter 7, we will see that when the UNCLS algorithm implemented in conjunction with NCLS in Chapter 3 for partial unsupervised subpixel detection, 34 target signatures are also required. The only difference between these two cases is the selection of the initial target signature. In UNCLS-OSP the initial target signatures was selected by UNCLS algorithm while the initial target signatures for UNCLS-UNCLS was selected by prior knowledge. Nevertheless, their results are very close.

The above experiments demonstrate that UVQ-OSP was not as effective as UTGP-OSP and UNCLS-OSP in unsupervised subpixel detection. This is because the target information produced by UVQ is mainly based on sample correlation, i.e. nearest neighboring rule. It did not take advantage of spectral properties as did UTGP and the UNCLS algorithm which are primarily designed for spectral analysis. More studies on the UNCLS algorithm and UTGP can be found in Chapters 10 and 13 respectively.

5.6 CONCLUSIONS

The OSP studied in Chapter 3 requires complete target knowledge. Insufficient target knowledge generally results in degradation in detection performance. In order to remedy this dilemma, this chapter presents three unsupervised algorithms, UVQ, UTGP and UNCLS algorithm that are developed for this purpose. They can be used to obtain additional target information directly from the image data and further improve detection performance. By means of these algorithms, supervised subpixel detection can be extended to unsupervised subpixel detection so that the necessary target knowledge can be generated directly from image data in an unsupervised fashion. However, this also requires the knowledge about the number of target signatures to generate, which must be provided *a priori*. This issue will be discussed in Chapter 17. In order to evaluate the effectiveness of these unsupervised algorithms, a comparative study is conducted through a series of experiments. The experimental results show that UTGP and UNCLS algorithm are more effective than UVQ.

6

AUTOMATIC SUBPIXEL DETECTION: ANOMALY DETECTION

One type of automatic target detection, unsupervised subpixel detection has been considered in Chapter 5. This chapter considers another type of automatic target detection, anomaly detection. Unlike the unsupervised subpixel detection, which requires generating *a posteriori* target information, anomaly detection does not need any target information at all. The purpose of anomaly detection is to locate and search for targets which are generally unknown, but relatively small with low probabilities of occurrence in an image scene. These anomalous targets cannot be identified by prior knowledge. Two approaches are considered in this chapter, the RX algorithm developed by Reed and Yu (1990) and the low probability detection (LPD) proposed in Harsanyi's dissertation (1993), both of which implement a same form as does CEM described in Chapter 4. The only difference is the matched target signature used in their filter design. The matched target signature used in CEM is the desired target signature $\mathbf{d}$, whereas RXD and LPD use the sample pixel vector $\mathbf{r}$ and the unity vector $\mathbf{1}$ as their respective matched signatures. Like CEM, anomaly detection can be also implemented in real time.

6.1 INTRODUCTION

Automatic subpixel detection can be considered in two aspects. One is to detect targets of interest in an unknown image scene from which the desired target knowledge can be generated directly from the scene in an unsupervised means. Another is to detect anomalous targets in a blind scene where the targets are generally unknown and small, but interesting. This type of targets generally has low probabilities of occurrence. The techniques developed for the former case are designed to find undesired or unwanted signatures and eliminate these target signatures subsequently before target detection takes place. So, a general approach is to produce a target signature matrix $\mathbf{M}$ that can well represent the entire image data including image background signatures. Because there is no sample spectral information considered, such subpixel target detection can be considered as pixel-level spectral target detection and was discussed in Chapter 5. On the contrary, the techniques developed for the latter case search unknown but interesting targets via suppression of image background. A commonly used method is to use the sample spectral correlation (or covariance) matrix for background suppression. Therefore, such detection can be viewed as sample correlation-based spectral target detection. An example of the relationship between these two approaches was given in Section 4.3 in Chapter 4 where the relationship between $P_{\mathbf{U}}^{\perp}$ used in OSP and $\mathbf{R}_{L\times L}^{-1}$ used in CEM was explored. Since the knowledge of the undesired target signature matrix $\mathbf{U}$ assumed to be known in $P_{\mathbf{U}}^{\perp}$ is not available in CEM, CEM must estimate $P_{\mathbf{U}}^{\perp}$ directly from the image data. One way to do so is to approximate the $P_{\mathbf{U}}^{\perp}$ in the sense of minimum least-squares error using the spectral correlation information provided by sample pixel vectors in the image data. If we compare $\delta_{\mathrm{OSP}}(\mathbf{r}) = \mathbf{d}^T P_{\mathbf{U}}^{\perp}\mathbf{r}$ specified by (3.9) against $\delta_{\mathrm{CEM}}(\mathbf{r}) = \left(\mathbf{d}^T\mathbf{R}_{L\times L}^{-1}\mathbf{d}\right)^{-1}\mathbf{d}^T\mathbf{R}_{L\times L}^{-1}\mathbf{r}$ specified by (4.8), we will discover that it is indeed the case, where that CEM uses the sample spectral information provided by $\mathbf{R}_{L\times L}$ to estimate the spectral information $\mathbf{U}$ used in OSP. More specifically, the undesired target spectral projector $P_{\mathbf{U}}^{\perp}$ is approximated by its minimum least-squares error based estimate $\kappa\mathbf{R}_{L\times L}^{-1}$ with $\kappa = \left(\mathbf{d}^T\mathbf{R}_{L\times L}^{-1}\mathbf{d}\right)^{-1}$ included to account for the estimation error (Chang, 1998). If the image data to be processed are assumed to be zero-mean and uncorrelated, the sample spectral correlation matrix $\mathbf{R}_{L\times L}$ is reduced to the identity matrix $\mathbf{I}_{L\times L}$ and $\mathbf{d}^T\mathbf{R}_{L\times L}^{-1}\mathbf{d} = \mathbf{d}^T\mathbf{d}$. As a result, CEM becomes a normalized spectral matched filter, that is, $\delta_{\mathrm{CEM}}(\mathbf{r}) = \left(\mathbf{d}^T\mathbf{d}\right)^{-1}\mathbf{d}^T\mathbf{r}$. On the other hand, if no sample spectral correlation is taken into account, i.e. $\mathbf{R}_{L\times L} = \mathbf{I}_{L\times L}$ and the desired target signature $\mathbf{d}$ is orthogonal to $\mathbf{U}$, i.e. $P_{\mathbf{U}}^{\perp}\mathbf{d} = \mathbf{d}$, then $\delta_{\mathrm{OSP}}(\mathbf{r}) = \mathbf{d}^T P_{\mathbf{U}}^{\perp}\mathbf{r} = \left(P_{\mathbf{U}}^{\perp}\mathbf{d}\right)^T\mathbf{r} = \mathbf{d}^T\mathbf{r}$. This implies that CEM is identical to OSP by a normalization constant $\kappa = \left(\mathbf{d}^T\mathbf{d}\right)^{-1}$. In other words, OSP can be considered a whitened version of CEM as illustrated in Section 4.3. However, as shown in Chang (1998), the constant κ is the result from a least-squares estimation error of α. It is bounded below by $\left(\mathbf{d}^T\mathbf{d}\right)^{-1}$ with $\left(\mathbf{d}^T\mathbf{d}\right)^{-1} \le \kappa = \left(\mathbf{d}^T P_{\mathbf{U}}^{\perp}\mathbf{d}\right)^{-1} \le \infty$. The optimal result can be achieved by the greatest low bound, $\kappa = \left(\mathbf{d}^T\mathbf{d}\right)^{-1}$. It makes sense since the sample spectral correlation matrix is an identity matrix and no portion of $\mathbf{d}$ will be projected onto $<\mathbf{U}>^{\perp}$ via $P_{\mathbf{U}}^{\perp}$. Therefore, the information provided by $\mathbf{d}$ is not deteriorated; thus the matched filter achieves the best performance. In this case, both OSP and CEM wind up an identical detector $\left(\mathbf{d}^T\mathbf{d}\right)^{-1}\mathbf{d}^T\mathbf{r}$. Such OSP is called *a posteriori* OSP in Chapter 8.

Generally speaking, when there is no prior information available for subpixel target detection, the required target information such as the desired target signature $\mathbf{d}$, target signature matrix $\mathbf{M}$, must be estimated from the image data. This information may not be accurate due to the noise and measurement error. One way to avoid this dilemma is to design a detector, which does not need any target information. In this chapter, we address this issue and consider subpixel detection with no required prior information. The detector of this type can be designed by a matched filter that uses the currently being visited image pixel vector $\mathbf{r}$ as the matched signature. Two detectors are of interest. One is RXD where the matched signature is the sample image pixel vector $\mathbf{r}$. Another is LPD with the constant unity vector $\mathbf{1}$ as its matched signature. As we will see, if the matched signature **r** used in RXD and the unity vector **1** used in LPD are replaced with the desired target signature **d**, both resulting RXD and LPD become an non-normalized CEM.

6.2 RXD

The main purpose of the unsupervised algorithms studied in Chapter 5 is to find an appropriate **U** that can characterize the image background. The role of the initial target $\mathbf{t}_0$ in these algorithms is simply used for initialization. How $\mathbf{t}_0$ is chosen is generally not important and does not affect the ultimate performance. Therefore, these algorithms are not particularly designed for anomaly detection. An anomaly detector does not extract targets randomly. It extracts unknown targets, which are supposed to be interesting and meaningful, particularly spectrally distinct from their surroundings. In this section, we describe two types of detectors that can be used to detect anomalous targets. Both are very closely related to CEM discussed in Chapter 4. It is interesting to see that these three perform a same functional form of a matched filter with using different matched signatures.

The RXD described in this chapter is referenced in Ashton and Schaum (1998) and Stellman et al. (2000) which was developed by Reed and Yu (1990). More specifically, for each image pixel vector **r**, RXD implements a filter specified by

$$\delta_{\mathrm{RXD}}(\mathbf{r}) = (\mathbf{r} - \mu)^T \mathbf{K}_{L\times L}^{-1} (\mathbf{r} - \mu) \tag{6.1}$$

where μ is the sample mean and $\mathbf{K}_{L\times L}$ is the sample spectral covariance matrix. The $\delta_{\mathrm{RXD}}(\mathbf{r})$ in (6.1) actually a measure that calculates the well-known Mahalanobis distance. It has a similar form to that of CEM specified by (4.8) where the sample spectral covariance matrix $\mathbf{K}_{L\times L}$ and the matched signal signature **d** are replaced with the sample spectral correlation matrix $\mathbf{R}_{L\times L}$ and the pixel vector $\mathbf{r} - \mu$ respectively, and the scale constant $\left(\mathbf{d}^T \mathbf{R}_{L\times L}^{-1} \mathbf{d}\right)^{-1}$ is discarded. Since there is no prior information required for RXD, it can be only used as an anomaly detector. Furthermore, as will be discussed in Chapter 7, the scale constant appearing in CEM but not showing in RXD is a crucial factor on detection performance because it reflects the information to be used in a detector. In RXD, there is no information used; thus the scale constant κ is missing in (6.1).

Mathematically, RXD can be considered as an inverse operation of the principal components analysis (PCA). PCA decorrelates the data matrix in such a manner that different amounts of the image information can be preserved in separate component images, each of which represents a different piece of uncorrelated image information. So, the strength of PCA is to compress most of significant image information into a few major principal components specified by the engenvectors of $\mathbf{K}_{L\times L}$ that correspond to large eigenvalues. So, it is not designed for detection or classification. On the contrary, if the image data contain interesting target samples which only occur with low probabilities in the data (i.e. the size of target samples is small), these targets will not be shown in major principal components, but rather in minor components specified by small eigenvalues. This phenomenon was observed and demonstrated in Chang and Heinz (2000b). It provides explanation of why RXD works for anomaly detection. As we recall (6.1), RXD looks for anomalous targets by calculating $(\mathbf{r} - \mu)^T \mathbf{K}_{L\times L}^{-1} (\mathbf{r} - \mu)$. If there is an anomaly in the data with relatively small energy, it will be specified by a small eigenvalue. In this case, a small eigenvalue will produce a large value of $(\mathbf{r}-\mu)^T\mathbf{K}_{L\times L}^{-1}(\mathbf{r}-\mu)$. This is equivalent to searching for minor components by finding smaller eigenvalues of $\mathbf{K}_{L\times L}$. As a consequence of this operation, RXD can be very effective in detecting small targets, which are retained in minor components. One problem with this approach is how to separate small eigenvalues from noise variance in the data. This is determined by the intrinsic dimesnionality of the data. It is a noise sensitivity issue, which will be discussed in Chapter 7. The explanation described above can be derived mathematically as follows using the spectral decomposition of a covariance matrix in Poor (1992), Stark and Woods (1991).

Assume that $\lambda_1 \geq \lambda_2 \geq \cdots \geq \lambda_L$ are the eigenvalues of $\mathbf{K}_{L\times L}$ with zero mean, $\mu = \mathbf{0}$ and $\{\mathbf{v}_1, \mathbf{v}_2, \cdots, \mathbf{v}_L\}$ are their associated orthogonal unit eigenvectors (i.e. the vector length is one) with $\mathbf{v}_l$ corresponding to λ_l. Let $\mathbf{A} = [\mathbf{v}_1\mathbf{v}_2 \cdots \mathbf{v}_L]$ be an eigen-matrix with the l-th column specified by the l-th eigenvector $\mathbf{v}_l$. Then the matrix $\mathbf{A}$ is a unitary matrix and can be used to decorrelate $\mathbf{K}_{L\times L}$ into a diagonal matrix $\Lambda = diag\{\lambda_1, \lambda_2, \cdots, \lambda_L\}$ such that $\mathbf{A}^T\mathbf{K}_{L\times L}\mathbf{A} = \Lambda$. If we let $\mathbf{y} = \mathbf{A}^T\mathbf{r}$, then

$$\mathbf{r}^T\mathbf{K}_{L\times L}\mathbf{r} = (\mathbf{A}\mathbf{y})^T\mathbf{K}_{L\times L}(\mathbf{A}\mathbf{y}) = \mathbf{y}^T\left[\mathbf{A}^T\mathbf{K}_{L\times L}\mathbf{A}\right]\mathbf{y}$$
$$= \mathbf{y}^T\mathbf{K}_{L\times L}\mathbf{y} = \Sigma_{l=1}^{L}\lambda_l y_l^2 \tag{6.2}$$

By means of (6.2) we can further represent RXD by

$$\mathbf{r}^T\mathbf{K}_{L\times L}^{-1}\mathbf{r} = \Sigma_{l=1}^{L}\lambda_l^{-1}y_l^2. \tag{6.3}$$

According to (6.2), the larger the eigenvalue is, the greater the value of $\mathbf{r}^T\mathbf{K}_{L\times L}\mathbf{r}$ is. So, using (6.2) we are able to compress data by retaining only a few principal components specified by the first few largest eigenvalues. By contrast, (6.3) allows RXD to detect anomalous targets with small energies that are represented by small eigenvalues. This is because the smaller the eigenvalue is, the greater the value of $\mathbf{r}^T\mathbf{K}_{L\times L}^{-1}\mathbf{r}$ is. In this case, it is crucial to determine p, the number of targets present in the image scene so that the eigenvalues beyond the first p largest values will be noise energies. Fig. 6.1(a-b) show the results of the LCVF in Fig. 1.6(a) produced by the first principal component of the standard PCA and (6.2) respectively, where Figs. 6.1(a) and 6.1(b) are comparable and both preserve vast information of the image scene. Fig. 6.1(c) was generated by RXD using (6.1) or (6.3) where RXD detected a two-pixel anomaly on the left upper edge of the drylake along with a small portion of vegetation on upper right corner shown in Fig. 6.1(c).

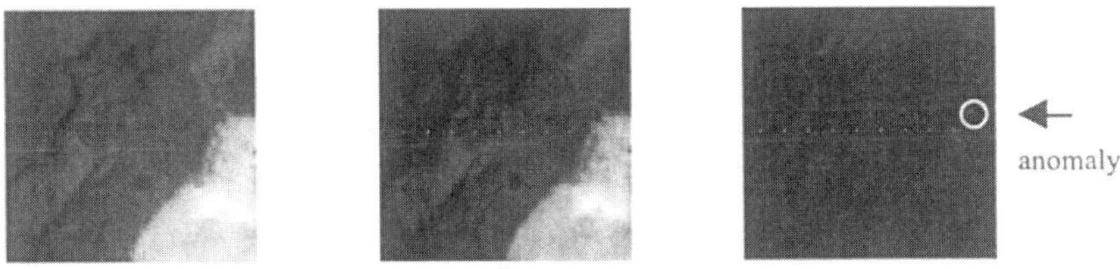

(a) PCA (b) $(\mathbf{r}-\mu)^T \mathbf{K}_{L\times L}(\mathbf{r}-\mu)$ (c) RXD

Figure 6.1. (a) PCA; (b) Results of using (6.2); (c) Detection results of using RXD

As noted in the introduction of this chapter, OSP, CEM and RXD operate some form of a matched filter, but their performance is closely related to the scale constant κ that appears in front of their matched filters. With this interpretation two variants of RXD, referred to as normalized RXD, denoted by $\delta_{\text{NRXD}}(\mathbf{r})$ and modified RXD, denoted by $\delta_{\text{MRXD}}(\mathbf{r})$ can be derived from (6.1) for anomaly detection as follows.

$$\begin{aligned}\delta_{\text{NRXD}}(\mathbf{r}) &= \frac{(\mathbf{r}-\mu)^T \mathbf{K}_{L\times L}^{-1}(\mathbf{r}-\mu)}{(\mathbf{r}-\mu)^T(\mathbf{r}-\mu)} \\ &= \left[(\mathbf{r}-\mu)^T(\mathbf{r}-\mu)\right]^{-1}(\mathbf{r}-\mu)^T \mathbf{K}_{L\times L}^{-1}(\mathbf{r}-\mu)\end{aligned} \tag{6.4}$$

which can be viewed as RXD with the scale constant $\kappa = \left[(\mathbf{r}-\mu)^T(\mathbf{r}-\mu)\right]^{-1}$ as well as a matched filter with the matched signature given by $\mathbf{K}_{L\times L}^{-1}(\mathbf{r}-\mu)$ with the scale constant $\kappa = \left[(\mathbf{r}-\mu)^T(\mathbf{r}-\mu)\right]^{-1}$. Or equivalently,

$$\begin{aligned}\delta_{\text{NRXD}}(\mathbf{r}) &= \left(\left[(\mathbf{r}-\mu)^T(\mathbf{r}-\mu)\right]^{-1/2}(\mathbf{r}-\mu)\right)^T \mathbf{K}_{L\times L}^{-1}\left(\left[(\mathbf{r}-\mu)^T(\mathbf{r}-\mu)\right]^{-1/2}(\mathbf{r}-\mu)\right) \\ &= \bar{\mathbf{r}}^T \mathbf{K}_{L\times L}^{-1}\bar{\mathbf{r}}\end{aligned} \tag{6.5}$$

with $\bar{\mathbf{r}} = \left[(\mathbf{r}-\mu)^T(\mathbf{r}-\mu)\right]^{1/2}(\mathbf{r}-\mu)$, which can be considered as normalized RXD (NRXD).

Another variant of RXD is

$$\begin{aligned}\delta_{\text{MRXD}}(\mathbf{r}) &= \left[(\mathbf{r}-\mu)^T(\mathbf{r}-\mu)\right]^{-1/2}(\mathbf{r}-\mu)^T \mathbf{K}_{L\times L}^{-1}(\mathbf{r}-\mu) \\ &= (\bar{\mathbf{r}})^T \mathbf{K}_{L\times L}^{-1}(\mathbf{r}-\mu)\end{aligned} \tag{6.6}$$

which can be considered as RXD with the scale constant $\kappa = \left[(\mathbf{r}-\mu)^T(\mathbf{r}-\mu)\right]^{-1/2}$ or interpreted as a matched filter with the matched signal given by $\mathbf{K}_{L\times L}^{-1}\bar{\mathbf{r}}$ with the scale constant $\kappa = 1$ where $\bar{\mathbf{r}}$ is the normalized image pixel vector of $\mathbf{r}-\mu$.

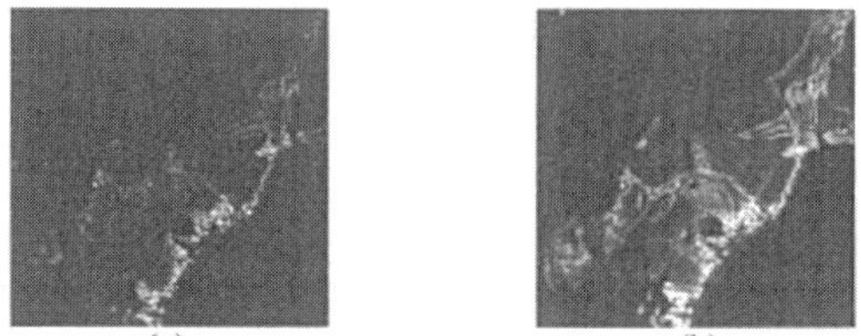

(a) (b)

Figure 6.2. (a) Detection results of $\delta_{\mathrm{NRXD}}(\mathbf{r})$; (b) Detection results of $\delta_{\mathrm{MRXD}}(\mathbf{r})$

Fig. 6.2(a-b) shows the results of $\delta_{\mathrm{NRXD}}(\mathbf{r})$ and $\delta_{\mathrm{MRXD}}(\mathbf{r})$. Interestingly, instead of detecting the anomaly as RXD did, $\delta_{\mathrm{NRXD}}(\mathbf{r})$ and $\delta_{\mathrm{MRXD}}(\mathbf{r})$ extracted different signatures such as background pixels and edge pixels of the dry lake.

6.3 LPTD AND UTD

Another type of anomaly detection was previously derived in Harsanyi's dissertation (1993). It is called the low probability target detector (LPTD) and given by

$$\delta_{\mathrm{LPTD}}(\mathbf{r}) = \mathbf{1}_{L\times 1}{}^{T}\mathbf{R}_{L\times L}^{-1}\mathbf{r} \tag{6.7}$$

which was based on the sample correlation matrix $\mathbf{R}_{L\times L}$. If we replace $\mathbf{R}_{L\times L}$ with the sample covariance matrix $\mathbf{K}_{L\times L}$, we can define an alternative LPTD using sample covariance matrix $\mathbf{K}_{L\times L}$, referred to as uniform target detector (UTD), $\delta_{\mathrm{UTD}}(\mathbf{r})$ which is given by

$$\delta_{\mathrm{UTD}}(\mathbf{r}) = \left(\mathbf{1}_{L\times 1} - \mu\right)^{T}\mathbf{K}_{L\times L}^{-1}\left(\mathbf{r} - \mu\right) \tag{6.8}$$

where $\mathbf{1}_{L\times 1} = \underbrace{(1,1,\cdots,1)}_{L}{}^{T}$ is the unity vector with ones in all the components. The UTD also uses the unity vector **1** as its matched signature. The reason of choosing the unity vector is the following. Because there is no prior information available, the best scenario is not to introduce any information into the detector. In this case, the anomalous targets are assumed to have radiance uniformly distributed over all the spectral bands. As a consequence , background signatures will be extracted as anomalies. However, if there is some partial knowledge available, the unity vector can be replaced by a certain specific vector. For example, if we are interested in short wavelengths, we can set 0's for all visible bands while assigning ones to all the near infrared bands. Fig. 6.3(a) shows the result of the LCVF in Fig. 3.1 produced by UTD in which it did not detect the anomaly shown in Fig. 6.1(c); instead, it detected most of the image background. This is exactly what we expect from the unity vector since the background signatures can be assumed to be uniformly distributed.

(a) (b)

Figure 6.3. (a) Detection results of $\delta_{\text{UTD}}(\mathbf{r})$; (b) Detection result of $\delta_{\text{RXD-UTD}}(\mathbf{r})$

It is interesting to note that the background subtraction could enhance RXD detection performance as shown in Ashton and Schaum (1998). This suggests that we can incorporate UTD into RXD to remove the image background as well as noise to improve the performance of RXD. This advantage enables us to design a new type of anomaly detector. Instead of following Ashton and Schaum's approach which used the constrained linear mixture model to estimate the image background, we design an anomaly detector, denoted by $\delta_{\text{RXD-UTD}}(\mathbf{r})$ by subtracting UTD from RXD as follows.

$$\delta_{\text{RXD-UTD}}(\mathbf{r}) = (\mathbf{r}-\mathbf{1})^T \mathbf{K}_{L\times L}^{-1}(\mathbf{r}-\mu). \tag{6.9}$$

Fig. 6.3(b) shows the result of $\delta_{\text{RXD-UTD}}(\mathbf{r})$ with little improvement by visual inspection. However, the magnitude of the anomaly was slightly improved. This is because the total energy of the background in LCVF scene is less than that of anomaly. In this case, the background subtraction using (6.8) did not improve much over the RXD.

As an another example, the 15-panel HYDICE scene was also used for experiments. Fig. 6.4 shows the results produced by PCA and (6.2). In Fig. 6.4(a), PCA retained most of image information in its first component, which included the forest on the left edge and some panels. So, from an information preservation viewpoint, PCA achieved its goal in terms of information preservation. However, if we look at Fig. 6.4(b), $(\mathbf{r}-\mu)^T \mathbf{K}_{L\times L}(\mathbf{r}-\mu)$ specified by (6.2) extracted only the forest. Therefore, from a classification point of view, PCA did not perform well since it could not separate the forest from the panels in one component image. It is interesting to note that compared to PCA, $(\mathbf{r}-\mu)^T \mathbf{K}_{L\times L}(\mathbf{r}-\mu)$ successfully separated the forest in an individual component image. This implies that $(\mathbf{r}-\mu)^T \mathbf{K}_{L\times L}(\mathbf{r}-\mu)$ performed better than PCA in terms of target detection.

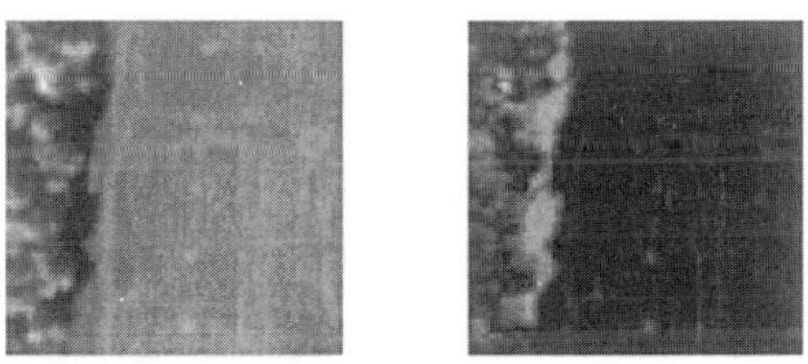

(a) PCA (b) $(\mathbf{r}-\mu)^T \mathbf{K}_{L\times L}(\mathbf{r}-\mu)$

Figure 6.4. (a) The first component of PCA; (b) Results of using (6.2)

Furthermore, Fig. 6.5 shows the detection results of the five anomaly detectors, $\delta_{\text{RXD}}(\mathbf{r})$, $\delta_{\text{NRXD}}(\mathbf{r})$, $\delta_{\text{MRXD}}(\mathbf{r})$, $\delta_{\text{UTD}}(\mathbf{r})$ and $\delta_{\text{RXD-UTD}}(\mathbf{r})$ where $\delta_{\text{RXD}}(\mathbf{r})$ and $\delta_{\text{RXD-UTD}}(\mathbf{r})$ performed nearly the same. They detected an interferer on the left upper corner and all the panels in the first column, but missed all the panels in the second and third columns as shown in Fig. 6.5(a,e) as opposed to $\delta_{\text{UTD}}(\mathbf{r})$ that extracted the image background as expected in Fig. 6.5(d). Apparently, $\delta_{\text{NRXD}}(\mathbf{r})$ and $\delta_{\text{MRXD}}(\mathbf{r})$ did not detect any panel in the scene.

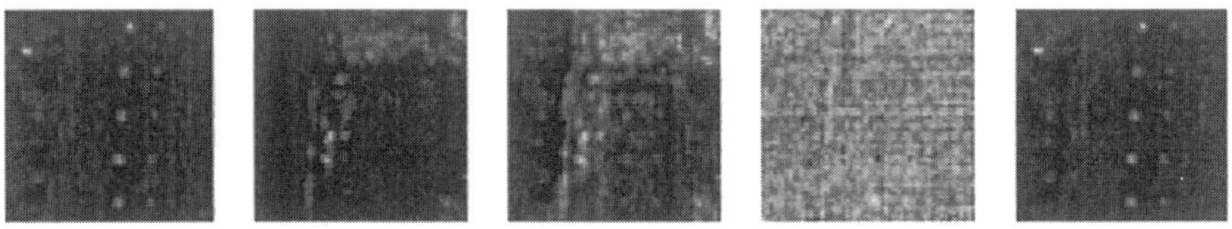

(a) $\delta_{\text{RXD}}(\mathbf{r})$ (b) $\delta_{\text{NRXD}}(\mathbf{r})$ (c) $\delta_{\text{MRXD}}(\mathbf{r})$ (d) $\delta_{\text{UTD}}(\mathbf{r})$ (e) $\delta_{\text{RXD-UTD}}(\mathbf{r})$

Figure 6.5. Detection results of the 15-panel HYDICE image using (a) $\delta_{\text{RXD}}(\mathbf{r})$; (b) $\delta_{\text{NRXD}}(\mathbf{r})$; (c) $\delta_{\text{MRXD}}(\mathbf{r})$; (d) $\delta_{\text{UTD}}(\mathbf{r})$; (e) $\delta_{\text{RXD-UTD}}(\mathbf{r})$

The hyperspectral image experiments conducted in Figs. 6.2 and 6.5 may lead to a conclusion that NRXD, MRXD and UTD could only perform background extraction. As will be shown in the following experiments, it is generally not the case. This is due to the fact that the information used in the detectors was $\mathbf{K}_{L\times L}$ which had removed the information of the first-order statistics. If the spectral properties of target signatures can be only characterized by the first-order statistics, these targets may not be detected by a detector designed solely based on the second-order statistics. This is particularly true for nonstationary images such as remotely sensed imagery. In order to account for both the first-order and second-order statistics, we substitute $\mathbf{R}_{L\times L}$ for $\mathbf{K}_{L\times L}$ and $\mathbf{r}$ for $\mathbf{r}-\mu$ in (6.1), (6.4-6.8). Figs. 6.6 and 6.7 show their detection results of the AVIRIS and HYDICE images respectively. In this case, UTD becomes LPTD. From these figures, we can see that the performance of RXD and RXD-UTD remained almost unchanged while UTD still detecting background and noise. To our surprise, NRXD and MRXD performed quite differently. Unlike Fig. 6.2, both NRXD and MRXD detected the shade in Fig. 6.6(b-c). In addition to the shade, MRXD also detected the anomaly in Fig. 6.6(c). Most interestingly is Fig. 6.7(b-c). Compared to Fig. 6.5(b-c) where NRXD and MRXD extracted only image background, Fig. 6.7(b-c) shows that NRXD and MRXD detected panels that were also detected by RXD. Besides, they both also extracted some tree signatures and interferers.

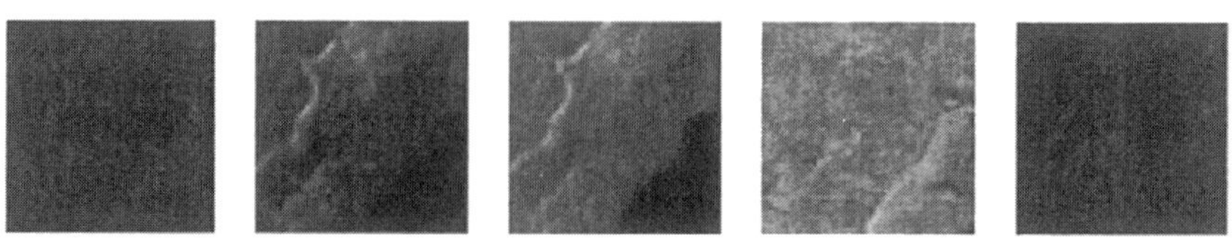

(a) $\delta_{\text{RXD}}(\mathbf{r})$ (b) $\delta_{\text{NRXD}}(\mathbf{r})$ (c) $\delta_{\text{MRXD}}(\mathbf{r})$ (d) $\delta_{\text{UTD}}(\mathbf{r})$ (e) $\delta_{\text{RXD-UTD}}(\mathbf{r})$

Figure 6.6. Detection results of the AVIRIS image using (a) $\delta_{\text{RXD}}(\mathbf{r})$; (b) $\delta_{\text{NRXD}}(\mathbf{r})$; (c) $\delta_{\text{MRXD}}(\mathbf{r})$; (d) $\delta_{\text{UTD}}(\mathbf{r})$; (e) $\delta_{\text{RXD-UTD}}(\mathbf{r})$

(a) $\delta_{\text{RXD}}(\mathbf{r})$ (b) $\delta_{\text{NRXD}}(\mathbf{r})$ (c) $\delta_{\text{MRXD}}(\mathbf{r})$ (d) $\delta_{\text{LPTD}}(\mathbf{r})$ (e) $\delta_{\text{RXD-LPTD}}(\mathbf{r})$

Figure 6.7. Detection results of the 15-panel HYDICE image using (a) $\delta_{\text{RXD}}(\mathbf{r})$; (b) $\delta_{\text{NRXD}}(\mathbf{r})$; (c) $\delta_{\text{MRXD}}(\mathbf{r})$; (d) $\delta_{\text{LPTD}}(\mathbf{r})$; (e) $\delta_{\text{RXD-LPTD}}(\mathbf{r})$

These two experiments demonstrated that when images are not stationary and cannot be characterized by only second-order statistics, using the sample correlation matrix $\mathbf{R}_{L\times L}$ is more appropriate and effective than using the sample covariance matrix $\mathbf{K}_{L\times L}$ to capture image characteristics.

6.4 RELATIONSHIP BETWEEN CEM AND RXD

As discussed in Section 4.3, $\delta_{\text{OSP}}(\mathbf{r})$ and $\delta_{\text{CEM}}(\mathbf{r})$ performed some sort of a matched filter with a different scale constant κ appearing in front of the matched filter. In order to investigate the relationship between CEM and RXD, we are interested in an RXD, which implements $\mathbf{r}$ and $\mathbf{R}_{L\times L}$ in place of (r-μ) and $\mathbf{K}_{L\times L}$ respectively. Such $\mathbf{R}_{L\times L}$-based RXD is referred to as R-RXD denoted by $\delta_{\text{R-RXD}}(\mathbf{r})$ afterwards. Unlike $\delta_{\text{CEM}}(\mathbf{r})$ which requires the prior knowledge of the desired target $\mathbf{d}$ as the matched signal, $\delta_{\text{R-RXD}}(\mathbf{r})$ does not have any prior target information and can be implemented with no prior knowledge. So, it is intuitive to choose the currently being processed image pixel $\mathbf{r}$ as its matched signature. Also noted, since anomaly detection primarily performs target detection not abundance estimation, the constant κ in $\delta_{\text{R-RXD}}(\mathbf{r})$ may not be a crucial parameter and can be set to $\kappa = 1$. So, if we replace the matched signature $\mathbf{d}$ in $\delta_{\text{CEM}}(\mathbf{r})$ with $\mathbf{r}$ and discard the constant $\kappa = \left(\mathbf{d}^T\mathbf{R}_{L\times L}^{-1}\mathbf{d}\right)^{-1}$ by setting $\kappa = 1$, $\delta_{\text{CEM}}(\mathbf{r})$ becomes $\delta_{\text{R-RXD}}(\mathbf{r})$.

In analogy with (6.4) and (6.6), we can also define normalized R-NRXD, $\delta_{\text{R-NRXD}}(\mathbf{r})$ and modified R-RXD, $\delta_{\text{R-MRXD}}(\mathbf{r})$ as follows.

$$\delta_{\text{R-NRXD}}(\mathbf{r}) = \left(\mathbf{r}/||\mathbf{r}||\right)^T \mathbf{R}_{L\times L}^{-1}\left(\mathbf{r}/||\mathbf{r}||\right) = \left(||\mathbf{r}||^2\right)^{-1} \mathbf{r}^T\mathbf{R}_{L\times L}^{-1}\mathbf{r} = \left(\mathbf{r}/||\mathbf{r}||^2\right)^T \mathbf{R}_{L\times L}^{-1}\mathbf{r} \quad (6.10)$$

$$\delta_{\text{R-MRXD}}(\mathbf{r}) = ||\mathbf{r}||^{-1}\mathbf{r}^T\mathbf{R}_{L\times L}^{-1}\mathbf{r} = \left(\mathbf{r}/||\mathbf{r}||\right)^T \mathbf{R}_{L\times L}^{-1}\mathbf{r} \quad (6.11)$$

where $||\mathbf{r}|| = \sqrt{\mathbf{r}^T\mathbf{r}}$ is the norm (vector length) of $\mathbf{r}$.

Following the same treatment for NRXD and MRXD, $\delta_{\text{R-NRXD}}(\mathbf{r})$ specified by (6.10) can be interpreted in three different ways. One is viewed as a normalized version of R-RXD. Another interpretation for $\delta_{\text{R-NRXD}}(\mathbf{r})$ is that it can be regarded as an R-RXD with the matched signature $\mathbf{d} = \mathbf{r}$ and a different scale constant $\kappa = ||\mathbf{r}||^{-2}$. Or equivalently,

$\delta_{\text{R-NRXD}}(\mathbf{r})$ can be thought of as a matched filter with the matched signature given by $\mathbf{d} = \mathbf{r}/||\mathbf{r}||^2$ with $\kappa = 1$. Similarly, the $\delta_{\text{R-MRXD}}(\mathbf{r})$ specified by (6.11) can be interpreted as an R-RXD with the constant $\kappa = ||\mathbf{r}||^{-1}$ or a matched filter with the matched signature $\mathbf{d} = \mathbf{r}/||\mathbf{r}||$ and $\kappa = 1$.

The $\delta_{\text{LPTD}}(\mathbf{r})$ and $\delta_{\text{UTD}}(\mathbf{r})$ can be also interpreted by a matched filter with the matched signature specified by an $L \times 1$ dimensional unity vector $\mathbf{1}$ with $\kappa = 1$. In analogy with $\delta_{\text{R-RXD}}(\mathbf{r})$, we can set the constant κ to 1 since $\delta_{\text{LPTD}}(\mathbf{r})$ and $\delta_{\text{UTD}}(\mathbf{r})$ are only used for target detection. The difference between $\delta_{\text{R-RXD}}(\mathbf{r})$ and $\delta_{\text{LPTD}}(\mathbf{r})$, $\delta_{\text{UTD}}(\mathbf{r})$ is that $\delta_{\text{R-RXD}}(\mathbf{r})$ uses $\mathbf{r}$ as its matched signature, while $\delta_{\text{LPTD}}(\mathbf{r})$ and $\delta_{\text{UTD}}(\mathbf{r})$ using $\mathbf{1}$ as their matched signatures. Because they are anomaly detectors not abundance estimators, there is no need of using the constant κ to account for estimation error.

In order to conduct a quantitative study among these $\mathbf{R}_{L\times L}$-based anomaly detectors, the data used in Section 4.5.1 were also used for computer simulations where 400 mixed pixel vectors, $\{\mathbf{r}_i\}_{i=1}^{400}$ were simulated and shown in Fig. 4.9. Fig. 6.8(a-d) shows anomaly detection of 5 creosote leaves pixels by the $\delta_{\text{R-RXD}}(\mathbf{r})$, $\delta_{\text{R-NRXD}}(\mathbf{r})$, $\delta_{\text{R-MRXD}}(\mathbf{r})$ and $\delta_{\text{LPTD}}(\mathbf{r})$ where $\delta_{\text{R-RXD}}(\mathbf{r})$ and $\delta_{\text{R-MRXD}}(\mathbf{r})$ performed better than $\delta_{\text{R-NRXD}}(\mathbf{r})$ in detection of creosote leaves, while $\delta_{\text{LPTD}}(\mathbf{r})$ basically extracted most background pixels. Comparing to the results in Figs. 4.10 and 4.11, all the four anomaly detectors cannot compete with $\delta_{\text{CEM}}(\mathbf{r})$ and $\delta_{\text{TCIMF}}(\mathbf{r})$ in detection of creosote leaves. This is because the latter requires the specific prior knowledge to detect targets as opposed to an anomaly detector which does not *have a priori* knowledge, but looks for targets with signatures spectrally distinct from neighboring pixels.

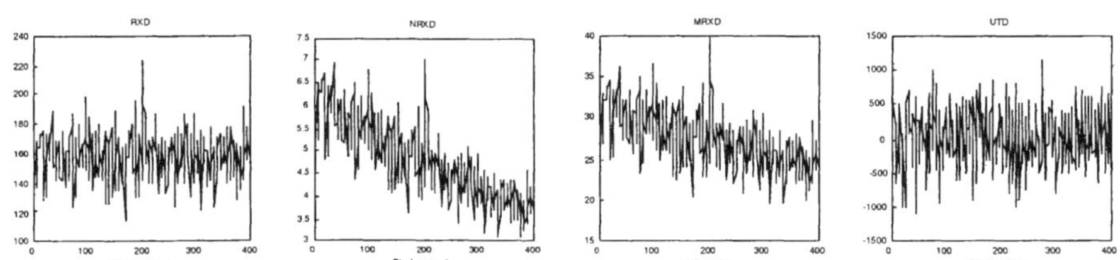

(a) $\delta_{\text{R-RXD}}(\mathbf{r})$ (b) $\delta_{\text{R-NRXD}}(\mathbf{r})$ (c) $\delta_{\text{R-MRXD}}(\mathbf{r})$ (d) $\delta_{\text{LPTD}}(\mathbf{r})$

Figure 6.8. Detection results of (a) $\delta_{\text{R-RXD}}(\mathbf{r})$; (b) $\delta_{\text{R-NRXD}}(\mathbf{r})$; (c) $\delta_{\text{R-MRXD}}(\mathbf{r})$; (d) $\delta_{\text{LPTD}}(\mathbf{r})$

Like Table 4.1, we summarize the relationship among $\delta_{\text{CEM}}(\mathbf{r})$, $\delta_{\text{R-RXD}}(\mathbf{r})$, $\delta_{\text{R-NRXD}}(\mathbf{r})$, $\delta_{\text{R-MRXD}}(\mathbf{r})$ and $\delta_{\text{LPTD}}(\mathbf{r})$ in Table 6.1 in terms of their matched signatures and their corresponding constants κ. It should be noted that all the four sample correlation matrix $\mathbf{R}_{L\times L}$-based anomaly detectors and $\delta_{\text{CEM}}(\mathbf{r})$ include $\mathbf{R}_{L\times L}^{-1}$ in their filter design.

Table 6.1. Relationship among $\delta_{\text{CEM}}(\mathbf{r})$, $\delta_{\text{R-RXD}}(\mathbf{r})$, $\delta_{\text{R-NRXD}}(\mathbf{r})$, $\delta_{\text{R-MRXD}}(\mathbf{r})$ and $\delta_{\text{LPTD}}(\mathbf{r})$

	κ	matched signatures	filter vector $\mathbf{w}$
$\delta_{\text{CEM}}(\mathbf{r})$	$\left(\mathbf{d}^T\mathbf{R}_{L\times L}^{-1}\mathbf{d}\right)^{-1}$	$\mathbf{d}$	$\mathbf{d}^T\mathbf{R}_{L\times L}^{-1}$ (4.8)
$\delta_{\text{R-RXD}}(\mathbf{r})$	1	$\mathbf{r}$	$\mathbf{r}^T\mathbf{R}_{L\times L}^{-1}$ (6.1)
$\delta_{\text{R-NRXD}}(\mathbf{r})$	1	$\mathbf{r}/\lVert\mathbf{r}\rVert^2$	$\left(\mathbf{r}/\lVert\mathbf{r}\rVert^2\right)^T\mathbf{R}_{L\times L}^{-1}$ (6.10)
$\delta_{\text{R-MRXD}}(\mathbf{r})$	1	$\mathbf{r}/\lVert\mathbf{r}\rVert$	$\left(\mathbf{r}/\lVert\mathbf{r}\rVert\right)^T\mathbf{R}_{L\times L}^{-1}$ (6.11)
$\delta_{\text{LPTD}}(\mathbf{r})$	1	$\mathbf{1}$	$\mathbf{1}^T\mathbf{R}_{L\times L}^{-1}$ (6.7)

6.5 REAL-TIME PROCESSING

The RXD defined by (6.1) cannot be implemented in real-time because the computation of $\mathbf{K}_{L\times L}^{-1}$ requires to calculate the sample mean of the entire image. On the other hand, it has been demonstrated in Chang et al. (2001) that the computation of the sample correlation matrix $\mathbf{R}_{L\times L}^{-1}$ can be implemented by a real-time process via a QR-decomposition (Golub and Van Loan, 1989). Comparing Fig. 6.1(c) against Fig. 6.6(a) and Fig. 6.5(a) against Fig. 6.7(a), there was no visible and appreciable difference for RXD to use $\mathbf{K}_{L\times L}^{-1}$ or $\mathbf{R}_{L\times L}^{-1}$. Fig. 6.9(a-b) shows their respective differential images obtained by taking the absolute difference between Fig. 6.1(c) and 6.6(a) for the AVIRIS image and the absolute difference between Fig. 6.5(a) and 6.7(a) for the HYDICE image. The plots next to the images show their corresponding abundance.

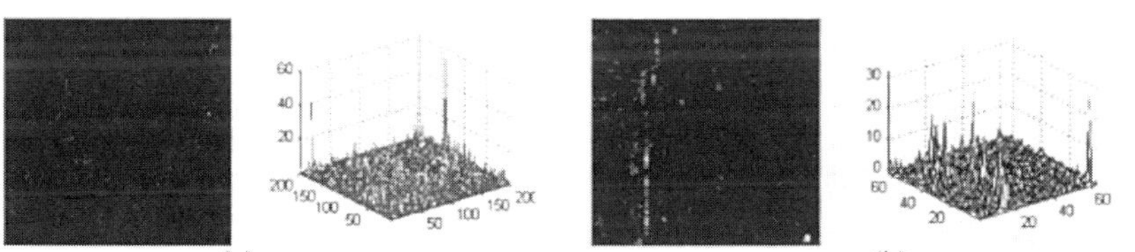

(a) (b)

Figure 6.9. (a) differential image between Fig. 6.1(b) and Fig. 6.6(a) (b) differential image between Fig. 6.5(a) and Fig. 6.7(a)

In Fig. 6.9(a), the only visible difference is the highest peak resulting from that both detectors detected different amounts of abundance in anomaly. In Fig. 6.9(b), both detectors detected nearly the same amount of abundance in all the panels, but different abundance fractions of background signatures, specifically, the one at the bottom right.

As also demonstrated in Figs. 6.2, 6.6(b-c), and Figs. 6.5(b-c), 6.7(b-c), it suggested that $\mathbf{R}_{L\times L}$ might be better than $\mathbf{K}_{L\times L}$ in characterizing spectral properties of non-natural targets. So, in order to implement RXD in real-time, it is advantageous to use RXD with $\mathbf{K}_{L\times L}^{-1}$ replaced by $\mathbf{R}_{L\times L}^{-1}$, i.e., R-RXD. The resulting R-RXD is referred to as causal R-RXD (CRXD). Since the information used in CRXD is updated in a causal manner, it may yield results different from that produced by R-RXD as demonstrated in the following experiments.

Analogous to the line-by-line real-time processing implemented for CEM in Section 4.6, we implemented the same line-by-line real-time processing for CRXD on the

AVIRIS image. The result is shown in Fig. 6.10 where the vegetation was first detected in the first 15 lines, then it disappeared as soon as the anomaly was detected at pixel 180. Finally, when the process was completed, the vegetation eventually vanished while the anomaly was enhanced.

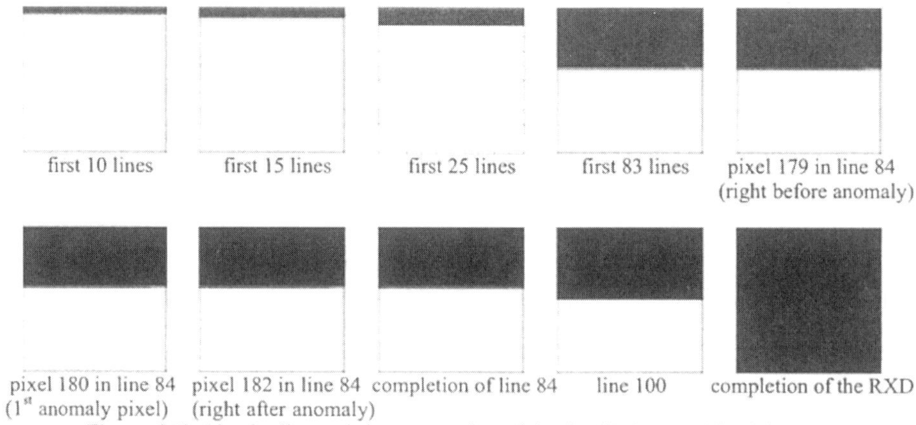

Figure 6.10. Line-by-line real-time processing of the CRXD for an AVIRIS image

A similar line-by-line real-time processing experiment for CRXD was also conducted for the 15-panel HYDICE scene as shown in Fig. 6.11.

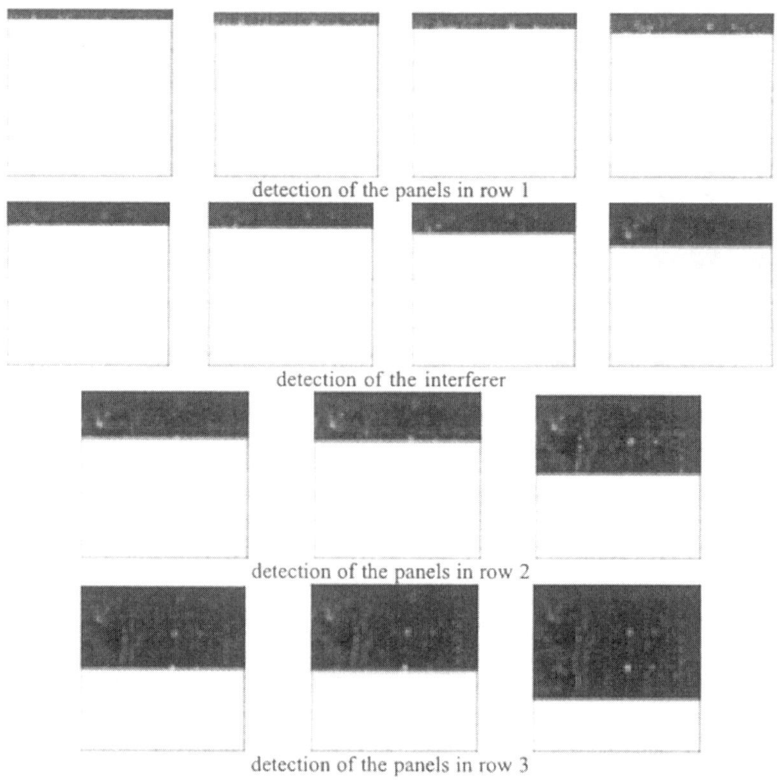

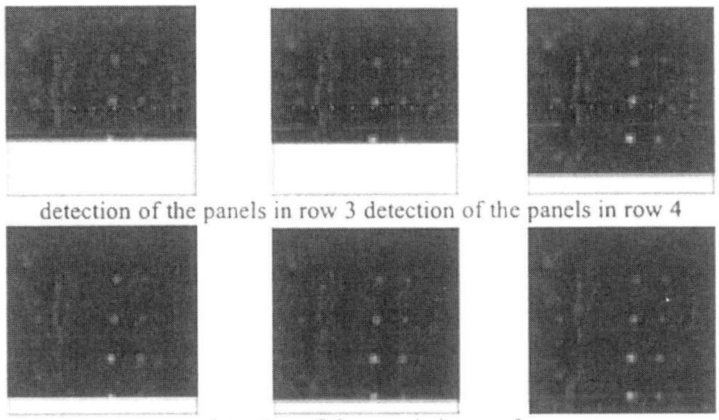

detection of the panels in row 3 detection of the panels in row 4

detection of the panels in row 5

Figure 6.11. Line-by-line real-time processing of CRXD for a 15-panel HYDICE image

Compared to Fig. 6.5(a), it is interesting to observe that the panel pixels in row 1 were first detected, then they faded away when an interferer was detected at the left upper corner. After this stage, their visibility was gradually diminished and eventually vanished when the process was completed. This experiment demonstrated a subtle difference between the line-by-line real-time processing of CRXD and RXD using $\mathbf{K}_{L\times L}^{-1}$ where the former used the causal information up to the line at which the pixel being processed, while the latter used noncausal information provided by the entire image.

Depending upon how the causal information used to calculate $\mathbf{R}_{L\times L}^{-1}$ a real-time implementation can be carried out in two different ways. One is the line-by-line real-time processing as described above. It used the line-by-line causal information that was provided by the pixels from the first line up to the line currently being examined by CRXD. Another is a pixel-by-pixel real-time processing which uses the pixel-by-pixel causal information that is provided by all pixels up to the one currently being examined by CRXD. Since CEM used the prior knowledge of the desired target signature, these two real-time processes resulted in very little difference for CEM. However, it is not true for CRXD because there is no assumed prior knowledge and the anomalies are generally sensitive to noise due to their small energies. Consequently, CRXD is sensitive to the information used in $\mathbf{R}_{L\times L}^{-1}$. Fig. 6.12 shows the experiment of the pixel-by-pixel real processing of CRXD for the 15-panel HYDICE image. If we compare Fig. 6.11 to Fig. 6.12, there were more background signatures detected in Fig. 6.11 than in Fig. 6.12, specifically, up to the detection of panels in row 3. However, it was found that for the AVIRIS image there was no visible difference between the line-by-line processing and pixel-by-pixel processing. This may be due to the low 20-m spatial resolution of the AVIRIS image and the pixel-by-pixel real-time processing was only sensitive to high pixel resolution. The results are not included here.

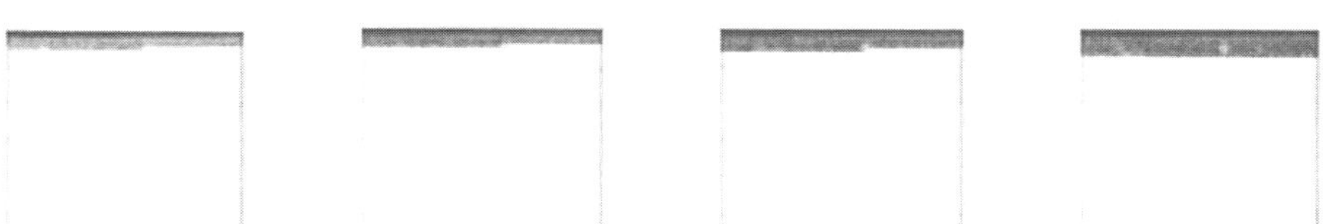

no target detected 1st panel pixel detected panel pixels detected completion of detection

(detection of panel pixels in row 1)

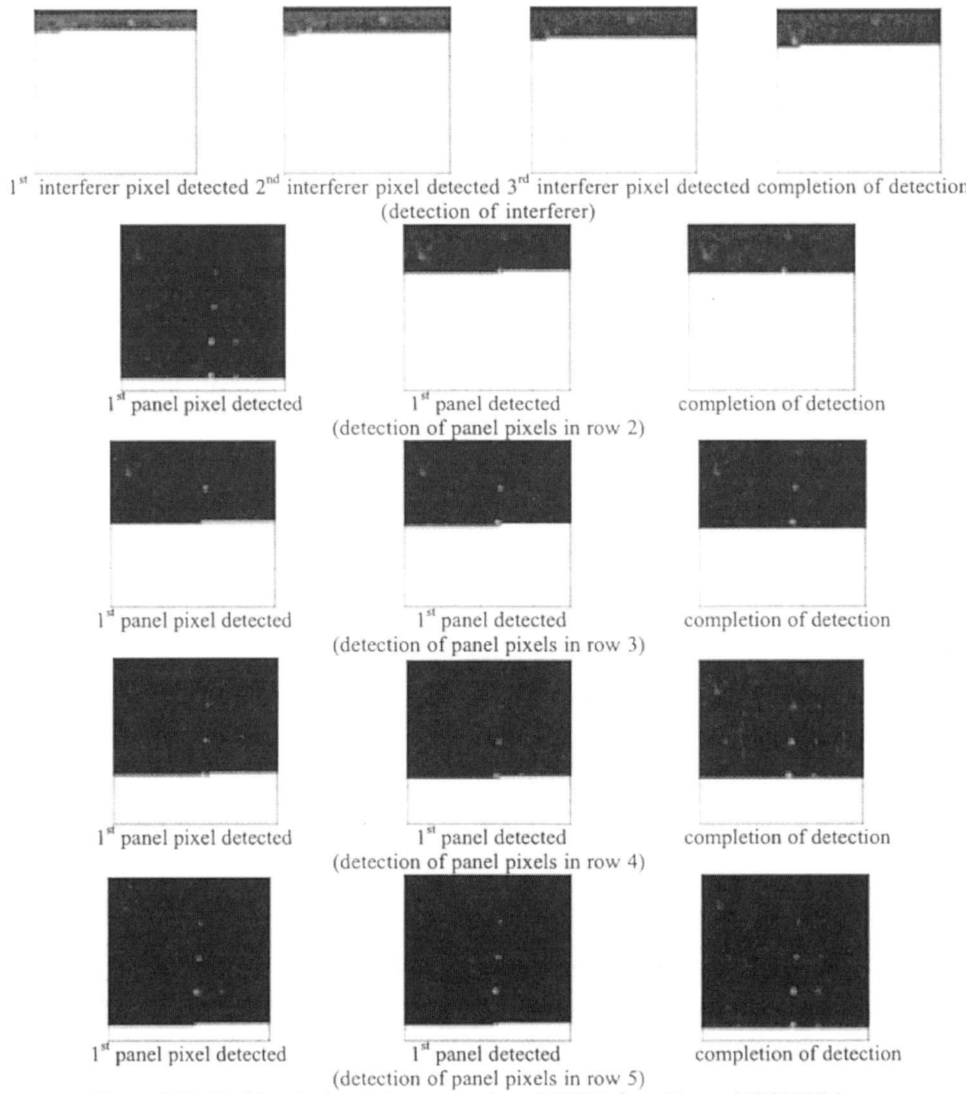

Figure 6.12. Pixel-by-pixel real-time processing of CRXD for a 15-panel HYDICE image

As a final comment, two significant advantages of using real-time processing are computational efficiency and detection of moving targets. This is particularly useful for hyperspectral imagery because the data cube is generally very large.

6.6 CONCLUSIONS

Despite that the unsupervised subpixel detection considered in Chapter 5 required no prior target knowledge, it did make use of target information generated by an unsupervised algorithm to perform supervised subpixel detection. Such obtained target information can be viewed as unsupervised target knowledge. This chapter considers anomaly detection, which does not require unsupervised target knowledge. In surveillance applications, we are often interested in anomalies that are relatively small objects with low probabilities in an image, but do not belong to the image background. These anomalous targets are generally unknown *a priori* and cannot be identified from an image scene visually. To meet this need, several anomaly detectors are developed to extract such targets. They all operate in a functional form of a matched filter with a different matched signature. Furthermore, they can be also implemented in a real-time process. Unfortunately, there is a disadvantage resulting from no prior knowledge. An anomaly detector cannot classify the targets it detects since it cannot discriminate its detected targets one from another. In order to extend anomaly detection to anomaly classification, it requires criteria for target discrimination. This issue will be addressed in Chapter 14.

7

SENSITIVITY OF SUBPIXEL DETECTION

In Chapter 3, we considered the target abundance-constrained subpixel detection (TACSD) and evaluated three techniques, OSP, SCLS and NCLS. A major limitation of these techniques is the requirement of complete target knowledge. In Chapter 4, we studied the target signature-constrained subpixel detection (TSCSD) and also evaluated two techniques, CEM and TCIMF. However, their performance is sensitive to the target knowledge used in their filter design and is also closely related to the intrinsic dimensionality (ID) of $\mathbf{R}_{L\times L}$ which is generally unknown and needs to estimate. Unfortunately, the accuracy of the ID estimation is determined by noise level. Similarly, the anomaly detection developed in Chapter 6 also requires the computation of $\mathbf{R}_{L\times L}^{-1}$ with their performance also affected by noise sensitivity. So, this chapter is devoted to a thorough investigation of two issues related to detection performance. One is the sensitivity of target knowledge. Another is the noise sensitivity to computation of $\mathbf{R}_{L\times L}^{-1}$ required for TSCSD. Because the noise variances are determined by eigenvalues, the performance of TSCSD will be evaluated based on the number of eigenvectors used to calculate $\mathbf{R}_{L\times L}^{-1}$. As will be demonstrated, this number plays a significant role in detection performance.

7.1 INTRODUCTION

In Chapters 3 and 4, we investigated TACSD and TSCSD respectively. Several distinctions between these two approaches are worth noting. The most prominent is that TACSD is basically a pixel-level spectral analysis technique, which operates on a pixel-by-pixel basis. It takes advantage of inter-band spectral information provided by individual bands within a pixel vector. In the OSP approach, such inter-band spectral information is provided by the knowledge of **d** and $P_{\mathbf{U}}^{\perp}$. In the PCLS methods, the spectral information specified by $(\mathbf{M}^T\mathbf{M})^{-1}\mathbf{M}^T$ in (3.10) is known. By contrast, TSCSD is a sample spectral correlation matrix-based analysis technique, which utilizes the sample spectral correlation matrix $\mathbf{R}_{L\times L}$ or the sample spectral covariance matrix $\mathbf{K}_{L\times L}$ to account for spectral correlation provided by sample pixel vectors. This can be seen from

(4.6), (4.10), (6.1), (6.4)-(6.7) where the detectors must calculate $\mathbf{R}_{L\times L}^{-1}$. Nevertheless, all these detectors result in a same functional form of a matched filter given by (3.6) with different matched signatures and scale constants κ that account for abundance estimation error. When a spectral matched filter is used for TACSD, the inter-band spectral information specified by $P_{\mathbf{U}}^{\perp}$ or $(\mathbf{M}^T\mathbf{M})^{-1}\mathbf{M}^T$ is used to preprocess a data vector by eliminating undesired signatures before a matched signature is applied. When a sample spectral correlation matrix-based matched filter is used for TSCSD, the sample spectral correlation/covariance matrix, $\mathbf{R}_{L\times L}^{-1}$ or $\mathbf{K}_{L\times L}^{-1}$ is used to whiten a data vector prior to application of a matched signature. It was shown in Chang (2002a) that OSP is a whitened version of CEM through TCIMF with an inclusion of a scale constant κ that accounts for estimation error. This is because OSP does not estimate the abundance fraction vector α, but CEM does. On the other hand, it was also shown in Chang and Heinz (2000) that the detection performance of TSCSD was also determined by q, the number of eigenvectors used in the computation of $\mathbf{R}_{L\times L}^{-1}$. This q was tied with the scale constant κ. Therefore, with respect to information being used in a matched filter, TACSD and TSCSD can be viewed as completely different approaches in the sense that the former uses $P_{\mathbf{U}}^{\perp}$ or $(\mathbf{M}^T\mathbf{M})^{-1}\mathbf{M}^T$ for information preprocessing as opposed to the latter using $\mathbf{R}_{L\times L}^{-1}$ or $\mathbf{K}_{L\times L}^{-1}$ for whitening the data. However, there is a close link between these two. As discussed in Section 5.1 of Chapter 5 as well as in Chang (2002b), a connection between TACSD and TSCSD is the relationship between $P_{\mathbf{U}}^{\perp}$ and $\mathbf{R}_{L\times L}^{-1}$. Basically, $P_{\mathbf{U}}^{\perp}$ requires the complete knowledge about the undesired target signature matrix $\mathbf{U}$ compared to $\mathbf{R}_{L\times L}^{-1}$ which can be computed directly from sample data vectors without prior target knowledge. In order to obtain the prior information $P_{\mathbf{U}}^{\perp}$, $\mathbf{R}_{L\times L}^{-1}$ makes use of spectral correlation among sample vectors to approximate $P_{\mathbf{U}}^{\perp}$ in the sense of minimum least squares error. As a result, how well $\mathbf{R}_{L\times L}^{-1}$ approximates $P_{\mathbf{U}}^{\perp}$ is determined by how accurate target information is represented by $\mathbf{R}_{L\times L}$. This is also in turn determined by the intrinsic dimensionality (ID), which turns out to be p, the number of target signatures used in the target signature matrix $\mathbf{M}$. According to this analysis, two major factors have substantial impact on detection performance of TACSD and TSCSD. One is the target knowledge, $\mathbf{m}_1, \mathbf{m}_2, \cdots, \mathbf{m}_p$, used in $\mathbf{M}$ and another is the number of target signatures, p which determines the number of eigenvectors, q to be used to computer $\mathbf{R}_{L\times L}^{-1}$. So, in this chapter we investigate these two sensitivity issues on subpixel target detection.

The first issue arises from the high spectral resolution of a hyperspectral sensor which can uncover many unknown target signatures in an image scene including background signatures, natural targets, interferers, clutter, etc. Under this circumstance, finding an appropriate $\mathbf{M}$ to well represent the image scene is crucial. But it is also very difficult because some of these target signatures cannot be identified *a priori*. In order to resolve this problem, three unsupervised algorithms developed in Chapter 5 can be used to find these target signatures in an unsupervised manner. As will be demonstrated in experiments, including these target signatures in $\mathbf{M}$ will significantly improve the performance of TACSD.

The second issue is to estimate an appropriate p, which determines the performance of TSCSD. But this is closely related to ID of the image data, viz. the number of image endmembers in the data. Unfortunately, ID is generally not known in practice. As a result, without knowing the exact value of q calculating $\mathbf{R}_{L\times L}^{-1}$ may be very tricky. This is due to the fact that small eigenvalues dominate $\mathbf{R}_{L\times L}^{-1}$ and differentiating small eigenvalues from noise variances is considered to be very challenging in signal processing. A similar problem also occurs in anomaly detection in Chapter 6 where the detection performance is also heavily affected by the number of eigenvectors used to calculate $\mathbf{R}_{L\times L}^{-1}$ or $\mathbf{K}_{L\times L}^{-1}$.

There is another major distinction between TACSD and TSCSD. TACSD requires a linear mixture model to estimate the abundance fractions of target signatures present in an image scene. Therefore, it needs the complete knowledge of target information in the image data. To the contrary, TSCSD is developed without knowing image background. So, no linear mixture model is required for TSCSD. However, both types of subpixel target detection do share the same difficulty, i.e. solving an inverse problem. To invert a linear mixing problem, TACSD requires that the number of target signatures present in an image scene cannot be greater than ID to avoid over-representing the image data. Correspondingly, TSCSD requires that the number of pixel samples be greater than ID to avoid singularity in computation of $\mathbf{R}_{L\times L}^{-1}$ or $\mathbf{K}_{L\times L}^{-1}$. In both cases, the problem winds up the determination of ID. In Chapter 17, we will introduce a new concept, called virtual dimensionality (VD), which is a little bit different from ID in the sense that VD is defined by a number of spectrally distinct signal sources in image data. The VD may provide a good estimate of how many eigenvectors, q required to calculate $\mathbf{R}_{L\times L}^{-1}$. In Chapter 17, this problem will be addressed and several methods to estimate VD will be proposed for this purpose.

7.2 SENSITIVITY OF TARGET KNOWLEDGE

The techniques presented in Chapters 3 and 4 required different degrees of target knowledge. In order to conduct a fair comparative analysis, the same target information must be made available to these techniques despite the fact that TSCSD requires only the knowledge of targets of interest. To meet this need, the UNCLS algorithm in Section 5.4 will be used to generate such unsupervised target information for a thorough comparative study where the five different subpixel target detection methods, OSP, TACSD (SCLS and NCLS) and TSCSD (CEM and TCIMF) are selected for evaluation. The term of "*unsupervised target information or knowledge*" is referred to as information or knowledge that is obtained directly from the data in an unsupervised means. The problem of sensitivity to target knowledge will be addressed first, then noise sensitivity to computation of $\mathbf{R}_{L\times L}^{-1}$. The following computer simulations are particularly designed to illustrate the sensitivity issue of target information used in TACSD and TSCSD.

The laboratory AVIRIS data set in Fig. 1.5 will be used to evaluate the performance of OSP, SCLS, NCLS, CEM and TCIMF. The data set contains five field reflectance spectra, dry grass, red soil, creosote leaves, blackbrush and sagebrush. In this case, the complete prior target knowledge is the target signature matrix $\mathbf{M} = [\mathbf{m}_1\, \mathbf{m}_2\, \mathbf{m}_3\, \mathbf{m}_4\, \mathbf{m}_5]$ consisting of these five spectral signatures with their associated abundance fractions given by $\alpha = (\alpha_1, \alpha_2, \alpha_3, \alpha_4, \alpha_5)^T$. Unlike the experiments conducted in Chapter 3, which was based on the complete knowledge of these five target signatures, the following experiments assume that the target knowledge is either partial available or has no a priori information at all.

Two examples are considered to illustrate how the target information affects the subpixel target detection performance. The first example assumes that there is partial but not complete target knowledge available prior to detection. In this case, the UNCLS algorithm will be used to generate the necessary target information that was required for OSP, SCLS and NCLS target detectors. The second example makes an assumption that no prior target knowledge is available and all the required target information must be obtained directly from the unknown image data. Once again, the UNCLS algorithm is also used to generate the necessary target information for all the five methods, OSP, SCLS, NCLS, CEM and TCIMF.

Example 7.1 Partial Target Knowledge

The data used in this example were the same used in Section 4.5.1 where only three target signatures, creosote leaves, dry grass and red soil were used for illustration. Assume that the only target knowledge available to us was the creosote leaves. We simulated 400 mixed pixel vectors, $\{\mathbf{r}_i\}_{i=1}^{400}$, as follows. We started the first pixel vector with 100% red soil and 0% dry grass, then began to increase 0.25% dry grass and decrease 0.25% red soil every pixel vector until the 400th pixel vector which contained 100% dry grass. We then added creosote leaves to pixel vector numbers 198-202 at abundance fractions 10% while reducing the abundance of red soil and dry grass accordingly. For example, after addition of creosote leaves, the resulting pixel vector 200 contained 10% creosote leaves, 45% red soil and 45% dry grass. White Gaussian noise was also added to each pixel vector to achieve a 30:1 signal-to-noise ratio as was defined as 50% reflectance divided by the standard deviation of the noise in Harsanyi and Chang (1994). The spectra of these 400 simulated pixel vectors shown in Fig. 4.9 are also reproduced in Fig. 7.1 for convenience where Fig. 7.1(a) shows the abundance of each pixel at particular band 30 and Fig. 7.1(b) shows the abundance of each pixel averaged over the total 158 bands.

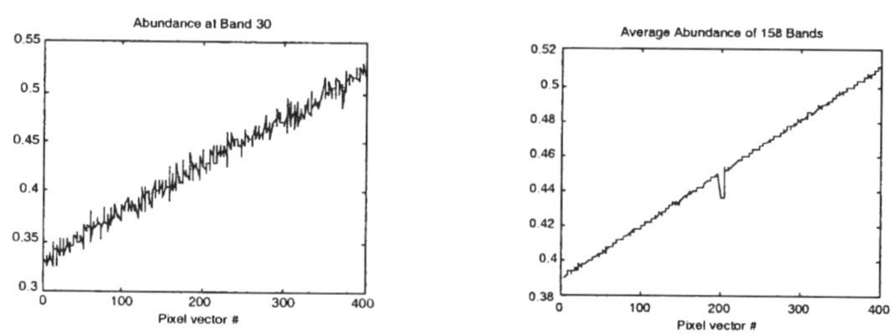

(a) Abundance of each pixel at band 30 (b) Abundance of each pixel averaged over 158 bands

Figure 7.1. 400 simulated mixed pixel vectors used in Example 7.1

Since the creosote leaves was the only target knowledge available to us, we used it as an initial target $\mathbf{t}_0$ to specify $\mathbf{m}_0$ used in the UNCLS algorithm. The estimated abundance fraction of $\mathbf{m}_0$ for each of 400 simulated pixel vectors is denoted by $\hat{\alpha}_0^{(1)}(\mathbf{r}_i)$

where $\mathbf{r}_i$ indicates the i-th simulated pixel vector with $i = 1,2,\cdots,400$. Using $\hat{\alpha}_0^{(1)}(\mathbf{r}_i)$ we calculated LSE between all simulated pixel vectors $\mathbf{r}_k$ for $1 \le k \le 400$ and $\hat{\alpha}_0^{(1)}(\mathbf{r}_i)\mathbf{m}_0$. Since the resulting maximum LSE was not below a prescribed threshold, UNCLS was continued to search for a pixel vector that yielded the maximum LSE. In this example, the seventh pixel vector with 98.5% red soil, $\mathbf{r}_7$ was selected as the first target, denoted by $\mathbf{t}_1 = \mathbf{r}_7$ with its signature specified by $\mathbf{m}_1$. The UNCLS algorithm was again used to estimate the abundance fractions of $\mathbf{m}_0$ and $\mathbf{m}_1$, denoted by $\hat{\alpha}_0^{(2)}(\mathbf{r}_i)$ and $\hat{\alpha}_1^{(2)}(\mathbf{r}_i)$ for each of 400 simulated pixel vectors with $i = 1,2,\cdots,400$. Using the estimated abundance fractions we calculated LSE of all simulated pixels vectors $\mathbf{r}_k$ for $1 \le k \le 400$ between the least squares linear mixture, $\hat{\alpha}_0^{(2)}(\mathbf{r}_i)\mathbf{m}_0 + \hat{\alpha}_1^{(2)}(\mathbf{r}_i)\mathbf{m}_1$. Because the resulting maximum LSE was still not below the prescribed threshold, the UNCLS algorithm was continued and the 400-th pixel with 100% dry grass, $\mathbf{r}_{400}$ was selected as a second target $\mathbf{t}_2 = \mathbf{r}_{400}$ with its signature specified by $\mathbf{m}_2$. After finding $\mathbf{t}_2$, the resulting maximum LSE was below the prescribed threshold and the UNCLS algorithm was terminated. At this point, two targets $\mathbf{t}_1 = \mathbf{r}_7 =$ red soil and $\mathbf{t}_2 = \mathbf{r}_{400} =$ dry grass were identified, which were not known *a priori*. Using these three target signatures $\mathbf{m}_0$, $\mathbf{m}_1$ and $\mathbf{m}_2$ to form a desired target signature matrix $\mathbf{M}$, Fig. 7.2(a-c) shows the detection results of OSP, SCLS, NCLS, CEM and TCIMF where creosote leaves, red soil and dry grass were used as the desired signatures and were also detected respectively.

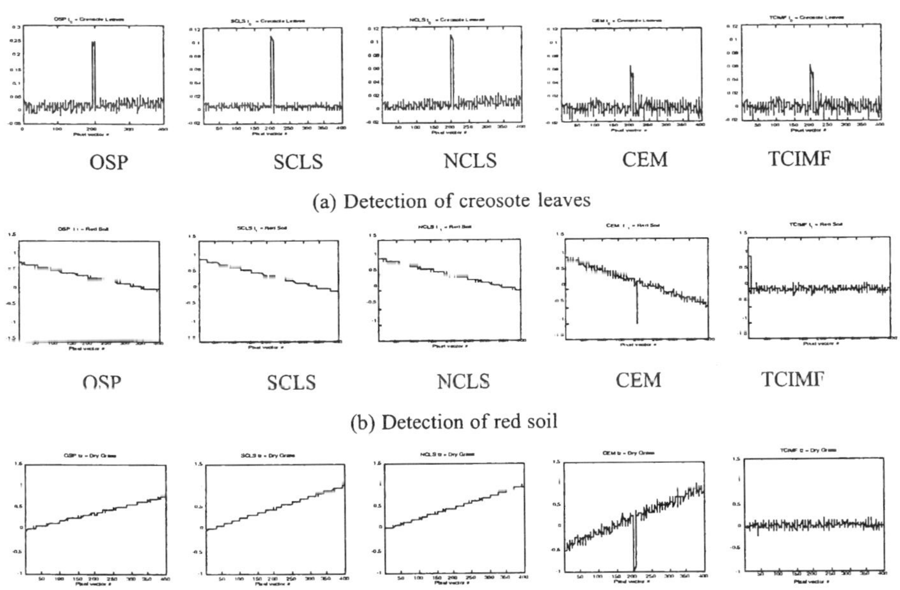

OSP SCLS NCLS CEM TCIMF

(a) Detection of creosote leaves

OSP SCLS NCLS CEM TCIMF

(b) Detection of red soil

OSP SCLS NCLS CEM TCIMF

(c) Detection of dry grass

Figure 7.2. Comparative results of OSP, SCLS, NCLS, CEM and TCIMF

As we can see from figures in Fig. 7.2(a-c), OSP and two PCLS detectors, SCLS and NCLS, performed similarly. NCLS was the best among these three detectors in the sense that the abundance fractions of the desired target signatures were detected very accurately. On the other hand, OSP was the worst because the detected abundance fractions of desired target signatures were not correct despite that it also detected the desired target signatures. CEM and TCIMF performed very differently from OSP, SCLS and NCLS. Both CEM and TCIMF correctly detected creosote leaves in Fig. 7.2(a), but they had completely different performance in detection red soil and dry grass. The amounts of red soil detected in Fig. 7.2(b) by CEM filter fluctuated and started from 1, then gradually decreased to -0.5. Interestingly, it also detected the creosote leaves at pixels 198-202 but with large negative amounts. A similar phenomenon in detection of dry grass was also observed in Fig.7.2(c) for CEM. The most intriguing case is the performance of TCIMF. Since TCIMF constrained the particular targets $\mathbf{r}_7$ and $\mathbf{r}_{400}$ while eliminating all other target pixels, it only detected a single red soil pixel $\mathbf{r}_7$ in Fig. 7.2(b) and a single dry grass pixel $\mathbf{r}_{400}$ in Fig. 7.2(c).

Example 7.2 No Target Knowledge Available *A Priori*

The only difference between this example and Example 7.1 is that no initial target signature $\mathbf{t}_0$ was given *a priori*. The target $\mathbf{t}_0$ must be generated from the data set. In this case, we selected the pixel vector with maximum length, which turned out to be the 400th pixel vector with 100% dry grass, $\mathbf{r}_{400}$. Using $\mathbf{t}_0 = \mathbf{r}_{400}$ to initialize the UNCLS algorithm and following the same procedure in Example 7.1 we found the fourth pixel vector, $\mathbf{r}_4$ to be the first target $\mathbf{t}_1 = \mathbf{r}_4 =$ red soil with 99.25% and the 200th pixel vector $\mathbf{r}_{200}$ with 10% creosote leaves to be the second target $\mathbf{t}_2 = \mathbf{r}_{200}$, and then the UNCLS algorithm was terminated. The signatures of three targets, $\mathbf{m}_0$, $\mathbf{m}_1$ and $\mathbf{m}_2$ were used to form a desired target signature matrix $\mathbf{M}$. It should be noted that targets generated by UNCLS, $\mathbf{r}_4$ and $\mathbf{r}_{200}$ in this example were different from $\mathbf{r}_7$ and $\mathbf{r}_{400}$ generated in Example 7.1.

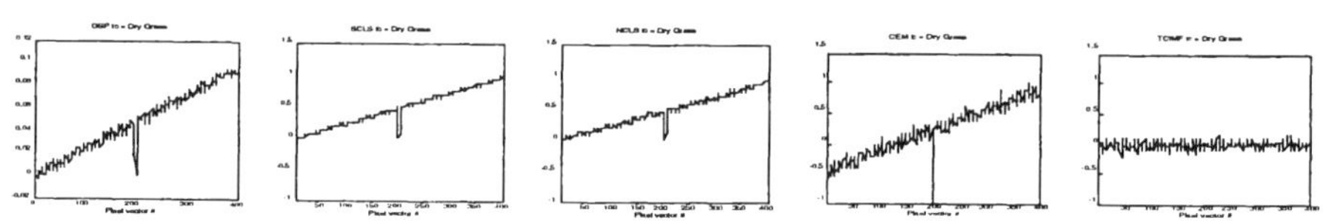

OSP SCLS NCLS CEM TCIMF

(a) Detection of the first found target, dry grass

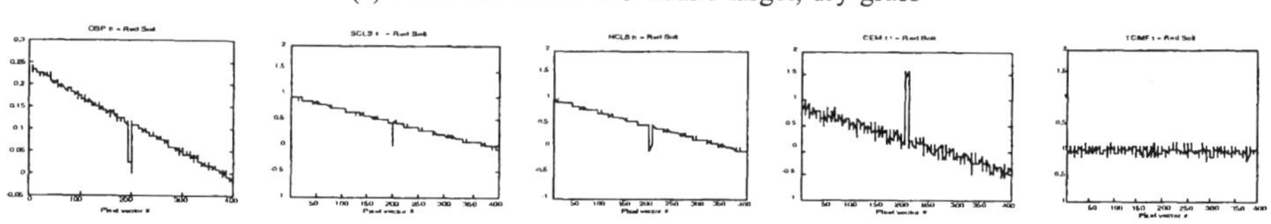

OSP SCLS NCLS CEM TCIMF

(b) Detection of the second found target, red soil

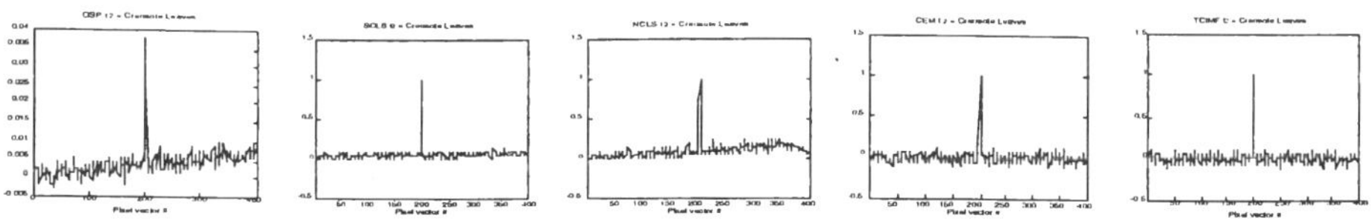

OSP SCLS NCLS CEM TCIMF

(c) Detection of the third found target, creosote leaves

Figure 7.3. Comparative results of OSP, SCLS, NCLS, CEM and TCIMF with no prior target knowledge

Fig. 7.3(a-c) shows the detection results produced by OSP, SCLS, NCLS, CEM and TCIMF where only the OSP-result was shown by a different scale. Once again, NCLS performed the best among all the five detectors, but it did not perform as well as it did in Fig. 7.2 in terms of estimating abundance fractions due to the lack of prior information about target signatures. Similar observations were also applied to OSP and SCLS. Under this circumstance OSP, SCLS and NCLS behaved more like a target detector rather than a target abundance estimator as shown in Fig. 7.2. Interestingly, the detection result of red soil in Fig. 7.3(b) produced by CEM looked the exactly upside down of that in Fig. 7.2(b) produced by CEM with slight different magnitudes. In detection of creosote leaves, the results produced by CEM in Figs. 7.2(a) and 7.3(c) looked similar but the detected abundance fractions were different. In detection of dry grass, CEM produced nearly same results in both cases. Like Example 7.1, TCIMF only detected single target pixels on which it constrained. If we compare the detection result of the creosote leaves produced by TCIMF in Fig. 7.2(a) to that in Fig. 7.3(c), TCIMF detected creosote leaves at pixel numbers 198-202 in Fig. 6(a) with TCIMF only detecting the creosote leaves at $\mathbf{r}_{200}$ in Fig. 7.3(c). This is because the former used the precise knowledge of the creosote leaves to produce the result in Fig. 7.2(a), while the latter used only $\mathbf{r}_{200}$ as the desired target knowledge to produce the result in Fig. 7.3(c).

7.3 SENSITIVITY OF NOISE

In this section, we investigate the noise sensitivity issue for TSCSD. Since this issue is closely related to the number of eigenvectors, q that is used to calculate $\mathbf{R}_{L\times L}^{-1}$, we will deal with the issue of the effect caused by q rather than noise directly in the following experiments. In order to complete characterize the impacts resulting from q, we also make an assumption that the complete target knowledge is known *a priori*.

7.3.1 TSCSD

According to (4.6) and (4.10), CEM and TCIMF share the same detector structure that calculates $\mathbf{R}_{L\times L}^{-1}$. As expected, their performance will be very similar. Only in some cases, TCIMF may take advantage of partial target knowledge to outperform CEM.

Example 7.3 Target Signatures with Relatively Large Abundance Fractions

Unlike two examples considered in Section 7.2, we simulated 400 mixed pixel vectors, $\{\mathbf{r}_i\}_{i=1}^{400}$, using all the five field reflectance spectra, dry grass, red soil, creosote leaves, blackbrush and sagebrush shown in Fig. 1.5. They were divided into four groups, each of which contains a hundred pixel vectors with the same mixture. The first group consisted of the first hundred pixel vectors with the mixture made up of 50% sagebrush and 50% dry grass. The second group consisted of the second hundred pixel vectors with the mixture made up of 50% sagebrush and 50% red soil. The third group consisted of the third hundred pixel vectors with the mixture made up of 50% sagebrush and 50% creosote leaves. The fourth group consisted of the fourth hundred pixel vectors with the mixture made up of 50% sagebrush and 50% blackbrush. More specifically, each of the 400 simulated pixel vectors was a two-component mixture with 50-50 split and all of the 400 pixel vectors share the same amount of sagebrush, that is, 50% sagebrush. White Gaussian noise was added to each pixel vector to achieve a 30:1 signal-to-noise ratio as defined in Example 7.1. Fig. 7.4 shows these 400 simulated pixels.

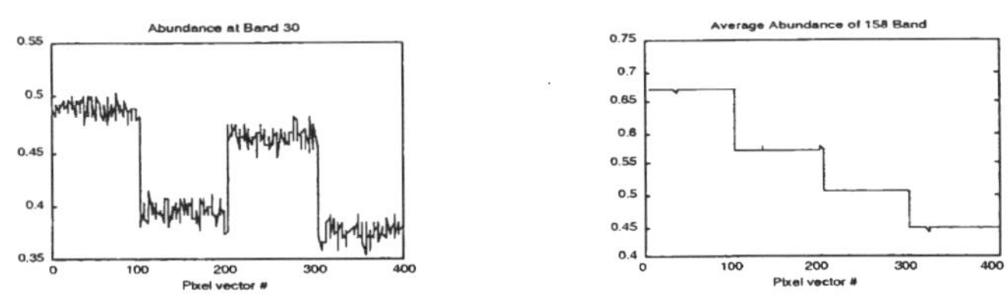

(a) Abundance of each pixel at band 30 (b) Abundance of each pixel averaged over 158 bands

Figure 7.4. 400 simulated mixed pixel vectors used in Example 7.3

Fig. 7.5(a-e) shows the results of OSP, SCLS, NCLS, CEM and TCIMF in detection of dry grass, red soil, creosote leaves, blackbrush and sagebrush respectively.

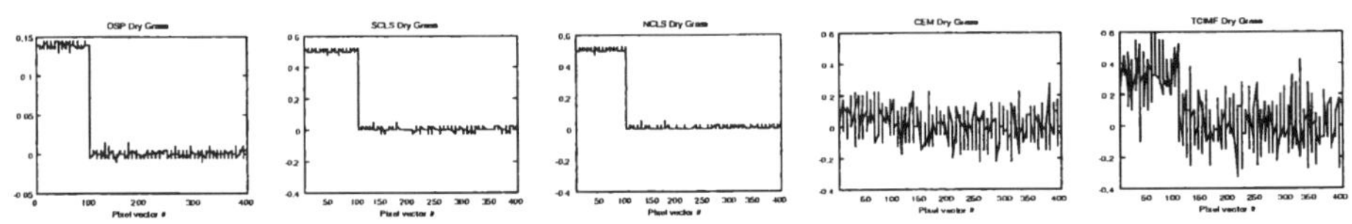

OSP SCLS NCLS CEM TCIMF

(a) Detection of dry grass

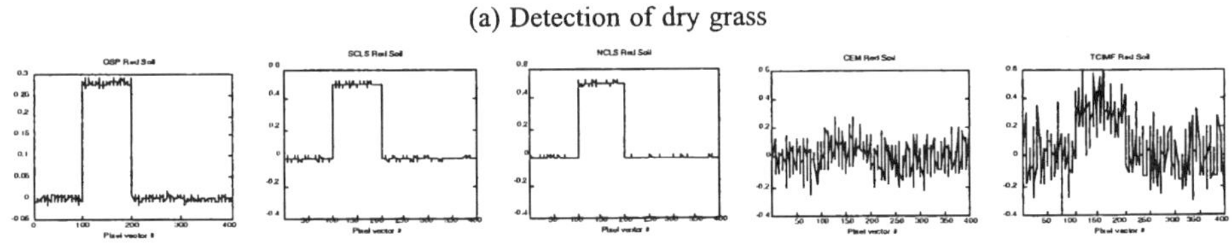

OSP SCLS NCLS CEM TCIMF

(b) Detection of red soil

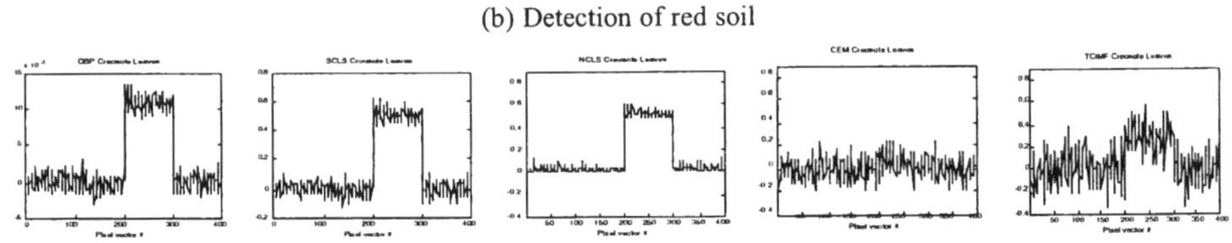

OSP SCLS NCLS CEM TCIMF

(c) Detection of creosote leaves

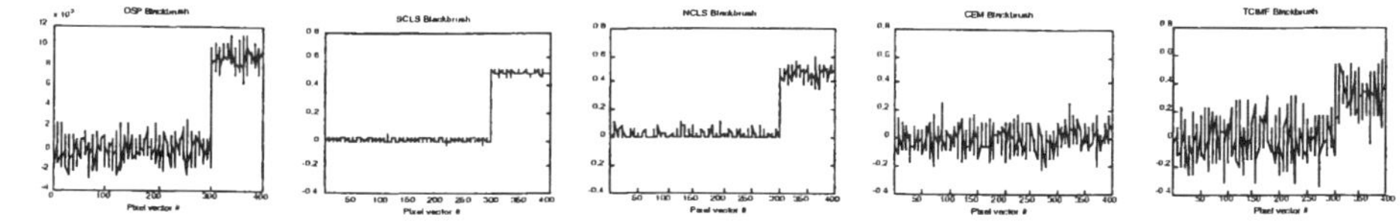

OSP SCLS NCLS CEM TCIMF

(d) Detection blackbrush

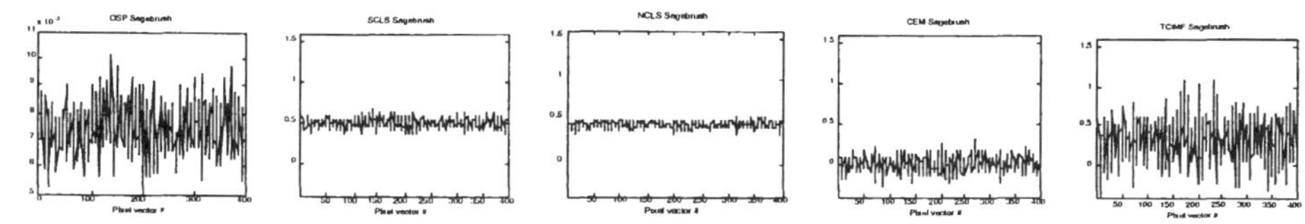

OSP SCLS NCLS CEM TCIMF

(e) Detection of sagebrush

Figure 7.5. Results of OSP, SCLS, NCLS, CEM and TCIMF in detection of dry grass, red soil, creosote leaves, blackbrush and sagebrush

Due to nature of unconstrained abundance, the abundance fractions detected in Fig. 7.5 had a wide range of values. In order to display the detection results for visual inspection, two different scales were used in Fig. 7.5, one for OSP, another for SCLS, NCLS, CEM, and TCIMF. As shown, OSP, SCLS and NCLS were able to detect all the five signatures, but OSP did not perform well in detection of sagebrush. However, if we examine the amounts of their detected target abundance fractions, the one detected by OSP did not reflect the true abundance, but those detected by SCLS and NCLS did. This is because OSP is unconstrained, and both SCLS and NCLS are at least partially constrained. Surprisingly, CEM, which was shown to be effective in Harsanyi (1993), Harsanyi et al. (1994), Farrand and Harsanyi (1997) and TCIMF performed poorly in this example. Similarly, TCIMF did not perform well either in detection of these five targets. Nevertheless, it did show high peaks when it detected desired targets, compared to CEM, which indicated no sign of detecting the five targets. This is because the q used to calculate $\mathbf{R}_{L\times L}^{-1}$ in (4.6) is crucial. So, in order to see how the q affected the detection performance, experiments were conducted for CEM and TCIMF using $q = 3, 5, 10, 60$ to detect blackbrush and the results are shown in Fig. 7.6(a-b). As we can see from Fig. 7.6, CEM detected the blackbrush when $q = 5$ and 10. TCIMF also detected blackbrush at $q = 60$ and performed better than the case that it used all full bands in Fig. 7.5(d). Interestingly, when q was too small such as 2, 3, both CEM and TCIMF detected wrong targets whose mixtures were creosote leaves and sagebrush resulted in larger values than those whose mixtures were blackbrush and sagebrush. This experiment demonstrated the significance of the q in computation of $\mathbf{R}_{L\times L}^{-1}$.

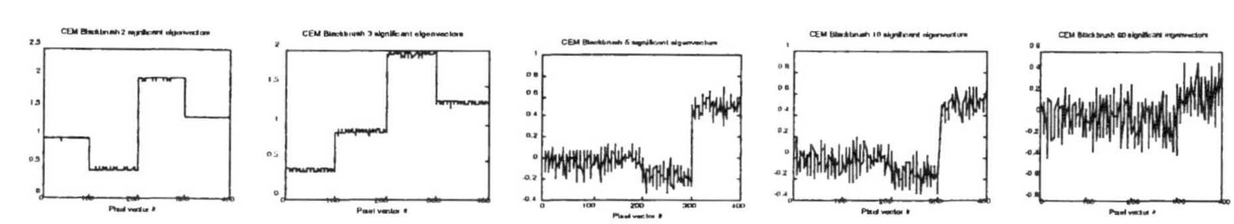

$q = 2$ $q = 3$ $q = 5$ $q = 10$ $q = 60$

(a) CEM

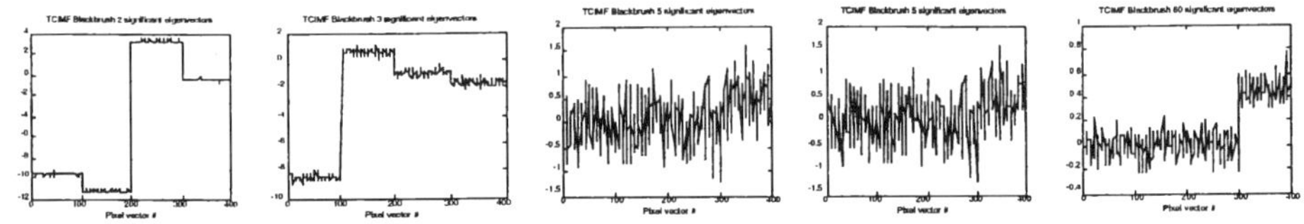

$q = 2$ $q = 3$ $q = 5$ $q = 10$ $q = 60$

(b) TCIMF

Figure 7.6. Results of CEM and TCIMF using q = 2, 3, 5, 10, 60 with blackbrush as the desired signature

Example 7.4 Target Signatures with Small Amount of Abundance Fractions

The experiments conducted in this example provide a different result for CEM and TCIMF. The used data were the same set of the 400 simulated pixels shown in Fig. 7.1 in Example 7.1, but the complete knowledge of creosote leaves, dry grass and red soil was assumed to be known. Fig. 7.7 shows the results of OSP, SCLS, NCLS, CEM and TCIMF in detection of creosote leaves (Note that only OSP-result is shown by a different scale) which were very close to those in Fig. 7.2(a). Unlike Example 7.3, this time all five methods, OSP, SCLS, NCLS, CEM and TCIMF produced comparable detection results. However, OSP was the worst in terms of detecting the true abundance fractions of the creosote leaves, whereas SCLS and NCLS were among the best.

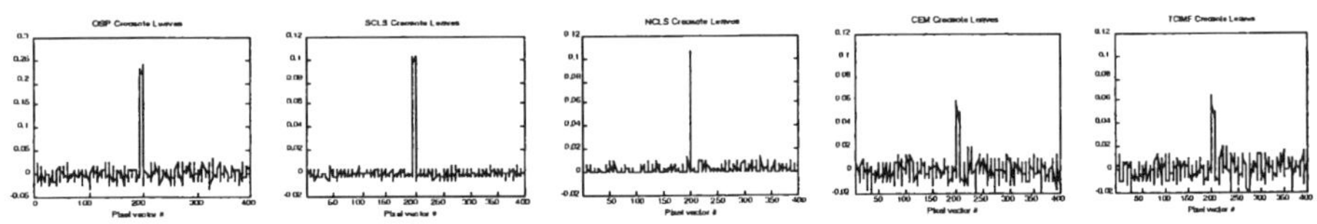

OSP SCLS NCLS CEM TCIMF

Figure 7.7. Results of OSP, SCLS, NCLS, CEM and TCIMF in detection of creosote leaves

In order to see the effects resulting from different values of q used to calculate $\mathbf{R}_{L\times L}^{-1}$ in CEM and TCIMF, Fig. 7.8(a-b) shows the results of CEM and TCIMF using q = 2, 3, 5, 10, 60. Compared to the results produced using q = 158 in Fig. 7.7, the abundance of the creosote leaves detected in Fig. 7.8 for CEM and TCIMF was more accurate. Interestingly, when $q = 2$, TCIMF missed the target, while CEM extracted creosote leaves but also extracted dry grass across the entire range. For CEM, q = 60 yielded the best result where the estimated abundance of creosote leaves was nearly accurate. But even in this case, the result was still not as good as that produced by SCLS and NCLS in Fig. 7.7 in the sense of its performance in detecting abundance fractions of other pixels.

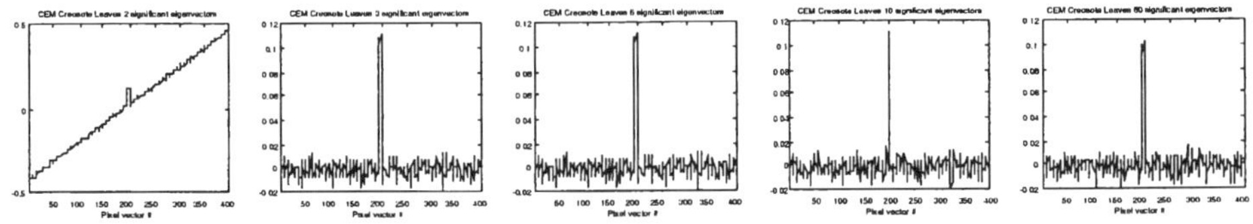

$q = 2$ $q = 3$ $q = 5$ $q = 10$ $q = 60$

(a) CEM

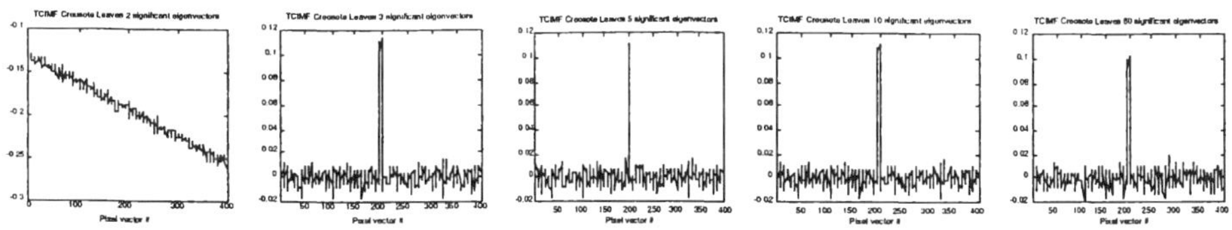

$q = 2$ $q = 3$ $q = 5$ $q = 10$ $q = 60$

(b) TCIMF

Figure 7.8. Results of (a) CEM and (b) TCIMF using the number of eigenvectors q = 2, 3, 5, 10, 60 in the detection of creosote leaves.

For TCIMF, it performed very well except q = 2. In particular, it detected nearly correct abundance fractions of the creosote leaves at q = 60.

While Example 7.3 shows one extreme case for CEM and TCIMF, Example 7.4 offers another extreme case for CEM and TCIMF. Both examples further demonstrate the crucial role the number of eigenvectors, q plays in the performance of CEM and TCIMF. If each eigenvector is interpreted as a piece of information, the larger the eigenvalue is, the more significant information it represents. So, these two examples suggest that when the desired target is small or occurs with low probability, the number of eigenvectors to be used for q is generally very high because targets with small abundance fractions may correspond to small eigenvalues and can be then viewed as insignificant targets. Under this circumstance, they may not be able to be detected by using only a few eigenvectors. Therefore, it requires a large set of eigenvectors to find these targets. This explains why CEM and TCIMF can be used to detect small targets so effectively using full bands. Conversely, if the desired targets are relatively large and wide-spread like the one studied in Example 7.3, a smaller q may be more appropriate to make CEM and TCIMF effective because the information provided by these targets can be well-represented by a few largest eigenvectors. In this case, a small set of large eigenvectors may be sufficient to detect these targets.

Example 7.5 Target Signatures Used as Interferers

The same simulated data used in Example 7.4 are also used in this example except that two more additional signatures, blackbrush and sagebrush were assumed to be present in the data even though they were actually not present. In this case, the signature matrix $\mathbf{M} = [\mathbf{m}_1 \mathbf{m}_2 \mathbf{m}_3 \mathbf{m}_4 \mathbf{m}_5]$ contained the five signatures, dry grass, red soil, creosote leaves, blackbrush and sagebrush. With this scenario, the blackbrush and sagebrush acted as interferers rather than target signatures. Fig. 7.9 shows the results of OSP, SCLS, NCLS, CEM and TCIMF in detection of creosote leaves.

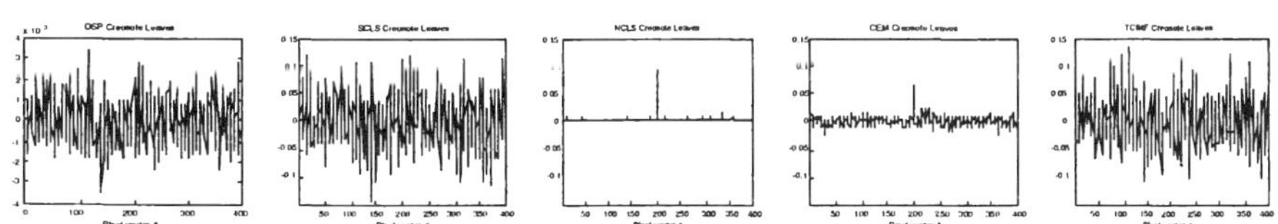

OSP SCLS NCLS CEM TCIMF

Figure 7.9. Results of OSP, CEM, NCLS, SCLS and TCIMF in detection of creosote leaves with blackbrush and sagebrush acted as interferers.

Unlike Fig. 7.7, the performances of the five detectors were quite different. The worst performance was produced by OSP. This is because the spectra of the two new added interferers, blackbrush and sagebrush signatures were very similar to that of creosote leaves and the detection capability of OSP was considerably reduced by the undesired signature annihilator $P_{\mathbf{U}}^{\perp}$. Similarly, SCLS and NCLS also suffered from the same problem. Nevertheless, NCLS still managed to perform reasonably well in spite of a slight degradation in detection of creosote leaves at pixel 200. Further, it actually did better than it did in Fig. 7.7 at other pixels by nulling out the abundance of creosote leaves. For CEM, the result was identical to that in Fig. 7.7 because the addition of blackbrush and sagebrush with zero abundance did not affect the output energy of CEM. But it did make difference for TCIMF since it eliminated the creosote leaves whose signature was similar to the two interferers. It performance was very similar to that produced by SCLS. However, if we conducted experiments using different values for q in TCIMF, the results shown in Fig. 7.10 were surprising. TCIMF actually detected the creosote leaves effectively for $q = 2, 3, 5, 10$ even for the case $q = 2$ where it detected the target with negative amounts. Fig. 7.10 further shows an advantage of TCIMF over CEM and a major difference between TACSD and TSCSD.

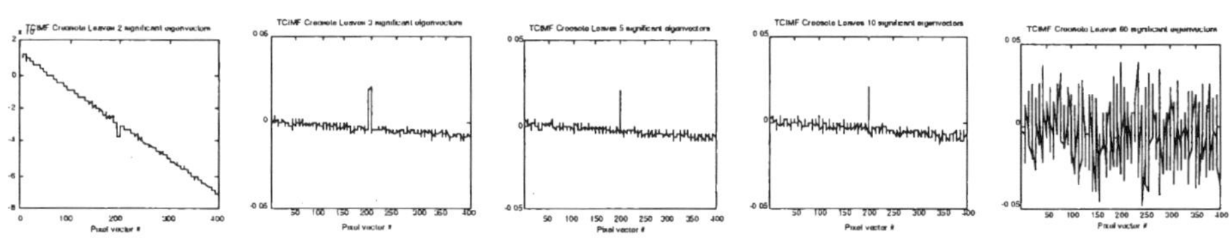

$q = 2$ $\quad$ $q = 3$ $\quad$ $q = 5$ $\quad$ $q = 10$ $\quad$ $q = 60$

Figure 7.10. Results of TCIMF using $q = 2$, 3, 5, 10, 60 in the detection of creosote leaves with blackbrush and sagebrush acted as interferers.

7.3.2 Hyperspectral Image Experiments

The computer simulations conducted in the previous section concluded that the sensitivity of target knowledge had significant impact on detection performance. In this section, real hyperspectral image experiments will provide further evidence on this issue.

7.3.2.1 AVIRIS Data

The hyperspectral data used in the following examples are AVIRIS data shown in Figure 1.6(a).

Example 7.6 (complete target knowledge)

In this example, the target knowledge is assumed to be known *a priori* where the signatures of the five targets, cinders, playa, rhyolite, shade and vegetation were directly extracted from the image scene as the same way conducted in Harsanyi and Chang (1994). Figure 7.11 shows the results of OSP, SCLS, NCLS, CEM and TCIMF using the entire eigenvectors, $q = 158$ where the $\mathbf{U}$ used in TCIMF was made of all other four target signatures except the desired target signature.

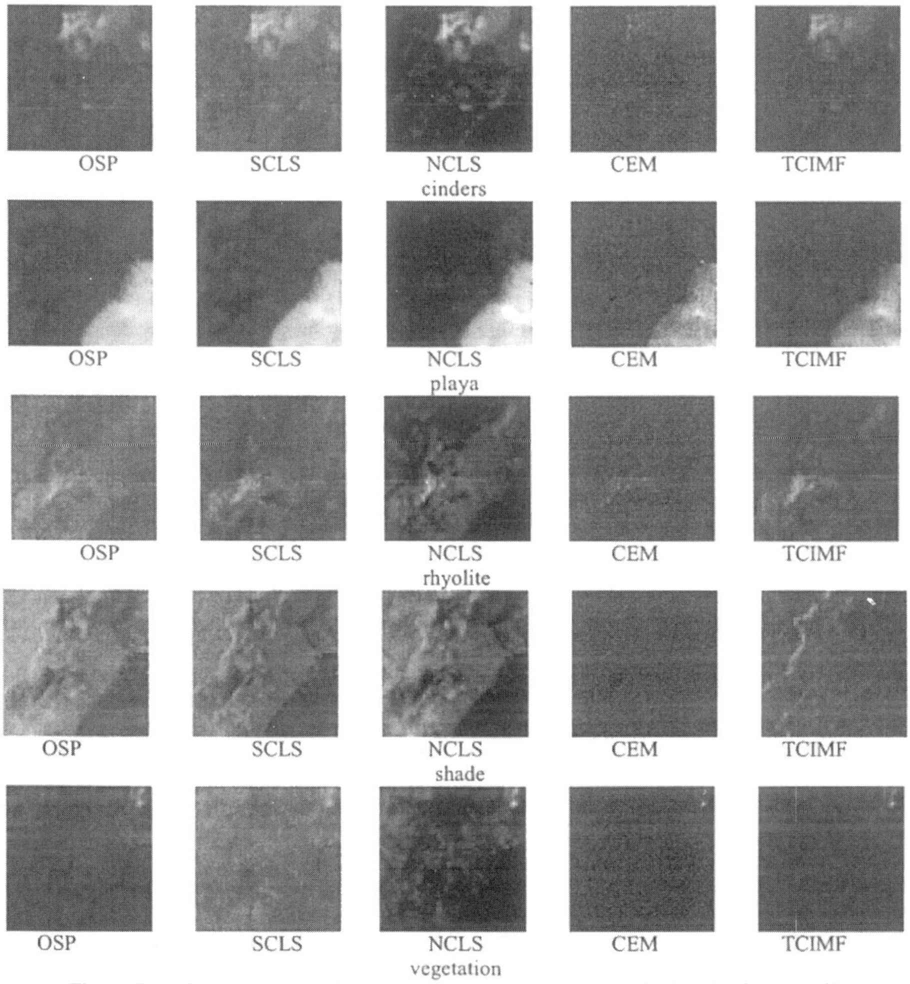

Figure 7.11. Comparative results of OSP, SCLS, NCLS, CEM and TCIMF using $q = 158$

From the images in Fig. 7.11, we see that NCLS performed best in detection of all the five targets. In order to see how the number of eigenvectors, q affected the performance of CEM and TCIMF, 5 different numbers, $q = 5, 10, 20, 40, 80$ were used in CEM and TCIMF implementation and the results are shown in Figs. 7.12 and 7.13.

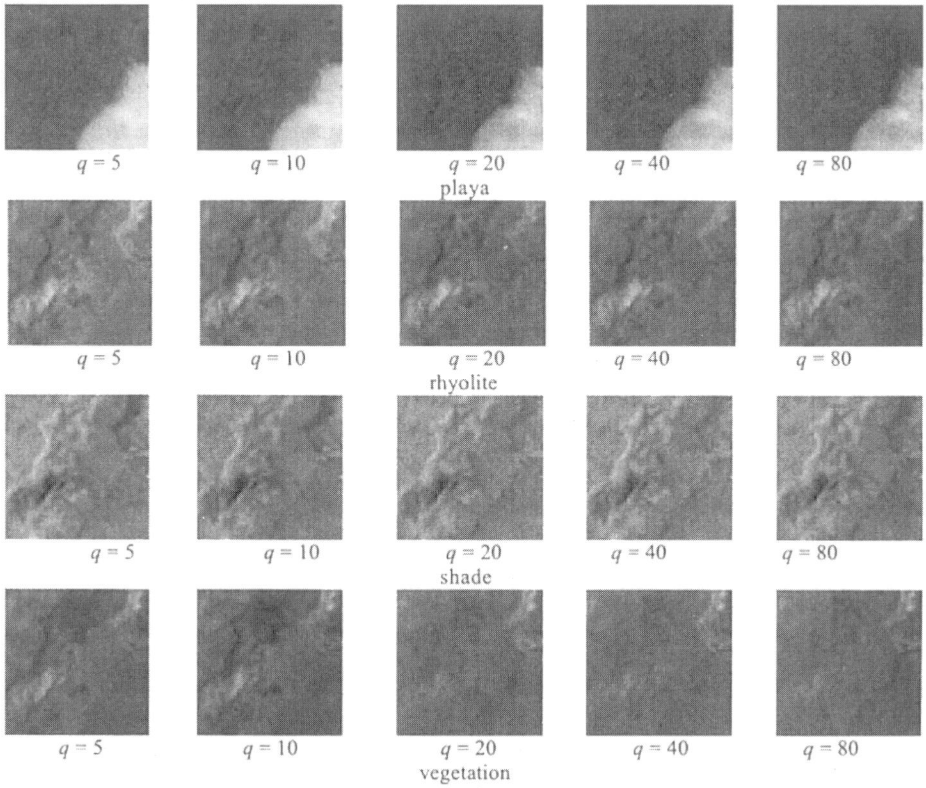

Figure 7.12. Comparative results of CEM using q = 5, 10, 20, 40, 80

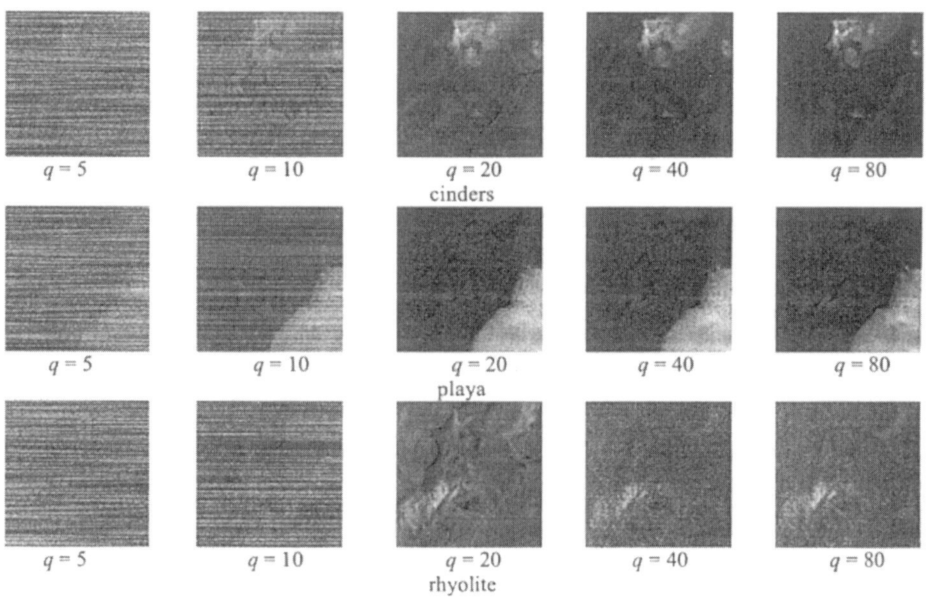

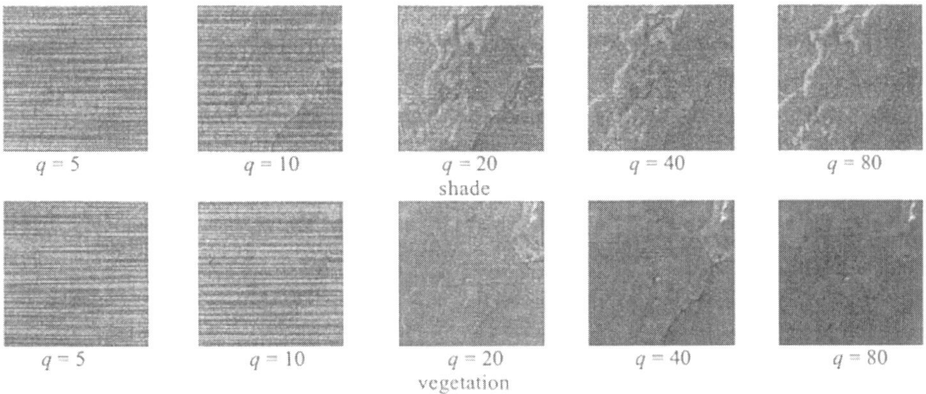

Figure 7.13. Comparative results of TCIMF using q = 5, 10, 20, 40 and 80

As shown in Figs. 7.12-7.13, TCIMF performed better than CEM in detection of all the five targets with $q = 80$. On the other hand, CEM seemed to do better than TCIMF when q was small. But in all both cases, they did not perform well and their performance was worse than that of CEM and TCIMF using $q = 158$ as shown in Fig. 7.11.

Example 7.7 (one target signature known *a priori*)

In Example 7.6, the complete knowledge of all the five target signatures was known *a priori*. In this example, we assumed that only partial target knowledge is available, which is the signature of the playa. The UNCLS algorithm was then used to generate five more unknown targets which has been identified as $\mathbf{t}_1$ = cinders, $\mathbf{t}_2$ = anomaly, $\mathbf{t}_3$ = vegetation, $\mathbf{t}_4$ = shade and $\mathbf{t}_5$ = rhyolite with $\mathbf{t}_0$ = playa. Fig. 7.14 shows the comparative results of OSP, SCLS, NCLS, CEM and TCIMF where the $\mathbf{U}$ used in TCIMF was made up of all the remaining five target signatures by excluding the desired target signature. Interestingly, the third detected target $\mathbf{t}_3$ was a playa pixel located at the upper left corner edge of the dry lakebed. This pixel vector was an anomaly, which was missed in Example 7.6 because it was very difficult to be identified by visual inspection from the image scene.

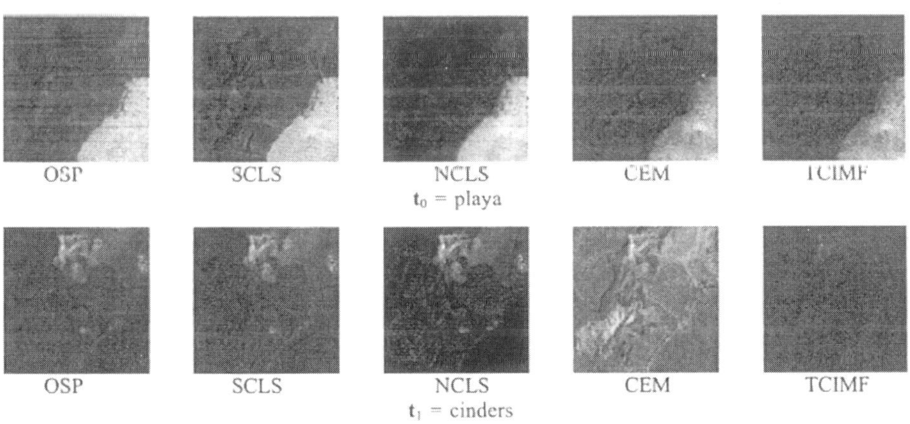

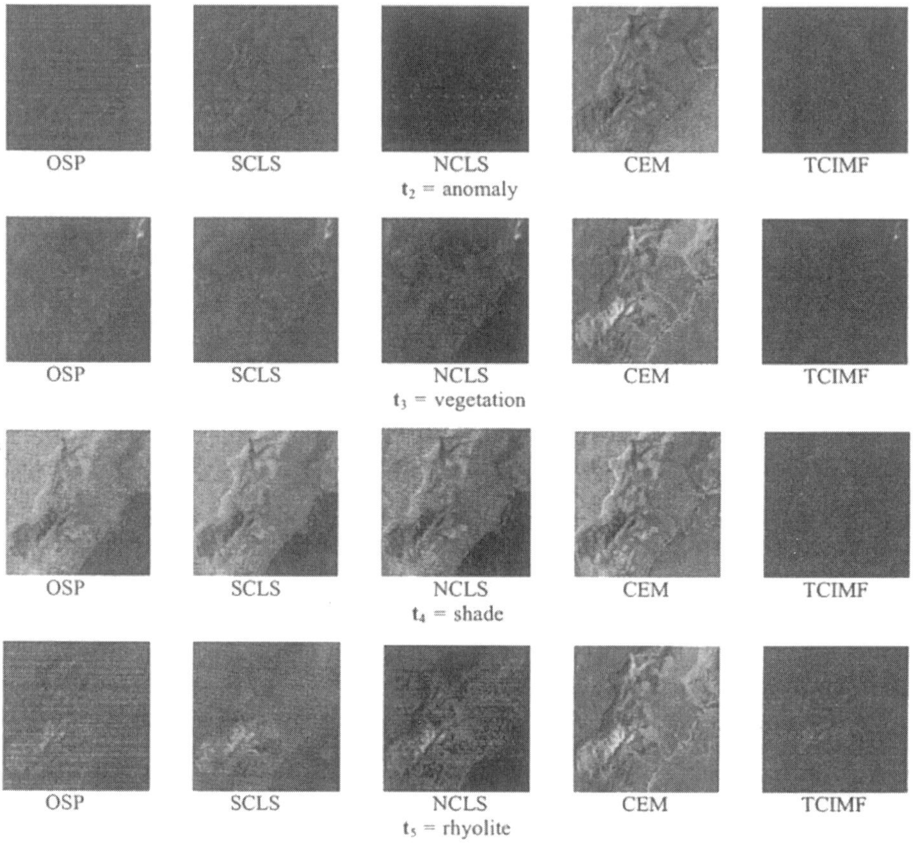

Figure 7.14. Comparative results of OSP, SCLS, NCLS, CEM and TCIMF with only one signature (t_0 = playa) known *a priori*

Comparing Fig. 7.14 to Fig. 7.11, all the five methods performed slightly worse than did their counterparts using full target knowledge. However, this experiment demonstrated a potential advantage of the UNCLS algorithm that can be used to detect anomalies.

Like Example 7.6, we also examine the effects of $q = 5, 10, 20, 40, 80$ for CEM and TCIMF. Figs. 7.15-16 show the results of CEM and TCIMF respectively.

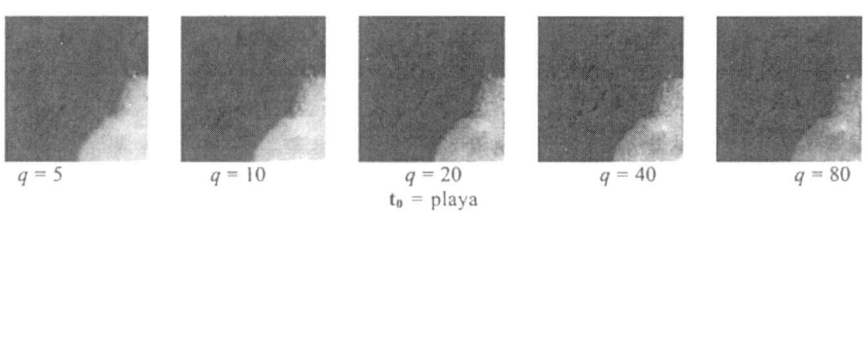

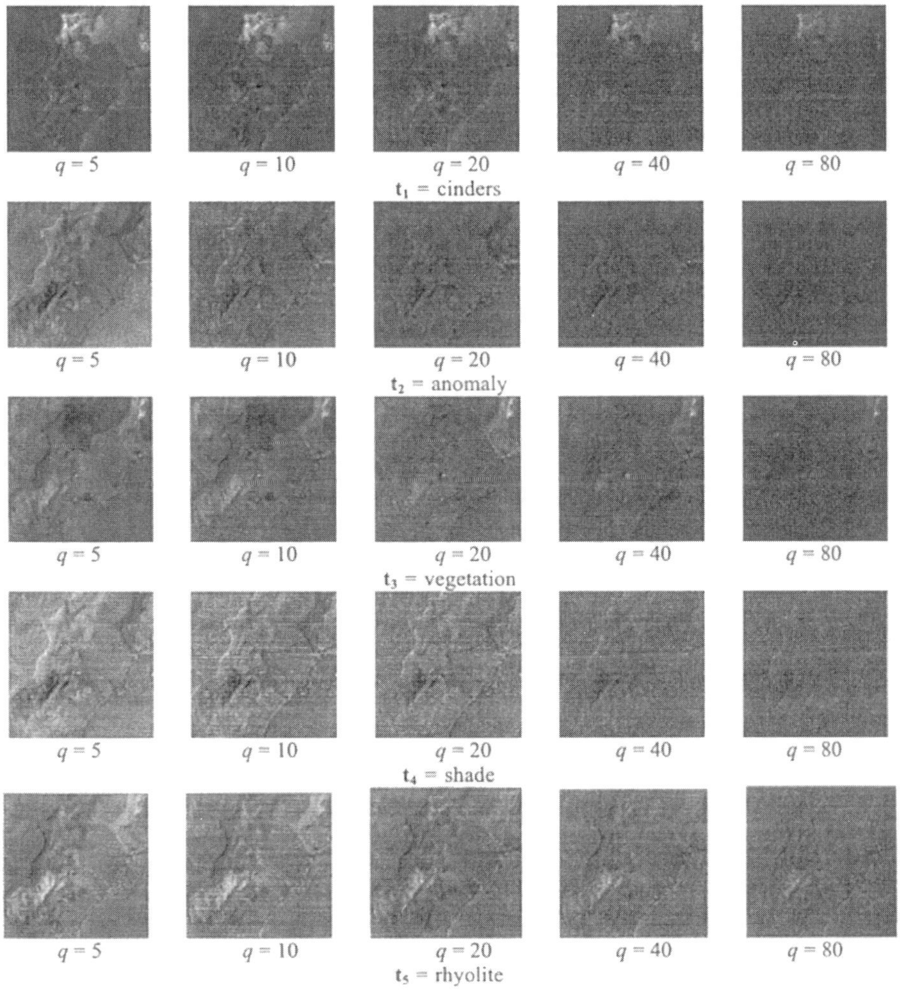

Figure 7.15. CEM using t_0 = playa as a known target signature for q = 5, 10, 20, 40, 80

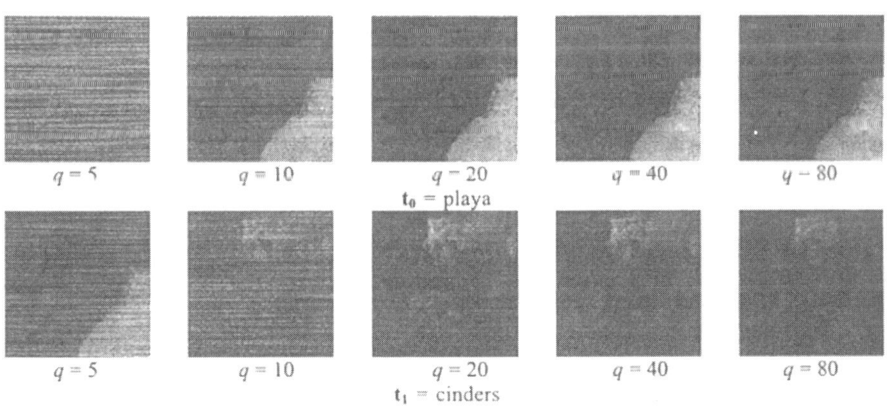

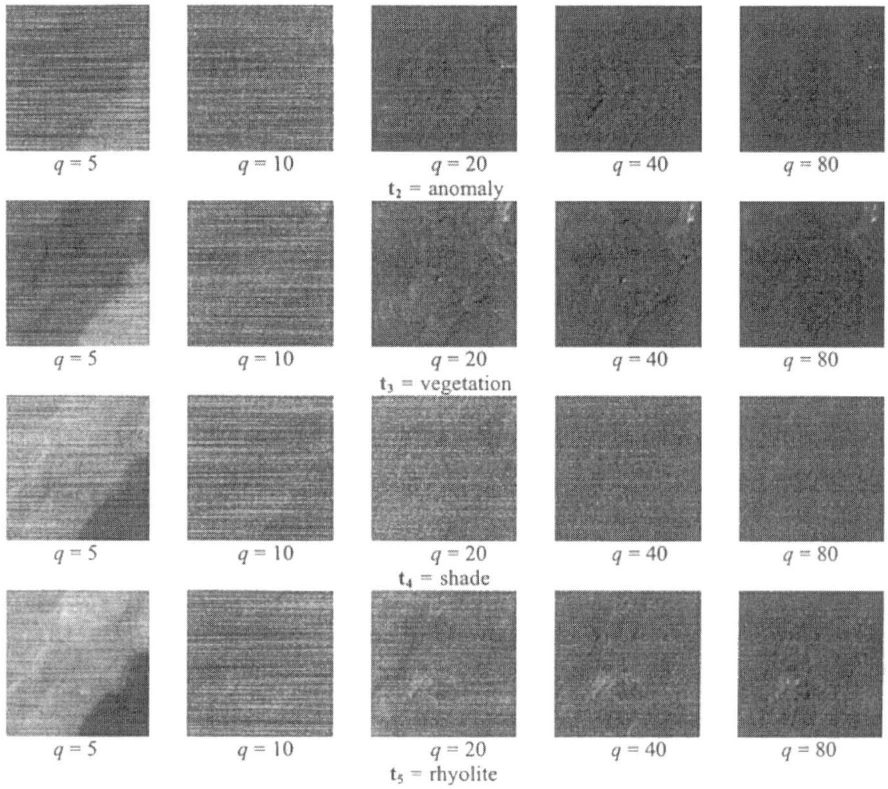

Figure 7.16. TCIMF using $\mathbf{t}_0$ = playa as a known target signature for q = 5, 10, 20, 40, 80

Example 7.8 (no *a priori* target knowledge)

Finally, we conclude this example by assuming that no prior target knowledge is given. In this particular situation, the UNCLS algorithm was used to generate 6 targets directly from the scene. They were identified as $\mathbf{t}_0 = \text{playa}$, $\mathbf{t}_1 = \text{cinders}$, $\mathbf{t}_2 = \text{anomaly}$, $\mathbf{t}_3 = \text{vegetation}$, $\mathbf{t}_4 = \text{shade}$ and $\mathbf{t}_5 = \text{rhyolite}$ where except for the initial target $\mathbf{t}_0$, the other five target pixels were the same pixels identified in Example 7.7. Fig. 7.16 show the results of OSP, SCLS, NCLS, CEM and TCIMF using these 6 targets as target information to detect cinders, anomaly, vegetation, shade and rhyolite in the image scene.

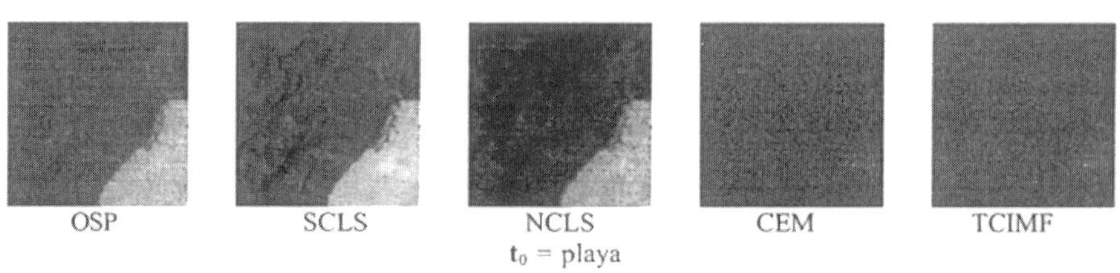

Figure 7.17. Comparative results of OSP, SCLS, NCLS, CEM and TCIMF with no a priori target knowledge using $q = 158$

Since no prior knowledge was assumed in this example, the initial target pixel $\mathbf{t}_0$ must be generated from the image scene. As can be expected, the $\mathbf{t}_0$ would not be the same initial target $\mathbf{t}_0$ used in Example 7.7 that was assumed to be known *a priori* and obtained by averaging a large area in the dry lake. It was a single playa pixel extracted from the dry lake. As a result, CEM only performed differently from that in Example 7.7 in detection of playa.

If we compare Fig. 7.17 to Fig. 7.11, OSP, SCLS and NCLS with no *a priori* target information performed slightly worse than their counterparts with complete target knowledge. But, their results were still comparable. However, CEM and TCIMF did not perform well in detection of cinders, playa, shade and rhyolite, but were effective in detection of vegetation and anomaly. This was because CEM and TCIMF were very sensitive to the target information to be used in these detectors. In Fig. 7.17, only single pixel for each target was used for information. So, CEM and TCIMF worked well for small targets such as vegetation and anomaly, but not for large areas such as of cinders, playa, shade and rhyolite which cover a wide range of spectral variation.

To evaluate effects of q on the performance of CEM and TCIMF, the experiments were also conducted for q = 5, 10, 20, 40, 80. Fig. 7.18 shows only results of playa since the information used for other five targets in CEM was the same used in Example 7.7. In comparison between Fig. 7.18 and Fig. 7.12(playa), the CEM using a single playa pixel performed much worse than the CEM using the information obtained by averaging a large number of pixels located in the dry lake.

$q = 5$ $q = 10$ $q = 20$ $q = 40$ $q = 80$

Figure 7.18. CEM with $\mathbf{t}_0$ = playa generated by the UNCLS algorithm for q = 5, 10, 20, 40, 80

Unlike CEM, the TCIMF in this example performed very differently from the TCIMF in Example 7.7 because the undesired target signature matrix **U** was made up of a single playa pixel not the averaged playa signature used in Example 7.7. Fig. 7.19 shows the results of TCIMF in detection of the 6 target signatures, playa, cinders, anomaly, vegetation, shade and rhyolite.

$q = 5$ $q = 10$ $q = 20$ $q = 40$ $q = 80$

$\mathbf{t}_0$ = playa

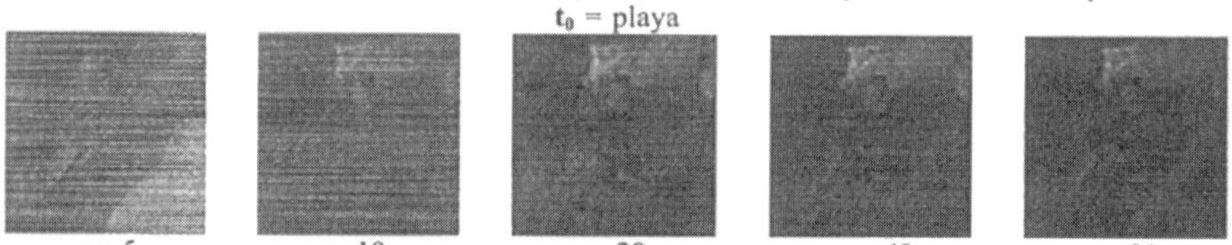

$q = 5$ $q = 10$ $q = 20$ $q = 40$ $q = 80$

$\mathbf{t}_1$ = cinders

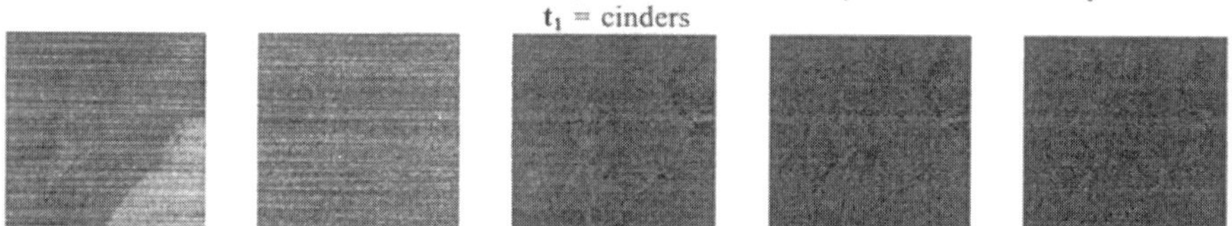

$q = 5$ $q = 10$ $q = 20$ $q = 40$ $q = 80$

$\mathbf{t}_2$ = anomaly

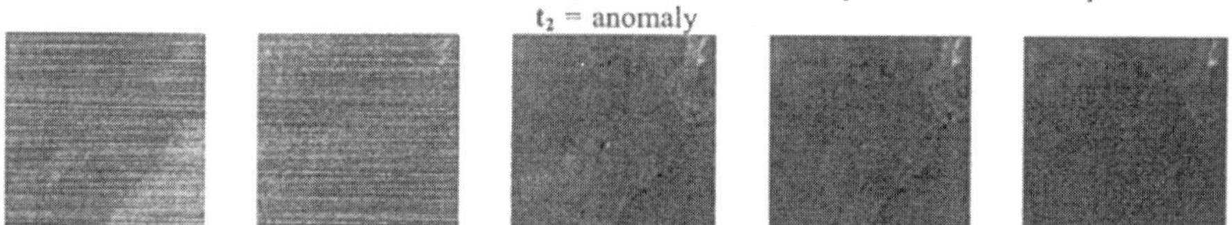

$q = 5$ $q = 10$ $q = 20$ $q = 40$ $q = 80$

$\mathbf{t}_3$ = vegetation

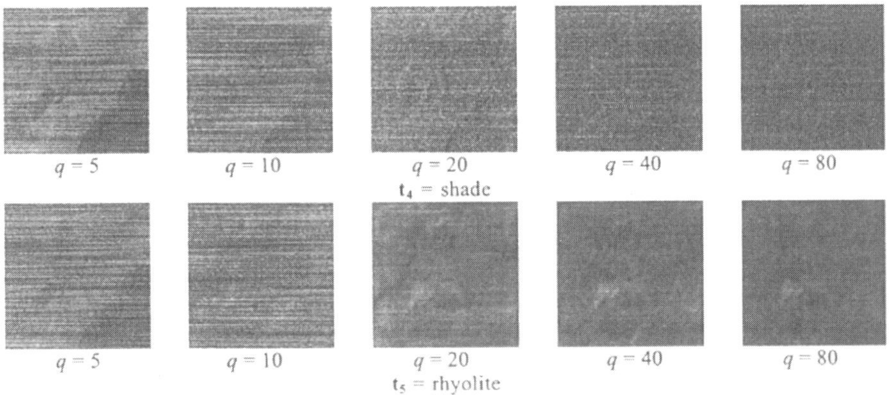

Figure 7.19. TCIMF with no *a priori* target knowledge for q = 5, 10, 20, 40, 80

From Fig. 7.19, the performance of TCIMF was very similar to that in Example 7.7 except the case of playa where TCIMF performed much worse that that in Example 7.7 due to insufficient target information represented by a single playa pixel.

7.3.2.2 HYDICE Data

In Chapters 3 and 4, the HYDICE experiments were conducted based on the complete knowledge of the 15 panels in the image scene in Fig. 1.7(a). This section investigates the issue of sensitivity to target knowledge with only partial or no prior knowledge available as well as the issue of noise sensitivity.

Example 7.9

In this example, we studied the case when the only available target knowledge was $\mathbf{t}_0$, which was chosen to be the first center panel pixel in row 1, p_{11}. Then the UNCLS algorithm in Section 5.4 was used to generate 34 unknown target pixels in the image scene among which 6 target pixels were identified by ground truth as $\mathbf{t}_0 = p_{11}$, $\mathbf{t}_1$ = interferer, $\mathbf{t}_4 = p_{51}$, $\mathbf{t}_5 = p_{31}$, $\mathbf{t}_{21} = p_{41}$, $\mathbf{t}_{33} = p_{21}$. Fig. 7.20(a-f) shows the detection results of these 6 targets using OSP, SCLS, NCLS, CEM and TCIMF. As we can see from the images, only CEM and TCIMF detected tiny amounts of p_{13}, while the other three missed it because of its small amount of abundance. In addition, NCLS, CEM and TCIMF were among the best and performed very similarly except that NCLS has some difficulty with classifying panels in rows 2 and 3, panels in rows 4 and 5 in Fig. 7.20(c,d,f). The OSP and the SCLS were the worst, but they still managed to classify panels in five rows except p_{13} and the interferer.

(a) $\mathbf{t}_0 = p_{11}$

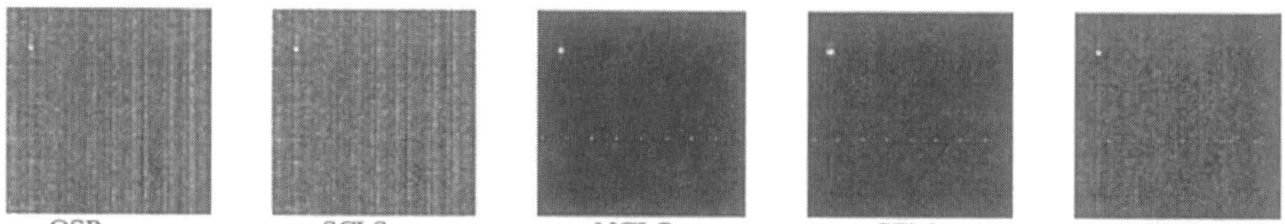

(b) $\mathbf{t}_1$ = interferer

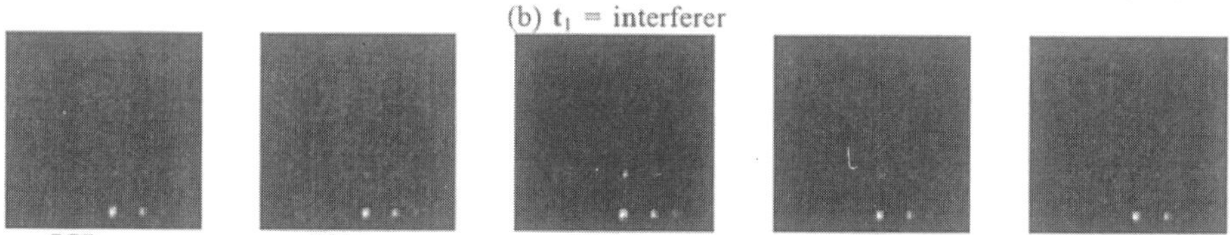

(c) $\mathbf{t}_4 = \mathrm{p}_{51}$

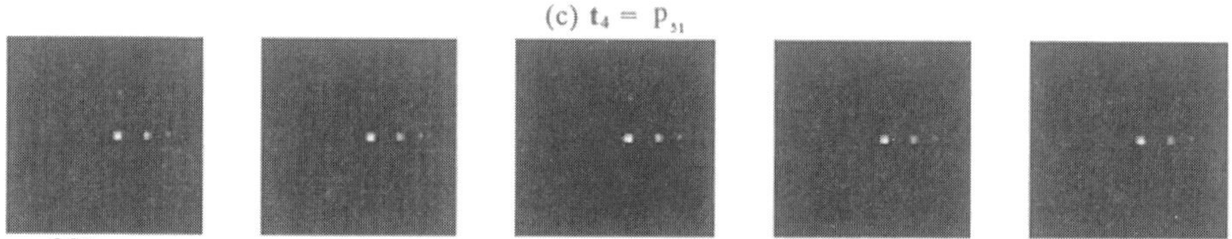

(d) $\mathbf{t}_5 = \mathrm{p}_{31}$

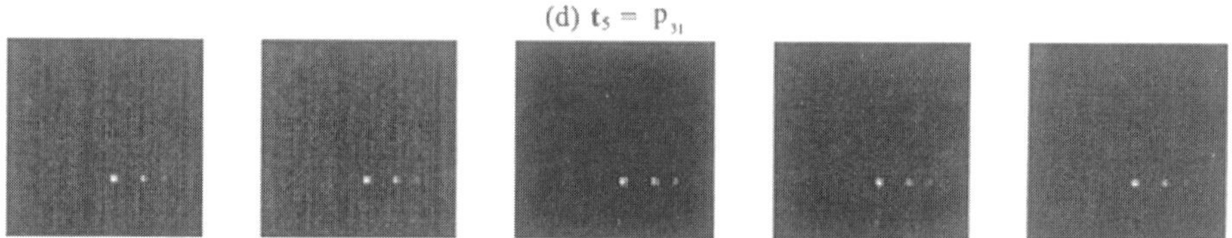

(e) $\mathbf{t}_{21} = \mathrm{p}_{41}$

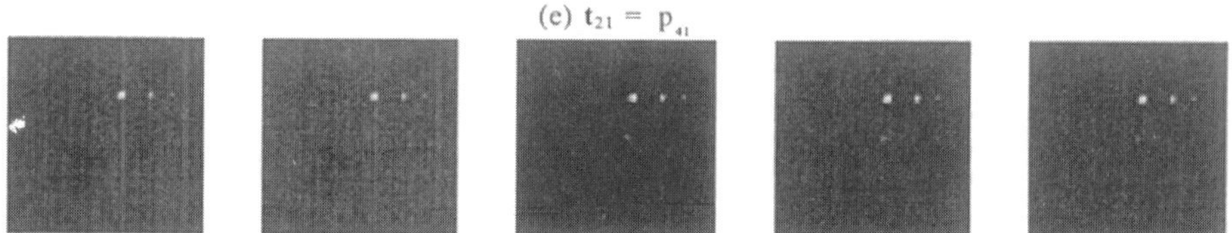

(f) $\mathbf{t}_{33} = \mathrm{p}_{21}$

Figure 7.20. Results of OSP, SCLS, NCLS, CEM and TCIMF using only the target information $\mathbf{t}_0 = \mathrm{p}_{11}$

In order to see how q affects the performance of CEM and TCIMF, Figs. 7.21 and 7.22 show the detection results of the same 6 targets obtained in Fig. 7.20 for CEM and TCIMF using q = 5, 10, 20, 40 and 80 respectively.

(a) $\mathbf{t}_0 = \mathrm{p}_{11}$

$q = 5$ $q = 10$ $q = 20$ $q = 40$ $q = 80$

(b) $\mathbf{t}_1$ = interferer

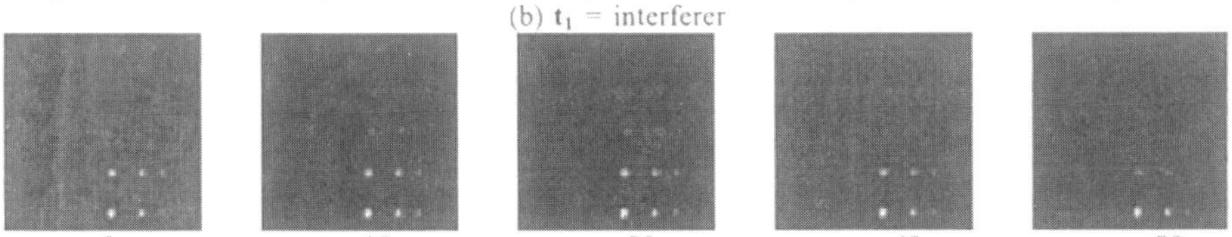

$q = 5$ $q = 10$ $q = 20$ $q = 40$ $q = 80$

(c) $\mathbf{t}_4 = \mathbf{p}_{51}$

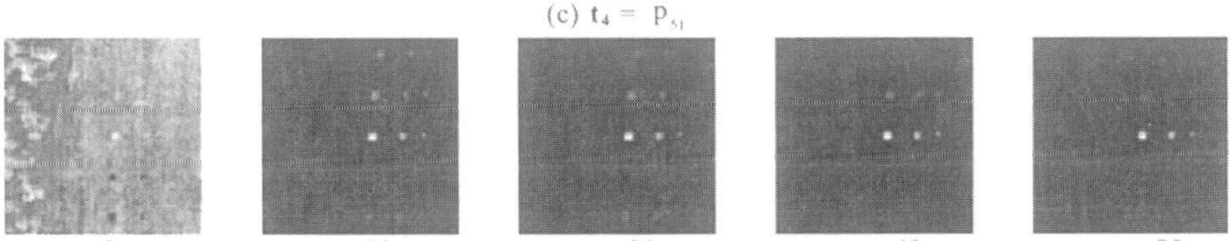

$q = 5$ $q = 10$ $q = 20$ $q = 40$ $q = 80$

(d) $\mathbf{t}_5 = \mathbf{p}_{31}$

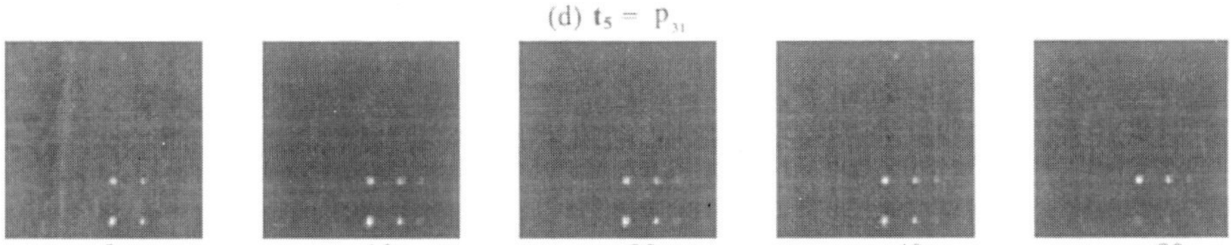

$q = 5$ $q = 10$ $q = 20$ $q = 40$ $q = 80$

(e) $\mathbf{t}_{21} = \mathbf{p}_{41}$

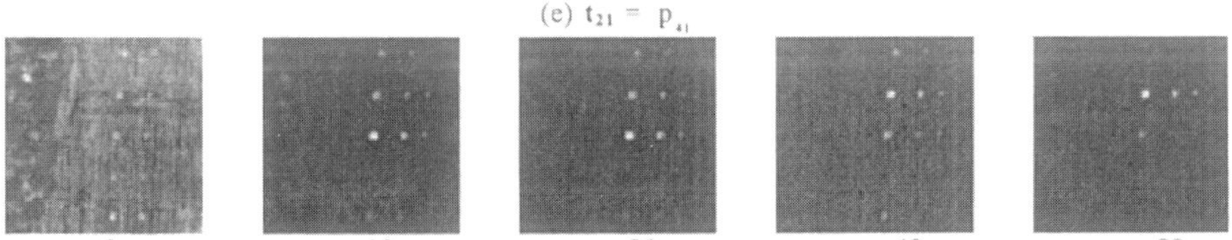

$q = 5$ $q = 10$ $q = 20$ $q = 40$ $q = 80$

(f) $\mathbf{t}_{33} = \mathbf{p}_{21}$

Figure 7.21. Performance of CEM using $q = 5, 10, 20, 40$, and 80 with $\mathbf{t}_0 = \mathbf{p}_{11}$

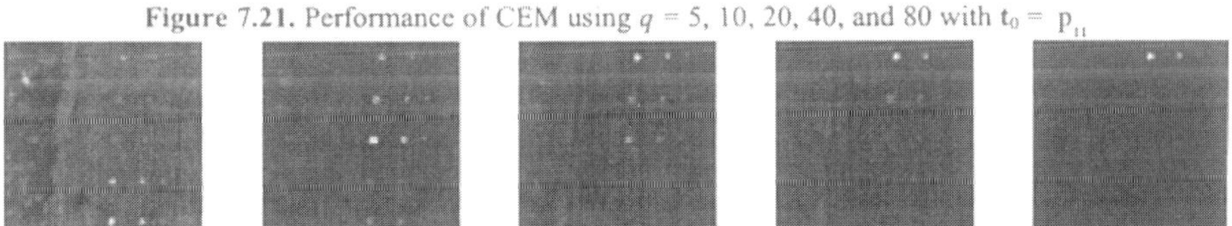

$q = 5$ $q = 10$ $q = 20$ $q = 40$ $q = 80$

(a) $\mathbf{t}_0 = \mathbf{p}_{11}$

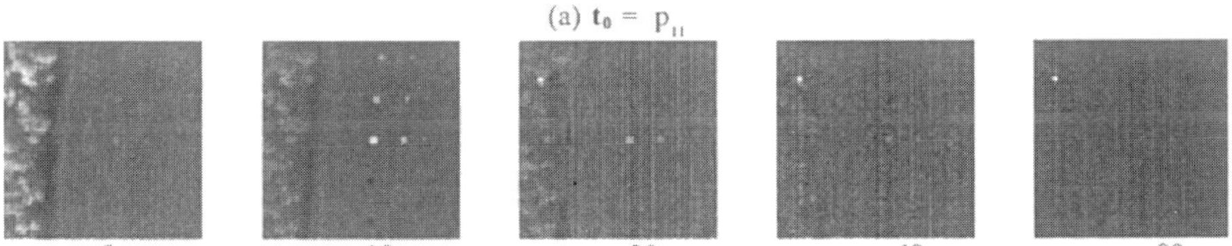

$q = 5$ $q = 10$ $q = 20$ $q = 40$ $q = 80$

(b) $\mathbf{t}_1$ = interferer

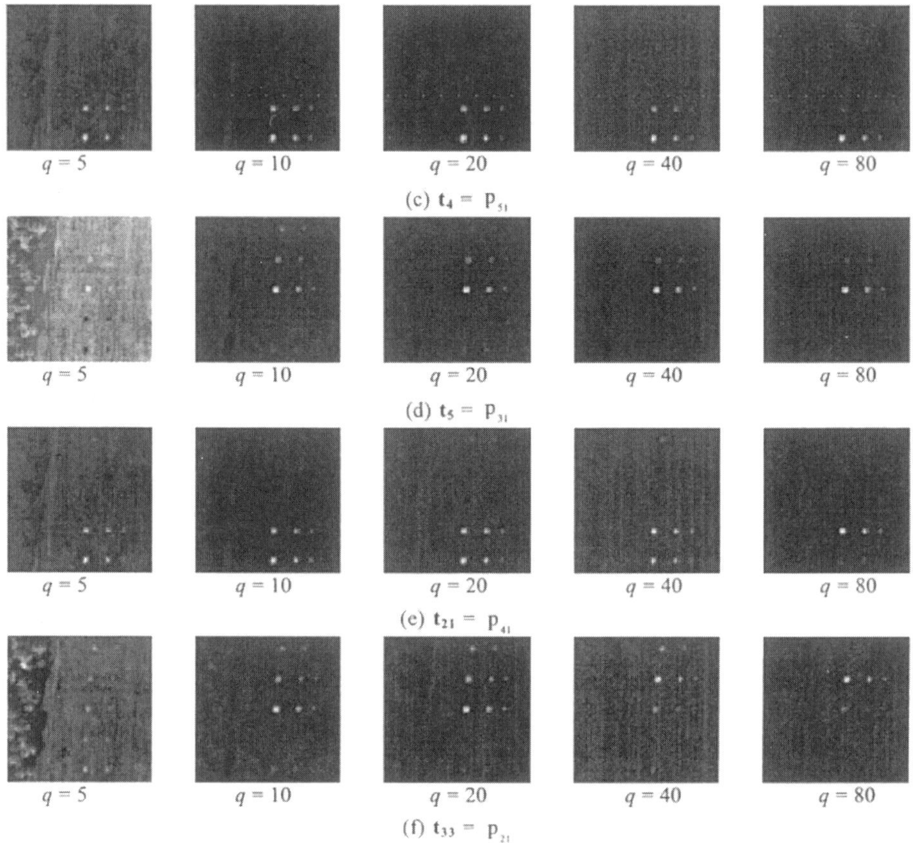

Figure 7.22. Performance of TCIMF using q = 5, 10, 20, 40, and 80 with $\mathbf{t}_0 = \mathrm{p}_{11}$

As demonstrated, the larger the q was, the better the detection was. Both CEM and TCIMF performed very closely except for the case of q = 80 in detection of panels in rows 3-5 shown in images labeled by (d-f) where TCIMF eliminated more interference than did CEM.

Example 7.10

Example 7.9 assumed that the initial target $\mathbf{t}_0$ was selected by partial knowledge. If we further assumed that no such partial knowledge was available, the UNCLS algorithm must be used to generate all necessary targets directly from the image scene including the initial target, $\mathbf{t}_0$. Once again, 34 unknown target pixels including $\mathbf{t}_0$ were generated. Among them six target pixels were identified by ground truth, which were $\mathbf{t}_0$ = interferer, $\mathbf{t}_1 = \mathrm{p}_{51}$ $\mathbf{t}_2 = \mathrm{p}_{31}$, $\mathbf{t}_3 = \mathrm{p}_{11}$, $\mathbf{t}_4 = \mathrm{p}_{41}$ and $\mathbf{t}_5 = \mathrm{p}_{21}$. Since these 6 target pixels were the same 6 target pixels generated in Example 7.8, the classification results of all the five methods were identical to those in Figs. 7.19-7.21. This example demonstrated that completely unsupervised classification may perform as well as supervised classification as long as the desired target information obtained from the image scene is sufficiently accurate. For more details of comparative analysis between NCLS and CEM, we refer to Chang and Heinz (2000a).

7.4 SENSITIVITY OF ANOMALY DETECTION

As demonstrated in Fig. 6.1(c), RXD was very effective to detect a single two-pixel anomalous target because the ratio of the target size to the entire image is 2:40000 and very small. However, if the anomaly is not small enough compared to its surroundings, RXD may not work well. In order to see that, we consider the following example.

Example 7.11

When RXD using $\mathbf{K}^{-1}_{L\times L}$ and $\mathbf{R}^{-1}_{L\times L}$ was applied to a smaller portion of the LCVF image scene with size from 60×60 down to 20×20. In this case, the ratio of the anomaly to the image is changed from 2:3600 to 2:400. Fig. 7.22(a-b) provides the detection results of RXD using $\mathbf{K}^{-1}_{L\times L}$ and $\mathbf{R}^{-1}_{L\times L}$ where the detection performance of the anomaly was gradually reduced.

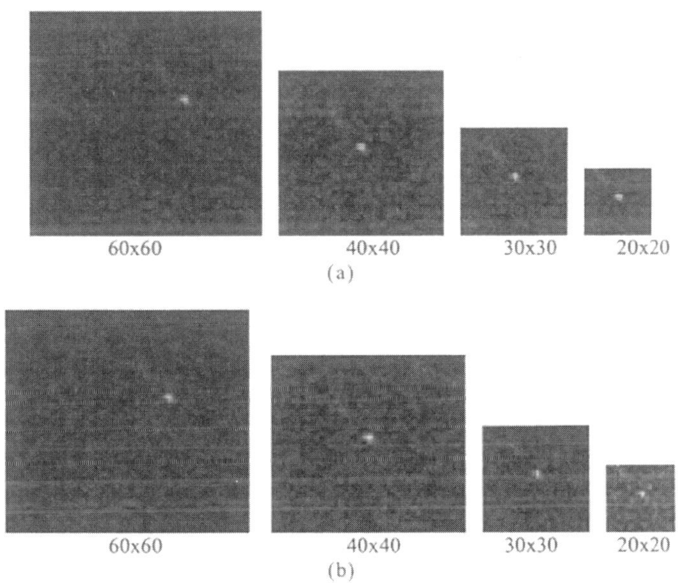

Figure 7.23. (a) Anomaly detection using $\mathbf{K}^{-1}_{L\times L}$, (b) Anomaly detection using $\mathbf{R}^{-1}_{L\times L}$

The detection performance of RXD was even worse when it was applied to the same 400-simulated data points considered in Example 7.1 with the creosote leaves as 5-pixel anomaly where the ratio of the anomaly to the data set is 5:400. As shown in Fig. 7.24, RXD completely failed to detect the creosote leaves.

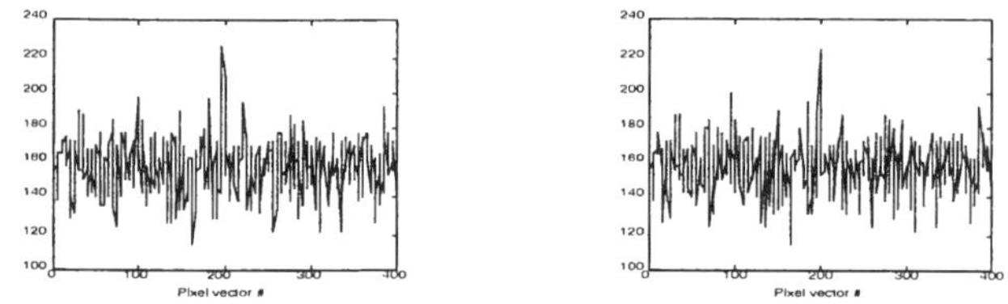

(a) (b)

Figure 7.24. (a) RXD using $\mathbf{K}_{L\times L}^{-1}$; (b) RXD using $\mathbf{R}_{L\times L}^{-1}$

The above experiments simply suggest that in order for RXD to be effective, the anomalous targets must sufficiently small. To illustrate this, we further expand the 400 simulated in Example 7.1 to 4000 mixed pixel vectors, $\{\mathbf{r}_i\}_{i=1}^{4000}$, as follows. We start the first pixel vector with 100% red soil and 0% dry grass, then began to increase 0.025% dry grass and decrease 0.025% red soil every pixel vector until the 400th pixel vector which contained 100% dry grass. We then added three anomalous targets to these 40000 simulated pixel vectors. The first anomalous target, creosote leaves was added to pixel vector numbers 1998-2002, $\{\mathbf{r}_i\}_{i=1998}^{2002}$ at abundance fractions 10% while reducing the abundance of red soil and dry grass evenly. A second anomalous target, blackbrush was added to pixel vectors 998-1002, $\{\mathbf{r}_i\}_{i=998}^{1002}$ at abundance fractions 5% while reducing the abundance fractions of red soil and dry grass evenly to make the total concentration of each pixel vector 100%. And a third anomalous target, sagebrush was added to pixel vectors 2998-3002, $\{\mathbf{r}_i\}_{i=2998}^{3002}$ at abundance fractions 5% while reducing the abundance fractions of red soil and dry grass evenly to make the total concentration of each pixel vector 100%. White Gaussian noise is also added to each pixel vector to achieve a 30:1 signal-to-noise ratio as was defined in (Harsanyi and Chang, 1994). The resulting band-30 spectra of the simulated 4000 pixel vectors are plotted in Fig. 7.25(a) which contained three types of anomalous targets, creosote leaves, blackbrush and sagebrush at pixel vectors $\{\mathbf{r}_i\}_{i=998}^{1002}$, $\{\mathbf{r}_i\}_{i=1998}^{2002}$ and $\{\mathbf{r}_i\}_{i=2998}^{3002}$ respectively. The result of detecting these three anomalous targets using RXD with $\mathbf{K}_{L\times L}^{-1}$ and $\mathbf{R}_{L\times L}^{-1}$ are shown in Figs. 7.25(b) and 7.25(c) respectively.

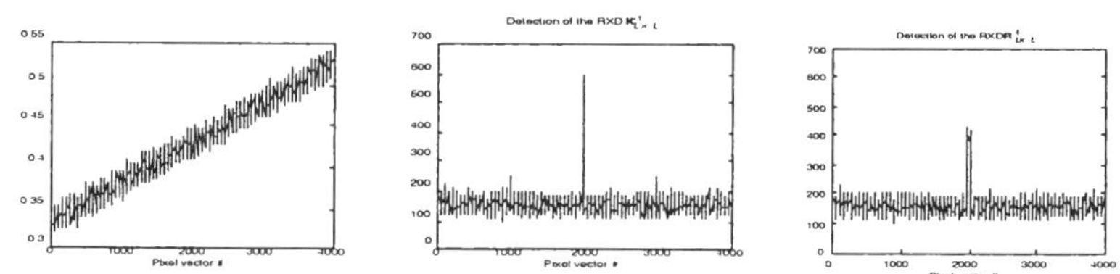

(a) (b) (c)

Figure 7.25. (a) Spectra of the 4000 simulated pixels (band 30); (b) RXD using $\mathbf{K}_{L\times L}^{-1}$; (c) RXD using $\mathbf{R}_{L\times L}^{-1}$

As we can see from Figs. 7.25(b) and 7.25(c), both versions of RXD detected the creosote leaves, but RXD using $\mathbf{K}_{L\times L}^{-1}$ was more effectively than RXD using $\mathbf{R}_{L\times L}^{-1}$. In detection of blackbrush and sagebrush, RXD using $\mathbf{K}_{L\times L}^{-1}$ extracted small fractions of their abundance, but it did better than RXD using $\mathbf{R}_{L\times L}^{-1}$, which missed both blackbrush and sagebrush. This may be due to the fact that the simulated data are stationary and the sample mean had more impact than the variance did in anomaly detection. Interestingly, when LPTD using $\mathbf{K}_{L\times L}^{-1}$ and $\mathbf{R}_{L\times L}^{-1}$ were applied to the same simulated data in Fig. 7.25(a), the detection results of LPTD using $\mathbf{K}_{L\times L}^{-1}$ shown in Fig. 7.26(a) were much better than that in Fig. 7.25(b) produced by RXD using $\mathbf{K}_{L\times L}^{-1}$ despite the fact that the abundance fractions of the three anomalies detected in Fig. 7.26(a) were negative. By contrast, LPTD using $\mathbf{R}_{L\times L}^{-1}$ performed very poorly by detecting very small amounts of creosote leaves and sagebrush. It should be noted that UTD is the LPTD using $\mathbf{K}_{L\times L}^{-1}$, in which the correlation matrix $\mathbf{R}_{L\times L}$ in LPTD is replaced with the covariance matrix $\mathbf{K}_{L\times L}$.

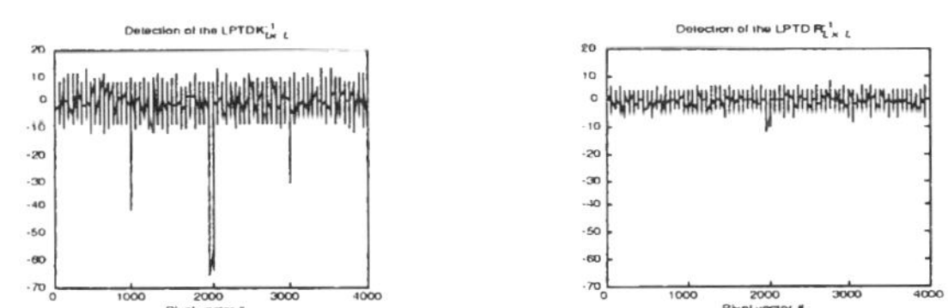

(a) (b)

Figure 7.26. (a) LPTD using $\mathbf{K}_{L\times L}^{-1}$; (b) LPTD using $\mathbf{R}_{L\times L}^{-1}$

As RXD-UTD and RXD-LPTD were further applied to the same data in Fig. 7.25(a), their detection results shown in Fig. 7.27(a-b) were very interesting, where the results produced by RXD-UTD in Fig. 7.27(a) were almost upside down of Fig. 7.26(a) with different abundance fractions. Once again, LPTD using $\mathbf{R}_{L\times L}^{-1}$ did not perform well in detecting blackbrush and sagebrush as shown Fig. 7.26(c).

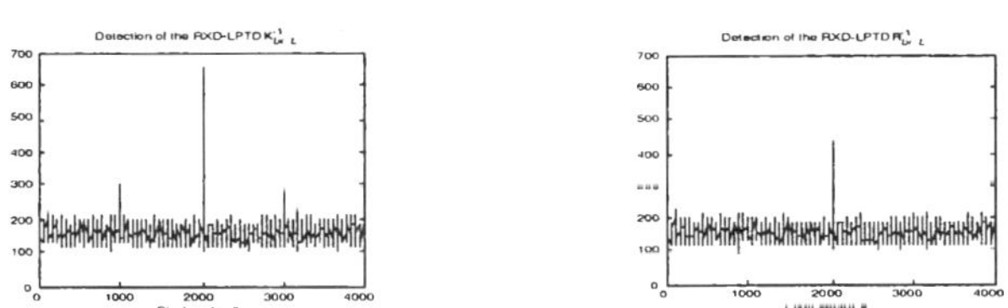

(a) (b)

Figure 7.27. (a) RXD-UTD; (b) RXD-LPTD

If we further compare the detection results in Fig. 7.27(a) to that in Fig. 7.25(b), RXD-UTD performed better than RXD in terms of the magnitude of detected anomalies.

Example 7.12

The reason that RXD did not detect the blackbrush and sagebrush as well as the creosote leaves was due to the fact that the fractions of their abundance contributed to the simulated data was only 5% which is relatively small. Under such circumstance, the noise begins to show its dominance in detection performance. In this example, we investigate this issue using the same data set in Example 7.11. Following a similar approach conducted in Section 7.3, the number of eigenvectors, q will be used to analyze the sensitivity of RXD to noise. Fig. 7.28 shows the results of RXD, NRXD, MRXD, UTD and RXD-UTD at q = 158. RXD detected the creosote leaves, but missed the blackbrush while barely detecting sagebrush. By contrast, UTD and RXD-UTD performed reasonably well in detecting the three anomalous targets at q = 158. On the other hand, the detection performance of NRXD and MRXD perform very poorly at q = 158.

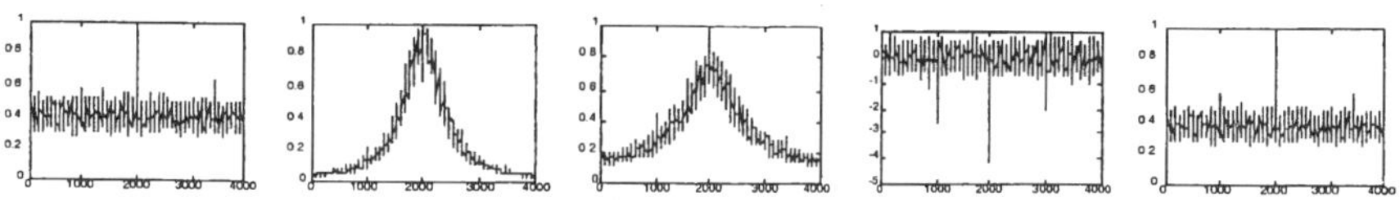

RXD NRXD MRXD UTD RXD-UTD

Figure 7.28. $q = L = 158$

However, the situation changed when q was small. Since there were only five material substances in the data set, the q cannot be too large. So, in this case, q was selected from 2 to 6 and 10 for experiments. Figs. 7.29-7.34 show the results of RXD, NRXD, MRXD, UTD and RXD-UTD respectively.

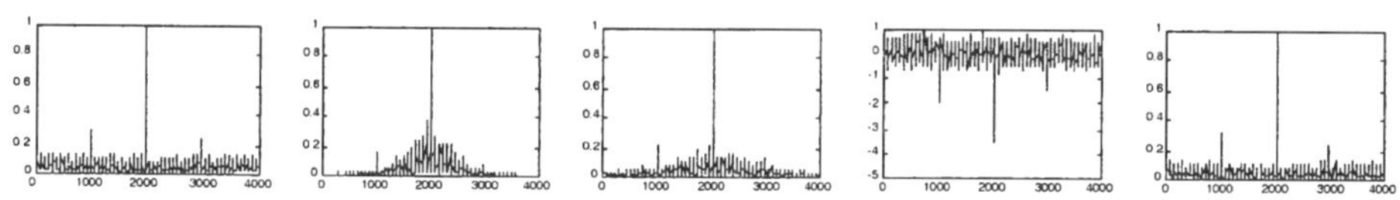

RXD NRXD MRXD UTD RXD-UTD

Figure 7.29. $q = 10$

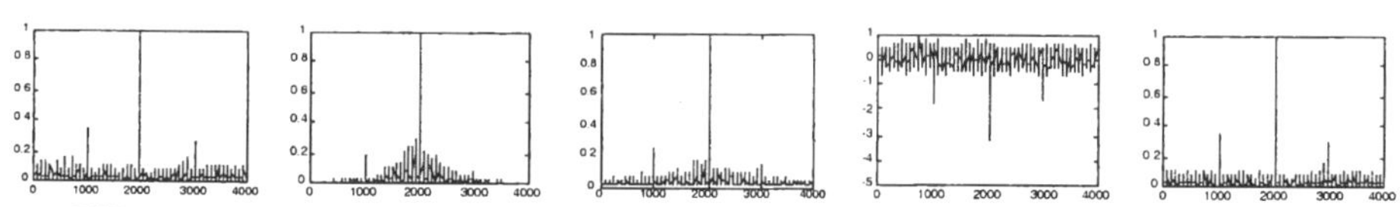

RXD NRXD MRXD UTD RXD-UTD

Figure 7.30. $q = 6$

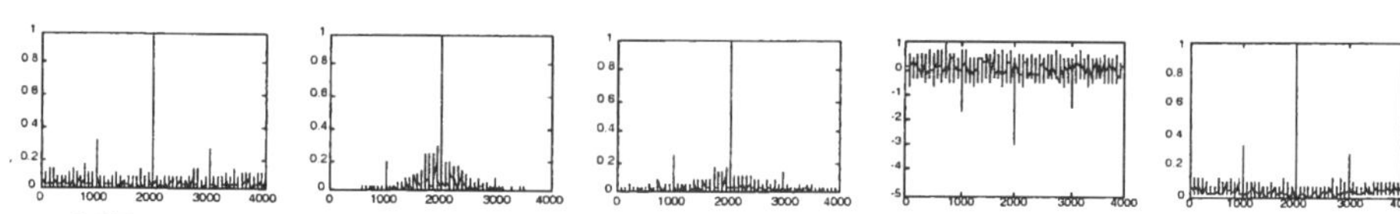

RXD NRXD MRXD UTD RXD-UTD

Figure 7.31. $q = 5$

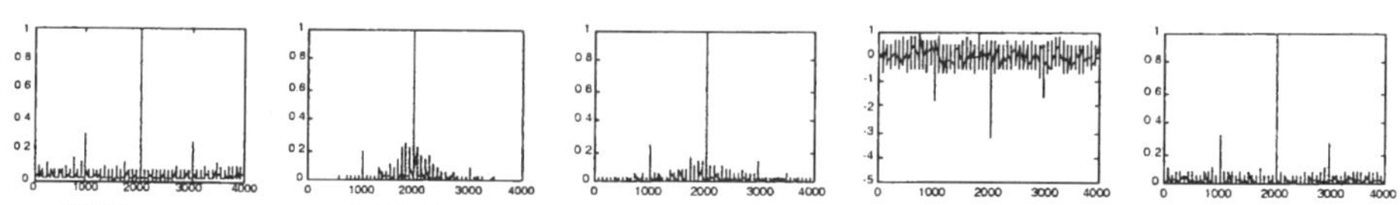

RXD NRXD MRXD UTD RXD-UTD

Figure 7.32. $q = 4$

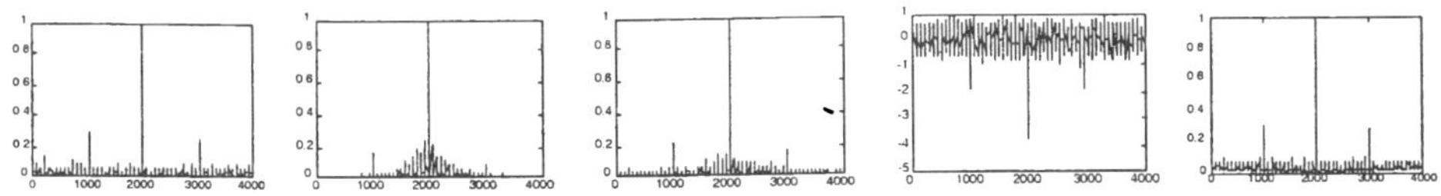

RXD NRXD MRXD UTD RXD-UTD

Figure 7.33. $q = 3$

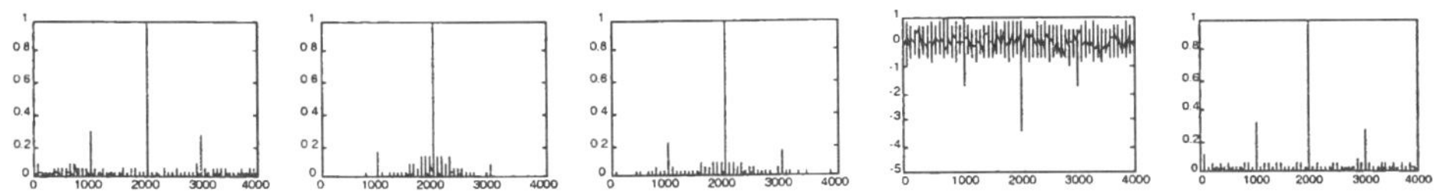

RXD NRXD MRXD UTD RXD-UTD

Figure 7.34. $q = 2$

From these figures, RXD, UTD and RXD-UTD performed better than they did at $q = 158$. Their performance was very stable at q = 2, 3, 4, 5 6, and 10. The only visible improvement resulting from using smaller q was more removal of interference resulting from other pixels. Interestingly, NRXD and MRXD performed much better than they did at $q = 158$ and their performance was improved as q was decreased from 10 to 2.

The above computer simulations demonstrated an important fact that the q played a crucial role in anomaly detection. In order for an anomaly detector to be effective, an appropriate q must be small. As mentioned previously, determination of the q is considered to be very difficult. Similar experiments were also conducted for RXD, NRXD, MRXD, LPTD and RXD-LPTD using $\mathbf{R}^{-1}_{L\times L}$. The results are not included here because they were comparable to those obtained in Figs. 7.28-7.34 using $\mathbf{K}^{-1}_{L\times L}$ with slight degradation in performance.

Example 7.13

In this example, the two hyperspectral images in Figs. 1.6(a) and 1.7(a) were used to investigate the effects of using different values of q on the detection performance of the RXD, NRXD, MRXD, UTD and RXD-UTD. In order to determine an appropriate q, some partial knowledge may be helpful. Since there are five known target signatures plus an anomaly in the AVIRIS image in Fig. 1.6(a), an adequate q may be in the range of 7. By using the results in Table 17.1 in Chapter 17 for VD, q was selected to be 4, 5, 8 for AVIRIS image experiments. Figs. 7.35-7.37 show the results of RXD, NRXD, MRXD, UTD and RXD-UTD obtained for q = 4, 5, 8 respectively. As we can see from these figures, they all performed very similarly for small q's. Compared to Figs. 6.2(a-b) and 6.3(a), NRXD, MRXD and UTD performed quite differently. In particular, UTD extracted the cinders along with the image background.

RXD NRXD MRXD UTD RXD-UTD

Figure 7.35. $q = 4$

RXD NRXD MRXD UTD RXD-UTD

Figure 7.36. $q = 5$

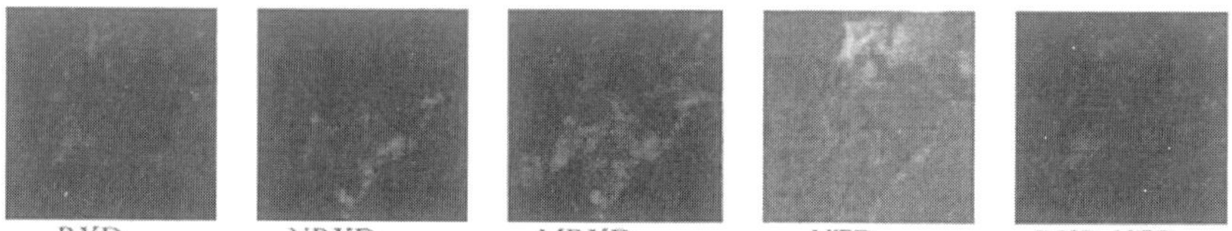

RXD NRXD MRXD UTD RXD-UTD

Figure 7.37. $q = 8$

Similar experiments were also conducted for the 15-panel HYDICE image where q was chosen to be 18, 20 and 22 according to Table 17.2 in Chapter 17. Figs. 7.38-7.40 show the results of RXD, NRXD, MRXD, UTD and RXD-UTD for q = 18, 20 and 22 respectively.

RXD NRXD MRXD UTD RXD-UTD

Figure 7.38. $q = 18$

RXD NRXD MRXD UTD RXD-UTD

Figure 7.39. $q = 20$

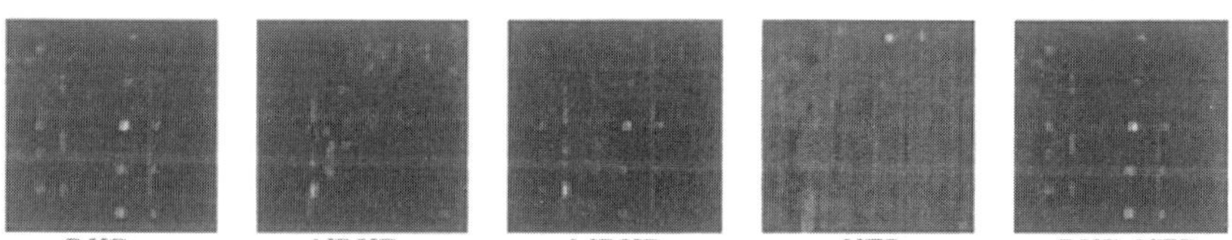

RXD NRXD MRXD UTD RXD-UTD

Figure 7.40. $q = 22$

As shown in these images, the results were very similar. Like AVIRIS image experiments, NRXD, MRXD and UTD performed differently from what they did at q = 169. It is also interesting to note that UTD could detect the panels in row 1 compared to Fig. 5.5(d) where it extracted image background. This example demonstrated that a small q could make a difference on the performance of NRXD, MRXD and LPTD, specifically, UTD. Same experiments were also conducted for RXD, NRXD, MRXD, LPTD and

RXD-LPTD using $\mathbf{R}_{L\times L}^{-1}$. The results showed that there was little difference for RXD-LPTD using $\mathbf{K}_{L\times L}^{-1}$ and using $\mathbf{R}_{L\times L}^{-1}$ when q was small.

Example 7.14

This example is particularly designed to study subtle difference in noise sensitivity resulting from RXD using $\mathbf{K}_{L\times L}^{-1}$ and $\mathbf{R}_{L\times L}^{-1}$. The 15-panel HYDICE scene can be used for this purpose because the ground truth map in Fig. 1.7(b) provides the precise locations of all panel pixels. The image in Fig. 7.41(a) shows the anomalous targets detected by RXD using $\mathbf{K}_{L\times L}^{-1}$ when full bands were used, i.e. $q = 169$. Next to the image is a plot that shows the abundance of targets detected in the image. The peak labeled by "*" in the plot is the maximum amount of abundance detected in the image. Similarly, Fig. 7.41(b) shows the results of RXD using $\mathbf{R}_{L\times L}^{-1}$.

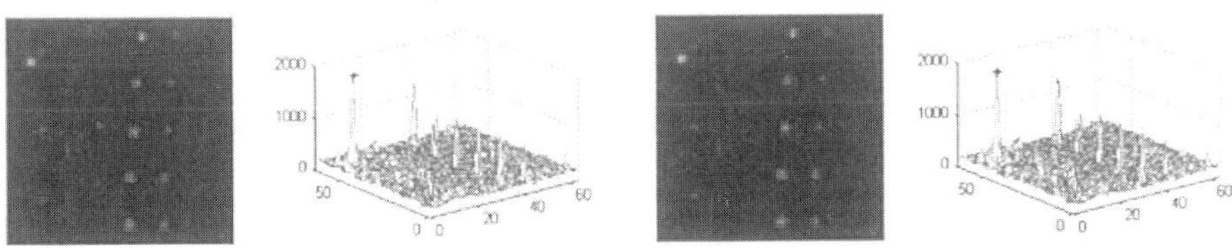

(a) maximum abundance = 1938.3390 at * using $\mathbf{K}_{L\times L}^{-1}$ (b) maximum abundance = 1934 at * using $\mathbf{R}_{L\times L}^{-1}$

Figure 7.41. Comparison between RXD using $\mathbf{K}_{L\times L}^{-1}$ and $\mathbf{R}_{L\times L}^{-1}$ at $q = 169$

As we can see from Fig. 7.41(a-b), there was no appreciable difference between RXD using $\mathbf{K}_{L\times L}^{-1}$ and $\mathbf{R}_{L\times L}^{-1}$. The highest peak, "*" appearing in both cases occurred at the interferer located at the top upper left corner in the image.

However, when the q was decreased from 169 to 80, the targets with "*" detected by RXD using $\mathbf{K}_{L\times L}^{-1}$ and $\mathbf{R}_{L\times L}^{-1}$ were different.

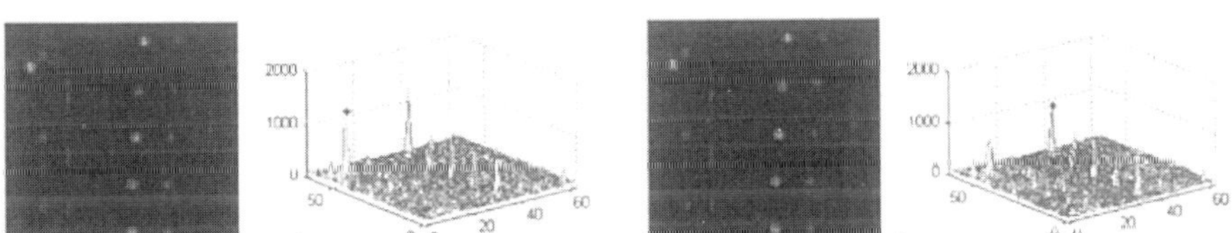

(a) maximum abundance = 1375.4855 at * using $\mathbf{K}_{L\times L}^{-1}$ (b) maximum abundance = 1372.3882 at * using $\mathbf{R}_{L\times L}^{-1}$

Figure 7.42. Comparison between RXD using $\mathbf{K}_{L\times L}^{-1}$ and $\mathbf{R}_{L\times L}^{-1}$ at $q = 80$

As shown in the plot of Fig. 7.42(a), the same interferer in Fig. 7.41(a) produced the highest peak. However, this is not true when $\mathbf{R}_{L\times L}^{-1}$ was used in RXD. The pixel that produced the highest peak in the plot of Fig. 7.42(b) was the panel pixel p_{11}, not the interferer in Fig. 7.42(a). But, when the q was further decreased to 60, both produced the same target pixel p_{11} with "*" as shown in Fig. 7.43(a-b).

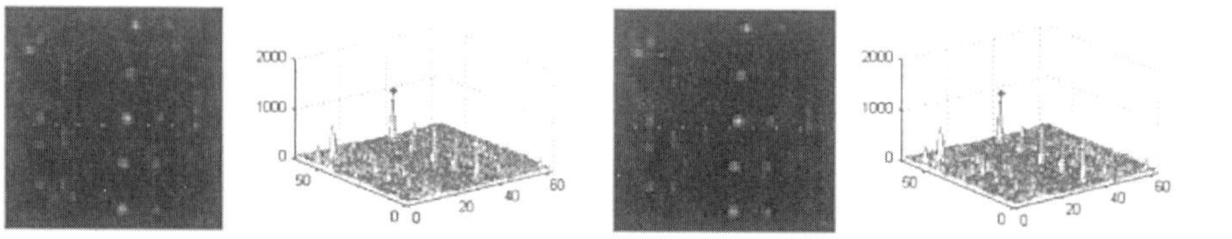

(a) maximum abundance = 1051.1350 at * using $\mathbf{K}_{L\times L}^{-1}$ (b) maximum abundance = 1001.7705 at * using $\mathbf{R}_{L\times L}^{-1}$

Figure 7.43. Comparison between RXD using $\mathbf{K}_{L\times L}^{-1}$ and $\mathbf{R}_{L\times L}^{-1}$ at $q = 60$ using K

Interestingly, as the q was reduced to 40, the image resulting from RXD using $\mathbf{K}_{L\times L}^{-1}$ still kept the same target pixel p_{11} with "*" in Fig. 7.44(a), but the target pixel with "*" detected by RXD using $\mathbf{R}_{L\times L}^{-1}$ in Fig. 7.44(b) was shifted to the panel pixel p_{31}.

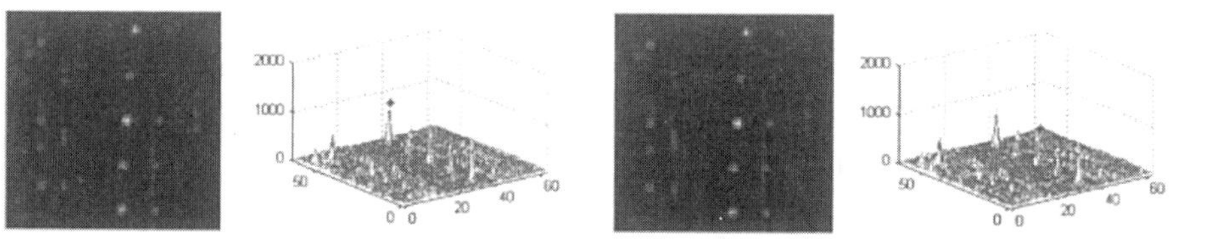

(a) maximum abundance = 857.2665 at * using $\mathbf{K}_{L\times L}^{-1}$ (b) maximum abundance = 814.0383 at * using $\mathbf{R}_{L\times L}^{-1}$

Figure 7.44. Comparison between RXD using $\mathbf{K}_{L\times L}^{-1}$ and $\mathbf{R}_{L\times L}^{-1}$ at $q = 40$

Now when the q was further reduced to 20, the panel pixel p_{31} remained the highest peak in the image generated by RXD using $\mathbf{R}_{L\times L}^{-1}$ shown in Fig. 7.45(a). But the target pixel with "*" detected by RXD using $\mathbf{K}_{L\times L}^{-1}$ was shifted from p_{11} to the panel pixel p_{51} shown in Fig. 7.45(b).

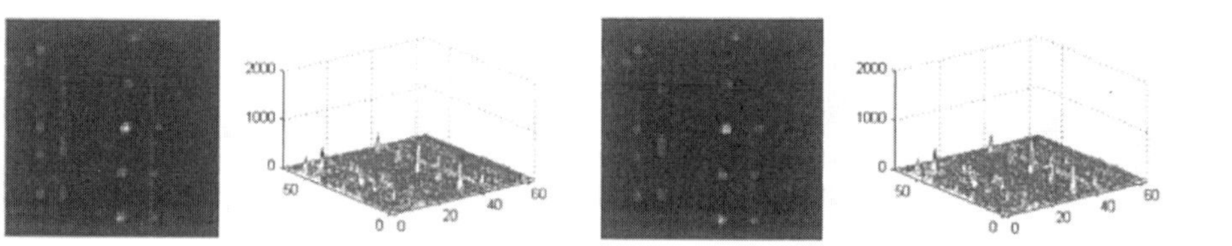

(a) maximum abundance = 711.3535 at * using $\mathbf{K}_{L\times L}^{-1}$ (b) maximum abundance = 687.9080 at * using $\mathbf{R}_{L\times L}^{-1}$

Figure 7.45. Comparison between RXD using $\mathbf{K}_{L\times L}^{-1}$ and $\mathbf{R}_{L\times L}^{-1}$ at $q = 20$

Finally, when the q was down to 10 and 5, the images in Figs. 7.46-7.47 produced by RXD using $\mathbf{K}_{L\times L}^{-1}$ and $\mathbf{R}_{L\times L}^{-1}$ were nearly the same with the same target pixel p_{51} having the highest peak.

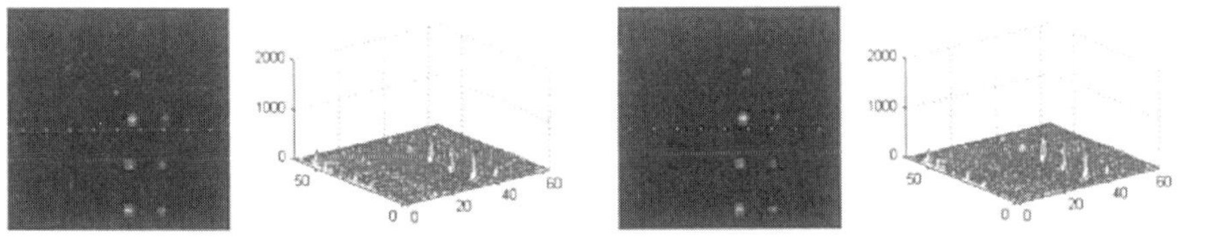

(a) maximum abundance = 609.9901 at * using $\mathbf{K}^{-1}_{L \times L}$ (b) maximum abundance = 618.2563 at * using $\mathbf{R}^{-1}_{L \times L}$

Figure 7.46. Comparison between RXD using $\mathbf{K}^{-1}_{L \times L}$ and $\mathbf{R}^{-1}_{L \times L}$ at $q = 10$

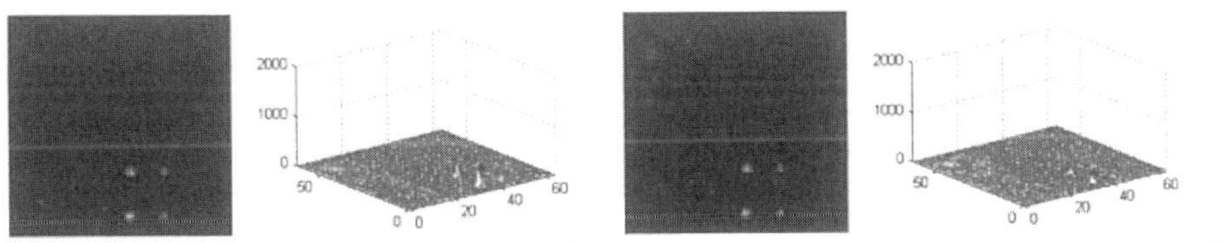

(a) maximum abundance = 384.2755 at * using $\mathbf{K}^{-1}_{L \times L}$ (b) maximum abundance = 190.6730 at * using $\mathbf{R}^{-1}_{L \times L}$

Figure 7.47. Comparison between RXD using $\mathbf{K}^{-1}_{L \times L}$ and $\mathbf{R}^{-1}_{L \times L}$ at $q = 5$ using K

Compared to Figs. 7.41-7.45, a noticeable difference was that the image background in Figs. 7.45-7.46 was almost nulled out. In particular, the strong interferer and the panel pixel p_{11} detected in an early stage at a larger q completely vanished. This implied that using a smaller q was more effective in background removal than using a larger q.

As a concluding remark, the study of noise sensitivity is a challenging issue that has been widely overlooked in the past. Using full bands is not always a good solution. Instead, finding an appropriate q is crucial to anomaly detection.

7.5 CONCLUSIONS

This chapter investigates two important issues arising in subpixel target detection, sensitivity of target knowledge and sensitivity of noise. Both play a significant role in the performance of TACSD, TSCSD and anomaly detection. In order to illustrate how these two issues affect the detection performance, a comprehensive experiment-based comparative analysis is studied for OSP, SCLS, NCLS, CEM and TCIMF. Since anomaly detection does not require *a priori* target knowledge, only the issue of noise sensitivity was investigated. A series of computer simulations is also conducted to analyze detection performance of the five anomaly detectors presented in Chapter 6. The experiments demonstrated that the noise has significant impact on anomaly detection due to small energy of anomalies. In order to reduce the noise sensitivity, the number of eigenvectors, q that is used to calculate the inverse of the sample covariance or correlation matrix must be small. How to select an appropriate q is generally a challenging problem. This issue will be addressed in Chapter 17.

III

UNCONSTRAINED MIXED PIXEL CLASSIFICATION

Classification in standard image processing is usually performed spatially on a pure pixel basis. The objects of interest are classified based on their spatial properties. In remote sensing image processing, image data are acquired by a number of spectral channels at different wavelengths; thus they are actually image cubes with the third dimension specified by spectral wavelengths. As a result, each image pixel is a column vector, of which each component is represented by a particular spectral band. The spectral information contained in such an image pixel vector cannot be generally explored by spatial-based classification methods. Mixed pixel classification (MPC) provides a solution. A general approach to MPC is linear unmixing which models the spectral signature of each pixel vector as a linear mixture of spectral signatures of targets present in the image. The classification is then performed based on the abundance fraction estimated for the spectral signature of each target resident in the pixel vector. As a result, MPC-generated images are generally gray scale and represent the abundance fractions of target signatures estimated from the image data. So, MPC is usually carried out by visual inspection. This approach is very different from the (0,1)-class membership assignment carried out by pure pixel classification (PPC) where a "1" implies that the object belongs to a specific class and a "0" indicates that the object is not in that particular class. There is no soft decision in PPC as is done in MPC with estimated abundance fractions. In Part III, least-squares based approaches to MPC are investigated, particularly, those without constraints imposed on abundance fractions. Chapter 8 describes a class of well-established OSP-based MPC methods that have shown success in hyperspectral image classification. It includes the OSP classifier first proposed by Harsanyi and Chang (1994), and several *a posteriori* OSP classifiers derived from Harsanyi-Chang's OSP classifier. Chapter 9 uses a standardized hyperspectral data to conduct a comparative analysis and quantitative study among several commonly used methods, specifically, the OSP-based MPC and Fisher's linear discriminant analysis-based PPC. A set of custom-designed criteria is also introduced for performance evaluation. In order to make comparison between MPC and PPC, the MPC-generated gray scale images must be converted to class-labeling images. Two thresholding criteria, winner-take-all and abundance percentage cutoff, are also developed for this purpose in Chapter 9 to convert a mixed pixel to a pure pixel. As a consequence of thresholding, the gray scale information produced by MPC for abundance fractions will be lost, which will result in performance degradation. A comprehensive experiment-based study on such mixed-to-pure pixel conversion is also conducted in Chapter 9 to demonstrate the significant importance of information provided by a mixed pixel through MPC.

8

UNCONSTRAINED MIXED PIXEL CLASSIFICATION: LEAST-SQUARES SUBSPACE PROJECTION

The orthogonal subspace projection (OSP) for hyperspectral image classification was first reported in Harsanyi and Chang (1994) and has been successfully applied to hyperspectral data exploitation since then. Its ability in subpixel detection was also demonstrated in Chapter 3. As we recall (3.9), the OSP-derived detector was given by $\delta_{\text{OSP}}(\mathbf{r}) = \mathbf{d}^T P_U^{\perp}(\mathbf{r})$ with the scale constant $\kappa = 1$. In Chapter 6, we have seen that this scale constant κ was actually determined by the *a posteriori* information that was used to estimate the unknown abundance fractions. Since OSP assumed the complete knowledge of the target signature matrix **M** and did not estimate the abundance vector α, the scale constant κ was absent in $\delta_{\text{OSP}}(\mathbf{r})$. As long as the abundance fractions detected for α provide sufficient amounts for target detection, it did not matter if α was estimated accurately. That was why OSP worked effectively for the real hyperspectral data experiments in Harsanyi and Chang (1994). However, this may not be true in terms of abundance estimation. So, in this chapter, the OSP in Chapter 3 is revisited for mixed pixel classification. It is then extended by three unconstrained least-squares subspace projection approaches, called signature subspace projection (SSP), target subspace projection (TSP) and oblique subspace projection (OBSP) where the abundance fractions of target signatures are not known *a priori*, but are required to be estimated from the data. The three subspace projection methods use their estimated signature abundance fractions to achieve target classification in a mixed pixel. As a result, they can be viewed as *a posteriori* OSP as opposed to the OSP in Chapter 3, which can be thought of as *a priori* OSP. In order to evaluate these three approaches, a least-squares estimation error is cast as a signal detection problem in the framework of the Neyman-Pearson detection theory so that the detection performance can be measured by the receiver operating characteristics (ROC) analysis.

8.1 INTRODUCTION

Image classification is a segmentation method, which aggregates image pixels into a finite number of classes by certain rules so that each class represents a distinct entity with specific properties. In multispectral/hyperspectral imagery, the spectral signature of a scene pixel is generally mixed by a number of target (or endmembers) spectral signatures. Consequently, such a scene pixel is considered as a mixed pixel. Two models have been proposed in the past to describe such mixing activities. One is the marcospectral mixture (Singer and McCord, 1979), which models a mixed pixel as a linear combination of target signatures resident in the pixel with relative concentrations. A second model suggested by Hapke (1981), called the intimate spectral mixture, is a nonlinear mixing of target signatures present within the pixel. Nevertheless, Hapke's model can be linearized by a method proposed by Johnson *et al.* (1983). In this chapter, only the linear spectral mixture model will be considered. By taking advantage of a linear mixture model, many image processing techniques are readily applied. Of particular interest is the principal components analysis (PCA) (Richards and Jia, 1999; Schowengerdt, 1997), also known as Karhunen-Loeve transformation, which is widely used to achieve data dimensionality reduction, feature extraction, etc. As a result of PCA, the data coordinates will be rotated in such a manner that the significant information of the data can be prioritized in accordance with the magnitude of the eigenvalues of the data covariance matrix. However, two disadvantages arise from the PCA approach. One is that the pixels in the PCA-transformed data are still a mixture of target signatures with unknown abundance fractions. So, the determination and identification of individual target spectral signatures remains unsolved. Malinowski (1977) and Heute (1986) proposed a solution. They first reconstructed the original data using the largest PCA-generated eigenvalue and measured the error between the raw data and the reconstructed data to see if the error falls within the prescribed tolerance. If not, they gradually added to data reconstruction by including additional eigenvalues in decreasing magnitude until the resulting error within the desired error level. A second disadvantage resulting from PCA is that PCA is only optimal in the sense of minimum mean squared error, but not necessarily optimal in terms of class discrimination and separability (Jenson and Waltz, 1979; Juang and Katagiri, 1992; Lee and Landgrebe, 1993) and signal-to-noise ratio (Green et al. 1981; Lee et al., 1990).

The OSP in Chapter 3 developed for subpixel detection can be also used for mixed pixel classification. It formulated an image classification problem as a generalized eigenvalue problem. The resulting classifier is an operator composed of two linear filters, a simultaneous diagonalization filter developed in Miller (1992) followed by a second filter, called matched filter derived from communication systems. However, as mentioned previously the model on which OSP was based assumed the complete target knowledge without estimating the abundance vector α. As a result, it can detect and classify correct target signatures, but cannot correctly estimate amounts of abundance fractions of target signatures. Despite the success of OSP in classification of AVIRIS data (Harsanyi and Chang, 1994; Farrand and Harsanyi, 1997), there is a lack of theory to explain how OSP works. In this chapter, we will revisit the OSP approach and offer a theoretical background for OSP from an estimation theory's point of view. The theory is derived based on unconstrained least-squares estimation and can be used to estimate target abundance fractions rather than detect target abundance fractions as does OSP. Accordingly, the approaches presented in this chapter can be referred to as *a posteriori* OSP, while the OSP in Chapter 3 (as also in Harsanyi and Chang, 1994) can be viewed as *a priori* OSP. A recent work in Settle (1996) also derived an unconstrained Gaussian maximum likelihood (GML) classifier which generated the same detector, $\delta_{\text{OSP}}(\mathbf{r}) = \mathbf{d}^T P_U^{\perp}(\mathbf{r})$ specified by (3.9) that were derived by Harsanyi and Chang (1994) with an additional constant $\kappa = \left(\mathbf{d}^T P_U^{\perp}\mathbf{d}\right)^{-1}$. Two important differences between the unconstrained GML classifier and OSP are significant. One is this constant κ that has been overlooked in the past. It has been considered as a normalization constant, which is immaterial to data processing. As noted in Chapters 5 and 6 this constant is indeed an important factor. It is determined by *a posteriori* information that can be used to estimate the abundance fractions. It is also very closely related to the accuracy of abundance estimation as demonstrated in Settle (1996) and Chang (1998). Another is the design rationale. The unconstrained GML classifier is an *a priori* approach, which maximizes the Gaussian conditional probability distribution of an unknown constant specified by target signature abundance fractions. Compared to the unconstrained GML classifier, OSP maximizes the signal-to-noise ratio based on Fisher's discriminant criterion, which only depends upon the noise second-order statistics. It can be derived as an *a posteriori* approach as will be shown in this chapter. When the noise is assumed to be Gaussian, the unconstrained GML classifier turns out to be *a posteriori* OSP.

This chapter presents three approaches that extend OSP in Chapter 3 to the case that the signature abundance fractions can be estimated directly from the image based on least-squares subspace projection. They all are unconstrained least-squares estimation methods. The first approach was proposed in Tu et al. (1998), which projected image pixel vectors into a signature space linearly spanned by target signatures in **M** so that the interfering effects caused by unwanted signatures can be suppressed by being mapped into a space orthogonal to the target signature subspace. The resulting projector is referred to as the signature subspace projection (SSP) classifier. Since the target signatures are those in which we are interested , a second approach is to project the observed pixels directly into a space that is linearly spanned only by the target signatures of interest rather than the entire target signatures in **M**. The resulting classifier is referred to as the target subspace projection (TSP) classifier. As we might expect, such TSP classifier should perform better than the SSP classifier in the sense that signatures other than the desired target signatures have been suppressed by the target subspace projection. Unfortunately, this is generally not the case. The drawback of the TSP classifier is that the undesired target signatures are not necessarily orthogonal to the desired target signatures, so the undesired target signatures may be scrambled into the signature space generated by the desired target signatures. As a result of such mixing, a bias becomes indispensable. In order to cope with this problem, a third approach based on the oblique subspace projection (Behrens and Scharf, 1994) is suggested to eliminate such a bias. It projects the desired target signatures and undesired target signatures into two separate orthogonal subspaces, range space and null space respectively. Since these two spaces are disjoint, no mixing will occur and nor will the bias. The resulting classifier will be referred to as an oblique subspace projection (OBSP) classifier. But, the OBSP classifier comes at a price. It is no longer an orthogonal classifier as are the SSP classifier and the TSP classifier. Nevertheless, it is still a projector. What it is interesting is that the SSP classifier can be actually decomposed into two oblique projectors, one of which happens to be the exactly the OBSP classifier. Moreover, we can also show that the OBSP classifier is identical to the unconstrained GML classifier derived in Settle (1996). In particular, the SSP classifier and the OBSP classifier are essentially equivalent in the sense of minimizing the least-squares error regardless of the fact that one is orthogonal and the other is not.

In order to evaluate the performance of these three classifiers, we model the least-squares based abundance estimation problem as a signal detection problem where the true target signature abundance is the desired signal and the estimation error is a result of the noise. By virtue of this detection model, the effectiveness of each classifier can be evaluated using the receiver operating characteristics (ROC) analysis. Interestingly, both the SSP classifier and the OBSP classifier generate an identical ROC curve. This implies that they basically are the same classifier. Surprisingly, the OSP classifier turns out to be the SSP classifier and the OBSP classifier with discarding the constant, $\kappa = \left(\mathbf{d}^T P_{\mathbf{U}}^{\perp}\mathbf{d}\right)^{-1}$. Since this constant affects the magnitude of the estimated abundance fractions of target signatures contained in a mixed pixel, the unconstrained GML classifier, the OSP classifier, the SSP classifier and the OBSP classifier can be considered as the same classifier with different classification powers. As for the TSP classifier, it produces a bias that deteriorates its performance. In order to resolve this problem, an unsupervised TSP classifier using the unsupervised algorithm, UVQ in Chapter 5 was developed in Brumbley and Chang (1999) where the bias was effectively eliminated by an orthogonalization algorithm developed in Chen et al. (1991).

8.2 *A POSTERIORI* OSP

In order to estimate $\alpha = \left(\alpha_1, \alpha_2, \cdots, \alpha_p\right)^T$ directly from the image data, several techniques have been developed (Settle, 1996; Tu et al., 1997; Chang, 1998; Chang et al. 1998b) based on *a posteriori* information obtained directly from the image scene. In this case, the model described by (3.1) in Chapter 3 can be expressed in terms of the *a posteriori* abundance estimation and the *a posteriori* noise estimation and given by

$$\begin{aligned}\mathbf{r} &= \mathbf{M}\hat{\alpha} + \hat{\mathbf{n}} \\ &= \mathbf{d}\hat{\alpha}_p + \mathbf{U}\hat{\gamma} + \hat{\mathbf{n}}\end{aligned} \tag{8.1}$$

where $\hat{\alpha}$, $\hat{\alpha}_p$ and $\hat{\gamma}$ are estimates of α, α_p and γ respectively based on the observed pixel vector, $\mathbf{r}$. Accordingly, the model specified by (8.1) is called an *a posteriori* model as opposed to the model by (3.1), which can be viewed as a Bayes or *a priori* model. For simplicity, the dependency on $\mathbf{r}$ will be dropped from all the notations of estimates throughout the rest of this chapter.

8.2.1 Signature Subspace Projection (SSP) Classifier

Using the least-squares error as an optimal criterion, the optimal least-squares estimate of α, $\hat{\alpha}_{LS}$ was given by (3.12) in Chapter 3. Substituting (3.12) for the estimate of α in model (8.1) results in

$$\mathbf{r} = \mathbf{M}\hat{\alpha}_{LS} + \hat{\mathbf{n}}_{LS} \tag{8.2}$$

where

$$\hat{\mathbf{n}}_{LS} = \mathbf{r} - \mathbf{M}\hat{\alpha}_{LS} = \mathbf{M}(\alpha - \hat{\alpha}_{LS}) + \mathbf{n}. \tag{8.3}$$

From (3.12) we define a signature space orthogonal projector by $P_{\mathbf{M}} = \mathbf{M}(\mathbf{M}^T\mathbf{M})^{-1}\mathbf{M}^T$ which projects $\mathbf{r}$ into the signature space $< \mathbf{M} >$. Applying $P_{\mathbf{M}}$ to model (8.2) yields

$$P_{\mathbf{M}}\mathbf{r} = P_{\mathbf{M}}\mathbf{M}\hat{\alpha}_{LS} + P_{\mathbf{M}}\hat{\mathbf{n}}_{LS} \tag{8.4}$$

$$= \mathbf{M}\hat{\alpha}_{LS} \tag{8.5}$$

where $P_{\mathbf{M}}\mathbf{M} = \mathbf{M}$ and the term $P_{\mathbf{M}}\hat{\mathbf{n}}_{\mathrm{LS}}$ vanishes in (8.5) since $P_{\mathbf{M}}$ annihilates $\hat{\mathbf{n}}_{\mathrm{LS}}$. Incorporating $P_{\mathbf{M}}$ into the OSP detector $\delta_{\mathrm{OSP}}(\mathbf{r})$ given by (3.9) results in a signature subspace projection (SSP) vector, $\mathbf{w}^{\mathrm{SSP}}$ given by

$$\left(\mathbf{w}^{\mathrm{SSP}}\right)^T = \mathbf{d}^T P_{\mathrm{U}}^{\perp} P_{\mathbf{M}}. \tag{8.6}$$

Now, if we apply $\mathbf{w}^{\mathrm{SSP}}$ to both *a priori* model (3.1) and *a posteriori* model (8.1), we obtain

$$\begin{aligned}\left(\mathbf{w}^{\mathrm{SSP}}\right)^T \mathbf{r} &= \mathbf{d}^T P_{\mathrm{U}}^{\perp} P_{\mathbf{M}}\left(\mathbf{M}\alpha + \mathbf{n}\right) \\ &= \mathbf{d}^T P_{\mathrm{U}}^{\perp} P_{\mathbf{M}}\mathbf{d}\alpha_p + \mathbf{d}^T P_{\mathrm{U}}^{\perp} P_{\mathbf{M}}\mathbf{n}\end{aligned} \tag{8.7}$$

and

$$\begin{aligned}\left(\mathbf{w}^{\mathrm{SSP}}\right)^T \mathbf{r} &= \left(\mathbf{w}^{\mathrm{SSP}}\right)^T (\mathbf{M}\hat{\alpha}_{\mathrm{LS}} + \mathbf{n}_{\mathrm{LS}}) \\ &= \mathbf{d}^T P_{\mathrm{U}}^{\perp} P_{\mathbf{M}}\mathbf{d}\hat{\alpha}_{\mathrm{SSP},p}\end{aligned} \tag{8.8}$$

where the estimate of α_p is denoted by $\hat{\alpha}_{\mathrm{SSP},p}$. Equating (8.7) and (8.8) gives rise to

$$\mathbf{d}^T P_{\mathrm{U}}^{\perp} P_{\mathbf{M}}\mathbf{d}\alpha_p + \mathbf{d}^T P_{\mathrm{U}}^{\perp} P_{\mathbf{M}}\mathbf{n} = \mathbf{d}^T P_{\mathrm{U}}^{\perp} P_{\mathbf{M}}\mathbf{d}\hat{\alpha}_{\mathrm{SSP},p}. \tag{8.9}$$

Dividing (8.9) by $\mathbf{d}^T P_{\mathrm{U}}^{\perp}\mathbf{d}$ and using the equality of $\mathbf{d}^T P_{\mathrm{U}}^{\perp} P_{\mathbf{M}}\mathbf{d} = \mathbf{d}^T P_{\mathrm{U}}^{\perp}\mathbf{d}$ result in

$$\hat{\alpha}_{\mathrm{SSP},p} = \alpha_p + \left(\mathbf{d}^T P_{\mathrm{U}}^{\perp}\mathbf{d}\right)^{-1}\left(\mathbf{w}^{\mathrm{SSP}}\right)^T \mathbf{n}. \tag{8.10}$$

Using (8.8) and (8.10) we can define a signature subspace classifier, $\delta_{\mathrm{SSP}}(\mathbf{r})$ by

$$\begin{aligned}\delta_{\mathrm{SSP}}(\mathbf{r}) &= \left(\mathbf{d}^T P_{\mathrm{U}}^{\perp}\mathbf{d}\right)^{-1}\left(\mathbf{w}^{\mathrm{SSP}}\right)^T \mathbf{r} - \left(\mathbf{d}^T P_{\mathrm{U}}^{\perp}\mathbf{d}\right)^{-1}\mathbf{d}^T P_{\mathrm{U}}^{\perp} P_{\mathbf{M}}\mathbf{r} \\ &= \hat{\alpha}_{\mathrm{SSP},p}\end{aligned} \tag{8.11}$$

The estimation error resulting from (8.11) and (8.10) is given by

$$\varepsilon_{\mathrm{SSP},p} = \hat{\alpha}_{\mathrm{SSP},p} - \alpha_p = \left(\mathbf{d}^T P_{\mathrm{U}}^{\perp}\mathbf{d}\right)^{-1}\left(\mathbf{w}^{\mathrm{SSP}}\right)^T \mathbf{n} \tag{8.12}$$

with $\mathrm{SNR}_{\mathrm{SSP,max}}$ defined by (3.6) or (3.10) as the maximum eigenvalue

$$\lambda_{\mathrm{SSP,max}} = \left(\alpha_p\sigma^{-1}\right)^2\left(\mathbf{d}^T P_{\mathrm{U}}^{\perp}\mathbf{d}\right)^{-1}\left[\mathbf{d}^T P_{\mathrm{U}}^{\perp} P_{\mathbf{M}}\mathbf{d}\right]^2 = \left(\alpha_p\sigma^{-1}\right)^2\mathbf{d}^T P_{\mathrm{U}}^{\perp}\mathbf{d}. \tag{8.13}$$

where σ^2 is assumed to be the variance of the noise $\mathbf{n}$.

Two comments are noteworthy.

1. Comparing (8.13) to (3.6), the maximum SNR (or eigenvalue) generated by $\delta_{\text{OSP}}(\mathbf{r})$ is exactly the same as that produced by $\delta_{\text{SSP}}(\mathbf{r})$ due to $\mathbf{d}^T P_{\mathbf{U}}^{\perp} P_{\mathbf{M}} \mathbf{d} = \mathbf{d}^T P_{\mathbf{U}}^{\perp} \mathbf{d}$. This shows that applying $P_{\mathbf{M}}$ does not increase the SNR. In other words, the maximum eigenvalue λ_{max} obtained by (3.6) remains unchanged after applying $P_{\mathbf{M}}$. This observation can be explained by Malinowski's error theory (Malinowski, 1977a) in which the eigenvalues are divided into two classes, a primary set containing larger eigenvalues and a secondary set of smaller eigenvalues. The former corresponds to imbedded errors that cannot be removed, whereas the latter represents experimental errors, which can be eliminated by designed techniques. According to this theory, the maximum eigenvalue is a primary eigenvalue, thus, it cannot be reduced by any estimation means. That (8.13) equals (3.6) illustrates this fact. A detailed error analysis resulting from the least-squares subspace approach was studied in Chang (1999).

2. Note that the magnitude of the SNR is determined by the quantity $\mathbf{d}^T P_{\mathbf{U}}^{\perp} \mathbf{d}$ in (8.13), namely, the degree of the target signature **d** correlated with the undesired signatures in $\mathbf{U}$. So, if **d** is very similar to one or more signatures in $\mathbf{U}$, $\mathbf{d}^T P_{\mathbf{U}}^{\perp} \mathbf{d}$ (i.e. the projection of **d** onto $< \mathbf{U} >^{\perp}$) will be small. This implies that the SNR will be low. As a result, the difficulty with the target signature discrimination is increased. Therefore, the magnitude of $\mathbf{d}^T P_{\mathbf{U}}^{\perp} \mathbf{d}$ can be used as a measure of the discrimination power of the SSP classifier. The larger the $\mathbf{d}^T P_{\mathbf{U}}^{\perp} \mathbf{d}$, the better the target discrimination. It should be also noted that the quantity of $\mathbf{d}^T P_{\mathbf{U}}^{\perp} \mathbf{d}$ is the inverse of the scale constant κ that determines the accuracy of abundance estimation (Chang 1998). This relationship further demonstrates how important role the κ plays in the performance of detection and classification.

8.2.2 Target Subspace Projection (TSP) Classifier

In the SSP classifier, $P_{\mathbf{M}}$ projects **r** onto the entire signature space $< \mathbf{M} >$. However, since we are only interested in classifying the target signature **d**, a natural approach is to project the observed pixel **r** onto the target signature space $< \mathbf{d} >$ rather than $< \mathbf{M} >$. This results in a second classifier, called target subspace projection (TSP) classifier, denoted by $\delta_{\text{TSP}}(\mathbf{r})$, given by

$$\delta_{\text{TSP}}(\mathbf{r}) = \mathbf{d}^T P_{\mathbf{U}}^{\perp} P_{\mathbf{d}} \mathbf{r} \tag{8.14}$$

where $P_{\mathbf{d}}$ is defined in the same fashion that $P_{\mathbf{M}}$ was defined, i.e., $P_{\mathbf{d}} = \mathbf{d}(\mathbf{d}^T\mathbf{d})^{-1}\mathbf{d}^T$.

Unfortunately, as shown in Chang et al. (1998b), the TSP classifier created an additional bias, which requires the complete knowledge of target signatures in the undesired target signature matrix $\mathbf{U}$. It was shown in Chang et al. (1998a) that the interference in hyperspectral imagery had significant impact on classification due to the sensors' high spatial and spectral resolution. Such intereferers are also considered to be undesired target signatures. They are generally unknown and cannot be identified by prior knowledge. Therefore, a direct application of the TSP classifier may not be a good approach. In order to mitigate this problem, Brumbley and Chang (1999) used an unsupervised vector quantization-based (UVQ) algorithm presented in Section 5.2 of Chapter 5 to find undesired target signatures and interferers in an image scene. Then they further used an algorithm proposed in Chen et al. (1991) to orthogonalize these undesired target signatures and interferers with respect to the desired target signature the desired target signature prior to detection of **d**. As a consequence, there is no bias created by such undesired mixing. The resulting classifier is referred to as unsupervised vector quantization-based target subspace projection (UVQTSP) classifier given by

$$\delta_{\text{UVQTSP}}(\mathbf{r}) = \mathbf{d}^T O_{\text{LS}}(\text{UVQ})(\mathbf{r}) \tag{8.15}$$

where O_{LS} is Chen's orthogonalization algorithm and the UVQ is the algorithm described in Section 5.2.

8.2.3 Oblique Subspace Projection (OBSP) Classifier

In the SSP classifier, the noise is first suppressed by $P_{\mathbf{M}}$, then followed by elimination of the undesired signatures in **U** by the projector $P_{\mathbf{U}}^{\perp}$. It would be convenient if we could have both two operations done in one step. One such operator, called an oblique subspace projection, was developed in Behren and Schlarf (1994). Let $<\mathbf{d}>$ and $<\mathbf{U}>$ denote the range space and the null space respectively. In this case, the oblique subspace projection is no longer orthogonal. Let $E_{\mathbf{XY}}$ be a projector with its range space **X** and null space **Y**. Then $P_{\mathbf{M}}$ can be decomposed as a sum of two oblique projectors and expressed by

$$P_{\mathbf{M}} = E_{\mathbf{dU}} + E_{\mathbf{Ud}} \tag{8.16}$$

where

$$E_{\mathbf{dU}} = \mathbf{d}\left(\mathbf{d}^T P_{\mathbf{U}}^{\perp}\mathbf{d}\right)^{-1}\mathbf{d}^T P_{\mathbf{U}}^{\perp} \tag{8.17}$$

$$E_{\mathbf{Ud}} = \mathbf{U}\left(\mathbf{U}^T P_{\mathbf{d}}^{\perp}\mathbf{U}\right)^{-1}\mathbf{U}^T P_{\mathbf{d}}^{\perp} \tag{8.18}$$

with $E_{\mathbf{dU}}\mathbf{d} = \mathbf{d}$ and $E_{\mathbf{dU}}\mathbf{U} = 0$.

In analogy with (8.6) an oblique subspace projection (OBSP) vector, $\mathbf{w}^{\text{OBSP}}$, can be obtained via (8.17) by

$$\left(\mathbf{w}^{\text{OBSP}}\right)^T\mathbf{r} = \mathbf{d}^T E_{\mathbf{dU}}\mathbf{r} = \left(\mathbf{d}^T\mathbf{d}\right)\left(\mathbf{d}^T P_{\mathbf{U}}^{\perp}\mathbf{d}\right)^{-1}\mathbf{d}^T P_{\mathbf{U}}^{\perp}\mathbf{r}. \tag{8.19}$$

Applying (8.19) to model (3.1) and model (8.1) results in

$$\left(\mathbf{w}^{\text{OBSP}}\right)^T\mathbf{r} = \mathbf{d}^T E_{\mathbf{dU}}\mathbf{r} = \mathbf{d}^T\mathbf{d}\alpha_p + \mathbf{d}^T E_{\mathbf{dU}}\mathbf{n} \tag{8.20}$$

and

$$\left(\mathbf{w}^{\text{OBSP}}\right)^T \mathbf{r} = \mathbf{d}^T E_{\mathbf{dU}} \mathbf{r} = \mathbf{d}^T \mathbf{d} \hat{\alpha}_{\text{OBCP},p} + \mathbf{d}^T E_{\mathbf{dU}} \hat{\mathbf{n}} = \mathbf{d}^T \mathbf{d} \hat{\alpha}_{\text{OBSP},p} \tag{8.21}$$

where $\mathbf{d}^T E_{\mathbf{dU}} \hat{\mathbf{n}} = 0$. Equating (8.20) and (8.21) yields

$$\hat{\alpha}_{\text{OBSP},p} = \alpha_p + \left(\mathbf{d}^T \mathbf{d}\right)^{-1} \mathbf{d}^T E_{\mathbf{dU}} \mathbf{n} \tag{8.22}$$

and

$$\hat{\alpha}_{\text{OBSP},p} = \alpha_p + \left(\mathbf{d}^T \mathbf{d}\right)^{-1} \mathbf{d}^T E_{\mathbf{dU}} \mathbf{n} = \alpha_p + \left(\mathbf{d}^T P_{\mathbf{U}}^{\perp} \mathbf{d}\right)^{-1} \mathbf{d}^T P_{\mathbf{U}}^{\perp} \mathbf{n} \tag{8.23}$$

So, from (8.22) and (8.23) we can define an oblique subspace classifier, $\delta_{\text{OBSP}}(\mathbf{r})$, by

$$\delta_{\text{OBSP}}(\mathbf{r}) = \left(\mathbf{d}^T \mathbf{d}\right)^{-1} \left(\mathbf{w}^{\text{OBSP}}\right)^T \mathbf{r} = \left(\mathbf{d}^T \mathbf{d}\right)^{-1} \mathbf{d}^T E_{\mathbf{dU}} \mathbf{r} = \left(\mathbf{d}^T P_{\mathbf{U}}^{\perp} \mathbf{d}\right)^{-1} \mathbf{d}^T P_{\mathbf{U}}^{\perp} \mathbf{r} \tag{8.24}$$

that produces the maximum SNR given by the maximum eigenvalue

$$\lambda_{\text{OBSP,max}} = \left(\alpha_p \sigma^{-1}\right)^2 \left[\mathbf{d}^T P_{\mathbf{U}}^{\perp} \mathbf{d}\right]^{-1}. \tag{8.25}$$

Interestingly, $\lambda_{\text{OBSP,max}}$ is also equal to $\lambda_{\text{SSP,max}}$ because (8.25) is identical to (8.13). Furthermore, the estimation error $\varepsilon_{\text{OBSP},p}(\mathbf{r})$ can be obtained from (8.23) as

$$\varepsilon_{\text{OBSP},p} = \hat{\alpha}_{\text{OBSP},p} - \alpha_p = \left(\mathbf{d}^T P_{\mathbf{U}}^{\perp} \mathbf{d}\right)^{-1} \mathbf{d}^T P_{\mathbf{U}}^{\perp} \mathbf{n}. \tag{8.26}$$

8.2.4 Unconstrained Maximum Likelihood Estimation Classifier

In the subspace projection approaches described in Sections 8.2.1-8.2.3 the only assumption made on the noise n was additive and white with variance given by $\sigma^2 \mathbf{I}_{L\times L}$. If the noise n is further assumed to be Gaussian, $p(\mathbf{r})$ in model (3.1) can be expressed as a Gaussian distribution with mean $\mathbf{M}\alpha$ and variance $\sigma^2 \mathbf{I}_{L\times L}$, i.e., $p(\mathbf{r}) \approx N(\mathbf{M}\alpha, \sigma^2 \mathbf{I}_{L\times L})$. In this case, the least-squares estimate of α for model (8.1) can be obtained by the unconstrained Gaussian maximum likelihood (GML) classifier (Settle, 1996; Chang et al., 1998b; Chang, 1998) which is given by

$$\delta_{\text{GML}}(\mathbf{r}) = \arg\{\max_{\mathbf{r}} p(\mathbf{r})\} = (\mathbf{M}^T \mathbf{M})^{-1} \mathbf{M}^T \mathbf{r}. \tag{8.27}$$

In particular, the estimate of the p-th abundance α_p, $\hat{\alpha}_{\text{GML},p}$ is given by

$$\begin{aligned}\hat{\alpha}_{\text{GML},p} &= \left(\mathbf{d}^T P_{\mathbf{U}}^{\perp}\mathbf{d}\right)^{-1}\left(\mathbf{d}^T P_{\mathbf{U}}^{\perp}\right)\mathbf{r} = \left(\mathbf{d}^T P_{\mathbf{U}}^{\perp}\mathbf{d}\right)^{-1}\delta_{\text{OSP}}(\mathbf{r}) \\ &= \left(\mathbf{d}^T P_{\mathbf{U}}^{\perp}\mathbf{d}\right)^{-1}\mathbf{d}^T P_{\mathbf{U}}^{\perp}\left(\mathbf{d}\alpha_p + \mathbf{n}\right) \\ &= \alpha_p + \left(\mathbf{d}^T P_{\mathbf{U}}^{\perp}\mathbf{d}\right)^{-1}\mathbf{d}^T P_{\mathbf{U}}^{\perp}\mathbf{n}\end{aligned} \tag{8.28}$$

and the associated estimation error is

$$\varepsilon_{\text{GML},p} = \hat{\alpha}_{\text{GML},p} - \alpha_p = \left(\mathbf{d}^T P_{\mathbf{U}}^{\perp}\mathbf{d}\right)^{-1}\mathbf{d}^T P_{\mathbf{U}}^{\perp}\mathbf{n}. \tag{8.29}$$

So, using (8.28) we can define an unconstrained GML classifier, $\delta_{\text{GML}}(\mathbf{r})$ by

$$\delta_{\text{GML}}(\mathbf{r}) = \left(\mathbf{d}^T P_{\mathbf{U}}^{\perp}\mathbf{d}\right)^{-1}\mathbf{d}^T P_{\mathbf{U}}^{\perp}\mathbf{r} = \hat{\alpha}_{\text{GML},p} \tag{8.30}$$

which produces the unconstrained Gaussian maximum likelihood estimate of α_p for the desired target signature **d** based on an image vector **r**.

Comparing $\left(\mathbf{d}^T P_{\mathbf{U}}^{\perp}\mathbf{d}\right)^{-1}\mathbf{d}^T P_{\mathbf{U}}^{\perp} P_{\mathbf{M}}\mathbf{n}$ in (8.12) which is the error estimate resulting from the SSP classifier, it is different from the estimation error resulting from $\delta_{\text{GML}}(\mathbf{r})$, $\left(\mathbf{d}^T P_{\mathbf{U}}^{\perp}\mathbf{d}\right)^{-1}\mathbf{d}^T P_{\mathbf{U}}^{\perp}\mathbf{n}$ in (8.29). However, if we further compare (8.26) to (8.29), we discover that both equations are identical. This implies that the unconstrained GML classifier is indeed the OBSP classifier provided the noise is white Gaussian. In this case, the unconstrained GML classifier can be replaced with the OBSP classifier in mixed pixel classification.

Several observations are also worth noting. Comparing (8.23) and (8.28) to the $\hat{\alpha}_p$ obtained in Settle (1996), the three abundance estimates (OBSP, unconstrained GML classifier, Settle's unconstrained GML classifier) are identical. Nevertheless, Settle provided an alternative approach to derive the unconstrained GML classifier $\hat{\alpha}_p$ (Settle, 1996). This implies that when the GML classifier and OBSP classifier are applied to linear spectral mixing problems specified by (3.1), they both arrive at the same identical classifier. Now comparing (8.24) and (8.30) to (3.9), there is an extra constant, $\left(\mathbf{d}^T P_{\mathbf{U}}^{\perp}\mathbf{d}\right)^{-1}$ appearing in (8.24) and (8.30). This constant is actually the same scale constant discussed in Chapter 7 and results from abundance estimation error. Since model (3.1) assumes the complete knowledge of the abundance vector α, there is no need of estimating the abundance vector α in model (3.1). Consequently, the constant was absent from (3.9). This demonstrates the significance of the constant $\left(\mathbf{d}^T P_{\mathbf{U}}^{\perp}\mathbf{d}\right)^{-1}$ that plays the distinction between the *a priori* model (3.1) and *a posteriori* model (8.2).

More interestingly, if we examine (8.24) and (8.30), we can discover that the scale constant $\kappa = \left(\mathbf{d}^T\mathbf{R}_{L\times L}^{-1}\mathbf{d}\right)^{-1}$ appearing in CEM is exactly the same as the constant appearing in $\delta_{\text{OBSP}}(\mathbf{r})$ and $\delta_{\text{GML}}(\mathbf{r})$ only if $\mathbf{R}_{L\times L}^{-1}$ in CEM is replaced by $P_{\mathbf{U}}^{\perp}$. This implies that the *a posteriori* information generated by $\delta_{\text{CEM}}(\mathbf{r})$ is $\mathbf{d}^T\mathbf{R}_{L\times L}^{-1}\mathbf{d}$ that approximates the prior information specified by $\mathbf{d}^T P_{\text{U}}^{\perp}\mathbf{d}$ used in the two classifiers $\delta_{\text{OBSP}}(\mathbf{r})$ and $\delta_{\text{GML}}(\mathbf{r})$. This relationship provides evidence that the three classifiers $\delta_{\text{CEM}}(\mathbf{r})$, $\delta_{\text{OBSP}}(\mathbf{r})$ and $\delta_{\text{GML}}(\mathbf{r})$ are optimal least-squares linear classifiers in the sense that they all make full use of information, (i.e., *a priori* target information, $\mathbf{d}^T P_{\text{U}}^{\perp}\mathbf{d}$ for $\delta_{\text{OBSP}}(\mathbf{r})$ and $\delta_{\text{GML}}(\mathbf{r})$ *a posteriori* target information, $\mathbf{d}^T\mathbf{R}_{L\times L}^{-1}\mathbf{d}$ for $\delta_{\text{CEM}}(\mathbf{r})$) to achieve best classification. The difference is that $\delta_{\text{CEM}}(\mathbf{r})$ is a sample spectral correlation-based matched filter, whereas $\delta_{\text{OBSP}}(\mathbf{r})$ and $\delta_{\text{GML}}(\mathbf{r})$ are merely a pixel-level spectral matched filter. The advantage of $\delta_{\text{CEM}}(\mathbf{r})$ over $\delta_{\text{OBSP}}(\mathbf{r})$ and $\delta_{\text{GML}}(\mathbf{r})$ is that the former does not require the prior information of U as the latter does. When it comes to abundance estimation, the *a posteriori* OSP-based classifiers, $\delta_{\text{OBSP}}(\mathbf{r})$ and $\delta_{\text{GML}}(\mathbf{r})$ include $\kappa = \left(\mathbf{d}^T\mathbf{R}_{L\times L}^{-1}\mathbf{d}\right)^{-1}$ in their filter design to account for abundance estimation error. If we further examine (4.15), $\delta_{\text{TCIMF}}(\mathbf{r})$ specified by the filter vector $\mathbf{w}^{\text{TCIMF}}_{\mathbf{R}_{L\times L}=\mathbf{I}_{L\times L}}$ is exactly identical to $\delta_{\text{OBSP}}(\mathbf{r})$ and $\delta_{\text{GML}}(\mathbf{r})$. This is not a surprise since these three filters are designed based on the same least-squares error criterion and they all generate the same abundance estimation constant $\kappa = \left(\mathbf{d}^T\mathbf{R}_{L\times L}^{-1}\mathbf{d}\right)^{-1}$.

8.3 ESTIMATION ERROR EVALUATED BY ROC ANALYSIS

In the previous section, three estimation errors were derived for the SSP classifier, the OBSP classifier and the MLE classifier. These errors are results of inaccurate abundance estimation of unknown signatures. In order to evaluate the error performance of these classifiers, we cast their associated estimation errors as a standard signal detection problem (Poor, 1994) where $\hat{\alpha}_p$ is viewed as a true target signature abundance fraction α_p corrupted by the noise that is represented by their estimation errors ε_p. By virtue of this formulation, these three classifiers can be also interpreted as subpixel target detectors, which can be used to detect the presence of a target signature in a mixed pixel. The effectiveness of these detectors depends upon the accuracy of the abundance estimate, $\hat{\alpha}_p$ and can be evaluated by the receiver operating characteristic (ROC) analysis via the Neyman-Pearson detection theory. An ROC curve is a graph plotted by the detection power versus the false alarm probability. Instead of using ROC curves as a performance criterion, we define a measure, called detection rate (DR) which calculates the area under an ROC curve to evaluate the effectiveness of the detector. Obviously, DR always lies between one and 1/2. The worst case occurs when $\text{DR} = 1/2$, i.e., the detection power is equal to the false alarm probability that implies that the detector is useless. On the other hand, the best case occurs only when $\text{DR} = 1$, namely, the detection power is always one regardless of the false alarm probability. Such ROC analysis has been widely used in diagnostic imaging (Metz, 1978; Swets and Pickett, 1982) for evaluation of computer-aided diagnostic methods where the detection power is measured by the true-positive probability and the false alarm probability is represented by the false-positive probability.

Assume that z is the projection resulting from applying a classifier $\delta(\mathbf{r})$ to the observed pixel $\mathbf{r}$. A signal detection model based on z can be described by a test of two hypotheses as follows.

$$H_0: \quad z = \delta(\mathbf{n}) \cong p_0(z)$$

$$versus \qquad (8.31)$$

$$H_1: \quad z = \alpha_p + \delta(\mathbf{n}) \cong p_1(z)$$

where the null hypothesis H_0 and the alternative hypothesis H_1 represent the case of noise alone, $z = \delta(\mathbf{n})$ (i.e., estimation error), and the case of the true target signature α_p present in the z respectively. The Neyman-Pearson detector for (8.31) is given by

$$\text{NPD}(z) = \begin{cases} 1; & z = \delta(\mathbf{n}) \geq \tau \\ 0; & z = \delta(\mathbf{n}) < \tau \end{cases}. \qquad (8.32)$$

From (8.32), we can also define the false alarm probability and detection power (detection probability) (Poor, 1994) as follows:

$$P_F = \int_\tau^\infty p_0(x)dx \qquad (8.33)$$

$$P_D = \int_\tau^\infty p_1(x)dx. \qquad (8.34)$$

Using (8.34), the detection power specifies the capability of the NPD in detecting the true target signature α_p. Therefore, the higher the detection power, the smaller the estimation error, the better classification the classifier.

8.3.1 Signature Subspace Projection (SSP) Classifier

Substituting $\delta_{\text{SSP}}(\mathbf{r})$ specified by (8.11) for q^T in (8.33) results in a subpixel target detection model given by

$$H_0: \quad z = \delta_{\text{SSP}}(\mathbf{n}) = \hat{n}_{\text{SSP}} \cong p_0(z)$$

$$versus \qquad (8.35)$$

$$H_1: \quad z = \alpha_p + \hat{n}_{\text{SSP}} \cong p_1(z)$$

where the noise $\hat{n}_{\text{SSP}}$ is generated by the estimation error produced by the SSP classifier. The hypothesis H_0 in (8.33) represents the case that the mixed pixel does not contain the target signature $\mathbf{d}$ while H_1 indicating the presence of $\mathbf{d}$ in the mixed pixel.

Based on (8.13)-(8.15), we obtain the error covariance matrix Σ_{SSP} for the estimation error $\hat{n}_{\text{SSP}}$ resulting from (8.14) as follows.

$$\Sigma_{\text{SSP}} = E[\varepsilon_{\text{SSP}}\varepsilon_{\text{SSP}}^T] = E[\hat{n}_{\text{SSP}}\hat{n}_{\text{SSP}}^T]$$
$$= \sigma^2 \frac{\mathbf{d}^T P_{\mathbf{U}}^{\perp} P_{\mathbf{M}} P_{\mathbf{M}} P_{\mathbf{U}}^{\perp} \mathbf{d}}{\mathbf{d}^T P_{\mathbf{U}}^{\perp} P_{\mathbf{M}} \mathbf{d}\mathbf{d}^T P_{\mathbf{U}}^{\perp} P_{\mathbf{M}} \mathbf{d}} = \frac{\sigma^2}{\mathbf{d}^T P_{\mathbf{U}}^{\perp} P_{\mathbf{M}} \mathbf{d}} = \left(\mathbf{d}^T P_{\mathbf{U}}^{\perp} \mathbf{d}\right)^{-1} \sigma^2. \tag{8.36}$$

Now, assume that $\mathbf{n}$ is a white Gaussian noise with zero mean and covariance matrix $\sigma^2 \mathbf{I}_{L\times L}$. Substituting (8.36) into (8.35) yields

$$p_0(z) \approx N\left(0, \left(\mathbf{d}^T P_{\mathbf{U}}^{\perp} \mathbf{d}\right)^{-1} \sigma^2\right)$$
$$p_1(z) \approx N\left(\alpha_p, \left(\mathbf{d}^T P_{\mathbf{U}}^{\perp} \mathbf{d}\right)^{-1} \sigma^2\right) \tag{8.37}$$

and the threshold τ_{SSP} given by

$$\tau_{\text{SSP}} = \sigma^{-1} \left(\mathbf{d}^T P_{\mathbf{U}}^{\perp} \mathbf{d}\right)^{1/2} \Phi^{-1}(1 - P_F) \tag{8.38}$$

where $\Phi(x)$ is the cumulative distribution function of the standard Gaussian random variable given by

$$\Phi(x) = (1/2\pi)^{1/2} \int_{-\infty}^{x} e^{-t^2/2} dt. \tag{8.39}$$

The desired detection power $P_{\text{SSP,D}}$ can be derived using (8.39) as follows

$$P_{\text{SSP,D}} = 1 - \Phi\left(\Phi^{-1}(1 - P_F) - \left(\sigma^{-2} \mathbf{d}^T P_{\mathbf{U}}^{\perp} \mathbf{d}\right)^{1/2} \alpha_p\right) \tag{8.40}$$
$$= 1 - \Phi\left(\Phi^{-1}(1 - P_F) - \sqrt{\lambda_{\text{SSP,max}}}\right). \tag{8.41}$$

In (8.40), $P_{\text{SSP,D}}$ is expressed in terms of $\mathbf{d}^T P_{\mathbf{U}}^{\perp} \mathbf{d}$ which indicates the degree of the correlation between $\mathbf{d}$ and the projection $P_{\mathbf{U}}^{\perp} \mathbf{d}$. On the other hand, (8.41) illustrates that the detection power is measured by the magnitude of SNR or equivalently, the maximum eigenvalue. So, both (8.40) and (8.41) can be used to evaluate the performance of the SSP classifier $\delta_{\text{SSP}}(\mathbf{r})$.

It is important to notice that for a fixed false alarm probability P_F, (8.40) shows that the discrimination power of $\delta_{\text{SSP}}(\mathbf{r})$ is proportional to the value of $\mathbf{d}^T P_{\mathbf{U}}^{\perp} \mathbf{d}$, whereas (8.41) suggests that the discrimination power of $\delta_{\text{SSP}}(\mathbf{r})$ is proportional to the magnitude of the maximum SNR or $\lambda_{\text{SSP,max}}$. This makes perfect sense since the former indicates the degree of similarity between the target signature $\mathbf{d}$ and the undesired signatures in $\mathbf{U}$, and the latter demonstrates that SNR determines the detection performance. More interestingly, if we interpret $\mathbf{d}^T P_{\mathbf{U}}^{\perp} \mathbf{d}$ as an inner product of $\mathbf{d}$ and the projection $P_{\mathbf{U}}^{\perp} \mathbf{d}$,

$\mathbf{d}^T P_{\mathbf{U}}^{\perp}\mathbf{d}$ measures how much projection of $\mathbf{d}$ is projected onto, $<\mathbf{U}>^{\perp}$, the orthogonal complement space. The more the projection, the less similarity between $\mathbf{d}$ and $\mathbf{U}$, thus the better the discrimination. Additionally, the detection power given by (8.40) is also a function of α_p. The higher the α_p, the better the detection of $\mathbf{d}$. Despite the fact that α_p is not present in (8.41), it is implicitly included in $\lambda_{\text{SSP,max}}$ given by (8.13). So, both (8.40) and (8.41) are also determined by the abundance fraction of α_p. All of these relationships can be well explained by the ROC analysis.

8.3.2 Oblique Subspace Projection (OBSP) Classifier

Analogous to (8.35), the oblique classifier $\delta_{\text{OBSP}}(\mathbf{r})$ given by (8.24) generates the following subpixel target detection problem.

$$\begin{aligned} &H_0: \quad z = \delta_{\text{OBSP}}(\mathbf{n}) = \hat{n}_{\text{OBSP}} \cong p_0(z) \\ &\textit{versus} \\ &H_1: \quad z = \alpha_p + \hat{n}_{\text{OBSP}} \cong p_1(z) \end{aligned} \tag{8.42}$$

with the noise covariance matrix $\Sigma_{\text{OBSP}}(\mathbf{r})$ given by

$$\begin{aligned} \Sigma_{\text{OBSP}} &= E[\varepsilon_{\text{OBSP}}\varepsilon_{\text{OBSP}}^T] = E[\hat{n}_{\text{OBSP}}\hat{n}_{\text{OBSP}}^T] \\ &= \left(\mathbf{d}^T P_{\mathbf{U}}^{\perp}\mathbf{d}\right)^{-1}\mathbf{d}^T P_{\mathbf{U}}^{\perp} E[\mathbf{n}\mathbf{n}^T]\left(\mathbf{d}^T P_{\mathbf{U}}^{\perp}\right)^T\left(\mathbf{d}^T P_{\mathbf{U}}^{\perp}\mathbf{d}\right)^{-1} = \left(\mathbf{d}^T P_{\mathbf{U}}^{\perp}\mathbf{d}\right)^{-1}\sigma^2 \end{aligned}. \tag{8.43}$$

Substituting (8.43) into (8.42) we obtain the same probability density functions $p_0(z)$ and $p_1(z)$ given by (8.37) for the detection problem described by (8.42). As a result, the threshold τ_{OBSP} is equal to the τ_{SSP} given by (8.38) and the desired detection power $P_{\text{OBSP,D}}$ given by

$$P_{\text{OBSP,D}} = 1 - \Phi\left(\Phi^{-1}(1 - P_F) - \sigma^{-1}\left(\mathbf{d}^T P_{\mathbf{U}}^{\perp}\mathbf{d}\right)\alpha_p\right) \tag{8.44}$$

$$= 1 - \Phi\left(\Phi^{-1}(1 - P_F) - \sqrt{\lambda_{\text{SSP,max}}}\right) \tag{8.45}$$

is also identical to the detection power $P_{\text{SSP,D}}$. This implies that $\delta_{\text{OBSP}}(\mathbf{r})$ is essentially equivalent to $\delta_{\text{SSP}}(\mathbf{r})$ in the sense that they both generate identical ROC curves.

8.4 COMPUTER SIMULATIONS AND HYPERSPECTRAL IMAGE EXPERIMENTS

In this section, computer simulations and the AVIRIS data in Fig. 1.5 will be used for performance evaluation. Since OSP does not use *a posteriori* target knowledge, its performance was not evaluated here. Furthermore, the OBSP classifier is essentially equivalent to the SSP classifier in terms of ROC analysis, only experiments were conducted for the SSP classifier in this section.

8.4.1 Computer Simulations

In the following simulations, two laboratory data sets are used to form the target signature matrix is $\mathbf{M} = [\mathbf{m}_1\, \mathbf{m}_2\, \mathbf{m}_3]$ with the abundance vector given by $\alpha = (\alpha_1\, \alpha_2\, \alpha_3)^T$. We also let $\mathbf{d} = \mathbf{m}_3$ be the target signature specified by abundance fraction α_3 and $\mathbf{U} = [\mathbf{m}_1\, \mathbf{m}_2]$ be the matrix made up of undesired signatures with abundance fractions given by $\alpha_1 . \alpha_2$. Data set 1 contains three field reflectances, dry grass, red soil and creosote leaves shown in Figure 1.5 with creosote leaves designated as $\mathbf{d}$. Data set 2 consists of sagebrush, blackbrush and creosote leaves with $\mathbf{U}$ made up of sagebrush, blackbrush and $\mathbf{d}$ = creosote leaves. The difference between data set 1 and data set 2 is that the spectrum of the target signature in data set 2 is very similar to that of sagebrush, while all the three spectra in data set 1 are distinguishable. There are 50 mixed pixels simulated with abundances in accordance with Table 8.1. Here, we only consider the case that the undesired signature vectors share their abundances evenly for illustrative purpose.

Table 8.1. 50 simulated pixels with assigned abundance fractions for data sets 1 and 2

	Pixel no. 1-10	Pixel no. 11-20	Pixel no. 21-30	Pixel no. 31-40	Pixel no. 41-50
Target signature	1%	5%	10%	15%	20%
undesired signature 1	49.5%	48.5%	45%	42.5%	40%
Undesired signature 2	49.5%	48.5%	45%	42.5%	40%

For the case of uneven abundances in $\mathbf{U}$, we refer to Zhao (1996) which shows no appreciable difference in the experiments, particularly for the desired signature with high abundance. In addition, different Gaussian noise levels are also simulated and added to generate SNR, 50:1, 30:1 and 10:1 with the SNR defined in Harsanyi and Chang (1994) as 50% reflectance divided by the standard deviation of the noise. Figs. 8.1 and 8.2 show the performance of the SSP classifier data set 1 and 2 respectively where figures labeled by (a), (b) and (c) are results obtained based on SNR, 50:1, 30:1 and 10:1 respectively.

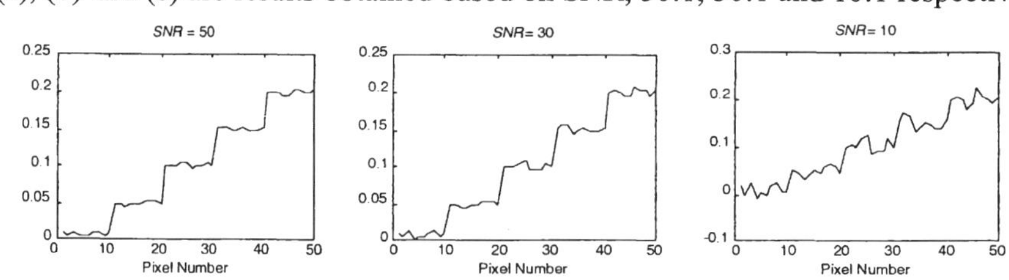

(a) SNR = 50:1 (b) SNR = 30:1 (c) SNR = 10:1

Figure 8.1. The performance of SSP for data set 1

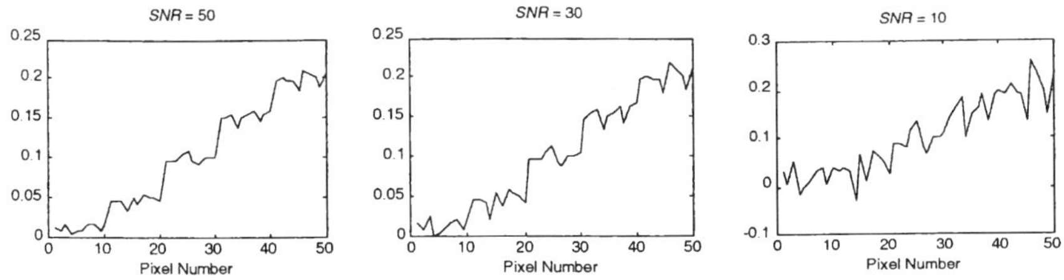

(a) SNR = 50:1 (b) SNR = 30:1 (c) SNR = 10:1

Figure 8.2. The performance of SSP for data set 2

Comparing Fig. 8.2 to Fig. 8.1, the SNR and spectral similarity among signatures are significant factors of detection performance.

In order to evaluate the estimation error performance of the SSP classifier, the ROC curves is plotted in Figs. 8.3 and 8.4 for data sets 1 and 2 with the same three signatures used for Figs. 8.1 and 8.2. Figures labeled by (a), (b) and (c) are results generated by α_3 / σ = 0.5, 0.3 and 0.1 respectively.

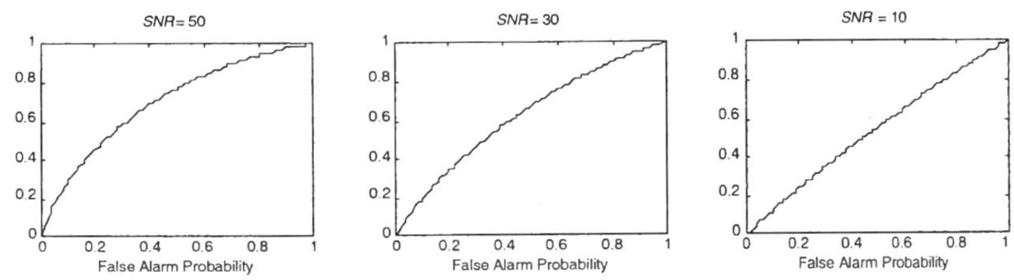

Figure 8.3. ROC of SSP data set 1

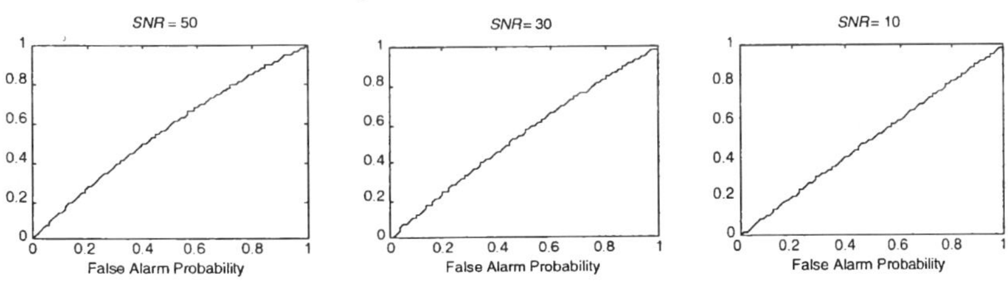

Figure 8.4. The ROC of SSP data set 2

It should be noted that each ROC curve is generated by one mixed pixel with a given desired signature abundance to noise ratio (ANR), α_3 / σ. For instance, if α_3 / σ = 0.5 and α_3 = 0.3, then the noise level will be σ = 0.6 and the other two undesired signatures will evenly split the remaining abundance 0.8. Their detection rates (DRs) are tabulated in Tables 8.2 for data set 1 and date set 2 respectively, which measure effects of different ANRs and spectral similarity.

Table 8.2. Detection rates produced by the SSP classifier for data sets 1 and 2

	DR (data set 1)	DR (data set 2)
$\alpha_3 / \sigma = 0.5$	0.71	0.57
$\alpha_3 / \sigma = 0.3$	0.63	0.54
$\alpha_3 / \sigma = 0.1$	0.55	0.52

8.4.2 Hyperspectral Data

A theoretical study on comparative analysis among subspace projection methods was investigated in Chang et al. (1998b). An experiment-based comparison among the OSP classifier, the OBSP classifier, UGMLC and the SSP classifier using the same AVIRIS image in Figure 1.6(a) was investigated in Chang et al. (1998b) and Chang and Ren (2000). In the following experiments, the standardized HYDICE data shown in Figure 1.7(a) will be used for comparative analysis. Since both the OBSP classifier and GML classifier generate an identical estimation error given by (8.24) and (8.30), we will only focus our experiments on the OSP classifier, the OBSP classifier and the SSP classifier. It is interesting to note that if we apply *a posteriori* OSP, $\left(\mathbf{d}^T P_{\mathrm{U}}^{\perp}\mathbf{d}\right)^{-1}\delta_{\mathrm{OSP}}(\mathbf{r})$ to model (3.2), it results in the same equations given by (8.26) and (8.31) with both $\hat{\alpha}_{\mathrm{OBSP},p}$ and $\hat{\alpha}_{\mathrm{UGMLC},p}$ replaced by α_p. This implies that if the knowledge about the abundance vector α is given *a priori*, then the OBSP classifier and UGMLC are reduced to OSP. On the other hand, if the abundance vector α is not known and needs to be estimated by $\hat{\alpha}$, then the OBSP classifier and GML classifier will be used to replace the OSP classifier. Consequently, OSP can be viewed as the *a priori* version of the OBSP classifier and GML classifier, while the OBSP classifier and GML classifier can be thought of as *a posteriori* OSP classifiers. Interestingly, despite the fact that the OSP used in Harsanyi and Chang (1994) was considered to be *a priori* OSP, the conducted experiments were actually based on *a posteriori* information where the target signatures, $\mathbf{m}_1, \mathbf{m}_2, \cdots, \mathbf{m}_p$ used in (3.1) were directly extracted from the image scene and treated as if they were true target signatures.

As noted, the constant $\left(\mathbf{d}^T P_{\mathrm{U}}^{\perp}\mathbf{d}\right)^{-1}$ accounts for the error of estimating the target abundance fractions present in classified pixels. This constant does not have impact on images displayed on computer because the abundance fractional images generated by OSP, the SSP and the OBSP classifiers for computer display are all scaled to 256 gray levels. In this case, the constant $\left(\mathbf{d}^T P_{\mathrm{U}}^{\perp}\mathbf{d}\right)^{-1}$ is absorbed for computer display. So, from a display point of view, they all produce identical results as shown in Fig. 8.5(a-c).

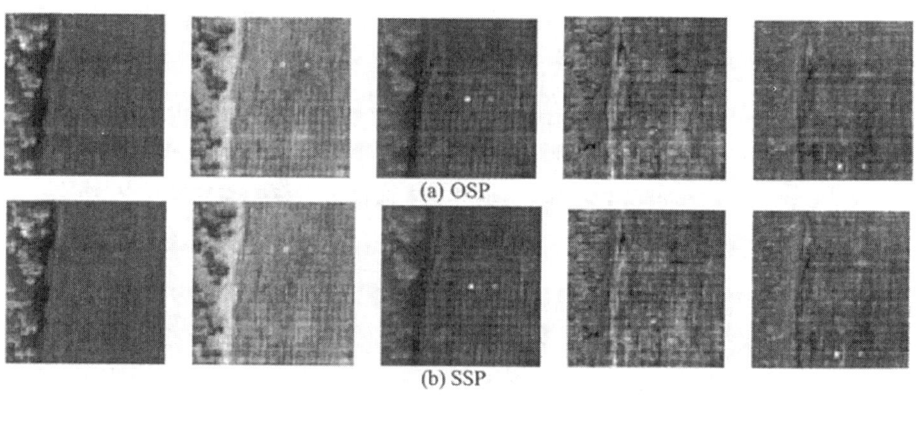

(a) OSP

(b) SSP

(c) OBSP

Figure 8.5. Classification results of OSP, SSP and OBSP

In order to substantiate the difference between the abundance fractions generated by OSP, the SSP and the OBSP classifiers, we can take the absolute differences between these three classifiers and display their error images in 256 gray scales in Fig. 8.6(a-c). If two classifiers generate identical results, their absolute difference should be 0 and their corresponding error images should be all black. Obviously, this is not true as shown in Figure 8.6(a-c). This justifies the subtle difference among these three classifiers.

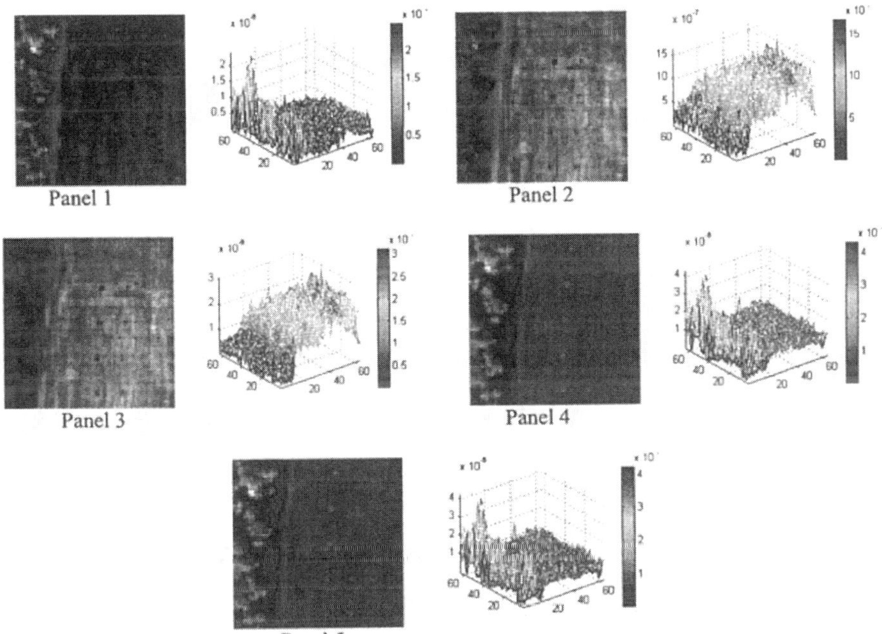

(a) Error images between OSP and SSP

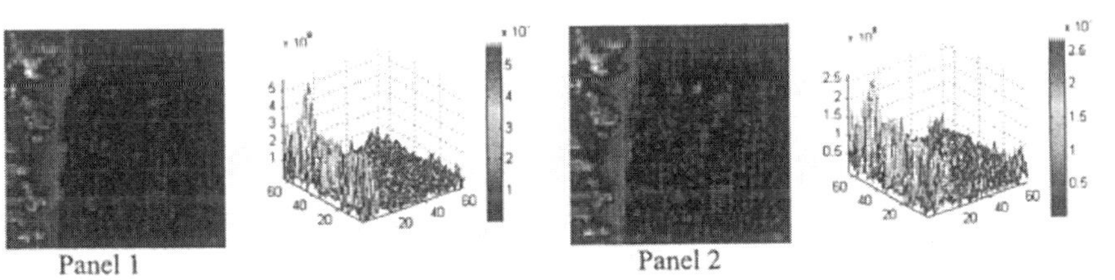

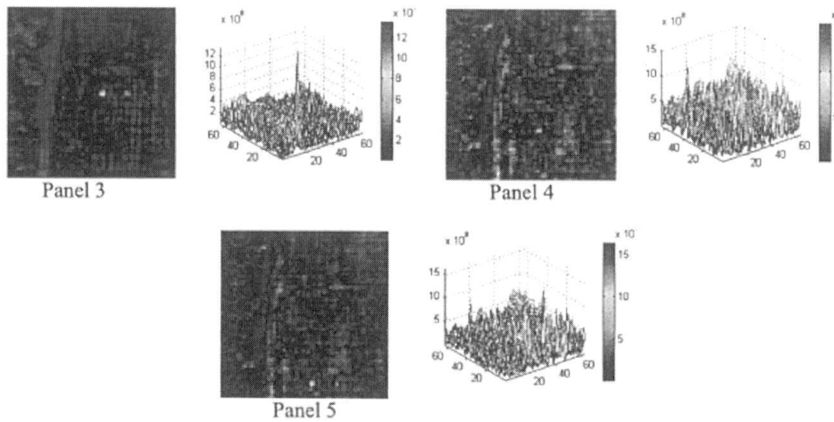

(b) Error images between OSP and OBSP

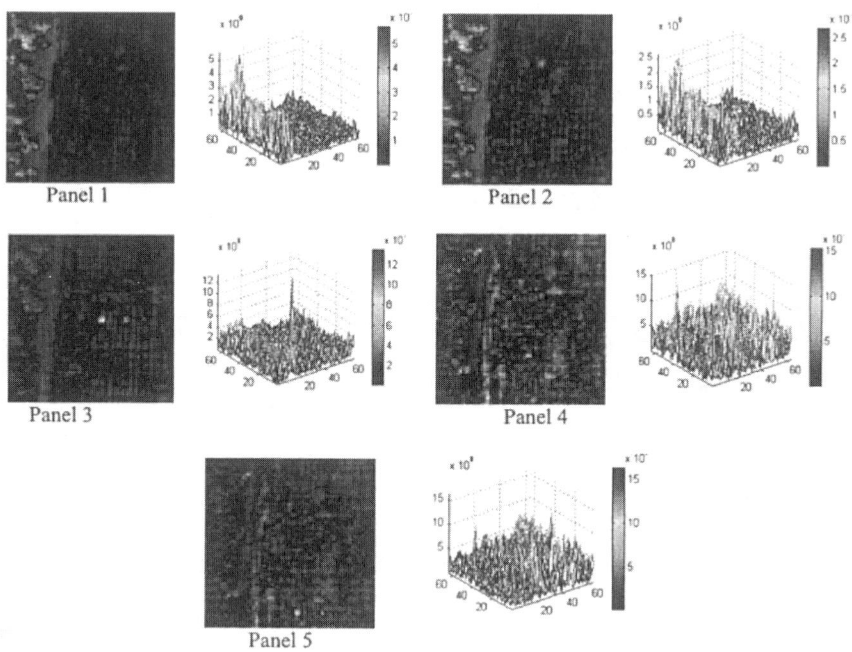

(c) Error images between OBSP and SSP

Figure 8.6. Error images among OSP-, SSP- and OBSP-generated images

If we further compare Figs. 8.6(b) and 8.6(c) to Fig. 8.6(a), the error images generated by Figs. 8.6(b) and 8.6(c) were very close, but very different from Fig. 8.6(a). This implies that OSP and the OBSP classifier performed nearly the same except that they did different amounts of target abundance fractions. On the other hand, the SSP classifier performed quite differently from OSP in that the SSP classifier includes an additional signature subspace projector $P_{\mathbf{M}}$ and the constant $\left(\mathbf{d}^T P_{\mathbf{U}}^{\perp}\mathbf{d}\right)^{-1}$ in its classifier. As a result, the estimation error generated by the SSP classifier using (8.11) is very different from that by OSP using (3.9). This fact was reflected in Fig. 8.6(a). It was shown in Tu et al. (1997) via ROC analysis that the SSP classifier greatly improved the OSP classifier in terms of SNR if the additive noise is assumed to be Gaussian. An error theory using ROC analysis for *a posteriori* OSP and OSP is further investigated in Chang (1999).

8.5 CONCLUSIONS

This chapter extends the orthogonal subspace projection (OSP) detector in Chapter 3 to a family of unconstrained *a posteriori* OSP classifiers which perform abundance fraction estimation rather than abundance fraction detection as does OSP. Three such *a posteriori* OSP classifiers, SSP classifier, TSP classifier and OBSP classifier, were introduced, which are all based on a least-squares subspace projection approach. The SSP classifier estimates the target signature by projecting a mixed pixel into the signature subspace from which the target signature can be extracted by a matched filter. Rather than mapping a mixed pixel into the entire signature subspace, the TSP classifier projects the pixel directly into the target subspace space. Unfortunately, it does not produce satisfactory performance due to the fact that the undesired signatures are also projected and mixed into the target subspace space. As a consequence, an unknown signature bias is created. In order to eliminate this bias, two approaches are considered. One is the unsupervised vector quantization-based TSP classifier (Brumbley and Chang, 1999). Another is the OBSP classifier that projects the desired target signatures into its range space while eliminating the undesired target signatures by mapping them into its null space. As a consequence, the OBSP classifier is no longer an orthogonal projection classifier. Nevertheless, the experiments demonstrated that both the SSP classifier and the OBSP classifier performed almost the same in terms of target detection. In particular, they generate identical ROC curves. However, as will be demonstrated in Chapter 9, we should not draw the same conclusion on their ability in abundance estimation. It is also interesting to note that if the noise in (3.1) is assumed to be additive Gaussian noise, the OBSP classifier becomes an unconstrained GML classifier, which is actually an *a priori* estimator. In this case, the unconstrained GML classifier can be considered as a special case of the OBSP classifier, which is indeed an *a posteriori* estimator.

9

A QUANTITATIVE ANALYSIS OF MIXED-TO-PURE PIXEL CONVERSION (MPCV)

Over the past years many algorithms have been developed for multispectral and hyperspectral image classification. Due to a lack of standardized data, these algorithms have not been rigorously compared within a unified framework. In this chapter, we present a comparative study of several popular classification algorithms through a standardized HYDICE data set with custom-designed detection and classification criteria. The algorithms to be considered for this study are those developed in Chapter 8, viz. the orthogonal subspace projection (OSP), unconstrained Gaussian maximum likelihood (GML) classifier, minimum distance, and Fisher's linear discriminant analysis (LDA). In order to compare mixed pixel classification (MPC) algorithms against pure pixel classification (PPC) algorithms, a mixed pixel is converted to a pure pixel via a mixed-to-pure pixel converter (MPCV). The standardized HYDICE data are then used to evaluate the performance of various PPC and MPC algorithms. Since the precise spatial locations of all the targets in the standardized HYDICE data are available, the candidate algorithms can be evaluated by tallying the number of targets detected and classified for quantitative analysis.

9.1 INTRODUCTION

Mixed pixel classification (MPC) becomes increasingly important in hyperspectral image analysis and provides an effective means for detection, classification, discrimination, quantification, etc. New algorithms have been reported every year. Due to a lack of standardized data, it is difficult to substantiate each of these algorithms. In addition, no unified criterion has been accepted for rigorous and impartial comparison. This chapter develops a set of custom-designed criteria for MPC and conducts a comparative study among several well-known classification algorithms. We confine our study to two types of classification, MPC and pure pixel classification (PPC) are considered. A general approach to MPC is to estimate the abundance fraction of a target signature of interest present in an image pixel and the estimated abundance fraction is then used to classify the pixel. One such method is spectral unmixing discussed in Chapter 8. However, the images generated by MPC are generally gray scale with gray level values representing the amounts of the estimated abundance fractions present in pixels. Consequently, it requires visual interpretation of these gray scale abundance fractional images. Such human intervention is rather subjective and may not be reliable or repeatable. With no availability of standardized data or objective criteria a quantitative analysis for mixed pixel classification is nearly impossible. By contrast, PPC does not have such a problem. Unlike MPC it does not require estimation of signature abundance fractions for class-membership assignment. Its performance is completely determined on a pure pixel basis by criteria used for classification. So, two major issues are addressed in this chapter. One establishes a link between MPC and PPC via a mixed-to-pure pixel converter (MPCV). Another develops a set of acceptable criteria to conduct a comparative and quantitative analysis among commonly used MPC and PPC algorithms. The selection of candidate algorithms to be compared in this chapter is made for the following three major reasons. (1) As shown in Chapter 8, if the noise in a linear mixing problem is white Gaussian, the Gaussian maximum likelihood (GML) is essentially equivalent to OSP and identical to OBSP. So, this allows us to restrict MPC algorithms to a family of OSP-based classifiers. (2) Additionally, if we also assume that the sample spectral correlation/covariance matrix is stationary, the GML classifier is then further reduced to a minimum distance classifier. In this case, only minimum distance classifiers are considered. (3) Fisher's linear discriminant analysis (LDA) has been shown very successful in pattern classification because of its criterion designed based on maximization of class separability. When PPC is considered, Fisher's LDA must be included for study. The difference between OSP and the other approaches (i.e., minimum distance, Fisher's LDA) is that OSP is primarily designed for MPC, whereas the latter is developed for PPC. Nevertheless, both can be used for MPC and PPC. As will be shown, MPC can be reinterpreted by imposing appropriate constraints on the abundance fractions and further reduced to PPC by a mixed-to-pure pixel conversion (MPCV). Through an MPCV a direct comparison between an MPC algorithm and a PPC algorithm is possible.

9.2 CONVERSION OF MPC TO PPC

Most MPC algorithms estimate the abundance fraction vector $\alpha = (\alpha_1\ \alpha_2 \cdots \alpha_p)^T$ present in a pixel vector $\mathbf{r}$ using a linear mixture model given by (3.1). Since model specified by (3.1) is not used to estimate α for OSP, it is considered as *a priori* model. On the other hand, the *a posteriori* model described by (8.1) requires an estimation of α, which leads to an *a posteriori* OSP approach. In both cases, the images generated by MPC algorithms are referred to as abundance fractional images, which are gray scale. The gray level values of the abundance fractional images are specified by the abundance fractions of $\hat{\alpha}_p$ estimated for the desired target signature $\mathbf{d}$ present in each mixed pixel vector $\mathbf{r}$ in the images. The only difference between the *a priori* model and the *a posteriori* model lies on the fact that the abundance fractions generated by the *a priori* model do not reflect true amounts of abundance fractions resident in the image pixel. Thus, these abundance fractions can be only used for detection and classification but not quantification. On the other hand, the images generated by the *a posteriori* model are estimates of true amounts of abundance fractions. Therefore, they can be better used for target classification and material quantification.

As noted, MPC is usually carried out by visual interpretation based on the amounts of abundance fractions estimated form image pixels. So, technically speaking, OSP and *a posteriori* OSP perform more than classification. They are essentially signature abundance estimation methods. In order for these algorithms to be used as classifiers without human interpretation, we need a mechanism, called a mixed-to-pure pixel converter (MPCV) that can threshold the estimate of an abundance fraction to a class-membership decision. This MPCV is very similar to an operator that has been widely used in communications and signal processing, called an analog-to-digital converter (A/D converter). Two versions of MPCV are suggested in this chapter, which are winner-take-all (WTA) and abundance percentage thresholding methods. The former is good for the case that the estimates of the abundance fractions of all the target signatures present in **r** are available, such as those produced by OSP in Chapters 3 and 8. The latter is generally applied to the case that only the estimates of target signatures of interest are available, e.g. those generated by LCMV developed in Chapter 4.

9.2.1 Mixed-to-Pure Pixel Converter (MPCV)

In order to compare PPC to MPC, we need to interpret an MPC problem in the context of PPC. One way to convert abundance estimation of a mixed pixel to classification of a pure pixel is to consider model (3.1) as a constrained problem with some specific restrictions imposed on the estimated abundance vector $\hat{\alpha}$.

Assume that the abundance vector α in model (3.1) satisfies constraints, $\alpha_j \geq 0$ for all $1 \leq j \leq p$ and $\sum_{j=1}^{p} \alpha_j = 1$. Additionally, the estimate $\hat{\alpha}$ is constrained to a set of p-dimensional vectors with one in only one component and zeros in the remaining p-1 components. Such vectors will be referred to as p-dimensional unit vectors. Let $\mathbf{u}_j$ be a p-dimensional vector with 1 in its j-th component and 0's in all other remaining components, i.e., $\mathbf{u}_j = (0,\cdots,0,\underset{j}{\underset{\downarrow}{1}},0,\cdots,0)^T$, then $\mathbf{u}_j$ is called the j-th p-dimensional unit vector. In this case, the estimated abundance vector $\hat{\alpha}$ must be forced to a pure abundance vector and there are only p choices for $\hat{\alpha}$. In other words, $\hat{\alpha}$ can be assigned to one and only one from p classes. By virtue of this constraint an MPC problem is reduced to a p-class classification problem. As a result, it can be solved by PPC techniques. Imposing a constraint on p-dimensional unit vectors the *a posteriori* model (8.1) becomes

$$\chi_{\text{MPCV}}(\mathbf{r}) = \mathbf{M}\mathbf{u}_j = \mathbf{m}_j \text{ for some } 1 \leq j \leq p \tag{9.1}$$

where $\chi_{\text{MPCV}}(\mathbf{r})$ is called a mixed-to-pure pixel converter (MPCV) operating on a pixel vector **r** that assigns **r** to signature $\mathbf{m}_j$ for some j. It should be noted that the estimated noise $\hat{\mathbf{n}}$ in model (8.1) has been absorbed into $\mathbf{u}_j$ to account for misclassification error. So, if we reinterpret model (3.1) using (9.1), each target signature vector in M represents a distinct class and an image pixel vector r will be assigned to one of the signatures in M via the MPCV specified by (9.1). The image resulting from MPCV is a binary image, which shows only target pixels. An important but difficult task is to design an effective MPCV for (9.1), which must preserve as much information as possible contained in mixed pixels during MPCV.

One method was recently proposed in Ren (2000) which is based on *winner-take-all* (WTA) thresholding criterion. It is very similar to the winner-take-all learning algorithm used in neural networks (Haykin, 1994). This WTA thresholding criterion can be used for (9.1) to convert a mixed pixel to a pure pixel.

Assume that there are p target signatures $\{\mathbf{m}_j\}_{j=1}^p$ where $\mathbf{m}_j$ is the j-th signature. Let $\mathbf{r}$ be a mixed pixel vector to be classified and $\alpha(\mathbf{r}) = (\alpha_1(\mathbf{r})\ \alpha_2(\mathbf{r}) \cdots \alpha_p(\mathbf{r}))^T$ be its estimated p-dimensional abundance vector. Suppose that $\hat{\alpha}_j(\mathbf{r})$ is the estimated abundance fraction of $\mathbf{m}_j$ contained in $\mathbf{r}$ that is produced by an MPC classifier. We then compare all estimated abundance fractions $\{\hat{\alpha}_1(\mathbf{r}), \hat{\alpha}_2(\mathbf{r}), \cdots, \hat{\alpha}_p(\mathbf{r})\}$ and find the one with the maximum abundance fraction that will be assigned to $\mathbf{r}$. Using such a WTA thresholding criterion (9.1) becomes

$$\chi_{\text{WTAMPCV}}(\mathbf{r}) = \mathbf{M}\mathbf{u}_{j^*} = \mathbf{m}_{j^*} \tag{9.2}$$

with $j^* = \arg\{\max_{1 \le j \le p}\{\hat{\alpha}_j(\mathbf{r})\}\}$. In this case, $\hat{\alpha}_{j^*}(\mathbf{r}) = 1$ and $\hat{\alpha}_j(\mathbf{r}) = 0$ for $j \ne j^*$. The MPCV defined by (9.2) is called WTAMPCV. As a result of the WTA criterion, the mixed abundance vector $\hat{\alpha}$ is then converted to a pure abundance vector, the j^*-th p-dimensional unit vector $\mathbf{u}_{j^*} = (0, \cdots, 0, \underset{j^*}{1}, 0, \cdots, 0)^T$ with $j^* = \arg\{\max_{1 \le j \le p}\{\hat{\alpha}_j(\mathbf{r})\}\}$.

Another simple MPCV is to use the abundance percentage as a cut-off threshold value. If the estimated abundance fraction $\hat{\alpha}_p$ of $\mathbf{m}_j$ goes beyond a certain percentage within an image pixel vector $\mathbf{r}$, the $\mathbf{r}$ will be assigned to $\mathbf{m}_j$. More precisely, assume that a% is the abundance percentage used as a desired cut-off threshold value. Like (9.2), we can use a% as a thresholding value to define an a% MPCV via (9.1) by

$$\chi_{a\%\text{MPCV}}(\mathbf{r}) = \mathbf{M}\mathbf{u}_j = \mathbf{m}_j \text{ if } \hat{\alpha}_j(\mathbf{r}) \ge a \times 10^{-2}. \tag{9.3}$$

Using (9.3), the mixed abundance vector $\hat{\alpha}$ is converted to a pure abundance vector, which is a p-dimensional unit vector $\mathbf{u}_j$. As a consequence, a mixed pixel resulting from a%MPCV may be classified into multiple target classes if two or more estimated abundance fractions exceed the threshold value a%. This is not the case for WTAMPCV, which assigns one and only one p-dimensional unit vector $\mathbf{u}_j$ to $\mathbf{r}$. Nevertheless, a%MPCV has an advantage over WTAMPCV in the sense that WTAMPCV requires the complete target knowledge of an image scene. So, WTAMPCV is suitable for linear unmixing methods such as the OSP-based classifiers. On the contrary, a%MPCV does not need full knowledge of target information. Therefore, it is good for detectors and classifiers such as the ones discussed in Chapter 4 where a%MPCV was used to produce Tables 4.1-4.4. However, there is also a disadvantage of a%MPCV, which is the selection of an appropriate percentage threshold value to make a%MPCV effective.

9.2.2 Minimum Distance-Based Classification

In Section 9.2.1, we described WTAMPCV that directly converted abundance estimation of a mixed pixel to classification of a pure pixel. In the following two subsections, we use (9.1) as a vehicle to reinterpret two commonly used pure pixel classification methods, minimum distance-based classification and Fisher's linear discriminant analysis, in the context of constrained MPC.

As noted in (9.1), there is no noise term present in the equation. This is because the noise is already included as part of a misclassification error. As shown in Chapter 8, when the noise in model (3.1) was assume to be white Gaussian, the mixed pixel classifiers, OSP and *a posteriori* OSP, became unconstrained Gaussian maximum likelihood (GML) classifier, namely,

$$\hat{\alpha}_{\text{GML}} = \arg\left\{\max_{\hat{\alpha}\in\Delta} p(\mathbf{r})\right\}. \tag{9.4}$$

In this case, it should be noted that the Δ in (9.4) is confined to the set $\Delta = \left\{\mathbf{u}_j\right\}_{j=1}^{p} = \left\{(1,0,\cdots,0)^T,(0,1,\cdots,0)^T,\cdots,(0,0,\cdots,1)^T\right\}$ that contains only p unit vectors, i.e., $\Delta = \left\{\hat{\alpha} = (\hat{\alpha}_1,\cdots,\hat{\alpha}_p)^T \middle|\ \hat{\alpha}_j = 1 \text{ for some } j,\ \hat{\alpha}_i = 0 \text{ for } i \neq j\right\}$. In other words, the estimated abundance vector $\hat{\alpha}$ in (9.4) must be a p-dimensional unit vector. . Since there are p components, only p options are in Δ for $\hat{\alpha}_{\text{GML}}$. If we further assume that the sample covariance matrix used in $\hat{\alpha}_{\text{GML}}$ is constant for all sample pixel vectors (i.e., the sample covariance matrix is independent of sample vectors), the classifier using (9.4) is then simplified to a classifier that is based on the minimum distance of r from class means $\left\{\mathbf{m}_j\right\}_{j=1}^{p}$. As a result, an image pixel vector $\mathbf{r}$ is assigned to $\mathbf{m}_{j^*}$ according to

$$\chi_{\text{GML}}(\mathbf{r}) = \mathbf{M}\mathbf{u}_{j^*} = \mathbf{m}_{j^*}. \tag{9.5}$$

where $j^* = \arg\left\{\min_{1\leq j\leq p} \|\mathbf{r} - \mathbf{m}_j\|\right\}$.

More generally, assume that $\mathbf{x} = (x_1,\cdots,x_L)^T$ is an image pixel vector to be classified in a hyperspectral image. Let $\{\omega_1,\omega_2,\cdots,\omega_p\}$ be a set of classes of interest and ω_j be the class represented by the j-th signature $\mathbf{m}_j = (m_{j1},\cdots,m_{jL})^T$. In our case, classes $\{\omega_j\}_{j=1}^{p}$ are target classes of interest and $\mathbf{m}_j$ is the j-th target signature that belongs to class ω_j. Assume that $\mathbf{x}_i^j$ is the i-th sample vector in class ω_j and $\Xi = \left\{\mathbf{x}_i^j\right\}_{j=1,i=1}^{p,N_j}$ is the set of sample vectors needed to be classified where N_j is the number of sample vectors in class ω_j and $N = N_1 + \cdots + N_p$ is the total number of sample vectors. Two types of distance-based classifiers can be considered depending upon the used sample statistics.

1) The first-order statistics classifier:

Minimum distance classifier:

(a) Euclidean distance

$$\text{ED}(\mathbf{x},\mathbf{m}_j) = \sqrt{(\mathbf{x}-\mathbf{m}_j)^T(\mathbf{x}-\mathbf{m}_j)} = \sqrt{\sum_{l=1}^{L}\left(x_j - m_{jl}\right)^2} \tag{9.6}$$

Since the quadratic term in $\mathbf{x}$ of (9.6) is independent of class ω_j, the Euclidean distance-based minimum distance classifier is a linear classifier.

(b) City block distance:

$$\text{CBD}(\mathbf{x},\mathbf{m}_j) = \Sigma_{l=1}^{L}\left|x_l - m_{jl}\right| \tag{9.7}$$

(c) Tchebyshev (maximum) distance (TD):

$$\text{TD}(\mathbf{x},\mathbf{m}_j) = \max_{1\le l\le L}\left|x_l - m_{jl}\right| \tag{9.8}$$

2) Second-order statistics classifiers:

(a) Mahalanobis distance classifier (Fukunaga, 1991)

$$\text{MD}(\mathbf{x},\mathbf{m}_j) = (\mathbf{x}-\mathbf{m}_j)^T\left(\Sigma_j\right)^{-1}(\mathbf{x}-\mathbf{m}_j) \tag{9.9}$$

In general, the Mahalanobis distance classifier is a quadratic classifier. However, when $\Sigma_j = \Sigma_0$ is the same for all classes, then the Mahalanobis classifier is reduced to the Euclidean distance classifier.

(b) Bhattacharyya distance classifier (Fukunaga, 1991)

$$\begin{aligned} \text{B}_{ij} &= \frac{1}{8}(\mathbf{m}_i - \mathbf{m}_j)^T\left(\left(\Sigma_i + \Sigma_j\right)/2\right)^{-1}(\mathbf{m}_i - \mathbf{m}_j) \\ &+ \frac{1}{2}\ln\left(\frac{\left|\left(\Sigma_i + \Sigma_j\right)/2\right|}{\sqrt{|\Sigma_i||\Sigma_j|}}\right) \end{aligned} \tag{9.10}$$

In a special case that $\Sigma_i = \Sigma_j$ for classes ω_i and ω_j, then the Bhattacharyya distance classifier is reduced to the Mahalanobis distance classifier.

In case the covariance matrices Σ in (9.8) and (9.9) are not of full rank, their inverses will be replaced by their pseudo-inverses, $\Sigma^{\#} = \left(\Sigma^T\Sigma\right)^{-1}\Sigma^T$.

9.2.3 Fisher's Linear Discriminant Analysis (LDA)

Let R^d and $R^{\tilde{d}}$ be d-dimensional and $\tilde{d}$-dimensional vector spaces with $\tilde{d}$ usually less than d, $\tilde{d} < d$ where R^d and $R^{\tilde{d}}$ are considered to be the original data space and feature space respectively. Let $\{\omega_j\}_{j=1}^{c}$ denote c classes of interest and $\omega_j = \left\{\mathbf{x}_1^j, \mathbf{x}_2^j, \cdots, \mathbf{x}_{N_j}^j\right\} = \left\{\mathbf{x}_i^j\right\}_{i=1}^{N_j}$ be the j-th class containing N_j samples where

$\mathbf{x}_i^j = (x_{i1}^j, x_{i2}^j, \cdots, x_{id}^j)^T$ is a d-dimensional vector in the space R^d and the i-th sample in class ω_j. Let $N = N_1 + \cdots + N_c$ be the total number of training samples. From Fisher's discriminant analysis (Duda and Hart, 1973), we can form total, between-class and within-class scatter matrices as follows.

Let $\mu = (1/N)\Sigma_{j=1}^{c}\Sigma_{i=1}^{N_j}\mathbf{x}_i^j$ be the global mean and $\mu_j = (1/N_j)\Sigma_{i=1}^{N_j}\mathbf{x}_i^j$ the mean of class ω_j. Then

$$\mathbf{S}_\mathrm{T} = \Sigma_{j=1}^{c}\Sigma_{i=1}^{N_j}(\mathbf{x}_i^j - \mu)(\mathbf{x}_i^j - \mu)^T \tag{9.11}$$

$$\mathbf{S}_\mathrm{W} = \Sigma_{j=1}^{c}\Sigma_{i=1}^{N_j}(\mathbf{x}_i^j - \mu_j)(\mathbf{x}_i^j - \mu_j)^T \tag{9.12}$$

$$\mathbf{S}_\mathrm{B} = \Sigma_{j=1}^{c}N_j(\mu_j - \mu)(\mu_j - \mu)^T. \tag{9.13}$$

From (9.11)-(9.13), $\mathbf{S}_\mathrm{T} = \mathbf{S}_\mathrm{W} + \mathbf{S}_\mathrm{B}$.

Assume that $\Xi = \{\mathbf{x}_1, \mathbf{x}_2, \cdots, \mathbf{x}_N\} = \{\mathbf{x}_i^j\}_{i=1,j=1}^{N_j,c}$ is a set of data training samples. Fisher's discriminant analysis finds a $d \times \tilde{d}$ weight matrix $\mathbf{W} = [\mathbf{w}_1\,\mathbf{w}_2 \cdots \mathbf{w}_{\tilde{d}}]$ that projects the i-th sample $\mathbf{x}_i^j$ in class ω_j in R^d onto $\mathbf{y}_i^j$ which is in a low dimensional feature space $R^{\tilde{d}}$ in such a manner that the resulting projected data samples $\{\mathbf{y}_i^j\}_{i=1,j=1}^{N_j,c}$ yield the best possible class separability by the following matrix transformation

$$\mathbf{y}_i^j = \mathbf{W}^T\mathbf{x}_i^j, \tag{9.14}$$

where

$$y_{ik}^j = \mathbf{w}_k^T\mathbf{x}_i^j \quad \text{for } 1 \le k \le \tilde{d}. \tag{9.15}$$

and $\mathbf{w}_k$ is the k-th column vector with dimensionality $d \times 1$ in $\mathbf{W}$ and $\mathbf{y}_i^j = (y_{i1}^j, y_{i2}^j, \cdots, y_{i\tilde{d}}^j)^T$.

Using (9.11)-(9.14) we can define similar within-class and between-class scatter matrices for the projected samples $\mathbf{y}_i^j = (y_{i1}^j, y_{i2}^j, \cdots, y_{i\tilde{d}}^j)^T$ by

$$\tilde{\mathbf{S}}_\mathrm{W} = \Sigma_{j=1}^{c}\Sigma_{i=1}^{N_j}(\mathbf{y}_i^j - \mu_j)(\mathbf{y}_i^j - \mu_j)^T \tag{9.16}$$

$$\tilde{\mathbf{S}}_\mathrm{B} = \Sigma_{j=1}^{c}N_j(\tilde{\mu}_j - \tilde{\mu})(\tilde{\mu}_j - \tilde{\mu})^T. \tag{9.17}$$

where $\tilde{\mu} = (1/N)\Sigma_{j=1}^{c}\Sigma_{i=1}^{N_j}\mathbf{y}_i^j$ and $\tilde{\mu}_j = (1/N_j)\Sigma_{i=1}^{N_j}\mathbf{y}_i^j$.

Substituting (9.14) and (9-15) into (9.11)-(9.13) results in

$$\tilde{\mathbf{S}}_{\mathrm{W}} = \mathbf{W}^T \mathbf{S}_{\mathrm{W}} \mathbf{W} \tag{9.18}$$

$$\tilde{\mathbf{S}}_{\mathrm{B}} = \mathbf{W}^T \mathbf{S}_{\mathrm{B}} \mathbf{W}. \tag{9.19}$$

In order to find an optimal linear transformation matrix $\mathbf{W}$ of size $d \times \tilde{d}$ in terms of maximizing class separability, we use Fisher's discriminant function ratio, called Raleigh quotient which is the ratio of the between-class scatter matrix to the within-class scatter matrix given by

$$J(\mathbf{W}) = \frac{|\tilde{\mathbf{S}}_{\mathrm{B}}|}{|\tilde{\mathbf{S}}_{\mathrm{W}}|} = \frac{|\mathbf{W}^T \tilde{\mathbf{S}}_{\mathrm{B}} \mathbf{W}|}{|\mathbf{W}^T \tilde{\mathbf{S}}_{\mathrm{W}} \mathbf{W}|} \tag{9.20}$$

where $|\cdot|$ is the determinant of a matrix. The optimal solution to (9.20), denoted by

$$\mathbf{W}^* = \left[\mathbf{w}_1^* \, \mathbf{w}_2^* \cdots \mathbf{w}_{\tilde{d}}^*\right] \tag{9.21}$$

can be found by solving following generalized eigenvalue problem

$$\mathbf{S}_{\mathrm{B}} \mathbf{w}_k^* = \lambda_k \mathbf{S}_{\mathrm{W}} \mathbf{w}_k^*, \text{ for } 1 \le k \le \tilde{d}. \tag{9.22}$$

with $\mathbf{w}_k^*$ corresponding to the eigenvalue λ_k generated by (9.22). These $\tilde{d}$ eigenvectors, $\left\{\mathbf{w}_k^*\right\}_{k=1}^{\tilde{d}}$ obtained from (9.22) form a set of Fisher's linear discriminants to yield a new set of feature vectors

$$\mathbf{y}_i^j = \left(\mathbf{W}^*\right)^T \mathbf{x}_i^j \text{ for } 1 \le k \le \tilde{d}. \tag{9.23}$$

where $y_{ik}^j = \left(\mathbf{w}_k^*\right)^T \mathbf{x}_i^j$ is given by (9.15).

Since there are c classes, only $c-1$ eigenvalues, denoted by $\{\lambda_k\}_{k=1}^{c-1}$ are nonzeros where each eigenvalue λ_k produces its own eigenvector $\mathbf{w}_k^*$. By means of these eigenvectors $\left\{\mathbf{w}_k^*\right\}_{k=1}^{c-1}$ we can define a Fisher's discriminant analysis-based optimal linear transformation ϕ^* via (9.14), (9.20)-(9.23) by

$$\mathbf{y} = \phi^*(\mathbf{x}) = \left(\mathbf{W}^*\right)^T \mathbf{x} \tag{9.24}$$

where the size of $\mathbf{W}^*$ is $d \times (c-1)$ with the k-th column specified by the **k**-th eigenvector $\mathbf{w}_k^*$. The feature vector $\mathbf{y}_i^j$ derived by (9.14) is a feature representation of $\mathbf{x}_i^j$ that will be used for classification. It should be noted that there is no direct relationship between $\tilde{d}$, the dimensionality of the feature space $R^{\tilde{d}}$ and c, the number of classes.

However, it is often the case that $\tilde{d} < c$. For more details, we refer to Duda and Hart (1973).

According to the linear mixture model described by (3.1), there are p target signatures to be classified. Consequently, there are only p-1 nonzero eigenvalues. Assume that $\lambda_1 \geq \lambda_2 \geq \cdots \geq \lambda_{p-1} > 0$ are arranged in decreasing order of magnitude. Their corresponding eigenvectors $\{\mathbf{w}_j^*\}_{j=1}^{p-1}$ resulting from (9.22) are called Fisher's discriminant vectors. For instance, $\mathbf{w}_1^*$ corresponding to λ_1 is the first Fisher's discriminant vector, $\mathbf{w}_2^*$ corresponding to λ_2 is the second Fisher's discriminant vector, etc. Using these p-1 Fisher's discriminant vectors $\{\mathbf{w}_j^*\}_{j=1}^{p-1}$, we can construct an eigen-matrix $\mathbf{W}^*$ given by $\mathbf{W}^* = \left[\mathbf{w}_1^* \mathbf{w}_2^* \cdots \mathbf{w}_{p-1}^*\right]$ to project a pixel vector $\mathbf{x}$ onto its feature vector $\mathbf{y} = \mathbf{Wx}$ in a new space $\mathbf{Z}$ that is linearly spanned by $\{\mathbf{w}_j^*\}_{j=1}^{p-1}$. The Fisher LDA classification is then carried out in the space $\mathbf{Z}$ using a classification measure such as the minimum distance measures given by (9.5)-(9.9).

9.2.4 Unsupervised Classification

Although the distance-based classifiers described above are supervised and require a set of training samples, they can be extended to unsupervised classifiers by including a clustering process such as the nearest neighboring rule (NNR) described in Section 5.2 or a neural network-based self organization algorithm (Haykin, 1994). For example, ISODATA is has been widely used in pattern classification (Duda and Hart, 1973), which implements the minimum distance classifier in conjunction with an NNR clustering process.

9.3 CRITERIA FOR TARGET DETECTION AND CLASSIFICATION

In this section, we develop a set of custom-designed criteria for target detection and classification. Using the 15-panel HYDICE scene in Fig. 1.7(a) as an example, a typical masked target of size 4×4 is shown in Fig. 9.1 where an R pixel (i.e., a red pixel in Fig. 1.7(b)) is considered to be a target center pixel and a Y pixel (i.e., a yellow pixel in Fig. 1.7(b)) surrounding R pixels is a target pixel that may be either a target boundary pixel or a target pixel mixed with background pixels. With this ground truth map given in Fig. 1.7(b) we can actually tally panels pixels that are detected and classified by a specific algorithm.

Y	Y	Y	Y
Y	R	R	Y
Y	R	R	Y
Y	Y	Y	Y

Figure 9.1. A typical masked target

Here, we make a subtle distinction between a target *detected* and a target *hit*. When a target is detected, it means that at least one R target pixel is detected. When a target is hit, it means that at least either one R or one Y pixel is detected. As long as one of these R or Y pixels is detected, we declare that the target is hit. So, by way of this definition, a detected target always implies a hit target, but not vice versa.

The criteria that we develop in this chapter are (1) how many target R pixels are detected; (2) how many target Y pixels are detected; (3) how many pixels are detected as false alarms for a target in which case neither an R pixel or a Y pixel is detected; (4) how many target R pixels are missed. For example, suppose that the shaded pixels in Fig. 1 are those detected by a detection algorithm. We declare that the target is detected with one R pixel as well as hit with one R and two Y pixels. There are no false alarm pixels, but have three R pixels that are missed by detection. In order to quantitatively study target detection performance, the following definitions are introduced. Let

N = total number of sample pixel vectors
$\mathbf{t}$ = specific target to be detected
$N_{R+Y}(t)$ = total number of R plus Y pixels specified by $\mathbf{t}$
$N_R(\mathbf{t})$ = total number of R pixels specified by $\mathbf{t}$
$N_Y(\mathbf{t})$ = total number of Y pixels specified by $\mathbf{t}$
$N_{(R+Y)D}(\mathbf{t})$ = total number of either R or Y pixels detected as $\mathbf{t}$
$N_{RD}(\mathbf{t})$ = total number of R pixels detected as $\mathbf{t}$
$N_{YD}(\mathbf{t})$ = total number of Y pixels detected as $\mathbf{t}$
$N_{FA}(\mathbf{t})$ = total number of false alarms pixels, i.e., total number of pixels, which are neither R nor Y pixels, detected as $\mathbf{t}$
$N_M(\mathbf{t}) = N_{R+Y}(\mathbf{t}) - N_{(R+Y)D}(\mathbf{t})$ = total number of R and Y pixels that are specified by $\mathbf{t}$ missed

Using the above notations, we can further define the R-detection rate $R_{RD}(\mathbf{t})$ for target $\mathbf{t}$ by

$$R_{RD}(\mathbf{t}) = N_{RD}(\mathbf{t})/N_R(\mathbf{t}) \tag{9.25}$$

and the Y-detection rate $R_{YD}(\mathbf{t})$ for target $\mathbf{t}$ by

$$R_{YD}(\mathbf{t}) = N_{YD}(\mathbf{t})/N_Y(\mathbf{t}). \tag{9.26}$$

Since R pixels represent target center pixels and Y pixels are target boundary pixels or pixels mixed with background pixels, a good detection algorithm must have a higher target R-detection rate, $R_{RD}(\mathbf{t})$. On the other hand, detecting a Y pixel does not necessarily imply a target detected. Nevertheless, in this case, target $\mathbf{t}$ can be declared to be hit. For this purpose, we define the target hit rate $R_H(\mathbf{t})$ for target $\mathbf{t}$ by

$$R_H(\mathbf{t}) = N_{(R+Y)D}(\mathbf{t})/N_{R+Y}(\mathbf{t}). \tag{9.27}$$

So, from (9.27) a higher target hit rate $R_H(\mathbf{t})$ may not mean a higher target R-detection rate $R_{RD}(\mathbf{t})$, or vice versa. This is because the number of Y pixels are generally much greater than the number of R pixels. In most cases, Y pixels may even determine the performance of $R_H(\mathbf{t})$. As will be shown in the experiments, a detection algorithm may detect all R pixels but miss all Y pixels. In this case, this algorithm achieves 100% target R-detection rate, $R_{RD}(\mathbf{t}) = 1$ but target Y-detection rate, $R_{YD}(\mathbf{t}) = 0$. As a result, its target hit rate $R_H(\mathbf{t})$ is very small because $R_{YD}(\mathbf{t}) = 0$. On the other hand, if the target hit rate $R_H(\mathbf{t}) = 1$, it implies that all the R and Y pixels are detected. In this case, even a target is hit, we may still have difficulty with locating where the target is. So, the target R-detection rate, $R_{RD}(\mathbf{t})$ is more important than $R_H(\mathbf{t})$ since it provides the information about the exact location of the target **t**.

In addition to (9.25)-(9.27), we are also interested in target false alarm rate $R_{FA}(\mathbf{t})$ and target miss rate $R_M(\mathbf{t})$. They can be defined as follows.

$$R_{FA}(\mathbf{t}) = N_{FA}(\mathbf{t})/(N-N_{R+Y}(\mathbf{t})) \tag{9.28}$$

$$R_M(\mathbf{t}) = 1- R_H(\mathbf{t}) = N_M(\mathbf{t})/N_{R+Y}(\mathbf{t}) = (N_{R+Y}(\mathbf{t}) - N_{(R+Y)D}(\mathbf{t}))/N_{R+Y}(\mathbf{t}) \tag{9.29}$$

If there are p targets, $\mathbf{T} - \{\mathbf{t}_k\}_{k=1}^{p}$ to be classified, the overall detection rate $R_{OD}(\mathbf{T})$ for a class of targets, **T** can be defined as

$$R_{OD}(\mathbf{T}) = \sum_{k=1}^{p} p(\mathbf{t}_k)R_{RD}(\mathbf{t}_k) \tag{9.30}$$

where

$$p(\mathbf{t}_k) = N(\mathbf{t}_k) / \sum_{k=1}^{p} N(\mathbf{t}_k) \text{ for } 1 \le i \le p. \tag{9.31}$$

As will be seen in the following experiments, a higher $R_{OD}(\mathbf{T})$ does not imply higher classification accuracy because it may happen that several targets are detected in one single image due to their similar signature spectra, but may be difficult to discriminate one from another. As a result, it may yield poor classification. In order to account for this phenomenon we define the classification rate for a specific target $\mathbf{t}_k$, $R_C(\mathbf{t}_k)$ as

$$R_C(\mathbf{t}_k) = N_{RD}(\mathbf{t}_k) / \left(N_R(\mathbf{t}_k) + N_{FA}(\mathbf{t}_k)\right) \tag{9.32}$$

and the overall classification rate as

$$R_{OC}(\mathbf{T}) = \sum_{k=1}^{p} p(\mathbf{t}_k)R_C(\mathbf{t}_k) \tag{9.33}$$

where $p(\mathbf{t}_k)$ and $R_C(\mathbf{t}_k)$ are defined by (9.31) and (9.32) respectively. Now using (9.32)-(9.40) as criteria, we can evaluate detection and classification performance of various algorithms.

9.4 COMPARATIVE PERFORMANCE ANALYSIS

This section presents a thorough quantitative and comparative analysis among MPC and PPC algorithms described in this chapter. In order to validate such a study, the 15-panel HYDICE image scene in Fig. 1.7(a) is used for experiments. The spatial positions of these 15 target panel pixels are precisely located at pixel resolution by the ground truth map in Fig. 1.7(b) where the panel pixels are designated as either center pixels or masking pixels. A center pixel is generally considered as a pure pixel, while a masking pixel is referred to as a pixel, which may be a boundary pixel or a panel pixel mixed with background signatures. Using this data scene along with the designed criteria given in Section 9.3 a comparative analysis on classification accuracy becomes possible. The significance of this chapter is to offer a rigorous and objective means to evaluate the performance of each candidate algorithm on the same common ground. The target signature matrix **M** used for the 15 panels is formed by the five spectral signatures, P1, P2, P3, P4 and P5 in Fig. 1.8.

Example 9.1 (MPC comparative analysis)

In this example, an experiment-based quantitative study is conducted among three MPC classifiers, OSP, SSP and OBSP. In Chapter 7, it has shown that the GML classifier was identical to the OBSP classifier (see (7.26) and (7.31)). In this case, only the OBSP classifier was included for experiments. Fig. 9.2(a-c) shows the classification results of the 15 panels produced by OSP, SSP and OBSP respectively.

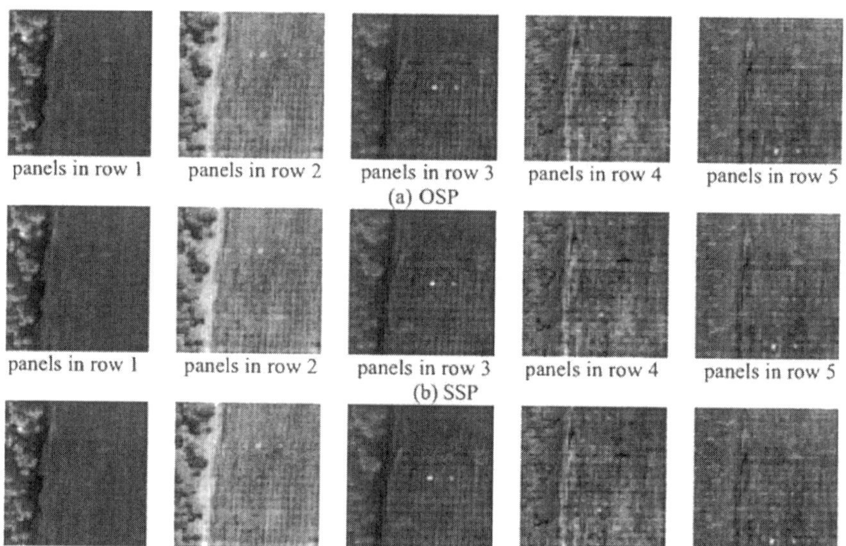

panels in row 1 panels in row 2 panels in row 3 panels in row 4 panels in row 5
(a) OSP
panels in row 1 panels in row 2 panels in row 3 panels in row 4 panels in row 5
(b) SSP
panels in row 1 panels in row 2 panels in row 3 panels in row 4 panels in row 5
(c) OBSP

Figure 9.2. Classification results of OSP, SSP and OBSP

The above experimental results may lead to a conclusion that the three classifiers are essentially the same classifier. Unfortunately, this is not case as shown in the following example.

Example 9.2

In this example, 250 mixed pixels were simulated according to Table 9.1 using the five panel signatures in Fig. 1.8.

Table 9.1. Abundance fraction assignment used to create 250 simulated pixels

Pixel #	P1	P2	P3	P4	P5	Tree	Grass
1 – 10	10 %	0 %	0 %	0 %	0 %	45 %	45 %
11 – 20	20 %	0 %	0 %	0 %	0 %	40 %	40 %
21 – 30	40 %	0 %	0 %	0 %	0 %	30 %	30 %
31 – 40	60 %	0 %	0 %	0 %	0 %	20 %	20 %
41 – 50	80 %	0 %	0 %	0 %	0 %	10 %	10 %
51 – 60	0 %	10 %	0 %	0 %	0 %	45 %	45 %
61 – 70	0 %	20 %	0 %	0 %	0 %	40 %	40 %
71 – 80	0 %	40 %	0 %	0 %	0 %	30 %	30 %
81 – 90	0 %	60 %	0 %	0 %	0 %	20 %	20 %
91 – 100	0 %	80 %	0 %	0 %	0 %	10 %	10 %
101 – 110	0 %	0 %	10 %	0 %	0 %	45 %	45 %
111 – 120	0 %	0 %	20 %	0 %	0 %	40 %	40 %
121 – 30	0 %	0 %	40 %	0 %	0 %	30 %	30 %
131 – 40	0 %	0 %	60 %	0 %	0 %	20 %	20 %
141 – 50	0 %	0 %	80 %	0 %	0 %	10 %	10 %
151 – 60	0 %	0 %	0 %	10 %	0 %	45 %	45 %
161 – 70	0 %	0 %	0 %	20 %	0 %	40 %	40 %
171 – 180	0 %	0 %	0 %	40 %	0 %	30 %	30 %
181 – 190	0 %	0 %	0 %	60 %	0 %	20 %	20 %
191 – 200	0 %	0 %	0 %	80 %	0 %	10 %	10 %
201 – 210	0 %	0 %	0 %	0 %	10 %	45 %	45 %
211 – 220	0 %	0 %	0 %	0 %	20 %	40 %	40 %
221 – 230	0 %	0 %	0 %	0 %	40 %	30 %	30 %
231 – 240	0 %	0 %	0 %	0 %	60 %	20 %	20 %
241 – 250	0 %	0 %	0 %	0 %	80 %	10 %	10 %

As an example, the first 50 pixels were simulated by the panel signature P1 with the pixel numbers 1-10 containing 10% P1 signature, pixel numbers 11-20 containing 20% P1 signature, pixel numbers 21-30 containing 40% P1 signature, 31-40 containing 60% P1 signature and pixel numbers 41-50 containing 80% P1 signature. In order to make the abundance fractions sum to one, we added two background signatures, tree and grass extracted from the scene in Fig. 1.7(a), which evenly split the remaining abundance. For instance, the first 10 pixels contained 10% P1 signature plus 45% tree and 45% grass, while the last 10 pixels (i.e., pixel numbers 241-250) contained 80% P5 signature, 10% tree and 10% grass. Additionally, a Gaussian noise was also added to achieve a 30:1 signal-to-noise ratio as defined in Harsanyi and Chang (1994). The resulting 250 simulated pixels are shown in Fig. 9.3.

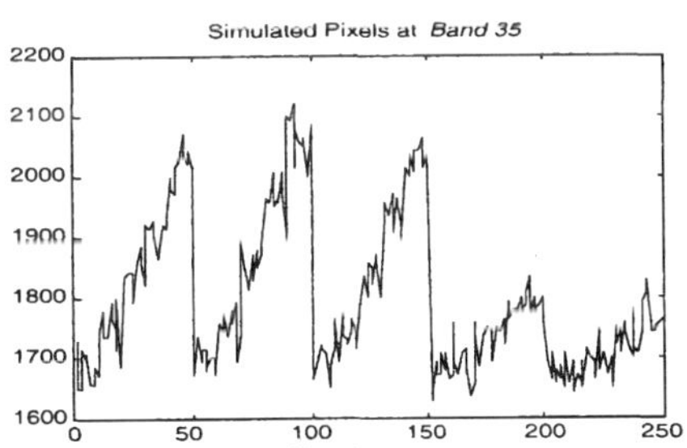

Figure 9.3. 250 simulated pixels

The classification results of OSP, SSP and OBSP classifiers are shown in the first, third and fourth columns of Fig. 9.4.

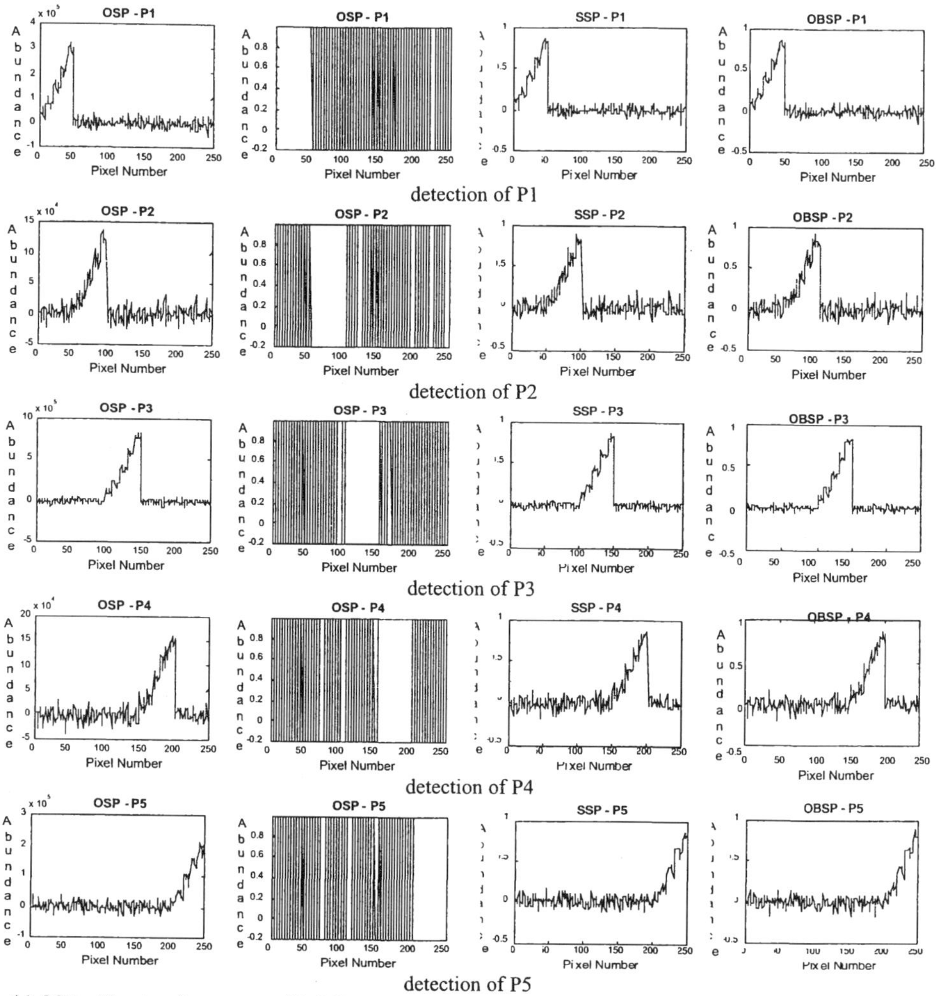

(a) OSP without scaling (b) OSP with scaling (c) SSP (d) OBSP

Figure 9.4. Classification results of OSP, SSP and OBSP

As we can see from Fig. 9.4, OSP performed as well as did SSP and OBSP, where a different scale was used to plot the OSP results. However, if we scaled the detected abundance fractions of the results in Fig. 9.4 to the same range [-0.2,1] that was used for SSP and OBSP in Fig. 9.4, OSP performed very poorly as shown in the second column of Fig. 9.4.

Fig. 9.4 demonstrates the significant difference between *a priori* classifier, viz., OSP classifier and *a posteriori* OSP classifier such as SSP and OBSP. As pointed out in Chapter 8, OSP does not include the scale constant $\kappa = \left(\mathbf{d}^T P_{\mathbf{U}}^{\perp} \mathbf{d}\right)^{-1}$ that appears in the front of SSP and OBSP. This constant may not alter the detection and classification results as demonstrated in Figs. 9.2 and 9.4, but it does have significant impact on the estimated target abundance fractions since the constant is generally determined by the accuracy of abundance fraction estimation (Chang, 1998a).

Example 9.3

In the previous two examples, comparisons were made among OSP-based classifiers. The abundance fractional images shown in Fig. 9.2(a-c) are gray scale and can be only interpreted by visual inspection. In order to avoid human intervention, these abundance fractional images must be thresholded by an automatic method. WTAMPCV and a%MPCV described by (9.2) and (9.3) can be used for this purpose. Using the criteria defined in Section 9.3 we tallied results by applying WTAMPCV and a%MPCV to the abundance fractional images in Fig. 9.2(a-c). Tables 9.2(a-b) tabulate detection and classification rates resulting from WTAMPCV for all the three classifiers, OSP, SSP and OBSP respectively. It should be noted that SSP and OBSP produced the same results in Table 9.2(b) despite that they differ by implementing a signature subspace projector P_M in SSP given by (8.12), which is absent in (8.27).

Table 9.2(a). Results of OSP resulting from WTAMPCV

	N_R	N_Y	N_{R+Y}	N_{RD}	N_{YD}	$N_{(R+Y)D}$	N_{FA}	N_M	R_{RD}	R_{YD}	R_H	R_{FA}	R_M
P1	3	47	50	3	27	30	1213	20	1.00	0.58	0.60	0.30	0.40
P2	4	41	45	4	8	12	1010	33	1.00	0.20	0.27	0.25	0.73
P3	4	37	41	4	32	36	1042	5	1.00	0.87	0.88	0.26	0.12
P4	4	40	44	4	10	14	348	30	1.00	0.25	0.32	0.09	0.68
P5	4	39	43	4	3	7	384	36	1.00	0.08	0.16	0.10	0.84
	19	204	223	19	80	99	3997	124	1.00	0.39	0.44	0.20	0.56

Table 9.2(b). Results of SSP and OBSP resulting from WTAMPCV

	N_R	N_Y	N_{R+Y}	N_{RD}	N_{YD}	$N_{(R+Y)D}$	N_{FA}	N_M	R_{RD}	R_{YD}	R_H	R_{FA}	R_M
P1	3	47	50	3	19	22	1136	28	1.00	0.40	0.44	0.28	0.56
P2	4	41	45	4	11	15	1210	30	1.00	0.27	0.33	0.30	0.67
P3	4	37	41	4	16	20	351	21	1.00	0.43	0.49	0.09	0.51
P4	4	40	44	4	18	22	778	22	1.00	0.45	0.50	0.20	0.50
P5	4	39	43	3	7	10	532	33	0.75	0.18	0.23	0.14	0.77
	19	204	223	18	71	89	4007	134	0.95	0.35	0.40	0.20	0.60

As a comparison, we also applied a%MPCV to the abundance fractional images in Fig. 9.2(a-c). Tables 9.3-9.4 are produced by 50%MPCV and 25%MPCV respectively.

Table 9.3. Results of OSP, SSP and OBSP resulting from 50%MPCV

	N_R	N_Y	N_{R+Y}	N_{RD}	N_{YD}	$N_{(R+Y)D}$	N_{FA}	N_M	R_{RD}	R_{YD}	R_H	R_{FA}	R_M
P1	3	47	50	0	0	0	76	50	0.00	0.00	0.00	0.02	1.00
P2	4	41	45	3	2	5	352	40	0.75	0.05	0.11	0.09	0.89
P3	4	37	41	3	2	5	0	36	0.75	0.05	0.12	0.00	0.88
P4	4	40	44	3	3	6	74	38	0.75	0.08	0.14	0.02	0.86
P5	4	39	43	3	2	5	2	38	0.75	0.05	0.12	0.00	0.88
	19	204	223	12	9	21	504	202	0.63	0.05	0.10	0.03	0.90

Table 9.4. Results of OSP, SSP and OBSP resulting from 25%MPCV

	N_R	N_Y	N_{R+Y}	N_{RD}	N_{YD}	$N_{(R+Y)D}$	N_{FA}	N_M	R_{RD}	R_{YD}	R_H	R_{FA}	R_M
P1	3	47	50	0	0	0	267	50	0.00	0.00	0.00	0.07	1.00
P2	4	41	45	4	6	10	799	35	1.00	0.15	0.22	0.20	0.78
P3	4	37	41	4	4	8	145	33	1.00	0.11	0.20	0.04	0.80
P4	4	40	44	4	19	23	552	21	1.00	0.48	0.52	0.14	0.48
P5	4	39	43	3	4	7	273	36	0.75	0.10	0.16	0.07	0.84
	19	204	223	15	33	48	2036	175	0.79	0.18	0.23	0.11	0.77

Interestingly, unlike WTAMPCV, all the three classifiers, OSP, SSP and OBP produced the same results for each case. These experiments imply that classification results can be altered by a different MPCV.

Since the $a\%$ used in $a\%$MPCV can go from 100% down to 0%, we can actually compute an ROC curve for $a\%$MPCV. However, in order for an ROC curve to be meaningful, the sample pool must be large to produce reliable statistics. In the 15-panel HYDICE image scene, there are only a total of 19 target R center pixels shown in the scene. It does not make lots of sense to plot an ROC only based on these 19 R pixels. In this case, tallying the detection results as we did Tables 9.2-9.4 is not an effective means to show detection performance. As an alternative, we replace the target detection probability by the target hit probability defined by (9.27) where there is a sufficient number of Y pixels (a total of 204 in the image scene) to generate a reasonable ROC curve. As a result, each $a\%$ generates a target hit probability and a false alarm probability. Then a 3-dimensional (3-D) ROC curve can be plotted based the generated target hit probability versus its false alarm probability via $a\%$MPCV where the x, y axes are specified by false alarm probability and abundance percentage % with the target hit probability represented by z-axis. Because our experiments showed that three classifiers, OSP, SSP and OBSP generated the same ROC curve, Fig. 9.5(a) only shows the 3-D curve of the 15-panel HYDICE image scene produced by SSP using $a\%$MPCV. Fig. 9.5(b) is its corresponding 2-D ROC curve plotted by the target hit probability versus the false alarm probability. As shown in Fig. 9.5, the results were not good compared to the results obtained in Fig. 11.5 for other classifiers, which will be discussed in Chapter 11.

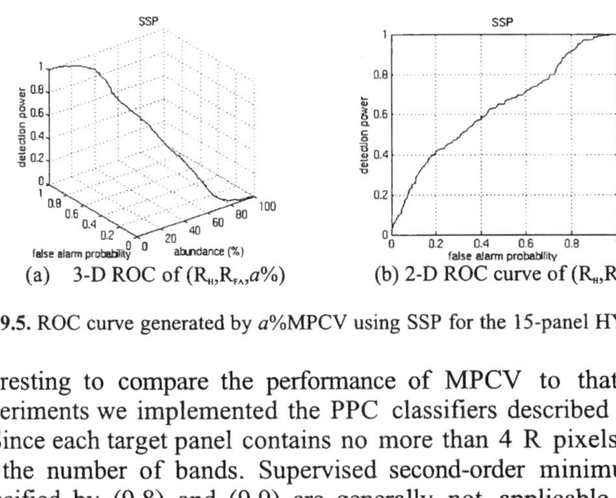

(a) 3-D ROC of $(R_H, R_{FA}, a\%)$ (b) 2-D ROC curve of (R_H, R_{FA})

Figure 9.5. ROC curve generated by $a\%$MPCV using SSP for the 15-panel HYDICE scene

It is interesting to compare the performance of MPCV to that of PPC. In the following experiments we implemented the PPC classifiers described in Section 9.3 for comparison. Since each target panel contains no more than 4 R pixels whose number is far less than the number of bands. Supervised second-order minimum distance-based classifiers specified by (9.8) and (9.9) are generally not applicable because the used covariance matrices will be ill-ranked. Under this circumstance, the number of target R pixels was used as the rank to calculate the covariance matrix. In order to further simplify experiments, only ED and MD were used for comparison because they are representatives of the first-order and second-order minimum distance-based classifiers. Tables 9.5-9.8 tally the detection and classification results produced by ED, LDAED (LDA using ED), MD and LDAMD (LDA using MD) respectively. From these four tables, LDAED performed the best among the four distance-based pure pixel classifiers. The reason why MD-based classifiers did not perform well was that there were no sufficient target samples of which they could take advantage for sample spectral correlation. In our experiments, the lack of target samples caused a difficulty of calculating the sample covariance matrix since it was not of full rank. One way to alleviate this problem was suggested in Ren (1998) where a new set of extra samples could be generated by nonlinear functions such as auto-correlation, cross-correlation from the original target pixels.

Table 9.5. Results of ED for the 15-panel HYDICE scene in Fig. 1.7(b)

	N_R	N_Y	N_{R+Y}	N_{RD}	N_{YD}	$N_{(R+Y)D}$	N_{FA}	N_M	R_{RD}	R_{YD}	R_H	R_{FA}	R_M
P1	3	47	50	2	2	4	120	46	0.67	0.04	0.08	0.03	0.92
P2	4	41	45	3	11	14	376	31	0.75	0.27	0.31	0.09	0.69
P3	4	37	41	3	2	5	0	36	0.75	0.05	0.12	0.00	0.88
P4	4	40	44	0	2	2	67	42	0.00	0.05	0.05	0.02	0.95
P5	4	39	43	3	1	4	65	39	0.75	0.03	0.09	0.02	0.91
	19	204	223	11	18	29	628	194	0.58	0.09	0.13	0.03	0.87

Table 9.6. Results of LDAED for the 15-panel HYDICE scene in Fig. 1.7(b)

	N_R	N_Y	N_{R+Y}	N_{RD}	N_{YD}	$N_{(R+Y)D}$	N_{FA}	N_M	R_{RD}	R_{YD}	R_H	R_{FA}	R_M
P1	3	47	50	3	8	11	1	39	1.00	0.17	0.22	0.00	0.78
P2	4	41	45	4	3	7	0	38	1.00	0.07	0.16	0.00	0.84
P3	4	37	41	4	4	8	0	33	1.00	0.11	0.20	0.00	0.80
P4	4	40	44	4	4	8	0	36	1.00	0.10	0.18	0.00	0.82
P5	4	39	43	4	5	9	0	34	1.00	0.13	0.21	0.00	0.79
	19	204	223	19	24	43	1	180	1.00	0.11	0.19	0.00	0.81

Table 9.7. Results of MD for the 15-panel HYDICE scene in Fig. 1.7(b)

	N_R	N_Y	N_{R+Y}	N_{RD}	N_{YD}	$N_{(R+Y)D}$	N_{FA}	N_M	R_{RD}	R_{YD}	R_H	R_{FA}	R_M
P1	3	47	50	0	0	0	10	50	0.00	0.00	0.00	0.00	1.00
P2	4	41	45	1	1	2	78	43	0.25	0.02	0.04	0.02	0.96
P3	4	37	41	3	10	13	464	28	0.75	0.27	0.32	0.11	0.68
P4	4	40	44	1	0	1	7	43	0.25	0.00	0.02	0.00	0.98
P5	4	39	43	2	1	3	2	40	0.50	0.03	0.07	0.00	0.93
	19	204	223	7	12	19	561	204	0.37	0.07	0.09	0.03	0.91

Table 9.9. Results of LDAMD for the 15-panel HYDICE scene in 1.7(b)

	N_R	N_Y	N_{R+Y}	N_{RD}	N_{YD}	$N_{(R+Y)D}$	N_{FA}	N_M	R_{RD}	R_{YD}	R_H	R_{FA}	R_M
P1	3	47	50	1	1	2	0	48	0.33	0.02	0.04	0.00	0.96
P2	4	41	45	3	2	5	0	40	0.75	0.05	0.11	0.00	0.89
P3	4	37	41	4	2	6	0	35	1.00	0.05	0.15	0.00	0.85
P4	4	40	44	2	2	4	0	40	0.50	0.05	0.09	0.00	0.91
P5	4	39	43	2	2	4	0	39	0.50	0.05	0.09	0.00	0.91
	19	204	223	12	9	21	0	202	0.63	0.05	0.10	0.00	0.90

If we further compare the results of LDAED to that of SSP, it found that LDAED performed better than SSP in terms of R-pixel detection but worse than SSP in Y-detection rate as well as target hit rate. However, both produced very close false alarm and target miss rates. More examples and experiments can be also found in Chang and Ren (2000).

9.5 CONCLUSIONS

Many hyperspectral target detection and image classification algorithms have been proposed in the literature. Comparing one to another is very challenging due to a lack of standardized data. Another difficulty also arises from the fact that there are no rigorous criteria to substantiate an algorithm. This chapter reinterprets MPC from a view point of

PPC by imposing constraints on target signature abundance fractions. As a result, the classes of target detection and classification algorithms can be reduced to three categories: OSP-based mixed pixel classifiers, minimum distance-based pure pixel classifiers and Fisher's LDA-based classifier. In addition, two MPCV are developed to convert an MPC problem into a PPC problem so that conventional PPC techniques could be readily applied. According to our experiments, MPCV may perform better than the minimum distance-based PPC. It unfortunately does not work as well as the class separability-based LDA due to the loss of gray level information about abundance fractions resulting from MPCV. The information, provided by the abundance-based gray scale images generated by MPC algorithms, contains very useful visual features, which can substantially improve image interpretation of classification results. The PPC algorithms cannot obtain such gray level information. Although we have made effort to make comparative analysis as comprehensive and rigorous as possible, completion is not claimed. In particular, the WTA-based converter used in this chapter for tallying target pixels is a simple thresholding technique and may not necessarily be optimal. There may exist an effective MPCV, which can produce better pure pixel classification performance. A thresholding technique using probability distribution in Chapter 16 and a Neyman-Pearson detection theory-based thresholding method in Chapter 17 are also promising approaches to MPCV. Finally, it should be noted that all the algorithms considered in this chapter are unconstrained in the sense that no constraints are imposed on signature abundances. Investigation of constrained mixed pixel classification problems will be discussed in Chapter 10.

IV

CONSTRAINED MIXED PIXEL CLASSIFICATION

The unconstrained mixed pixel classification (unconstrained MPC) discussed in PART III estimates abundance fractions of target signatures present in a pixel vector via a linear mixture model without imposing any constraint on the target signatures. The estimated abundance fractions are then used to classify an image pixel into one of the targets. Like subpixel detection, the performance of unconstrained MPC can be further improved by imposing constraints on target signatures or target abundance fractions. So, in PART IV two approaches to constrained MPC will be investigated, target abundance-constrained MPC (TACMPC) and target signature-constrained MPC (TSCMPC). Chapter 10 presents two least-squares based techniques, a fully constrained least-squares (FCLS) MPC and modified FCLS (MFCLS). Both extend the partially constrained least squares methods, SCLS and NCLS developed for subpixel detection in Chapter 3. One most significant and important benefit resulting from fully constrained MPC is mixed pixel quantification, a task that unconstrained MPC cannot achieve. The techniques in Chapter 10 provide solutions to target quantification at subpixel level. As mentioned in PART II, the vector direction may be as an important feature as abundance fractions to characterize a vector. In some cases such as target detection, target abundance fractions may not be of main concern. Instead, the directions of target signature vectors may provide more crucial information than target abundance fractions in image analysis. This was demonstrated in Chapter 4. So, in contrast to TACMPC, Chapters 11 and 12 consider TSCMPC, which constrains the vector directions of target signatures rather than abundance fractions. As noted in spectral angle mapper (SAM), two pixel vectors pointing to similar directions may result in similar spectral properties where the magnitude of a pixel is determined by its vector length, which represent a certain amount of abundance fraction contained in the pixel vector. Chapter 11 extends the LCMV subpixel detectors, CEM and TCIMF in Chapter 4 to various LCMV classifiers where the feasibility of real-time implementation is also discussed. Chapter 12 develops another new approach, linearly constrained discriminant analysis (LCDA) that is derived from Fisher's linear discriminant analysis considered in Chapter 9. It constrains the directions of targets of interest aligned with predetermined coordinates so that the targets to be classified can be separated in a way that we desire. The derived LCDA classifier turns out to be a constrained version of Harsanyi-Chang's OSP classifier in Chapter 8. Recently, a filter-vectors (FV) method was developed by Bowles et al. (Bowles et al., 1995) for mixed pixel classification. Surprisingly, Bowles et al.'s FV method can be shown to be a special version of LCMV and LCDA in Chapters 11-12. In particular, like OSP which can be considered as a whitened CEM, the FV method can be interpreted as a whitened background-removed LCMV (BRLCMV) classifier .

10

TARGET ABUNDANCE-CONSTRAINED MIXED PIXEL CLASSIFICATION (TACMPC)

The mixed pixel classification (MPC), which was considered in Chapters 8 and 9 is unconstrained with no constraint imposed on the target signature abundance fractions. Consequently, the resulting abundance estimates do not necessarily reflect their true amounts of abundance. In this case, these estimates can be only used for the purpose of target detection, discrimination and classification, but not for target quantification. In order for MPC to perform mixed pixel quantification, we need to consider a fully constrained mixed pixel classification problem, which imposes two constraints on the abundance fractions of target signatures. They were described in Chapter 3, (a) abundance sum-to-one constraint, referred to as ASC, $\sum_{j=1}^{p}\alpha_j = 1$ and (b) abundance nonnegativity constraint, $\alpha_j \geq 0$ for all $1 \leq j \leq p$, referred to as ANC. Since no closed form can be derived for a fully constrained mixed pixel classification problem, two least-squares approaches are presented in this chapter. One is referred to as fully constrained least-squares method (FCLS), which develops an efficient algorithm to yield least-squares optimal solutions. Another is referred to as modified fully constrained least-squares method (MFCLS), which applies the ASC, but replaces ANC with the absolute abundance sum-to-one constraint (AASC), i.e., $\sum_{j=1}^{p}|\alpha_j| = 1$, that allows one to derive an analytical form for optimal solutions. Both approaches arrive at nearly the same results. Additionally, both FCLS and MFCLS can be further extended to unsupervised methods by incorporating unsupervised algorithms presented in Chapter 5.

10.1 INTRODUCTION

Spectral mixture analysis (SMA) has been widely used in remote sensing for versatile applications such as target discrimination, detection and classification. A wealth of SMA techniques has been reported in the literature. For the purpose of mathematical tractability, most SMA-based techniques are developed based on a mixture model, which assumes to be linear and unconstrained, for example, OSP in Chapter 3 and *a posteriori* OSP, Gaussian maximum likelihood classifiers in Chapters 8 and 9. Unfortunately, such unconstrained methods generally do not take full advantage of a linear mixture model that can be implemented by ASC and ANC to improve solutions, particularly, to provide solutions to mixed pixel quantification problems. On the other hand, a linear constrained mixed pixel classification problem involves solving a set of inequality constraints, which usually does not have analytic solutions. In this case, we must rely on numerical algorithms to render optimal solutions. Over the past years some efforts were devoted to solving fully constrained linear mixing problems; however, they were designed mainly for a small number of target signatures. For instance, in Shimabukuro and Smith (1991) they considered several constrained least-squares mixing models and obtained constrained least-squares solutions by solving an overdetermined system that consisted of n equations with n unknowns and $n < m$ where m is the number of spectral channels and n is the number of target signatures. Since there are no closed-form solutions, one must exhaust all possible solutions in a feasible region bounded by ASC and ANC. The use of quadratic programming techniques to impose the ASC and ANC was previously investigated (Shimabukuro 1987, Boardman, 1990, Settle and Drake, 1993), but these methods are computationally expensive. Another method presented in Ashton and Schaum (1998) also suffered from excessive computational complexity as the number of target signatures increases.

In this chapter, we consider a fully constrained linear mixing problem for mixed pixel quantification in hyperspectral imagery. Because there are no analytical solutions, two least-squares-based approaches are developed. One is FCLS, which makes use of an efficient numerical algorithm to generate optimal solutions. The proposed FCLS method is based on a least-squares approach (Scharf, 1991) in conjunction with OSP in Chapter 3. A second method is MFCLS, which modifies the fully constrained linear mixing problem by replacing ANC, $\alpha_j \geq 0$ for all $1 \leq j \leq p$ with AASC, $\sum_{j=1}^{p}|\alpha_j| = 1$ while still implementing ASC, $\sum_{j=1}^{p}\alpha_j = 1$. The advantage of the proposed MFCLS method is to convert a set of inequality constraints to an equality constraint so that a closed-form solution can be found through Largrange multipliers. Both approaches take advantage of its SCLS solution derived in Chapter 3 and use it as a vehicle to bridge the gap between unconstrained and fully constrained solutions. As shown in Section 3.3, the SCLS solution can be obtained by an unconstrained least-squares solution plus an error correction term. The unconstrained least-squares solution in the SCLS solution is identical to the least-squares orthogonal subspace projection solution specified by (3.9) in Chapter 3. From here MFCLS solves the FCLS problem with an additional constraint, AASC using Largrange multipliers. Unlike MFCLS, FCLS implements the NCLS algorithm developed in Section 3.3 coupled with the SCLS solution. It is also a quadratic programming technique, but the algorithm used to simultaneously implementing both ASC and ANC is more computationally efficient. The significant saving in computational cost becomes more evident when the FCLS method is extended to an unsupervised FCLS method.

In order to further extend FCLS and MFCLS to unsupervised methods, the three unsupervised methods described in Chapter 5 can be used for this purpose. Of particular interest is the unsupervised FCLS method (UFCLS), which can be implemented by FCLS along with the unsupervised NCLS algorithm described in Section 5.4. One major advantage resulting from the developed UFCLS is its ability in identifying potential pure pixels in an unknown image scene. These pixels are then used to generate more accurate target signature information for supervised mixed pixel classification.

10.2 FULLY CONSTRAINED LEAST-SQUARES APPROACH

One simple approach to solving fully constrained linear mixing problems is to take advantage of partially constrained solutions. For an SCLS solution, we can simply throw out the target signatures with negative abundance fractions and normalize the abundance fractions of the remaining target signatures to unity. The resulting solution is called normalized SCLS (NSCLS) solution. For an NCLS solution we can normalize it to unity, which results in the normalized NCLS (NNCLS) solution. Unfortunately, as will be shown in the experiments, neither NSCLS nor NNCLS will yield optimal solutions since the two constraints, ANC and the ASC are carried out sequentially, not simultaneously. A method that simultaneously implemented ANC and ASC was recently proposed in Heinz et al. (1999). However, it still produced only a nearly optimal solution because it did not satisfy (3.24). In what follows, we will present an FCLS algorithm that will produce accurate abundance fractions of target signatures in a linear mixture of a mixed pixel. The proposed FCLS method is the same one considered in Haskell and Hanson (1981). It extended the nonnegative least-squares algorithm in Lawson and Hanson (1995) by including ASC.

10.2.1 Fully Constrained Least-Squares Method (FCLS)

In order to take care of ASC, we include ASC in the signature matrix $\mathbf{M}$ by introducing a new signature matrix, denoted by $\mathbf{N}$, which is defined by

$$\mathbf{N} = \begin{bmatrix} \delta\mathbf{M} \\ \mathbf{1}^T \end{bmatrix} \tag{10.1}$$

with $\mathbf{1} = (\underbrace{11\cdots 1}_{p})^T$, and a vector $\mathbf{s}$ by

$$\mathbf{s} = \begin{bmatrix} \delta\mathbf{r} \\ 1 \end{bmatrix}. \tag{10.2}$$

The utilization of δ in (10.1)-(10.2) controls the effect of ASC. Using these two equations an FCLS algorithm can be derived directly from the NCLS algorithm described in Section 5.4 by replacing $\mathbf{M}$ and $\mathbf{r}$ used in the NCLS algorithm with $\mathbf{N}$ and $\mathbf{s}$ respectively.

10.2.2 Unsupervised FCLS Method (UFCLS)

The FCLS requires a complete knowledge of the signature matrix $\mathbf{M}$. In order for it to apply to a situation where no *a priori* information is available, we need an unsupervised process to generate the required material information to apply FCLS. Since FCLS is a least-squares-based approach, the least-squares error (LSE)-based criterion described in Section 5.4 is used for the optimal criterion and the UNCLS algorithm is used to generate the desired target (material) information for the linear mixture model. This procedure is referred to as Unsupervised FCLS (UFCLS) algorithm, which can be summarized as follows.

Unsupervised FCLS (UFCLS) Algorithm

1) Initial condition:

   Select ε to be a prescribed error threshold and let $\mathbf{m}_0 = \arg\left\{\max_{\mathbf{r}}\left[\mathbf{r}^T\mathbf{r}\right]\right\}$ where $\mathbf{r}$ is run over all image pixel vectors. Let $k = 0$.

2) Find $\mathbf{m}_1$ that yields the largest $\text{LSE}^{(0)}(\mathbf{r}) = \left(\mathbf{r} - \mathbf{m}_0\right)^T\left(\mathbf{r} - \mathbf{m}_0\right)$, i.e., $\mathbf{m}_1 = \arg\left\{\max_{\mathbf{r}}\text{LSE}^{(0)}(\mathbf{r})\right\}$.

3) Let $k \leftarrow k+1$ and apply the FCLS algorithm with the signature matrix $\mathbf{M} = \left[\mathbf{m}_0\,\mathbf{m}_1\cdots\mathbf{m}_k\right]$, $\mathbf{N} = \begin{bmatrix}\delta\mathbf{M}\\ \mathbf{1}^T\end{bmatrix}$ and $\mathbf{s} = \begin{bmatrix}\delta\mathbf{r}\\ 1\end{bmatrix}$ to estimate the abundance fractions of $\mathbf{m}_0$, $\mathbf{m}_1$, $\cdots$, $\mathbf{m}_k$, $\hat{\alpha}_0^{(k)}(\mathbf{r})$, $\hat{\alpha}_1^{(k)}(\mathbf{r})$, $\cdots$, $\hat{\alpha}_{k-1}^{(k)}(\mathbf{r})$.

4) Find the least-squares error defined by

$$\text{LSE}^{(k)}(\mathbf{r}) = \left(\mathbf{r} - \left[\Sigma_{i=0}^{k}\hat{\alpha}_i^{(k)}(\mathbf{r})\mathbf{m}_i\right]\right)^T\left(\mathbf{r} - \left[\Sigma_{i=0}^{k}\hat{\alpha}_i^{(k)}(\mathbf{r})\mathbf{m}_i\right]\right) \tag{10.3}$$

   and check the error if $\text{LSE}^{(k)}(\mathbf{r}) < \varepsilon$ for all $\mathbf{r}$. If it is, the algorithm stops, otherwise continue.

5. Find $\mathbf{m}_{k+1} = \arg\left\{\max_{\mathbf{r}}\text{LSE}^{(k)}(\mathbf{r})\right\}$. Go to step 3.

It should be noted that $\left\{\text{LSE}^{(k)}(\mathbf{r})\right\}_k$ is a monotonically decreasing sequence at k, thus, it converges. Similarly, such an LSE-based unsupervised approach can be also used to extend LS specified by (3.12), SCLS, NSCLS, NCLS and NNCLS to Unsupervised LS, Unsupervised SCLS, Unsupervised NSCLS, Unsupervised NCLS and Unsupervised NNCLS.

10.3 MODIFIED FULLY CONSTRAINED LEAST-SQUARES (MFCLS) APPROACH

As mentioned previously, the main difficulty with solving constrained linear mixing problems is the constraint of abundance nonnegativity that prohibits one from using the Lagrange multiplier method to find solutions analytically. In this section, we propose an alternative approach. Instead of directly dealing with the inequalities, $\alpha_j \geq 0$ for each $1 \leq j \leq p$, we replace them with the absolute abundance sum-to-one constraint (AASC), $\Sigma_{j=1}^{p}\left|\alpha_j\right| = 1$. The advantage of AASC is that the Lagrange multiplier method is now applicable and can be used to derive an iterative algorithm that leads to a desired optimal constrained least-squares solution. Additionally, AASC enables us to exclude negative abundance fractions from solutions. In other words, the only solution to satisfy both constraints, sum-to-one ($\Sigma_{j=1}^{p}\alpha_j = 1$) and AASC ($\Sigma_{j=1}^{p}\left|\alpha_j\right| = 1$) is that all the abundance fractions $\left\{\alpha_j\right\}_{j=1}^{p}$ must be nonnegative. So, a modified least-squares linear mixing problem with constraints, ASC, $\Sigma_{j=1}^{p}\alpha_j = 1$ and AASC, $\Sigma_{j=1}^{p}|\alpha_j| = 1$ can be cast as follows

$$\min_{\alpha \in \Delta_f}\left\{(\mathbf{r}-\mathbf{M}\alpha)(\mathbf{r}-\mathbf{M}\alpha)^T\right\} \tag{10.4}$$

subject to

$$\Delta = \left\{\alpha \middle| \Sigma_{j=1}^{p}\alpha_j = 1 \text{ and } \Sigma_{j=1}^{p}|\alpha_j| = 1\right\} \tag{10.5}$$

Using a similar argument that was used to derive PCLS solutions in Chapter 3, an object function can be obtained by

$$J(\alpha) = (1/2)(\mathbf{r}-\mathbf{M}\alpha)(\mathbf{r}-\mathbf{M}\alpha)^T - \lambda_1\left(\Sigma_{j=1}^{p}\alpha_j - 1\right) - \lambda_2\left(\Sigma_{j=1}^{p}|\alpha_j| - 1\right). \tag{10.6}$$

Differentiating (10.6) with respect to α and setting to zero yields

$$\begin{aligned} \partial J(\alpha)/\partial\alpha|_{\hat{\alpha}_{\text{MCLS}}} = 0 &\Rightarrow \hat{\alpha}_{\text{MCLS}} = \left(\mathbf{M}^T\mathbf{M}\right)^{-1}\left[\mathbf{M}^T\mathbf{r} - \lambda_1\mathbf{1} - \lambda_2 sign(\alpha)\right] \\ &\Rightarrow \hat{\alpha}_{\text{MCLS}} = \hat{\alpha}_{\text{LS}} - \left(\mathbf{M}^T\mathbf{M}\right)^{-1}\left[\lambda_1\mathbf{1} + \lambda_2 sign(\alpha)\right] \end{aligned} \tag{10.7}$$

where $\hat{\alpha}_{\text{LS}} = \left(\mathbf{M}^T\mathbf{M}\right)^{-1}\mathbf{M}^T\mathbf{r}$ is the unconstrained least-squares estimate of α given by (3.12). We then substitute $\hat{\alpha}_{\text{LS}}$ for α in (10.7) using the following constraints

$$\Sigma_{j=1}^{p}\alpha_j = \mathbf{1}^T\alpha = 1 \tag{10.8}$$

$$\Sigma_{j=1}^{p}|\alpha_j| = sign(\alpha)^T\alpha = 1 \tag{10.9}$$

to compute λ_1 and λ_2. The $sign(\alpha)$ in (10.9) is defined by the sign function of α with the j-th component being the sign of α_j. More precisely, $sign(\alpha) = \left(\beta_1\,\beta_2\cdots\beta_p\right)^T$ is defined by

$$\beta_j = \begin{cases} \alpha_j / |\alpha_j|; & \text{if } \alpha_j \neq 0 \\ 0; & \text{if } \alpha_j = 0 \end{cases}. \tag{10.10}$$

Iteratively computing λ_1, λ_2 and $\hat{\alpha}_{\text{MCLS}}$ using (10.7)-(10.9) can obtain a solution to (10.4)-(10.5). The detail of implementing this algorithm is given below.

MFCLS Algorithm

1. Initialization: Set $\hat{\alpha}_{\text{MCLS}} = \hat{\alpha}_{\text{SCLS}}$ using (3.17)

2. Compute λ_1 and λ_2 using (10.8)-(10.9)
3. Compute $\hat{\alpha}_{MCLS} = \hat{\alpha}_{SCLS} - \left(\mathbf{M}^T\mathbf{M}\right)^{-1}\left[\lambda_1\mathbf{1} + \lambda_2 sign(\alpha)\right]$
4. If there exists one component in $\hat{\alpha}_{MCLS}$, which is negative, go to step 2. Otherwise, stop.

It should be noted that step 1 of the MFCLS algorithm is initialized by taking advantage of the SCLS solution $\hat{\alpha}_{SCLS}$ obtained by (3.17). The stopping criterion described in step 4 provides a rule of when the algorithm will be terminated, in which case all components must be nonnegative. In general, it requires a fair amount of computing time to reach this requirement. So, in real implementation, we suggest a simpler rule by checking $\sum_{j=1}^{p}\left|\alpha_j\right| - 1 < \varepsilon$ for a preselected threshold ε that will guarantee a quick termination.

10.4 COMPUTER SIMULATIONS AND REAL HYPERSPECTRAL IMAGE EXPERIMENTS

This section contains a series of computer simulations and experiments to evaluate comparative performance of LS, SCLS, NSCLS, NCLS, NNCLS, FCLS and their corresponding unsupervised versions. First, we conducted computer simulations to demonstrate advantages of FCLS and UFCLS. Then real hyperspectral image data were used to show superior performance of FCLS and UFCLS in comparison to other methods. In implementing FCLS and UFCLS, the value of δ used in (17) and (18) was fixed at $\delta = 1.0 \times 10^{-5}$ except for HYDICE experiments where various δ values were explored for performance analysis.

10.4.1 Computer Simulations

In the following simulations, two experiments were designed to demonstrate the performance of the LS, SCLS, NSCLS, NCLS, NNCLS, FCLS and MFCLS methods, (1) when the information of all signatures is completely known and (2) when some false information is used.

Example 10.1 Signature Matrix with Three Distinct Target Signatures

The same set of 400 simulated pixels, $\{\mathbf{r}_i\}_{i=1}^{400}$ used in Example 7.1 was used for experiments. They were simulated as follows. We started the first pixel vector with 100% red soil and 0% dry grass, then began to increase 0.25% dry grass and decrease 0.25% red soil every pixel vector until the 400th pixel vector which contained 100% dry grass. We then added creosote leaves to pixel vector numbers 198-202 at abundance fractions 10% while reducing the abundance of red soil and dry grass accordingly. For example, after addition of creosote leaves, the resulting pixel vector 200 contained 10% creosote leaves, 45% red soil and 45% dry grass. White Gaussian noise was also added to each pixel vectors to achieve a 30:1 signal-to-noise ratio. Fig. 10.1 shows the results of the LS, SCLS, NSCLS, NCLS, NNCLS, FCLS and MFCLS methods in detection of creosote leaves where all six methods performed similarly.

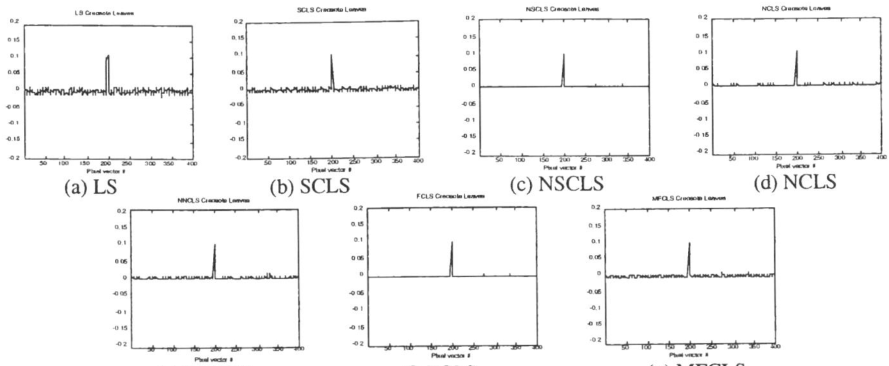

(a) LS (b) SCLS (c) NSCLS (d) NCLS

(e) NNCLS (f) FCLS (g) MFCLS

Figure 10.1. Classification results of LS, SCLS, NSCLS, NCLS, NNCLS, FCLS and MFCLS

In order to further compare their capabilities at quantifying creosote leaves, the squared errors between the actual and estimated abundance fractions of creosote leaves were averaged over 400 pixels. The resulting quantification errors for each method were 2.588×10^{-5} for LS, 7.850×10^{-6} for SCLS, 3.717×10^{-6} for NSCLS, 1.318×10^{-5} for NCLS, 1.303×10^{-5} for NNCLS, 3.715×10^{-6} for both FCLS and MFCLS. Obviously, FCLS produced the best quantification result in the sense of minimizing the average of squared errors. It is interesting to note that NSCLS, FCLS and MFCLS performed nearly the same. However, this will not be true in following simulation.

Example 10.2 Effects of Two Additional, Less Spectrally Distinct, Targets

The target signature matrix **M** was made up of the spectral signatures of all targets in the image scene. An image pixel may not be necessary to contain all these targets and may contain only a few of them. In order to demonstrate the effects of some targets used in **M** but absent in a pixel, the same simulated mixed pixel vectors used in Example 10.1. However, two additional signatures, blackbrush and sagebrush were added to the target signature matrix **M**. These signatures were not actually present in the pixel vectors, that is, the abundance fractions of blackbrush and sagebrush in these 400 simulated pixels are 0%. In this scenario, the signature matrix $\mathbf{M}=\lfloor \mathbf{m}_1\,\mathbf{m}_2\,\mathbf{m}_3\,\mathbf{m}_4\,\mathbf{m}_5 \rfloor$ was assume to consist of these five spectral signatures with abundance fractions given by $\alpha=(\alpha_1\,\alpha_2\,\alpha_3\,\alpha_4\,\alpha_5)^T$. Fig. 10.2 shows the results of LS, SCLS, NSCLS, NCLS, NNCLS, FCLS and MFCLS in detection of creosote leaves.

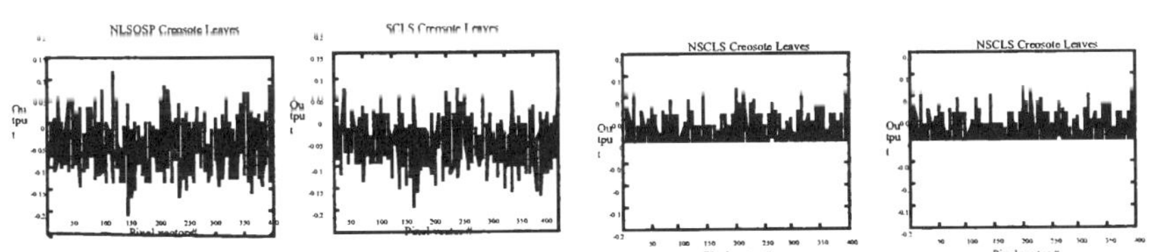

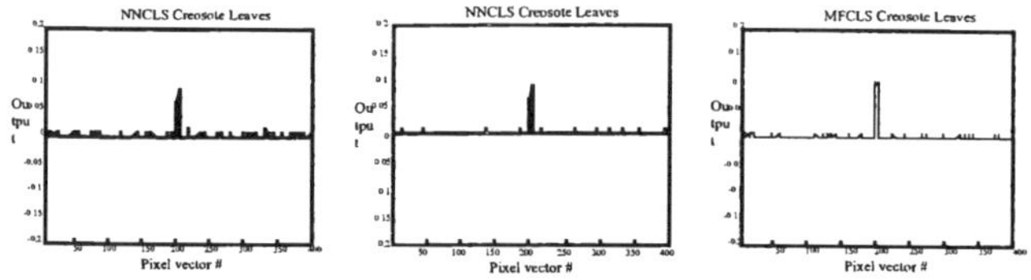

Figure 10.2. Classification results of LS, SCLS, NSCLS, NCLS, NNCLS, FCLS and MFCLS where **M** consisted of five endmembers: dry grass, red soil, creosote leaves, sagebrush and blackbrush

Unlike Fig. 10.1 the performance of LS, SCLS and NSCLS was very poor and their respective averaged squared quantification errors, 2.556×10^{-3} for LS, 1.945×10^{-3} for the SCLS, 7.960×10^{-4} for NSCLS were significantly worse than those produced by the other three methods, 4.823×10^{-5} for NCLS, 4.907×10^{-5} for NNCLS and 2.806×10^{-5} for FCLS. The detection performance of the LS was considerably decreased because the undesired signature annihilator $P_U^{\perp}$ used in LS nulled the undesired signatures, blackbrush and sagebrush whose spectra are similar to that of creosote leaves. The performance of SCLS and NSCLS was reduced because they assumed there were five signatures and their estimated abundance fractions must be summed to one. Since the spectrum of creosote leaves is close to those of blackbrush and sagebrush, the estimated abundance fraction of creosote leaves was forced to share with the inexistent blackbrush and sagebrush. As expected, SCLS and NSCLS would not perform well. By contrast, ANC significantly improved the performance of NCLS, NNCLS and FCLS. By implementing ANC these methods were able to effectively select an appropriate subset of material signatures for unmixing. This experiment demonstrated that NCLS, NNCLS and FCLS performed significantly better than LS, SCLS and NSCLS as the number and similarity of spectral signatures in the signature matrix **M** increases. In both examples, FCLS performed the best while LS was the worst.

10.4.2 AVIRIS Image Experiments

The data used in the following experiments are AVIRIS image in Fig. 1.6(a). For these experiments, representative pixels of each signature were manually extracted from the image scene and their average was used to represent the signature. For example, the signature for the playa was obtained by averaging 5033 pixels located at the bottom right corner of the image scene. The shade signature was generated by averaging pixels in a 5×5 square located in the darkest area of the scene. Each of the used signatures for cinders and rhyolite was produced by averaging four pixels in the scene. One pixel was extracted for vegetation signature. Three experiments were conducted using different degrees of prior information applied to signatures.

Example 10.3 Complete Prior Signature Information

With these five signatures manually selected as above, a signature matrix, **M** was formed. Fig. 10.3 shows the quantification results of LS, SCLS, NSCLS, NCLS, NNCLS, FCLS and MFCLS using cinders, playa, rhyolite, vegetation and shade as the desired signatures respectively.

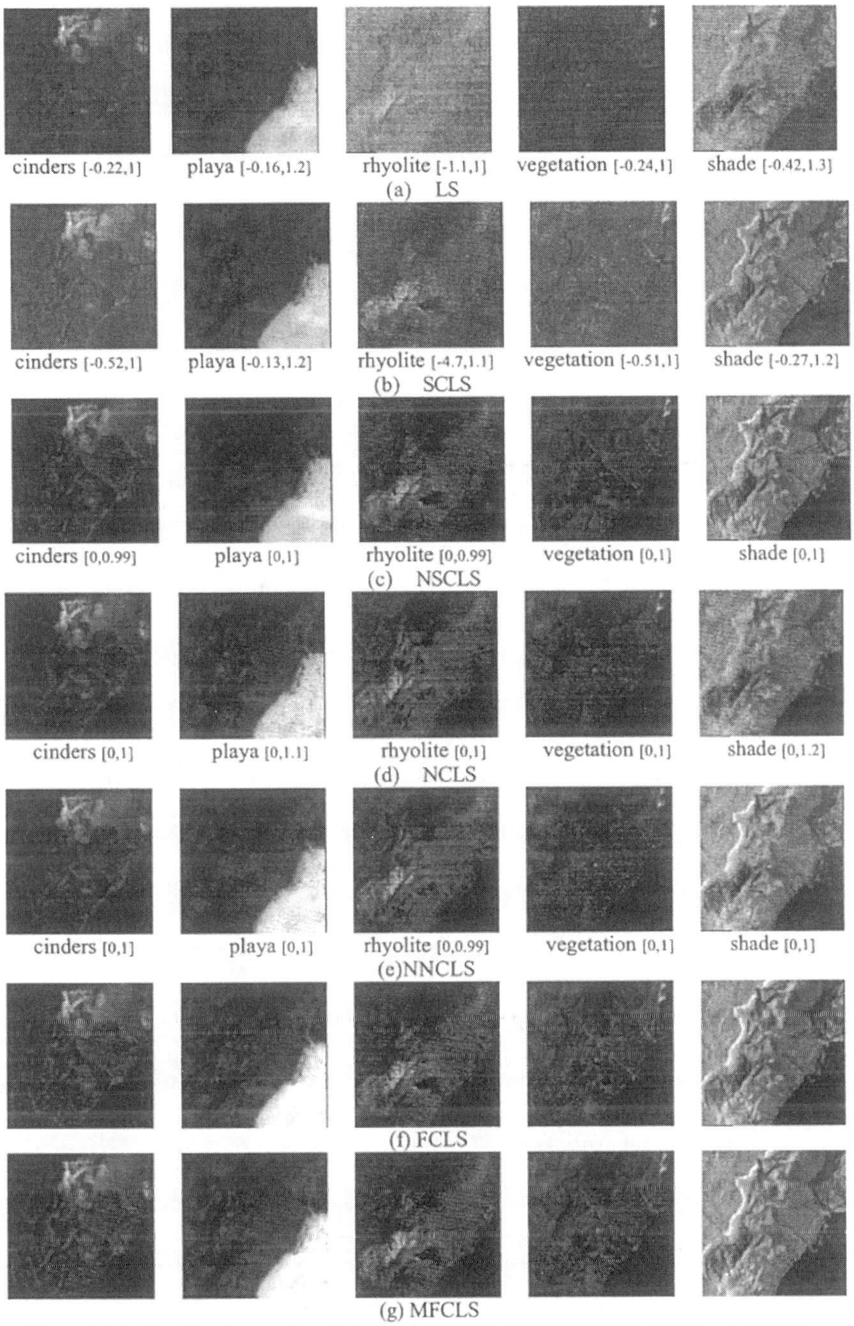

Figure 10.3. Classification results of LS, SCLS, NSCLS, NCLS, NNCLS, FCLS and MFCLS

The estimated abundance fraction values for each signature correspond directly to the gray scale values for each image in Fig. 10.3(a-f). Underneath each of the images generated by LS, SCLS, NSCLS, NCLS and NNCLS the ranges of abundance fraction values for their images are particularly specified since they are not really fully constrained methods. As we can see, from a detection point of view all the six methods performed similarly. However, in terms of quantification the performances of NSCLS, NCLS, NNCLS, FCLS and MFCLS are very close and better than LS and SCLS.

Example 10.4 Partial Prior Signature Information

In this experiment, only one of five signatures was assumed to be known *a priori*. In this case, the UFCLS algorithm was used to generate additional signatures to form a desired signature matrix **M** to quantify this particular signature. Fig. 10.4(a) shows the results of cinders after 16 iterations of the UFCLS algorithm. Similarly, Fig. 10.4(b)-(e) shows the results of playa, rhyolite, vegetation and shade after 3, 19, 25 and 5 iterations respectively.

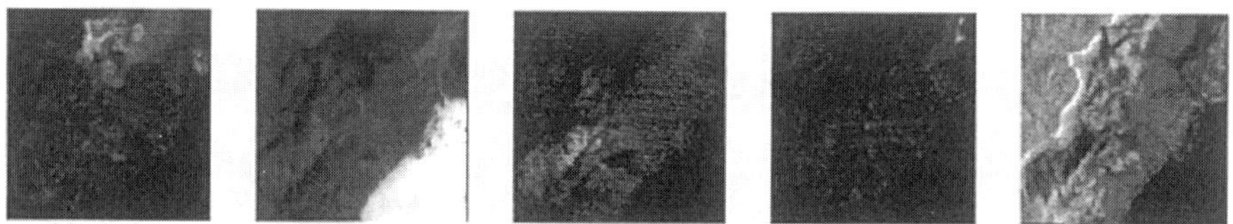

(a) cinders (b) playa (c) rhyolite (d) vegetation (e) shade

Figure 10.4. Results of UFCLS for (a) cinders, (b) playa, (c) rhyolite, (d) vegetation and (e) shade with partial knowledge, where for each image only signature of interest was known *a priori*.

Compared to Fig. 10.3, their results were very similar to those produced by NCLS, NNCLS, FCLS and MFCLS. This implies that the UFCLS using a single signature of interest performed as well as FCLS and MFCLS using all the five signatures.

Example 10.5 No Prior Signature Information

The following experiment is interesting and designed to explore the utility of UFCLS algorithm when no prior information is assumed. In order to initialize the algorithm, a prescribed error threshold ε and an initial material signature, $\mathbf{m}_0$ were required. To determine a value for ε we used the UFCLS algorithm and plotted the maximum least-squares error values resulting from (10.3), as shown in Fig. 10.5.

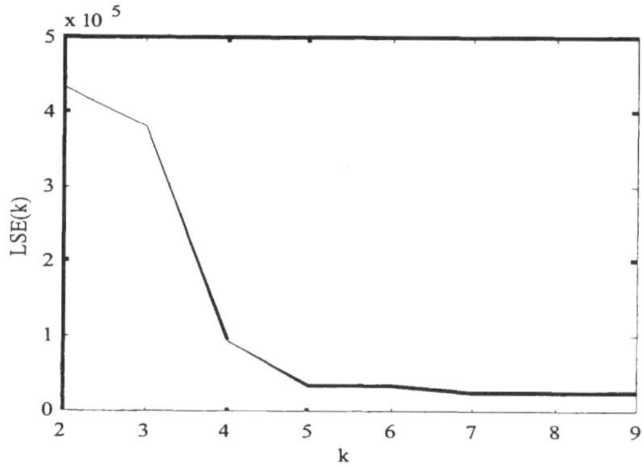

Figure 10.5. Plot of LSE values

Since the plot leveled off at the iteration $k = 5$, a threshold value of 5×10^4 was selected for ε. For the initial material signature $\mathbf{m}_0$, a pixel with the maximum vector length, i.e., a brightest pixel in the image scene was selected and turned out to be a pixel located in the center of the dry playa dry lake. Using the spectral signature of this pixel as the initial signature, $\mathbf{m}_0$, $\text{LSE}^{(0)}(\mathbf{r})$ was calculated for all pixels $\mathbf{r}$ in the image scene and the resulting image is shown in Fig. 10.6(a). The spectral signature of a shade image pixel that yielded the largest LSE in Fig. 10.6(a) was selected as the first signature $\mathbf{m}_1$. At this point, $\mathbf{m}_0$ and $\mathbf{m}_1$ represented two pixels that corresponded to the brightest and darkest pixels in the image scene. The UFCLS algorithm continued to calculate $\text{LSE}^{(1)}(\mathbf{r})$ as specified in step 4 and the result is shown in Fig. 10.6(b). A pixel containing cinders that produced the largest LSE in Fig. 10.6(b) was selected and its spectral signature was used as the second signature $\mathbf{m}_2$. Then following the same procedure the $\text{LSE}^{(2)}(\mathbf{r})$ image is shown in Fig. 10.6(c) where a pixel at the edge of the dry lakebed had the largest LSE and its spectral signature was selected as the third desired signature $\mathbf{m}_3$. Once again, using $\mathbf{m}_0$, $\mathbf{m}_1$, $\mathbf{m}_2$ and $\mathbf{m}_3$ $\text{LSE}^{(3)}(\mathbf{r})$ image was generated in Fig. 10.6(d) and the spectral signature of a vegetation pixel that generated the largest LSE was selected as the fourth desired signature $\mathbf{m}_4$. Interestingly, Figs. 10.6(c) and (d) are almost identical except for the pixel at the edge of the dry lakebed represented by $\mathbf{m}_3$. In this particular situation, UFCLS found an anomalous signature $\mathbf{m}_3$ that could be not be identified by visual inspection from the image scene. Accordingly, UFCLS can be used to detect anomalies. This task cannot be achieved by the supervised FCLS method. Then with $\mathbf{m}_0$, $\mathbf{m}_1$, $\mathbf{m}_2$, $\mathbf{m}_3$ and $\mathbf{m}_4$ $\text{LSE}^{(4)}(\mathbf{r})$ was generated in Fig. 10.6(e) where a rhyolite pixel produced the largest LSE and its spectral signature was designated as the fifth signature, $\mathbf{m}_5$. Since the largest LSE in the $\text{LSE}^{(5)}(\mathbf{r})$ image labeled by Fig. 10.6(f) was below the prescribed threshold $\varepsilon = 5 \times 10^4$, the UFCLS algorithm was terminated. At this point, the error image in Fig. 10.6(f) took on a chi-square distribution, except for a few pixels along the edge of the dry lakebed and a pixel in the upper left corner of the image. This chi-square distribution provided conformation that the six signatures $\{\mathbf{m}_j\}_{j=0}^{5}$ generated by the UFCLS algorithm represented the significant materials in the image scene.

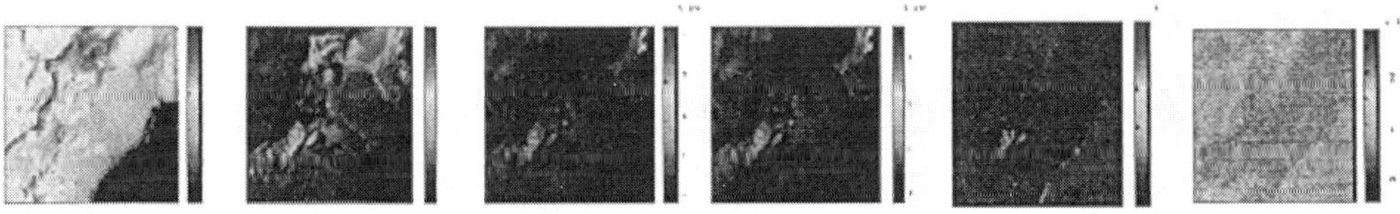

(a) (b) (c) (d) (e) (f)

Figure 10.6. Pixel detection of the UFCLS algorithm

The six signatures that were found in Fig. 10.6 were used to form a signature matrix, **M** to implement LS, SCLS, NSCLS, NCLS, NNCLS, FCLS and MFCLS. The results are shown in Fig. 10.7 where only the range of estimated abundance fractions for LS and SCLS were specified underneath the images they generated. This is because the abundance fractions estimated by all the other five classifiers were all in the range [0,1].

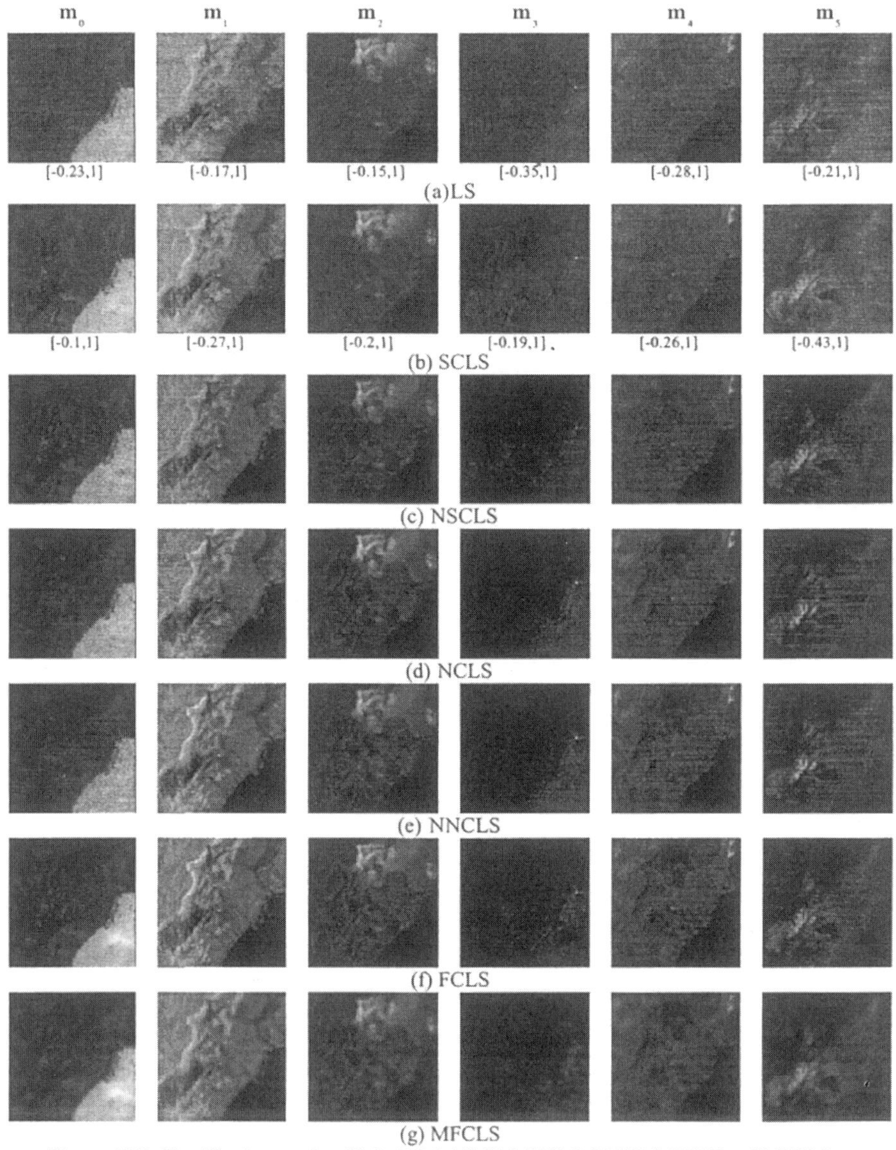

Figure 10.7. Classification results of LS, SCLS, NSCLS, NCLS, NNCLS, FCLS and MFCLS

Compared to the images in Fig. 10.3, the results in Fig. 10.7 were similar, but quantification results were worse, particularly for playa and shade. This is primarily due to the fact that only a single pixel was used to generate each of signatures in **M**. When a relatively large area such as dry lake and shade needs to be classified, a single pixel can not well represent the area to account for spectral variability. One way to resolve this dilemma is to add more sample pixels to generate a more robust signature. These additional sample pixels can be selected by using spectral measures such as, ED, SAM or SID discussed in Chapter 2.

10.4.3 HYDICE Image Experiments

The 15-panel HYDICE scene in Fig. 1.7(a) was used for experiments. Three examples were studied to demonstrate how different levels of signature knowledge affect the performance of constrained MPC.

Example 10.6 Complete Prior Signature Information

The five panel signatures, P1, P2, P3, P4 and P5 shown in Fig. 1.8 were used to form a desired signature matrix for **M**. In addition to the panel signatures, four more signatures were generated to represent the background. A tree signature was obtained by averaging 768 pixels in a large rectangle of pixels from the left side of the image. A grass signature was generated by averaging 1152 pixels from two large rectangles, one located between the trees and the panels, and the second at the right side of the image. Averaging 8 pixels from the gravel road and 7 pixels of shade located along the tree line produced road and shade signatures. Since most of the road was shaded, it is very difficult to see from Fig. 1.7(a). The spectra of these four background signatures are shown in Fig. 10.8.

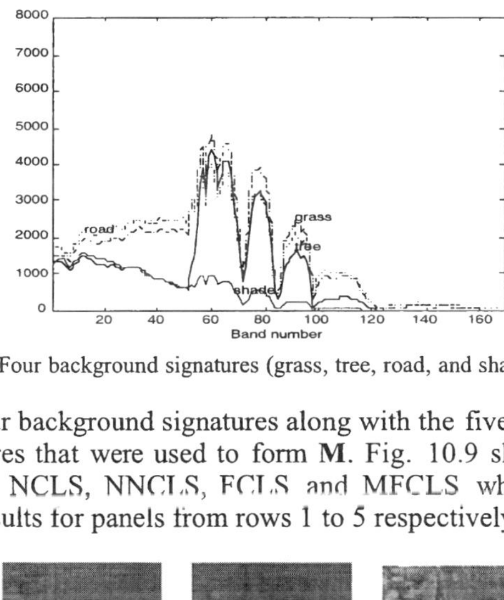

Figure 10.8. Four background signatures (grass, tree, road, and shade) in Fig. 1.7(a)

Including the four background signatures along with the five panel signatures gave a total of nine signatures that were used to form **M**. Fig. 10.9 shows the performance of LS, SCLS, NSCLS, NCLS, NNCLS, FCLS and MFCLS where figures from left to right are detection results for panels from rows 1 to 5 respectively.

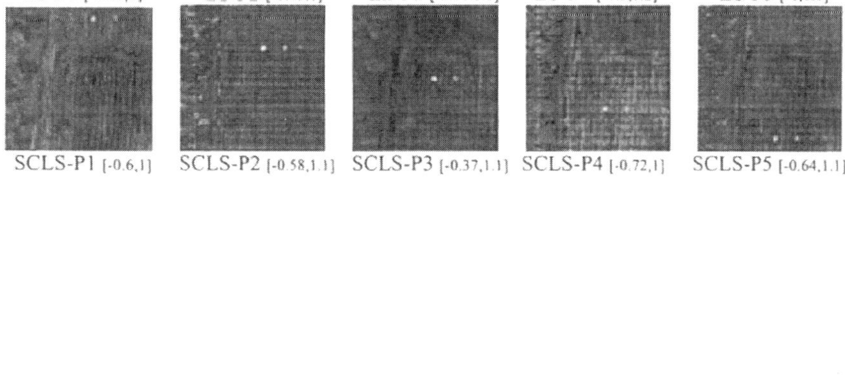

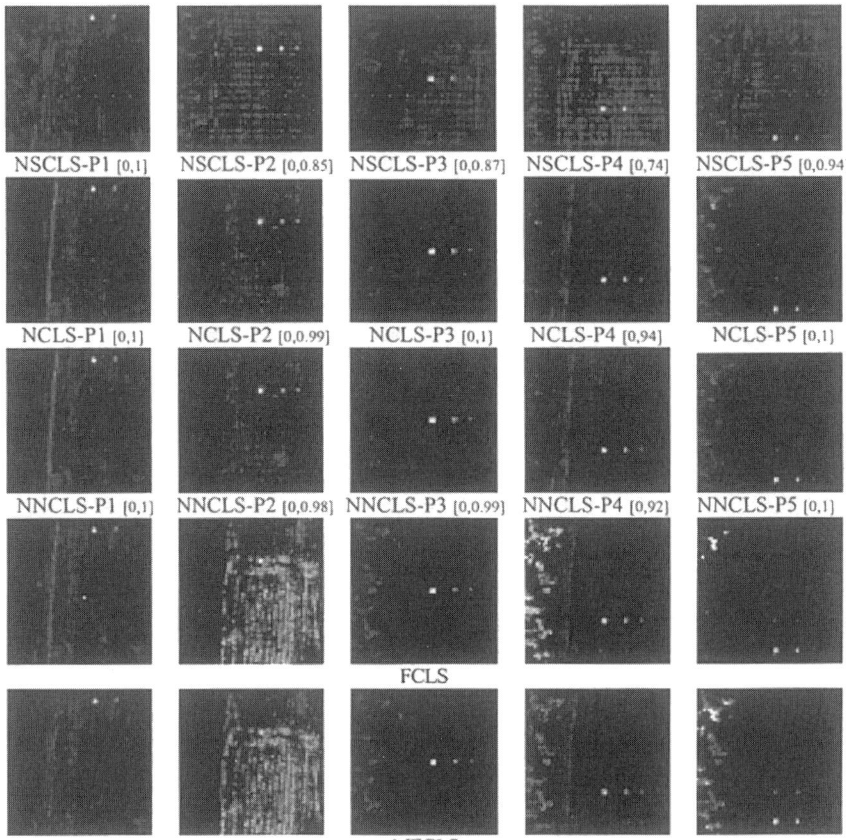

Figure 10.9. Classification results of LS, SCLS, NSCLS, NCLS, NNCLS, FCLS and MFCLS

Like Fig. 10.3, the range of abundance fractions estimated by LS, SCLS, NSCLS, NCLS, NNCLS are specified underneath of each image they generated. Apparently the best results were produced by NCLS. Interestingly, FCLS and MFCLS had the worst performance in detection of the panels in row 2, but performed reasonably well for other panels. This shows that fully constrained MPC does not necessarily perform better than unconstrained MPC or partially constrained MPC in target detection.

Until now we simply set the value of δ at $\delta = 1.0 \times 10^{-5}$ for use in FCLS and UFCLS. In order to demonstrate its effect on performance of FCLS, the same experiment was conducted with $\delta = 1.0 \times 10^{-4}$ and $\delta = 1.0 \times 10^{-2}$. These results are shown in Fig. 10.10(a-b) where the results of Fig. 10.10(b) were better than those in Fig. 10.10(a).

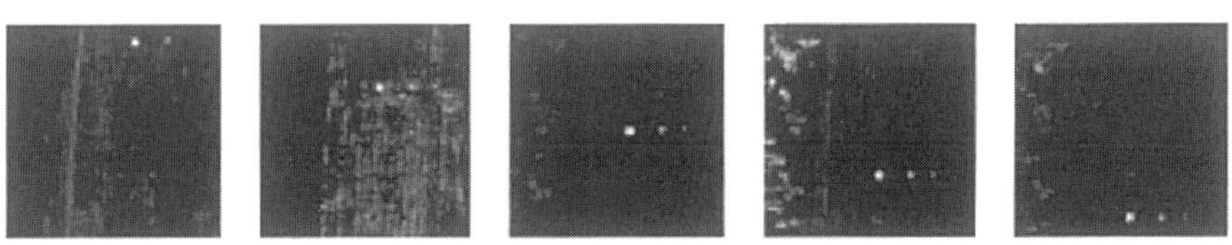

(a) $\delta = 1.0 \times 10^{-4}$

(b) $\delta = 1.0 \times 10^{-2}$

Figure 10.10. Results of $\delta = 1.0 \times 10^{-4}$ and $\delta = 1.0 \times 10^{-2}$

Comparison of these results in Fig. 10.10 to those of NCLS in Fig. 10.9 illustrates the increasingly similar performance as the value of δ was increased. This effect was expected, since δ controls the impact of ASC and a reduction of this impact should correspond to increased similarity between results of FCLS and NCLS.

An advantage of using panels for experiments is that ground truth provides their exact dimensions and can use this information to quantify the amount of each panel material present in the scene. Since each image pixel is approximately $1.5\,\text{m} \times 1.5\,\text{m}$, with area-per-pixel of 2.25m^2, we can determine the quantity of panel materials by dividing the area of the $3\text{m} \times 3\text{m}$, $2\text{m} \times 2\text{m}$ and $1\text{m} \times 1\text{m}$ panels by the area-per-pixel, which results in quantification values 4.0, 1.78 and 0.44 pixels respectively. Consequently, the image contains a quantity of approximately 6.22 pixels of each panel material. Table 10.1 contains the image quantification results for each of LS, SCLS, NSCLS, NCLS, NNCLS, FCLS and MFCLS. As we can see, each method performed poorly.

Table 10.1. Quantification of LS, SCLS, NSCLS, NCLS, NNCLS, FCLS and MFCLS

	LS	SCLS	NSCLS	NCLS	NNCLS	FCLS $\delta = 1 \times 10^{-5}$	FCLS $\delta = 1 \times 10^{-4}$	FCLS $\delta = 1 \times 10^{-2}$	MFCLS
P1	-73.97	-67.72	143.35	106.30	105.94	107.97	109.18	106.30	83.53
P2	60.05	53.66	152.40	67.29	64.31	608.29	355.33	67.33	623.83
P3	28.37	27.08	112.67	24.02	22.90	36.66	28.41	24.02	38.77
P4	137.45	129.06	215.44	44.34	45.48	169.11	123.00	44.34	120.09
P5	-101.32	-94.71	139.41	53.93	41.89	26.38	36.59	53.94	107.80

In order to determine how each of the methods performed in quantification of each of the three panel sizes, Table 10.2 contains quantification results for each of the methods using only the R and Y pixels of each panel.

Table 10.2. Quantification of results over only R and Y pixels for LS, SCLS, NSCLS, NCLS, NNSCLS and FCLS using manually selected background signatures

	LS	SCLS	NSCLS	NCLS	NNCLS	FCLS $\delta = 1 \times 10^{-5}$	FCLS $\delta = 1 \times 10^{-4}$	FCLS $\delta = 1 \times 10^{-2}$	MFCLS
P_{11}	5.99	4.26	3.70	4.08	4.08	4.11	4.11	4.08	4.04
P_{12}	3.01	2.09	1.83	1.70	1.69	1.71	1.70	1.70	1.70
P_{13}	0.50	-1.01	0.01	0.05	0.05	0.00	0.01	0.05	0.12
P_{21}	5.89	4.75	4.03	4.74	4.66	7.67	6.40	4.74	7.59
P_{22}	0.30	1.40	1.68	1.36	1.30	7.18	4.71	1.36	7.28
P_{23}	-0.02	0.96	0.78	0.73	0.71	2.49	1.75	0.73	2.36
P_{31}	4.39	4.61	3.62	4.59	4.53	3.87	4.14	4.59	3.80
P_{32}	3.86	3.80	2.40	1.76	1.71	1.01	1.25	1.76	0.97
P_{33}	1.28	1.44	0.95	0.58	0.56	0.34	0.39	0.58	0.29
P_{41}	4.53	5.89	4.03	3.16	3.05	3.49	3.62	3.16	3.37
P_{42}	8.66	5.91	3.78	1.46	1.44	1.50	1.62	1.46	1.19
P_{43}	3.48	3.59	2.05	0.44	0.43	0.46	0.43	0.44	0.59
P_{51}	-0.10	-0.07	3.10	4.12	3.94	3.67	3.61	4.12	4.11
P_{52}	0.25	0.75	1.08	1.82	1.81	1.70	1.82	1.82	1.50
P_{53}	-1.21	-0.88	0.13	0.41	0.40	0.26	0.35	0.41	0.19

For instance, the quantification results for p_{11} in row one of Table 10.2 were calculated using only the twenty-five pixels corresponding to p_{11}'s R and Y pixel mask. These results show that NCLS, NNCLS, FCLS ($\delta = 1.0 \times 10^{-2}$) and MFCLS performed better than other methods in terms of both detection and quantification. Since the large background field was not well represented by the four generated background signatures, all the methods performed poorly. As noted, the HYDICE sensor can extract objects less than 1 meter; thus, the sensor may also pick up many unknown targets. This is in shown in Fig. 10.9. Unfortunately, the knowledge of these unidentified materials cannot be obtained *a priori*. This is a situation where the UFCLS method finds its most useful applications.

Example 10.7 Partial Prior Signature Information

In this experiment, only the five panel signatures were assumed to be known *a priori*. In this case, the UFCLS algorithm was used to generate additional signatures to represent the background. The five panel signatures from Example 10.6 were selected as the initial material signature set to initialize the UFCLS algorithm. It is interesting to note that at iteration 11 the UFCLS algorithm selected the top R pixel for panel P5 as a new material signature. This indicated that one of the two R pixels was actually a mixed pixel. In order to determine which was the mixed pixel, UFCLS, with no prior signature knowledge was used. It selected a bottom R pixel; consequently, only the bottom R pixel was used to represent panel P5 in this experiment. As stated before, UFCLS requires a prescribed error threshold to determine how many targets will be included in the signature matrix **M**. In order to determine this value we used the UFCLS algorithm and plotted the maximum least squares error values resulting from (10.3). Fig. 10.11 shows the results for iterations of k from 10 to 49.

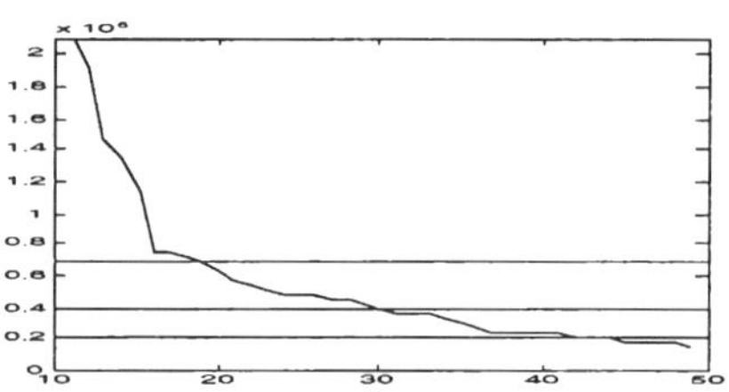

Figure 10.11. Results for iterations of k from 10 to 49

The plot starts to level off after about 18 iterations of the UFCLS algorithm. In order to see how the threshold value effects UFCLS, three threshold values (7.0×10^5, 4.0×10^5 and 2.0×10^5) were selected, each of which thresholded by the horizontal lines in Fig. 10.11. Fig. 10.12 shows the results of UFCLS using these three threshold values.

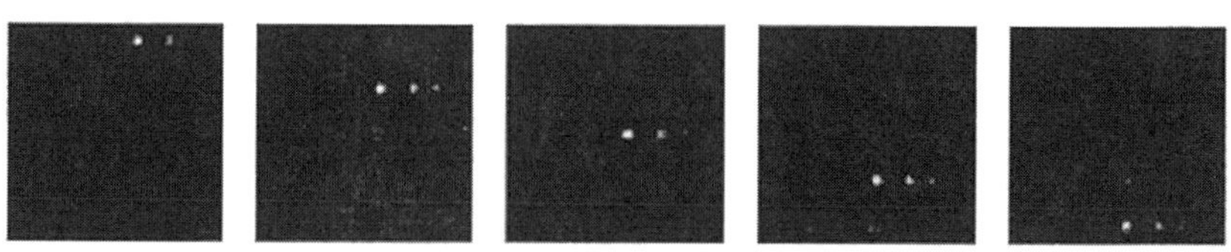

(a) $\varepsilon = 7.0 \times 10^5$

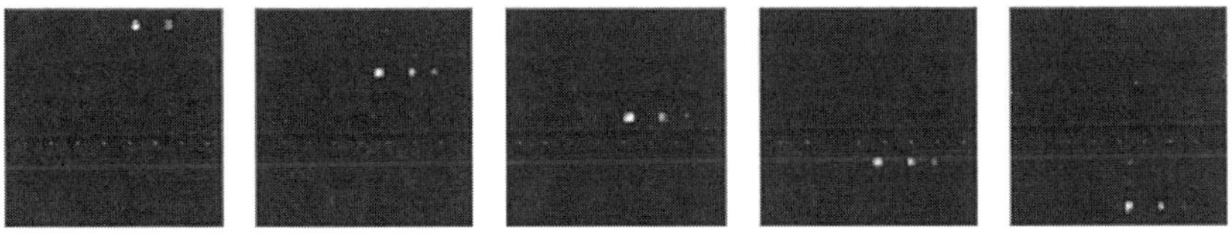

(b) $\varepsilon = 4.0 \times 10^5$

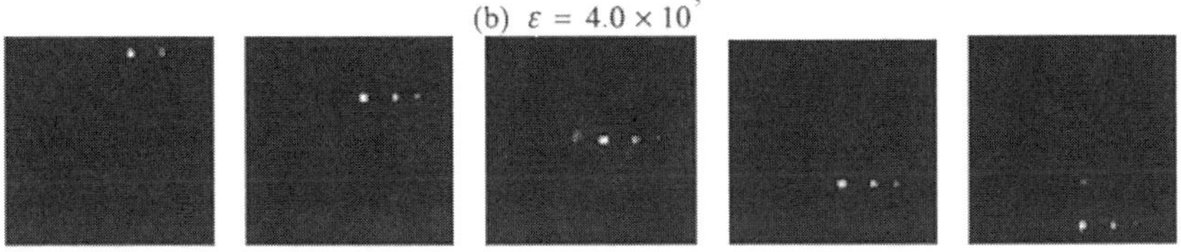

(c) $\varepsilon = 2.0 \times 10^5$

Figure 10.12. Results of UFCLS using these three threshold values

As we can see from Fig. 10.12(a), when the error threshold is too large it results in an insufficient number of signatures to represent the image background. Tables 10.3 and 10.4 contain quantification results for the entire image and for the R and Y pixels respectively.

Table 10.3. Quantification of quantification results over all image pixels using UFCLS method with partial prior signature knowledge for three different error threshold values.

	UFCLS $\varepsilon = 7.0 \times 10^5$	UFCLS $\varepsilon = 4.0 \times 10^5$	UFCLS $\varepsilon = 2.0 \times 10^5$
P1	19.7649	13.9114	7.3021
P2	109.4454	41.2200	19.9960
P3	20.3169	14.8010	23.0000
P4	15.2187	11.5537	9.4493
P5	21.5999	11.2960	10.2822

Table 10.4. Quantification of results over only R and Y pixels using UFCLS method with partial prior signature knowledge for three different error threshold values

	UFCLS $\varepsilon = 7.0 \times 10^5$	UFCLS $\varepsilon = 4.0 \times 10^5$	UFCLS $\varepsilon = 2.0 \times 10^5$
P_{11}	3.2691	3.1162	3.0897
P_{12}	1.3128	1.2508	1.2433
P_{13}	0.0413	0.0497	0.0526
P_{21}	4.9350	4.2786	4.2540
P_{22}	2.4201	1.8693	1.6412
P_{23}	1.1844	0.7295	0.5539
P_{31}	3.9199	4.0814	4.3011
P_{32}	1.7502	1.7754	2.0301
P_{33}	0.4203	0.4330	0.5630
P_{41}	2.9124	3.1241	2.9777
P_{42}	1.6915	1.5367	1.7529
P_{43}	0.4634	0.5104	0.4758
P_{51}	3.7763	3.8227	3.9045
P_{52}	1.8771	1.8153	1.8415
P_{53}	0.4937	0.5173	0.4756

As we can see from the quantification values in these two tables, UFCLS using partial prior material knowledge performed significantly better than FCLS using manually selected material signatures.

Example 10.8 No Prior Signature Information

The previous experiments assumed that *a priori* knowledge of the material signatures was available. In this final experiment we conclude by assuming that no prior material knowledge is given. As before, we must prescribe an error threshold value to terminate the UFCLS algorithm. Again, we used the UFCLS algorithm and plotted the maximum least squares error values resulting from (10.3), as shown in Fig. 10.13(a). In order to determine an appropriate threshold the difference between every fifth sample was calculated using the following equation

$$\Delta \mathrm{LSE}^{(k)} = \frac{\max_{\mathbf{r}}\left\{\mathrm{LSE}^{(k-5)}(\mathbf{r})\right\} - \max_{\mathbf{r}}\left\{\mathrm{LSE}^{(k)}(\mathbf{r})\right\}}{\max_{\mathbf{r}}\left\{\mathrm{LSE}^{(k-5)}(\mathbf{r})\right\}}. \tag{10.11}$$

This is the same method used to determine the number of targets or codewords for an unsupervised vector quantization method in Brumbley (1998). This function is plotted in Fig. 10.13(b) and shows a significant drop at values 11, 23, and 40.

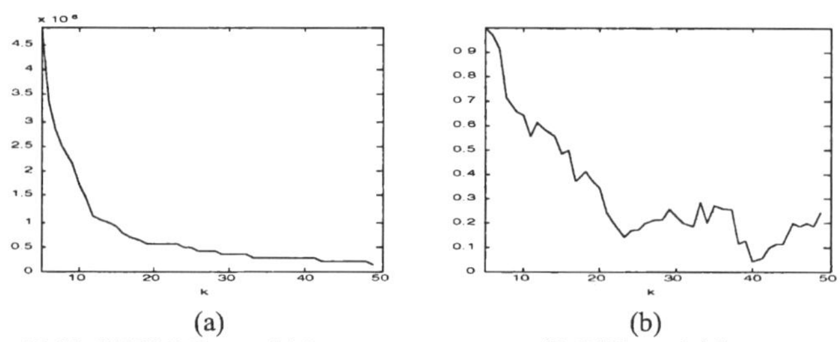

(a) (b)

Figure 10.13. UFCLS Error: (a) Least squares error, (b) Differential least squares error.

To begin this analysis we selected an error threshold value of $\varepsilon = 1.5 \times 10^6$ which resulted in termination of the UFCLS algorithm after 11 iterations. The resulting $\mathrm{LSE}^{(k)}(\mathbf{r})$ images are shown in Fig. 10.14.

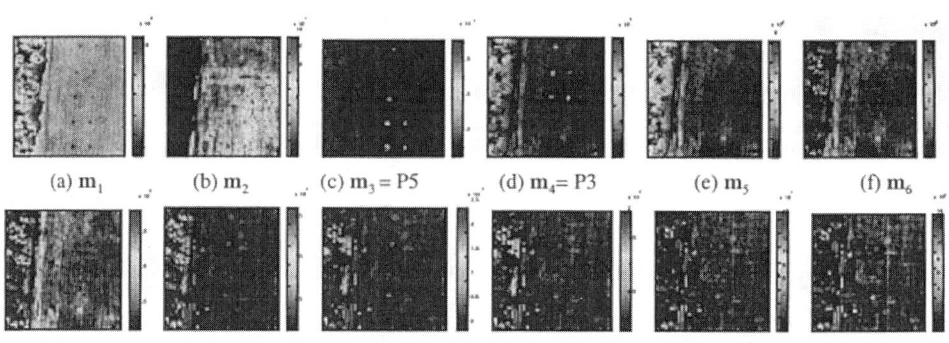

Figure 10.14. Results of UFCLS using these three threshold values

In Figs. 10.14(c), 10.14(d) and 10.14(i) R pixels for panel P5, P3 and P1 were selected as initial pixels respectively. However, unlike the AVIRIS data, the $\mathrm{LSE}^{(k)}(\mathbf{r})$

image in Fig. 10.14(l) was far from a chi-square distribution. Fig. 10.15(a)-(c) shows $\mathrm{LSE}^{(k)}(\mathbf{r})$ images for iterations, k equal to 23, 40 and 49 respectively.

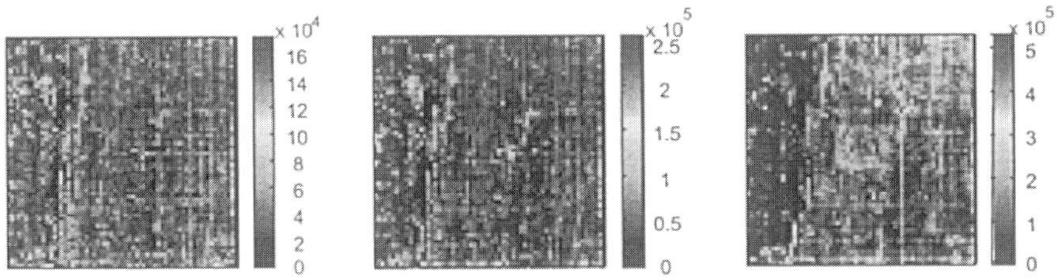

(a) $k = 23$ (b) $k = 40$ (c) $k = 40$

Figure 10.15. Results of $\mathrm{LSE}^{(k)}(\mathbf{r})$ images for iterations, k equal to 23, 40 and 49 respectively

Even after 49 iterations of the UFCLS algorithm, the error image still failed to achieve a chi-square distribution. This may result from the very fine spatial and spectral resolutions the HYDICE sensor, which can extract targets with size as small as 1-4 meters. Therefore, it may take a long iterative process for the UFCLS algorithm before it stops.

In order to compare the various methods a prescribed error threshold of $\varepsilon = 2.54 \times 10^5$ was selected for the UFCLS algorithm and resulted in the generation of 41 signatures. These 41 signatures used to form a signature matrix $\mathbf{M}$. Using this $\mathbf{M}$, LS, SCLS, NSCLS, NCLS, NNCLS, FCLS and MFCLS were implemented. Fig. 10.16 shows results for $\mathbf{m}_9$, $\mathbf{m}_{33}$, $\mathbf{m}_4$, $\mathbf{m}_{22}$, $\mathbf{m}_{35}$ and $\mathbf{m}_3$ where panel P1, P2, P3, P4, P4 and P5 were detected by various methods respectively.

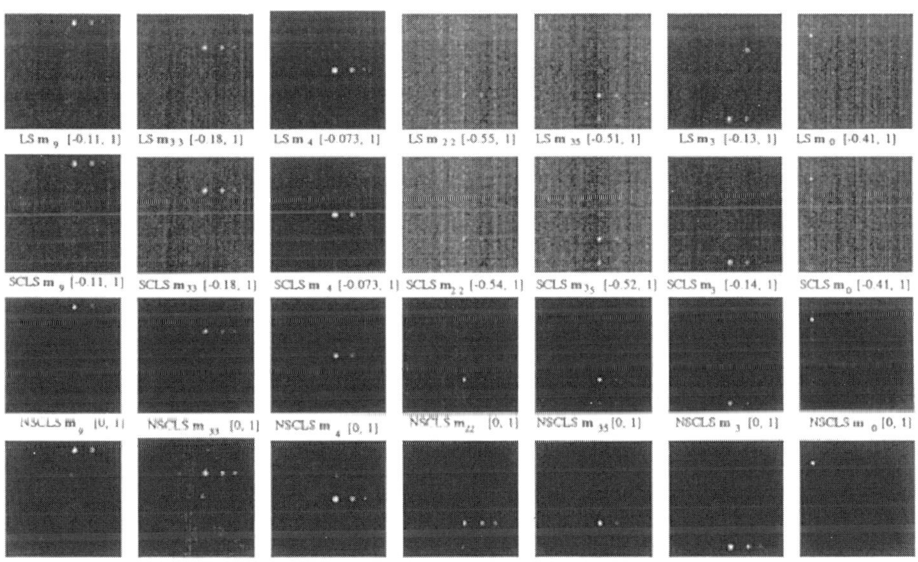

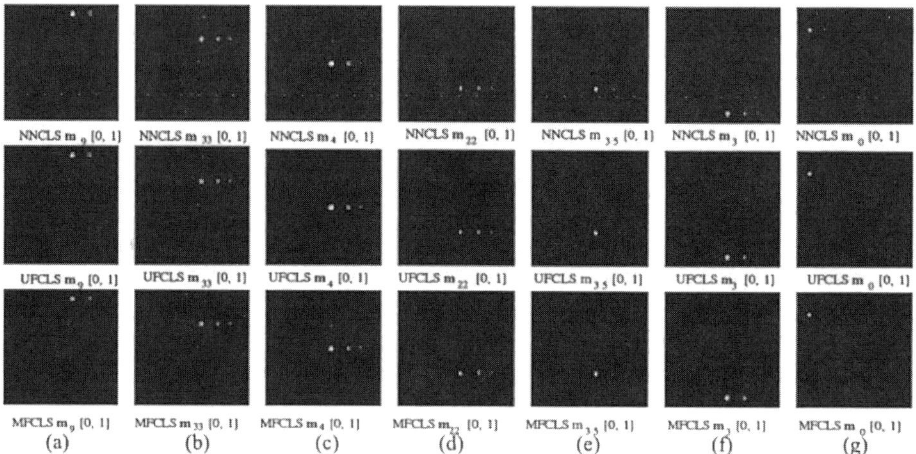

Figure 10.16. Results of LS, SCLS, NSCLS, NCLS, NNCLS, UFCLS and MFCLS using target signatures generated by the UFCLS algorithm

In each case, an R pixel was selected as a material signature, except for $\mathbf{m}_{33}$ which turned out to be a Y pixel next to the top R pixel of panel p_{21}. It should be noted that in this unsupervised case, the UFCLS algorithm actually detected the panels in row 4 in two separate images. This separation could have been avoided if the UFCLS algorithm was terminated at iteration 34 or 35. However, determining the most appropriate stopping threshold is generally difficult due to the lack of prior information. As we can see from these images, NCLS, NNCLS and UFCLS performed the best and significantly better than LS, SCLS and NSCLS. An anomaly was also observed in Fig. 10.16(g), where there was an unknown signature detected in the upper left corner of the tree line, $\mathbf{m}_0$. Tables 10.5 and 10.6 contain quantification results for the entire image and for the Y and R pixels respectively.

Table 10.5. Quantification results of LS, SCLS, NSCLS, NCLS, NNCLS, UFCLS and MFCLS for Y pixels

	LS	SCLS	NSCLS	NCLS	NNCLS	FCLS	MFCLS
p_{11}	3.21	3.14	1.74	3.14	3.12	2.96	2.94
p_{12}	1.02	0.97	0.49	1.29	1.28	1.26	1.23
p_{13}	-0.23	-0.23	0.06	0.09	0.09	0.05	0.05
p_{21}	5.21	4.96	2.37	3.70	3.69	3.37	3.28
p_{22}	1.20	1.30	0.56	1.20	1.17	1.43	1.43
p_{23}	0.47	0.52	0.23	0.57	0.56	0.61	0.68
p_{31}	4.01	4.01	1.91	4.09	4.10	4.03	4.06
p_{32}	2.52	2.52	0.86	1.95	1.94	1.96	1.91
p_{33}	0.77	0.77	0.25	0.49	0.49	0.50	0.45
p_{41}	4.70	4.79	2.18	1.94	1.90	1.97	1.91
p_{41}	-0.97	-1.01	1.20	1.57	1.54	2.05	1.95
p_{42}	-0.19	0.12	0.38	1.42	1.41	1.47	1.44
p_{42}	2.82	2.66	0.90	0.33	0.33	0.35	0.32
p_{43}	0.81	0.93	0.29	0.54	0.53	0.60	0.51
p_{43}	-0.12	-0.18	0.08	0	0	0	0.04
p_{51}	4.51	4.54	2.14	3.85	3.81	3.69	3.67
p_{52}	1.51	1.51	0.50	1.85	1.84	1.79	1.79
p_{53}	0.28	0.19	0.09	0.47	0.46	0.45	0.39

Table 10.6. Quantification results of LS, SCLS, NSCLS, NCLS, NNCLS, UFCLS and MFCLS for R pixels

	LS	SCLS	NSCLS	NCLS	NNCLS	UFCLS	MFCLS
$\mathbf{m}_9$	-49.2866	-53.4773	8.5554	9.0078	9.0037	9.5068	10.6713
$\mathbf{m}_{33}$	22.6381	37.7144	34.2775	34.4859	34.7202	28.3764	33.3212
$\mathbf{m}_4$	45.0890	45.0504	23.0562	22.9769	22.9950	24.9895	21.0518
$\mathbf{m}_{22}$	186.1792	202.9720	109.4225	7.0436	6.9863	8.7071	8.8039
$\mathbf{m}_{35}$	-135.5971	-143.8910	52.4539	3.2501	3.1987	6.5286	5.0979
$\mathbf{m}_3$	7.5093	-1.95281	9.8006	8.8974	8.8122	8.4374	9.2320

As we can see from the quantification values in these two tables, the NCLS, NNCLS and UFCLS performed significantly better than the other methods.

In the above experiment UFCLS was able to effectively detect and quantify materials in the image scene. However, since UFCLS was completely unsupervised, further analysis would be required to identify all the unknown materials. Calibrating the spectra of the extracted material signatures and comparing them against a reference database such as a spectral library can generally do this.

10.5 NEAR REAL-TIME IMPLEMENTATION

In general, a supervised algorithm that performs mixed pixel classification on a pixel-by-pixel basis can be implemented in real time such as OSP-based classifiers. This assumes that complete knowledge is provided *a priori*. It is also true for pixel-based constrained supervised algorithms such as NCLS and FCLS. The difficulty of real time implementation for unsupervised MPC arises from the case that there is no prior knowledge available about the image data. Under this circumstance the necessary knowledge must be obtained directly from image data. As described in Chapter 5, one way to obtain such information is to take advantage of the information provided by the image data such as sample correlation/covariance matrix used in CEM and TCIMF in Chapter 4 and anomaly detectors in Chapter 6.

An alternative approach is to design a two-pass algorithm where the first pass is designed to obtain desired image information, then followed by the second pass to perform subpixel detection and mixed pixel classification using the knowledge generated by the first pass. This section presents a near real time implementation approach. It is a two-pass process, but it makes two passes into one pass with negligible time delay. It only uses line-by-line causal information as the way implemented in CEM in Section 4.5 and RXD in Section 6.6.3.

Since the major strength of FCLS and UFCLS is quantification, they sometimes may not work as effective as NCLS and UNCLS in terms of target detection and classification. This is because the abundance fractions estimated by FCLS must satisfy ASC. When image data contains many signal sources such as 34 targets required for the 15-panel HYDICE scene to perform unsupervised subpixel detection and classification in Fig. 10.16, the abundance amounts generated by FCLS for each target signature become very small resulting from ASC. As expected, it may not perform as well as NCLS and UNCLS which do not impose ASC. So, in this section we present a near real-time implementation of UNCLS rather than UFCLS. Nevertheless, a near real-time implementation of UFCLS can be also implemented in exactly the same way as does UNCLS (Heinz, 2001). It is important to note that the detection and classification are conducted simultaneously on a pixel-by-pixel basis. The near real-time UNCLS algorithm can be described as follows.

Let L represent the number of spectral bands and let $\mathbf{r}_j$ be an $L \times 1$column pixel vector in a hyperspectral image. The first pixel vector $\mathbf{r}_1$, to be analyzed is selected as target $\mathbf{t}_1$ with its corresponding signature $\mathbf{m}_1$. Using $\mathbf{m}_1$, we form a target signature matrix $\mathbf{M} = [\mathbf{m}_1]$. This matrix is then used to estimate the abundance fraction $\hat{\alpha}_1(\mathbf{r}_2)$ in the second pixel vector $\mathbf{r}_2$. The LSE between $\mathbf{r}_2$ and its estimated linear mixture $\hat{\alpha}_1(\mathbf{r}_2)\mathbf{r}_2$ is then compared to a prescribed error threshold. If this error threshold is exceeded then $\mathbf{r}_2$ is selected as target $\mathbf{t}_2$ with signature $\mathbf{m}_2$, and a new target signature matrix $\mathbf{M} = [\mathbf{m}_1\,\mathbf{m}_2]$ is formed, otherwise, $\mathbf{M}$ remains unchanged. This same procedure is repeated on a pixel-by-pixel basis until all pixels have been processed.

Using the 15-panel HYDICE image scene in Fig. 1.7(a) as an example, the near real-time processing of the UNCLS algorithm was started at the upper-left pixel and continued across the top row on a pixel-by-pixel basis. As each new target signature was selected, a new classification image was displayed. After completion of the first row, each new row of data was processed using the same left to right pixel flow. After the first row was processed, there were seventeen targets $\mathbf{t}_1, \mathbf{t}_2, \cdots, \mathbf{t}_{17}$ with their corresponding signatures $\mathbf{m}_1, \mathbf{m}_2, \cdots, \mathbf{m}_{17}$ were detected. The classification results of the first 17 signatures are shown in Fig. 10.17.

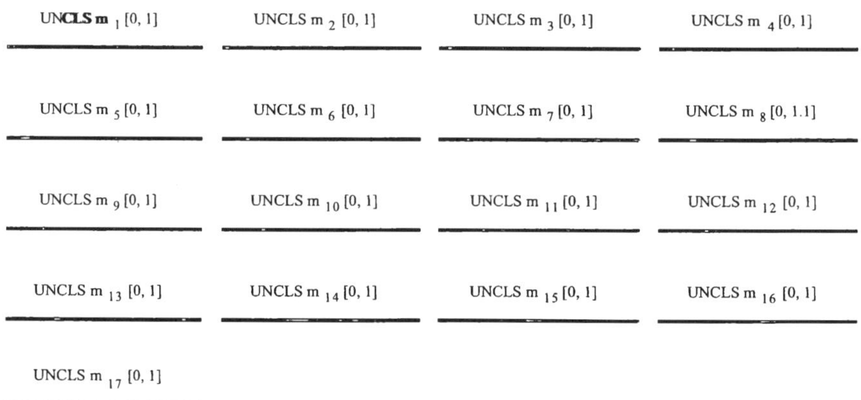

Figure 10.17. Classification results after the first row was processed in near real-time

Figure 10.18 shows classification results after completion of 20 rows in near real-time processing where 31 targets were generated.

In Fig. 10.18 the panel pixels p_{11}, p_{12} and p_{13} were classified as four separate targets (19, 20, 22 and 23). This resulted from the pixel-by-pixel real-time implementation which first detected the mixed pixel as targets 19 and 20 (mainly grass with some panel), before the more pure pixels, targets 22 and 23 were available. This separation also occurred for the second row of panels, with the panel pixels p_{21} and p_{22} being classified as four separate targets (27, 28, 29 and 30). One possible solution to this problem is to monitor the target signature matrix and remove mixed pixel targets upon detection of a more pure pixel target. However, this solution requires a technique for reclassification of pixels once the mixed pixel targets are removed. For instance, the detection of target 23 would cause the removal of targets 19, 20 and 22 from the target signature matrix and all pixels classified with targets 19, 20 and 22 in the target signature matrix would be reclassified. Another simpler method to avoid detection of mixed pixels is to implement a look-ahead procedure into the UNCLS algorithm. A two-row look-ahead is appropriate, since its goal is to avoid detection of mixed pixels at the transitions from one target to the next. In addition, this look-ahead procedure can also avoid display of discontinuities associated with the removal of targets.

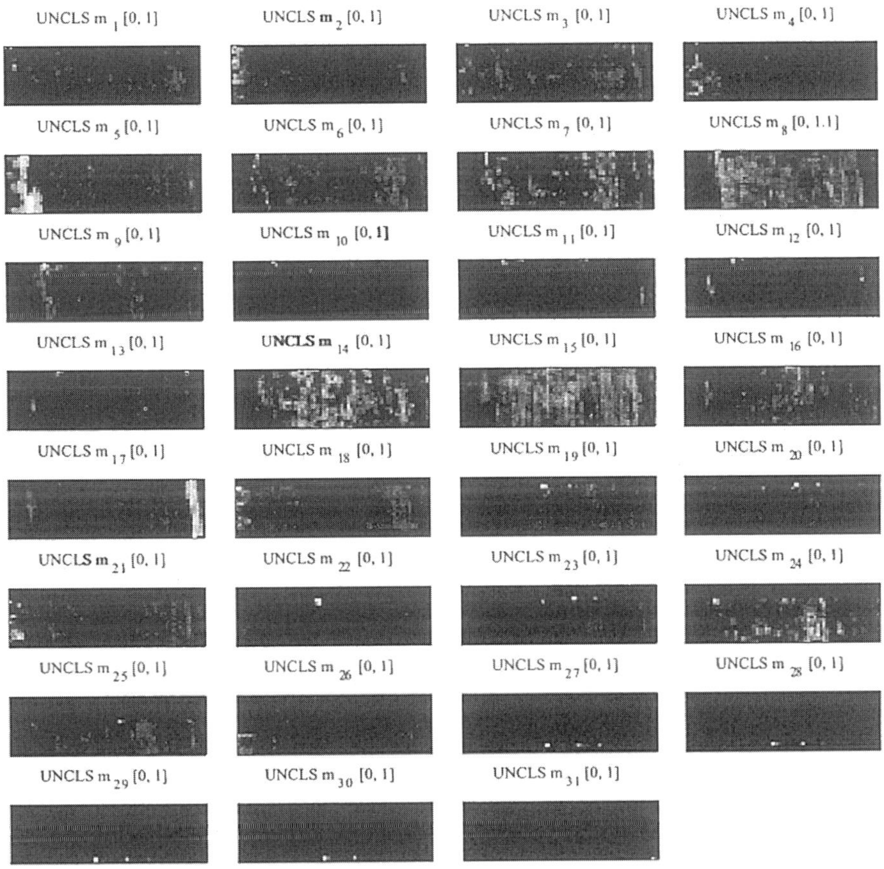

Figure 10.18. Classification results after 20 rows were processed in near real-time

However, even with this two-row look-ahead procedure there is still an additional problem with the current UNCLS algorithm in that once it selects a target it is not allowed to update this selection. Therefore, as a final modification to the near real-time UNCLS algorithm described above, a real-time target update was added. This update is based on least squares error and allows the processing algorithm to replace a target in the target signature matrix with the current pixel target, provided the least squares error is reduced by this replacement.

Fig. 10.19 shows the classification results after completion of the algorithm using the two-row look-ahead procedure and a real-time target update. Compared to 31 targets detected and classified in Fig. 10.18 without the real-time target update, only 19 targets detected and classified in Fig. 10.19. As we can see, all five rows of panels were detected and properly classified using these generated 19 targets. This is a significant reduction in target generation. Furthermore, by adding the two-row look-ahead procedure the mixed pixel target selections in the first two rows of panels were avoided.

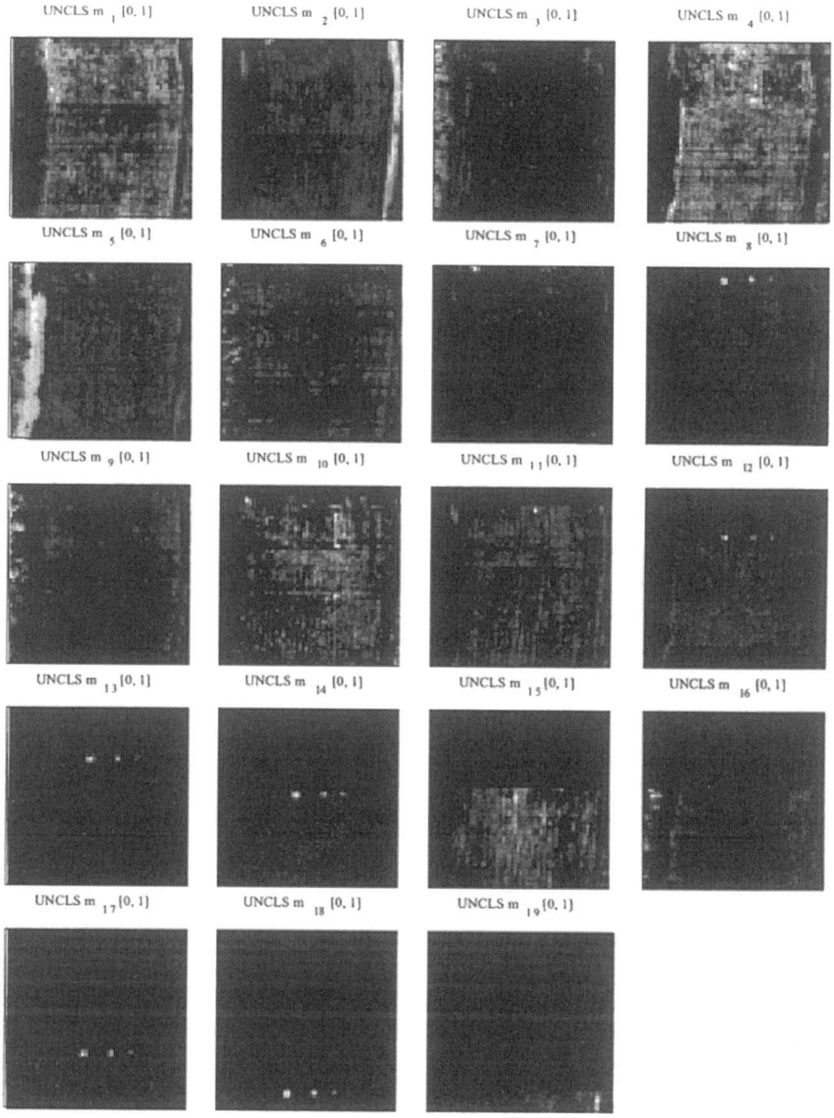

Figure 10.19. Classification results after all 64 rows were processed using a two-row look-ahead procedure and real-time target update

In order to illustrate the need of the real-time target update, Fig. 10.20 shows pixel locations (bright white pixels) where the updates to the road and shade targets (2 and 5) occurred. This experiment shows that processing without this target update would have resulted in separation of the road target into two classification images.

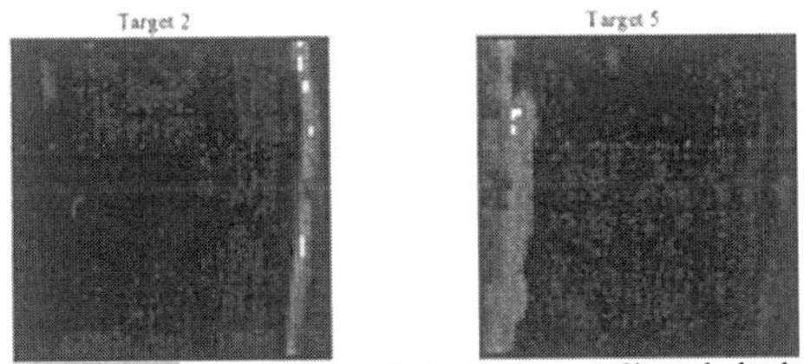

Figure 10.20. Pixel locations (bright white) with road pixels (target 2) and shade pixels (target 5) updated.

10.6 CONCLUSIONS

Unconstrained linear unmixing has been used for hyperspectral image classification due to its mathematical tractability. However, the resulting abundance fractions of target signatures may be negative and their sum may not be necessarily summed to one. Therefore, unconstrained linear unmixing generally cannot be used for target quantification. This chapter presents two least squares methods to solve fully constrained linear mixing problems. One is a fully constrained least squares (FCLS) method, which implements two constraints, ASC and ANC simultaneously. Since there is no closed form solved for fully constrained linear mixing problems, an efficient algorithm is further developed to generate the desired optimal solutions. A second method is a modified fully constrained least-squares (MFCLS) method, which also implements two constraints. One is ASC. Another is AASC, which replaces ANC in FCLS. Unlike FCLS, MFCLS takes advantage of AASC to derive an iterative algorithm to find optimal solutions. Since both FCLS and MFCLS require complete knowledge of target signatures, an unsupervised least squares error-based approach is also proposed to extend their capability to unsupervised methods, UFCLS and UMFCLS. Despite a slight degradation in quantification performance, the UFCLS and UMFCLS have a significant advantage that they can be used for finding and quantifying anomalies. Finally, like CEM and RXD UNCLS and UFCLS can be also implemented in real time. However, since UNCLS and UFCLS do not have complete target knowledge as required by a real-time implementation, a two-row look-ahead procedure and a real-time target update are included to make UNCLS and UFCLS process images in near real time. A similar concept of near real-time process will be discussed in Chapter 14 where on-line process will be used instead for distinction.

11

TARGET SIGNATURE-CONSTRAINED MIXED PIXEL CLASSIFICATION (TSCMPC): LCMV CLASSIFIERS

The target abundance-constrained mixed pixel classification that was considered in Chapter 10 imposed ASC and ANC on target abundance fractions. In this case, a linear mixture model is required and the target signature matrix **M** must be also known *a priori*. In this chapter, we consider the target signature-constrained mixed pixel classification which does not need a linear mixture model. Instead, it classifies a mixed pixel by constraining the spectral shapes or vector directions of target signatures rather than target abundance fractions. Such concept was explored in Chapter 4 and referred to as linearly constrained minimum variance (LCMV) approach. It was used to design target signature-constrained subpixel detectors, CEM and TCIMF. A subpixel detector can detect targets, but does not necessarily imply that it can also classify the targets it detected. It may occur that a detector can detect all targets of interest but cannot discriminate one from another. In this case, the detection rate can be as high as 100%, but the classification rate could as low as 0%. In order for LCMV-based detectors to also achieve target classification, they must be implemented one target at a time so that the detected targets can be classified in a separate image. This chapter extends LCMV-based detectors to LCMV classifiers. It develops an approach that expands the capability of the LCMV-based detectors in such a fashion that it can simultaneously detect and classify multiple targets in a single image where different colors are used to highlight distinct types of detected targets. In particular, such color assignment approach also allows us to extend a CEM-based detector in Chapter 4 to a CEM-based classifier. Despite that an LCMV classifier requires the prior knowledge of desired targets, it can take advantage of the unsupervised algorithms presented in Chapter 5 to generate the necessary target knowledge and make its classification unsupervised.

11.1 INTRODUCTION

In Chapter 4, a CEM-based detector detected a desired target $\mathbf{t}_p$ by constraining the spectral shape or vector direction of its signature vector $\mathbf{d} = \mathbf{m}_p$ to the scalar constraint constant 1. In this case, a CEM-based detector can only detect a single target. In order to detect multiple targets, a CEM-based detector was recently extended to an LCMV-based detector (Chang and Ren, 1999) where the scalar constant 1 was replaced by a unity vector $\mathbf{1} = (1,1,\cdots,1)^T$ as a constraint vector with each component "1" in $\mathbf{1}$ used to constrain a specific target of interest. More recently, this LCMV detector was further extended to TCIMF (Ren and Chang, 2001) described in Chapter 4. It can detect a set of desired targets while also eliminating a set of undesired targets. Their idea expands the unity vector $\mathbf{1}$ to a vector made up of a unity vector $\mathbf{1}$ and a zero vector $\mathbf{0}$, i.e. $\begin{pmatrix}\mathbf{1}\\ \mathbf{0}\end{pmatrix} = (1,1,\cdots,1,0,0,\cdots,0)^T$ such that a component "1" is used to constrain a specific desired target and a component "0" is used to annihilate an undesired target. The strength of TCIMF is to enhance detection performance by simultaneously eliminating undesired targets and minimizing the interfering effects resulting from unknown signal sources. A third version of an LCMV-based detector to be presented in this chapter also uses a unity vector $\mathbf{1}$ to remove the image background as was done by the LPTD in Section 5.5. Interestingly, this type of the background removed-LCMV approach includes as a special case the filter vectors (FV) developed by Bowles et al (1995). Although LCMV-based detectors can be also used for multiple-target classification, but they cannot detect and classify multiple targets at the same time. They must be carried out in a way that one type of targets is detected at a time, then different types of targets are classified in separate images. In order to resolve this problem, we need to find a mechanism for an LCMV-based detector to perform classification while the target detection takes place. One feasible way is to use specific colors to classify multiple targets when the targets are detected. During the course of the detection, targets of same type are specified by one particular color and targets of distinct types are discriminated by different colors. Using such color assignment we can also extend a CEM detector to a CEM classifier. The evolution from CEM through LCMV to TCIMF provides a clue for a possible solution. It started with a scalar constraint constant 1 for a CEM detector, then a vector constraint $\mathbf{1}$ for an LCMV detector, and finally arrived at a two-component constraint vector $(\mathbf{1},\mathbf{0})^T$ for a TCIMF detector. So, a natural extension of an LCMV detector may be one, which uses a constraint matrix. However, such approach is not trivial since we are now dealing with a matrix rather than a vector, in which case, we need to identify what roles the rows and columns of the constraint matrix will play. This chapter explores this approach to accomplish multiple-target detection and classification in one single operation.

11.2 LCMV CLASSIFIERS

In order to make an LCMV detector specified by (4.5) an LCMV classifier, we introduce a set of constraint vectors derived from $\mathbf{c}$, denoted by $\{\mathbf{c}_i\}_{i=1}^p$, where $\mathbf{c}_i = (0,\cdots,\underbrace{c_i}_{i},\cdots,0)^T$ is a p-dimensional column vector with "c_i" in the i-th components and "0"'s in all other components. Substituting $\{\mathbf{c}_i\}_{i=1}^p$ for the constraint $\mathbf{c}$ in (4.4) for each i, $1 \le i \le p$ results in the following LCMV classification problem

$$\min_{\mathbf{w}_i} \mathbf{w}_i^T \mathbf{R}_{L\times L} \mathbf{w}_i \text{ subject to } \mathbf{M}^T \mathbf{w}_i = \mathbf{c}_i \text{ for } 1 \le i \le p \quad (11.1)$$

where $\mathbf{w}_i$ is a weight vector used to classify the specific target $\mathbf{t}_i$ with its signature $\mathbf{m}_i$. The optimal solution to (11.1), $\mathbf{w}_i^{\text{LCMV}}$ can be obtained by

$$\mathbf{w}_i^{\text{LCMV}} = \mathbf{R}_{L\times L}^{-1}\mathbf{M}\left(\mathbf{M}^T\mathbf{R}_{L\times L}^{-1}\mathbf{M}\right)^{-1}\mathbf{c}_i \text{ for each } 1 \le i \le p. \tag{11.2}$$

A classifier that uses a set of $\left\{\mathbf{w}_i^{\text{LCMV}}\right\}_{i=1}^p$ specified by (11.2) to classify the p target signatures in $\mathbf{M}$ is called an LCMV classifier.

As noted in Section 5.5, $\delta_{\text{RXD-LPTD}}$ and $\delta_{\text{RXD-UTD}}$ improved its detection performance by background subtraction where a unity vector $\mathbf{1}$ was implemented by δ_{LPTD} and δ_{UTD} to extract the image background. The same idea can be also applied to an LCMV classifier to improve its classification performance. In doing so, we first augment the p-target signature matrix $\mathbf{M}$ to $\tilde{\mathbf{M}} = \left[\mathbf{m}_1\, \mathbf{m}_2 \cdots \mathbf{m}_p\, \mathbf{1}_{L\times 1}\right]$ which includes an L-dimensional unity vector $\mathbf{1}_{L\times 1} = \underbrace{(1,1,\cdots,1)}_{L}^T$ as the last column of $\mathbf{M}$ for the purpose of background annihilation.

Following a similar way that an LCMV classifier was derived in (11.2), we define $\tilde{\mathbf{c}}_i = (\mathbf{c}_i, 0) = (0,\cdots,\underbrace{c_i}_{i},\cdots,0,\underbrace{0}_{p+1})^T$ as a (p+1)-dimensional column vector with "c_i" in the i-the components and "0"'s in all other components. Substituting $\left\{\tilde{\mathbf{c}}_i\right\}_{i=1}^p$ for the constraint in (11.1) yields a background-removed LCMV classification problem

$$\min_{\mathbf{w}_i} \mathbf{w}_i^T\mathbf{R}_{L\times L}\mathbf{w}_i \text{ subject to } \tilde{\mathbf{M}}^T\mathbf{w}_i = \tilde{\mathbf{c}}_i \text{ for } 1 \le i \le p \tag{11.3}$$

where the optimal solution to (11.3), $\mathbf{w}_i^{\text{BRLCMV}}$ can be obtained by

$$\mathbf{w}_i^{\text{BRLCMV}} = \mathbf{R}_{L\times L}^{-1}\tilde{\mathbf{M}}\left(\tilde{\mathbf{M}}^T\mathbf{R}_{L\times L}^{-1}\tilde{\mathbf{M}}\right)^{-1}\tilde{\mathbf{c}}_i\,. \tag{11.4}$$

A classifier that uses the set of $\left\{\mathbf{w}_i^{\text{LCMV}}\right\}_{i-1}^p$ specified by (11.4) to classify the p target signatures in $\mathbf{M}$ is called a background-removed LCMV (BRLCMV) classifier.

11.3 BOWLES ET AL.'S FILTER VECTORS (FV) ALGORITHM

Recently, Bowles et al. developed a filter-vectors (FV) approach to hyperspectral data analysis (Bowles et al., 1995). Their idea is to construct a set of filter vectors to unmix targets of interest in an image scene, of which each filter vector is used to classify a specific target. Like linear unmixng methods studied in Chapter 7, the FV approach also requires a linear mixture model described by (3.1) to design a set of L-dimensional filter vectors, $\left\{\mathbf{w}_i\right\}_{i=1}^p$ that must satisfy the following three conditions,

$$\mathbf{w}_i^T \mathbf{m}_j = \delta_{ij} \tag{11.5}$$

$$\mathbf{w}_i \cdot \mathbf{1}_{L\times 1} = \mathbf{w}_i^T \mathbf{1}_{L\times 1} = 0 \quad \text{for each } 1 \le i \le p \tag{11.6}$$

$$\mathbf{w}_i^T \mathbf{w}_i \text{ is a minimum for each } 1 \le i \le p. \tag{11.7}$$

The solution derived by the FV approach consists of p filter vectors, $\left\{\mathbf{w}_i^{\mathrm{FV}}\right\}_{i=1}^{p}$, specified by the matrix $\mathbf{W}^{\mathrm{FV}} = \left[\mathbf{w}_1^{\mathrm{FV}}\, \mathbf{w}_2^{\mathrm{FV}} \cdots \mathbf{w}_p^{\mathrm{FV}}\right]$ that can be obtained by

$$\mathbf{W}^{\mathrm{FV}} = \mathbf{S}^{-1}\breve{\mathbf{M}} \tag{11.8}$$

where

$$\breve{\mathbf{M}} = \left[\breve{\mathbf{m}}_1\, \breve{\mathbf{m}}_2 \cdots \breve{\mathbf{m}}_p\right] \text{ with } \breve{\mathbf{m}}_i = \mathbf{m}_i - \left(\mathbf{1}_{L\times 1}^T \mathbf{m}_i / L\right)\mathbf{1}_{L\times 1} \tag{11.9}$$

$$s_{ij} = \breve{\mathbf{m}}_i^T \mathbf{m}_j \text{ and } \mathbf{S} = \left[s_{ij}\right]. \tag{11.10}$$

Solving (11.5)-(11.7) is equivalent to solving the following constrained optimization problem:

$$\min_{\mathbf{w}_i} \mathbf{w}_i^T \mathbf{w}_i \text{ for each } 1 \le i \le p \tag{11.11}$$

subject to

$$\mathbf{1}_{L\times 1}^T \mathbf{w}_i = 0 \text{ and } \mathbf{w}_i^T \mathbf{m}_j = \delta_{ij}. \tag{11.12}$$

Interestingly, if we examine closely the two constraints specified by (11.12), we will find that the first constraint (i.e. (11.5)) can be taken care of by a set of constraint vectors $\left\{\tilde{\mathbf{c}}_i\right\}_{i=1}^{p}$ given by (11.3) that is used in BRLCMV by letting $\mathbf{c}_i = 1$ for each $1 \le i \le p$. The second constraint in (11.12) (i.e. (11.6)) can be satisfied by including $\mathbf{1}_{L\times 1}$ in $\tilde{\mathbf{M}}$ with $c_{p+1} = 0$ for $1 \le i \le p$. Now if we further replace $\mathbf{R}_{L\times L}$ in (11.3) with the identity matrix $\mathbf{I}_{L\times L}$, the BRLCMV classification problem specified by (11.3) becomes

$$\min_{\mathbf{w}_i} \mathbf{w}_i^T \mathbf{w}_i \text{ subject to } \tilde{\mathbf{M}}^T \mathbf{w}_i = \tilde{\mathbf{c}}_i \text{ for each } 1 \le i \le p. \tag{11.13}$$

Compared the constrained classification problem specified by (11.13) to Bowles et al's filter vectors problem outlined by (11.5)-(11.7) (i.e., (11.11)-(11.12)), these two constrained optimization problems turn out to be identical. This implies that Bowles et al.'s FV classifier becomes a special case of the BRLCMV classifier where it can be considered a whitened BRLCMV with $\mathbf{R}_{L\times L} = \mathbf{I}_{L\times L}$, where the data are zero-mean and whitened. That is, Bowles et al.'s FV classifier operates pixel-by-pixel without accounting for spectral correlation among sample vectors, while (11.3) takes advantage of spectral correlation among samples by including $\mathbf{R}_{L\times L}$ into the optimization problem. As will be shown in experiments, the BRLCMV performs significantly better than the FV approach.

11.4 COLOR ASSIGNMENT OF LCMV CLASSIFIERS

The LCMV described in Chapter 4 can be used to detect targets of interest by simultaneously passing these targets through a filter constrained by (4.1). Unfortunately, it also suffered from a disadvantage that it could not differentiate these detected targets one another. To expand the capability of an LCMV detector as an LCMV classifier, we further augment the constraint vector **c** in (4.1) to a constraint matrix **C**.

More precisely, assume that the size of a constraint matrix **C** is $k \times p$ and denoted by $\mathbf{C} = \left[\mathbf{c}_1\ \mathbf{c}_2 \cdots \mathbf{c}_p\right]$. In this case, the weight vector **w** in (4.1) is also expanded by an $L \times p$ weight matrix $\mathbf{W} = \left[\mathbf{w}_1\ \mathbf{w}_2 \cdots \mathbf{w}_p\right]$, which is formed by p weight vectors $\mathbf{w}_1, \mathbf{w}_2, \cdots, \mathbf{w}_p$ and is represented by

$$\mathbf{M}^T\mathbf{W} = \mathbf{C} \tag{11.14}$$

The constraint matrix **C** plays a significant role in multiple-target classification. It makes use of each $k \times 1$ column vector $\mathbf{c}_i$ for $1 \le i \le p$ to detect a specific type of targets and each row vector of **C** to assign a particular color to the detected targets so as to achieve target classification. In this case, we are required to solve separately a set of the following constrained optimization problems

$$\min_{\mathbf{w}_i} \mathbf{w}_i^T\mathbf{R}_{L\times L}\mathbf{w}_i \text{ subject to } \mathbf{M}^T\mathbf{w}_i = \mathbf{c}_i \text{ for } 1 \le i \le p. \tag{11.15}$$

Using the constraint equation (11.14) and (11.2) which is a solution to (11.15), the optimal solution $\mathbf{W}^{\text{LCMV}} = \left[\mathbf{w}_1^{\text{LCMV}}\ \mathbf{w}_2^{\text{LCMV}} \cdots \mathbf{w}_p^{\text{LCMV}}\right]$ can be obtained by the following matrix form

$$\mathbf{W}_i^{\text{LCMV}} = \mathbf{R}_{L\times L}^{-1}\mathbf{M}\left(\mathbf{M}^T\mathbf{R}_{L\times L}^{-1}\mathbf{M}\right)^{-1}\mathbf{C}. \tag{11.16}$$

The use of the weight matrix $\mathbf{W}^{\text{LCMV}}$ in (11.16) is two-fold. It detects desired targets, and at the same time it also classifies targets with the pre-selected colors assigned by row vectors in the constraint matrix **C**. More specifically, the column dimensionality of **C**, p, specifies the number of colors to be used to classify targets, whereas the row dimensionality of **C**, k, is the number of targets of interest needed to be classified. For a TCIMF-based detector in Chapter 4 to be implemented as a TCIMF classifier, we only have to substitute $[\mathbf{DU}]$ in (4.9) for the **M** in (11.16).

To further illustrate the above idea, let us assume that there are k targets to be classified into p classes. In this case, the constraint matrix $\mathbf{C} = \left[\mathbf{c}_1\ \mathbf{c}_2 \cdots \mathbf{c}_p\right]$ given by

(11.14) must have size of $k \times p$ where each of column vectors $\mathbf{c}_i$ for $1 \le i \le p$ has same dimensionality, k and corresponds to the constraint vector of an LCMV classifier specified by (11.1). It should be noted that k is not necessarily equal to p. For example, if there are two targets of the same type which have slightly different but similar spectral signatures, they must be classified into the same class. So, in this case, $k < p$. On the other hand, if partial knowledge of natural background signatures is available, it generally presents interference to target detection and classification, and can be considered as undesired signatures. The elimination of such interference can enhance target detection and classification by virtue of TCIMF. These interfering signatures are annihilated by a zero column vector $\mathbf{0}$ in the constraint matrix $\mathbf{C}$. Since all the detected k targets must be classified into p distinct classes, it is impossible to show their classification with gray levels in one image. This difficulty can be resolved if the detected targets of distinct types are assigned by different colors. Surprisingly, the row vectors in the constraint matrix $\mathbf{C}$ can be used for this purpose as illustrated in the following.

Suppose that there are five targets, $\mathbf{t}_1,\mathbf{t}_2,\mathbf{t}_3,\mathbf{t}_4,\mathbf{t}_5$ which belong to three types of objects, denoted by O_1 ($\mathbf{t}_1$), O_2 ($\mathbf{t}_2,\mathbf{t}_3$), O_3 ($\mathbf{t}_4,\mathbf{t}_5$), and two background signatures $\mathbf{t}_6,\mathbf{t}_7$ which may be available from partial knowledge and can be used to represent undesired signatures. In this case, $k = 3$ and $p = 7$. If we assume that red, green, blue and black colors will be used to specify these three types of target classes and one class of background signatures, then the target signature matrix is formed by seven target signatures, $\mathbf{M} = \left[\mathbf{m}_1\ \mathbf{m}_2 \cdots \mathbf{m}_7\right]$ where $\mathbf{m}_1, \mathbf{m}_2, \cdots, \mathbf{m}_7$ are spectral signatures of the 7 targets, $\mathbf{t}_1, \mathbf{t}_2, \cdots, \mathbf{t}_7$, respectively. Then the constraint matrix $\mathbf{C}$ can be specified by

$$\mathbf{C}^T = \left[\mathbf{c}_1\ \mathbf{c}_2\ \mathbf{c}_3\right]^T = \begin{bmatrix} 1 & 0 & 0 & 0 & 0 & 0 & 0 \\ 0 & 1 & 1 & 0 & 0 & 0 & 0 \\ 0 & 0 & 0 & 1 & 1 & 0 & 0 \\ \underbrace{}_{\text{red}} & \underbrace{}_{\text{green}} & \underbrace{}_{\text{green}} & \underbrace{}_{\text{blue}} & \underbrace{}_{\text{blue}} & \underbrace{}_{\text{black}} & \underbrace{}_{\text{black}} \end{bmatrix}. \tag{11.17}$$

where $\mathbf{c}_1^T = (1, 0, 0, 0, 0, 0,0)^T$, $\mathbf{c}_2^T = (0, 1, 1, 0, 0, 0,0)^T$ and $\mathbf{c}_3^T = (0, 0, 0, 1, 1, 0,0)^T$ are constraint vectors to detect the 7 targets. The three colors, red, green and blue are used to specify three types of objects O_1, O_2 and O_3 respectively. The "1" in row 1 of $\mathbf{c}_1$ is used to detect $\mathbf{t}_1$, the two "1"'s in rows 2 and 3 of $\mathbf{c}_2$ are used to detect $\mathbf{t}_2,\mathbf{t}_3$ and the two "1"'s in rows 4 and 5 of $\mathbf{c}_3$ are used to detect $\mathbf{t}_4,\mathbf{t}_5$. As noticed in (11.17), the rows 6 and 7 in $\mathbf{C}$ (i.e. columns 6 and 7 in $\mathbf{C}^T$) are zero vectors which will be used to eliminate background signatures $\mathbf{t}_6,\mathbf{t}_7$ by the assigned black color. The reason we included background signatures $\mathbf{t}_6,\mathbf{t}_7$ in the above illustration is because in many practical applications, there always has some partial knowledge about image background. With the availability of this information we should take advantage of it. If there is no such knowledge available, the background signatures $\mathbf{t}_6,\mathbf{t}_7$ in (11.17) would be absent and $\mathbf{t}_6,\mathbf{t}_7$ would be considered as unknown interferers rather than undesired signatures. In this case, the size of the $\mathbf{C}$ would be reduced to $k = 3$ and $p = 5$. It is worth noting that for visual discrimination the colors chosen to classify targets should be as distinct as possible. Nevertheless, they can be arbitrary. More importantly, if the color of a pixel is a mixture of different colors, it may suggest that this particular pixel could be mixed by the materials represented by these colors. Such valuable color information of how a pixel is mixed cannot be provided by an LCMV detector.

11.5 EXTENSION OF CEM FILTER TO CLASSIFIERS

As noted previously, if a CEM detector is used as a classifier, it must be implemented for one target at a time so that distinct types of targets can be classified in separate images. In order to classify all different targets in a single image, it requires to combine the separate images produced by a set of CEM detectors into a single image where each CEM-detected target must be highlighted by a different color. Three ways of extending a CEM detector to a classifier can be derived by such target combination.

11.5.1 Winner-Take-All CEM (WTACEM) Classifier

Let $\mathbf{r}$ denote an image pixel and $\{\mathrm{CEM}_j(\mathbf{r})\}_{j=1}^{p}$ be p target abundance fractional images generated by p CEM detectors, each of which is used to detect one specific target. One CEM-derived classifier is the Winner-Take-All CEM classifier, referred to as WTACEM($\mathbf{r}$) which is similar to WTAMPC specified by (8.2). For each image pixel $\mathbf{r}$, it is defined by

$$\mathrm{WTACEM}(\mathbf{r}) = \max_{1 \le j \le p} \mathrm{CEM}_j(\mathbf{r}) \tag{11.18}$$

which uses the winner-take-all rule to determine the gray scale of $\mathbf{r}$.

11.5.2 Sum CEM (SCEM) Classifier

Another CEM-derived classifier is the Sum CEM classifier, referred to as SCEM($\mathbf{r}$), defined by

$$\mathrm{SCEM}(\mathbf{r}) = \Sigma_{j=1}^{p} \mathrm{CEM}_j(\mathbf{r}) \tag{11.19}$$

which simply sums up the p abundance fractions of each image pixel $\mathbf{r}$.

11.5.3 Multiple-Target CEM (MTCEM) Classifier

If the constraints $\mathbf{c}_i$ for $1 \le i \le p$ in (11.15) are replaced with the same scalar constant "1" for $1 \le i \le p$, the resulting LCMV, referred to as multiple-target CEM (MTCEM) classifier, can be viewed as a set of p CEM detectors implemented simultaneously by solving the following constrained optimization problem

$$\min_{\mathbf{w}_i} \mathbf{w}_i^T \mathbf{R}_{L \times L} \mathbf{w}_i \text{ subject to } \mathbf{M}^T \mathbf{w}_j = 1 \text{ for } 1 \le i \le p, \tag{11.20}$$

which is a special case of (11.15).

11.5.4 Target-Constrained Interference-Minimized (TCIM) Classifier

Like CEM, TCIMF can be also extended to a classifier, referred to as TCIM classifier. It implements a constraint matrix **C** in (11.14) to highlight different types of targets detected by TCIMF. The example using (11.17) for illustration in Section 11.4 is actually a color implementation of a TCIM classifier. Such TCIM classifier can be considered as a generalization of MTCEM, which includes a set of zero vectors to eliminate undesired target signatures.

It should be noted that the three CEM extensions, WTACEM, SCEM and MTCEM plus the TCIM classifier are all special cases of the LCMV classifier specified by (11.15). They all generate a single image to classify multiple targets using a color assignment to highlight detected targets. For the WTACEM classifier, the color to be used to highlight WTACEM(**r**) is the same color that was assigned to the $\mathrm{CEM}_{j^*}(\mathbf{r})$ detector where $j^* = \arg\left\{\max_{1\le j\le p} \mathrm{CEM}_j(\mathbf{r})\right\}$ is the solution to (11.18). For the SCEM classifier, the color of the SCEM(**r**) will be a mixture of the colors that are assigned to specify the p CEM-detected target signatures. One advantage of using SCEM is that its mixed color classification map provides different degrees of mixture in image pixel vectors. For instance, if a pixel is mixed by three target signatures with equal abundance fractions specified by red, green and blue, the resulting (red,green,blue)-mixed color of the pixel will be black. As another extreme, if a mixed pixel is a pure pixel, the color representing this pixel will be a pure color. For MTCEM and TCIM, their color assignment will be carried out by the procedure described in Section 11.4. In particular, when TCIMF is implemented with a color assignment for classification, it becomes a TCIM classifier and MTCEM classifier turns out to be its special version without including annihilation of undesired target signatures.

As a final concluding remark, it is worth noting that according to (11.13) Bowles et al.'s FV classifier discussed in Section 11.3 can be considered as a whitened TCIM classifier with $\mathbf{R}_{L\times L} = \mathbf{I}_{L\times L}$.

11.6 COMPUTER SIMULATIONS

In this section, computer simulations were conducted to demonstrate the performance of various LCMV classifiers. The used data set was the five AVIRIS reflectance signatures shown in Fig. 1.5. Since the spectral signatures of blackbrush, creosote leaves and sagebrush are similar, we chose these three targets to evaluate the detection performance of the LCMV classifier. A set of 400 mixed pixels was simulated. Red soil and dry grass were used to simulate background pixels with 50%-50% split. In addition, Gaussian noise was also added to each pixel to achieve a 30:1 signal-to-noise ratio. In order to implant three targets in these 400 pixels, we replaced pixel numbers 50th, 100th, 150th, 200th with 20%, 40%, 60%, 80% blackbrush pixels respectively and the remaining abundance split by red soil and dry grass evenly. For instance, the 100th pixel contains 40% blackbrush, 30% red soil and 30% dry grass. Similarly, the 250th, 300th, 350th, 400th pixels were replaced with 20%, 40%, 60%, 80% creosote leaves pixels respectively and the remaining abundance split by red soil and dry grass evenly. The

25th, 125th, 225th, 325th pixels were replaced with 20%, 40%, 60%, 80% sagebrush pixels respectively and the remaining abundance split by red soil and dry grass evenly.

Fig. 11.1 shows the detection and classification results of the LCMV, BRLCMV, FV and OSP classifiers with **M** = [blackbrush, creosote leaves, sagebrush].

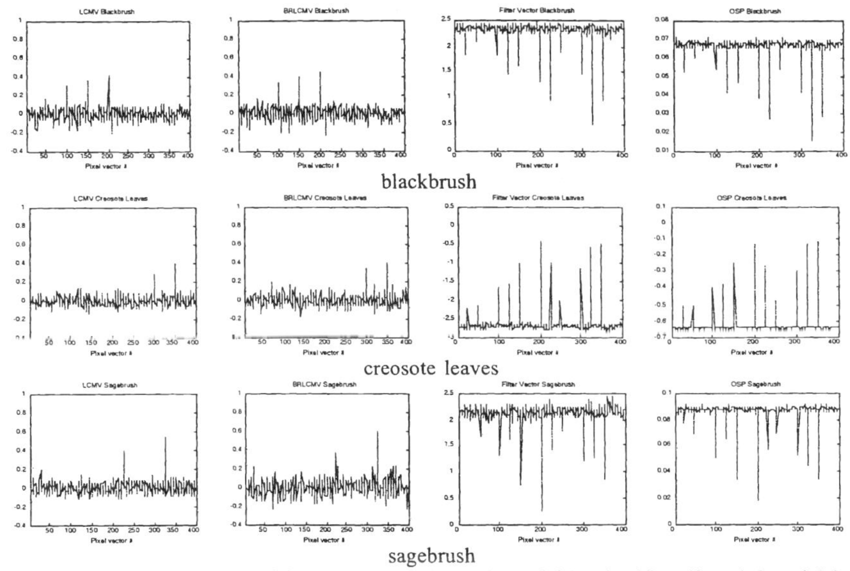

Figure 11.1. Classification results of the LCMV, BRLCMV, FV and OSP classifiers (from left to right) with **M** = **D** = [blackbrush, creosote leaves, sagebrush]

As shown in Fig. 11.1, the LCMV and the BRLCMV classifiers performed comparably while the FV and the OSP classifiers performed very similarly but completely failed in classification of the three targets. However, if we further included the background signatures in the target signature matrix **M** and used **U** = [dry grass, red soil] as the undesired target signatures for the LCMV and the BRLCMV classifiers, the performance of the FV and the OSP classifiers was significantly improved as shown in Fig. 11.2. Since the signatures of red soil and dry grass are different from those of three targets, removal of undesired target signatures did not have much impact on the performance of the LCMV and the BRLCMV classifiers. This is not true for the FV and OSP classifiers. Unlike Fig. 11.1, the FV and the OSP classifiers was able to detect all the three targets and even performed slightly better than the LCMV and the BRLCMV classifiers. The reason for this was because the FV and the OSP classifiers had complete target knowledge that was required for the linear mixture model. In Fig. 11.1, the background signatures, red soil and dry grass accounting for most target abundance were not included. As a result, the FV and the OSP classifiers performed poorly. By contrast, the FV and the OSP classifiers worked very effectively in Fig. 11.2 because the target signature matrix **M** provided complete target information.

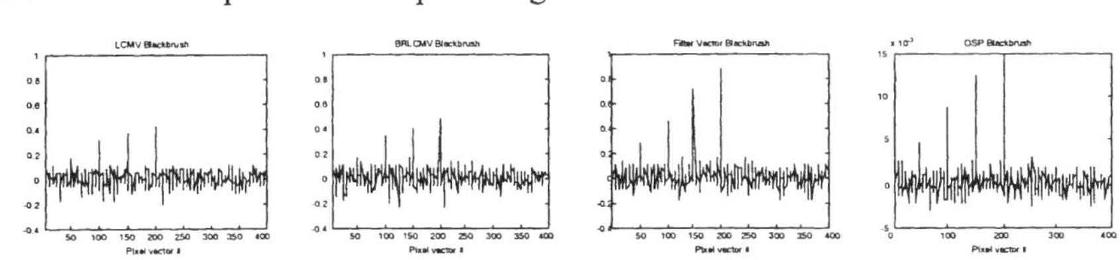

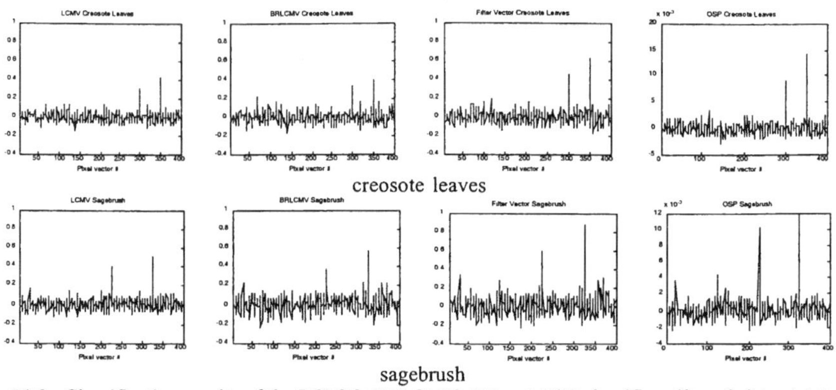

Figure 11.2. Classification results of the LCMV, BRLCMV, FV and OSP classifiers (from left to right) with $\mathbf{M} = [\mathbf{D}\ \mathbf{U}]$ = [blackbrush, creosote leaves, sagebrush, dry grass, red soil]

Despite that both the FV and the OSP classifiers performed very similarly in terms of target detection and classification, their detected abundance fractions were very different. It should be noted that a different scale was used to show the detectability of the OSP classifier in Fig. 11.2. Compared to the FV classifier, the OSP classifier did not accurately estimate abundance fractions for the targets it detected. As demonstrated in Chapters 6 and 7, in order for the OSP to produce more accurate abundance fraction estimates, the constant $\kappa = \left(\mathbf{d}^T \mathbf{R}_{L\times L}^{-1} \mathbf{d}\right)^{-1}$ must be included in OSP in (3.9). The resulting OSP classifier was referred to as, oblique subspace projection classifier (OBSP) or unconstrained Gaussian maximum likelihood estimation classifier (UGMLC) in Chapter 8. Fig. 11.3 shows the classification result of the same experiment done for Fig. 11.2 for the FV, the OSP and the OBSP classifiers at the same scale. As we can see from Fig. 11.3, the FV and the OBSP classifiers performed nearly the same while the OSP performed very poorly after the detected abundance fractions were scaled to the same range used for the FV and the OBSP classifiers. Interestingly, the experiment conducted in Fig. 11.3 also demonstrated that the FV classifier performed no better than the OBSP classifier, *a posteriori* OSP classifier discussed in Chapter 8.

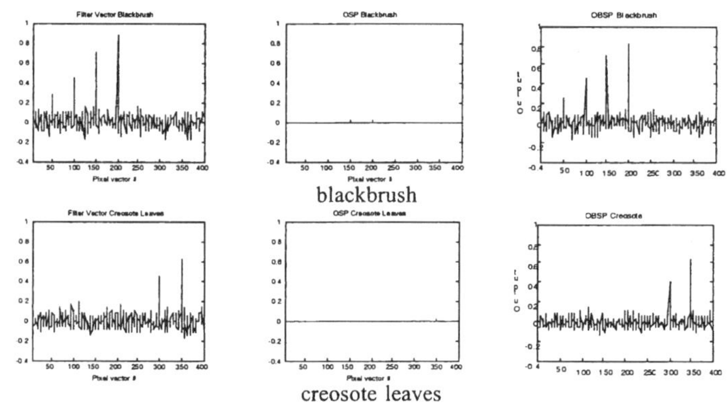

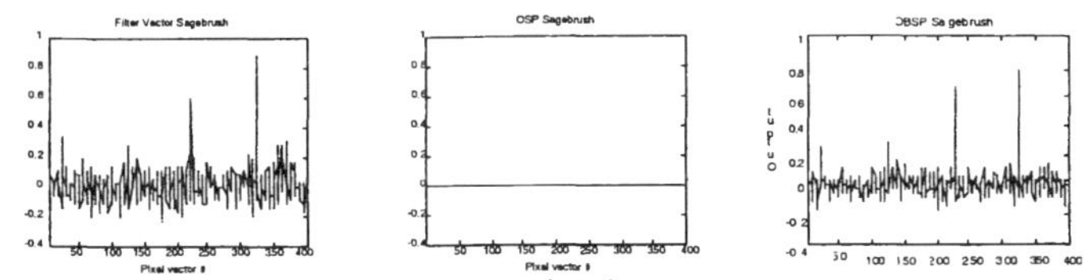

sagebrush

Figure 11.3. Classification results of the FV, OSP and OBSP classifiers (from left to right) with **M** = [blackbrush, creosote leaves, sagebrush, dry grass, red soil]

In order to see how an LCMV classifier classifies targets of interest in a single image, Fig. 11.4 shows the classification results produced by WTACEM, MTCEM, BRLCMV and LCMV. These results were obtained by combining the three figures in Fig. 11.1 into a single figure where red, green, and blue are used to highlight the three target signatures, blackbrush, creosote leaves and sagebrush. The SCEM result is not shown in Fig. 11.4 because its figure plots did not show its advantage as demonstrated in Fig. 11.9(e) where the mixed colors of an image pixel can really show the degrees of abundance fractions of targets detected in the pixel (see Fig. 11.9(e)).

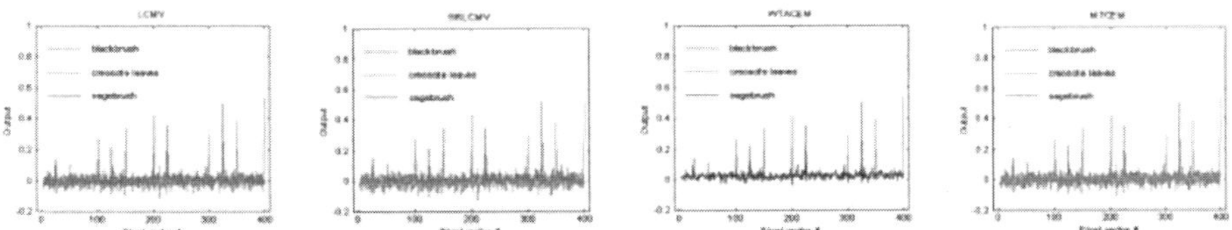

(a) LCMV (b) BRLCMV (c) WTACEM (d) MTCEM

Figure 11.4. Single-plot classifications of the LCMV, BRLCMV, WTACEM and MTCEM classifiers with **M** = **D** = [blackbrush, creosote leaves, sagebrush]

Similar experiments were also conducted for Fig. 11.2 which included two undesired target signatures, red soil and dry grass, Fig. 11.5 shows the classification results produced by the LCMV and BRLCMV classifiers by merging the three target signatures detected in Fig. 11.2 into one single figure, while eliminating background signatures. This resulting LCMV is actually the TCIM classifier. Since the WTACEM and MTCEM did not require knowledge of red soil and dry grass, their classification results in Fig. 11.4 remain unchanged and are not included in Fig 11.5

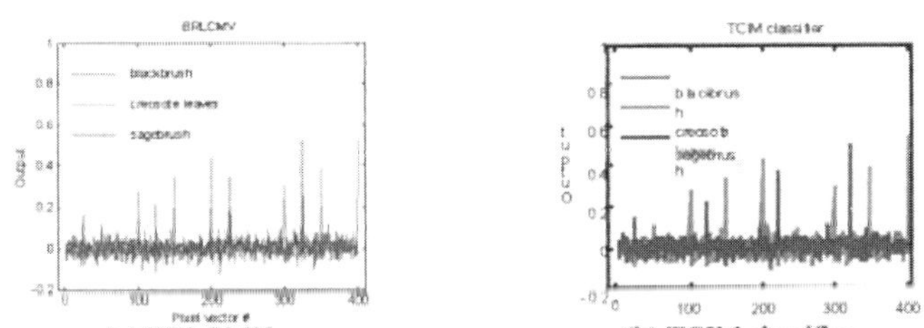

(a) BRLCMV (b) TCIM classifier

Figure 11.5. Single plots classifications of BRLCMV and TCIM classifiers with **M** = [**D U**] = [blackbrush, creosote leaves, sagebrush, dry grass, red soil]

11.7 HYPERSPECTRAL IMAGE EXPERIMENTS

Since there was no sample spectral correlation included in the conducted computer simulations, the results did not show the advantage of the sample correlation matrix $\mathbf{R}_{L\times L}$ used in the LCMV and BRLCMV classifiers (see (11.15)). In order to see how much improvement resulting from inclusion of sample spectral correlation, we conducted real image experiments using the AVIRIS image in Fig. 1.6(a) and the 15-panel scene in Fig. 1.7 for the LCMV, the BRLCMV, the FV and the OSP classifiers. As noted in Fig. 11.3, the OSP classifier did not detect correct amounts of target abundance. Nevertheless, as demonstrated in Chapter 8, the OSP and *a posteriori* OSP classifiers did produce very close results in classification because their generated abundance fractional images were scaled to 256 gray levels. So, only the OSP classifier was used for comparison in the following hyperspectral image experiments. Fig. 11.6 shows the AVIRIS classification results of cinders, rhyolite, playa, shade and vegetation.

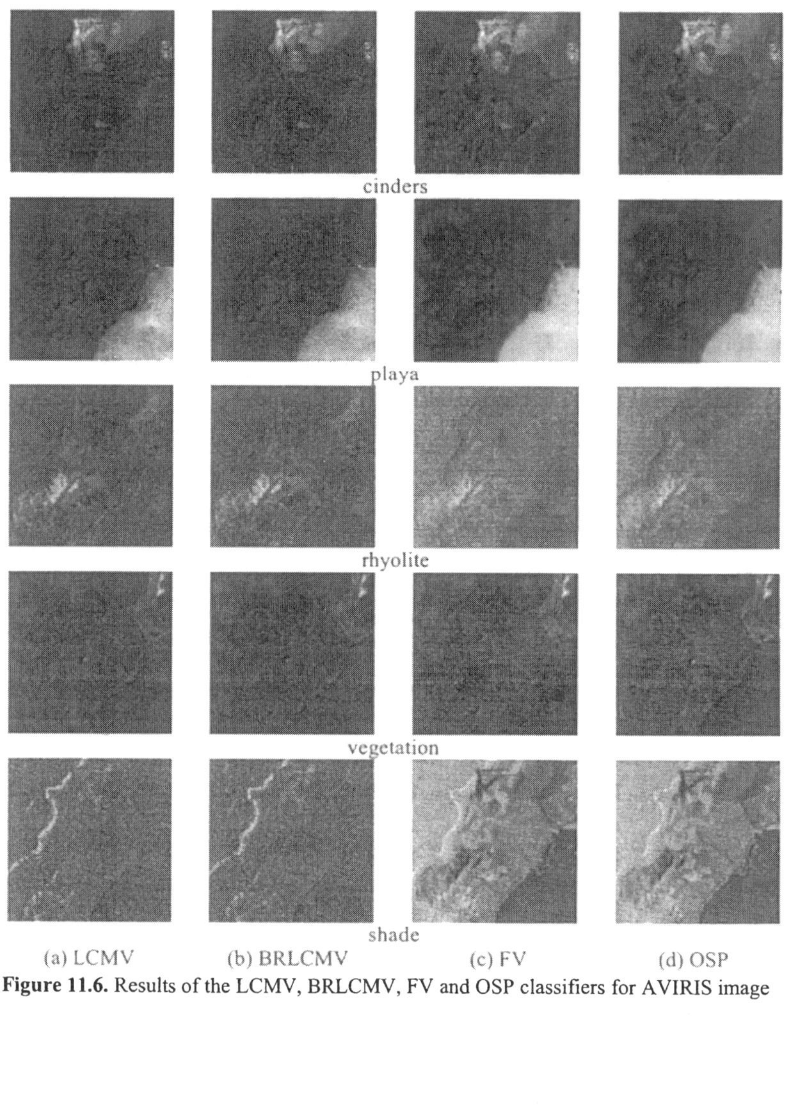

Figure 11.6. Results of the LCMV, BRLCMV, FV and OSP classifiers for AVIRIS image

From Fig. 11.6, we can see that when the LCVF AVIRIS data were used for the experiments, the LCMV and the BRLCMV classifiers performed slightly better than did the FV and the OSP, particularly, vegetation and shade. However, when the 15-panel HYDICE scene was used for experiments, the classification results changed significantly. Fig. 11.7 shows their classification results of the 15 panels. Fig. 11.7 shows the LCMV and the BRLCMV classifiers performed significantly better than did the FV and the OSP classifiers for all the 15 panels, particularly, panels in rows 1 and 4. This is because the former used $\mathbf{R}_{L\times L}$ to account for spatial correlation while the latter discarding it by operating individual image pixel independently, i.e., $\mathbf{R}_{L\times L} = \mathbf{I}_{L\times L}$. Since the HYDICE image has 1.5-m spatial resolution compared to 20-m spatial resolution of the AVIRIS data. In this case, the sample spectral correlation plays a crucial role in classification and the LCMV and the BRLCMV classifiers certainly took advantage of it.

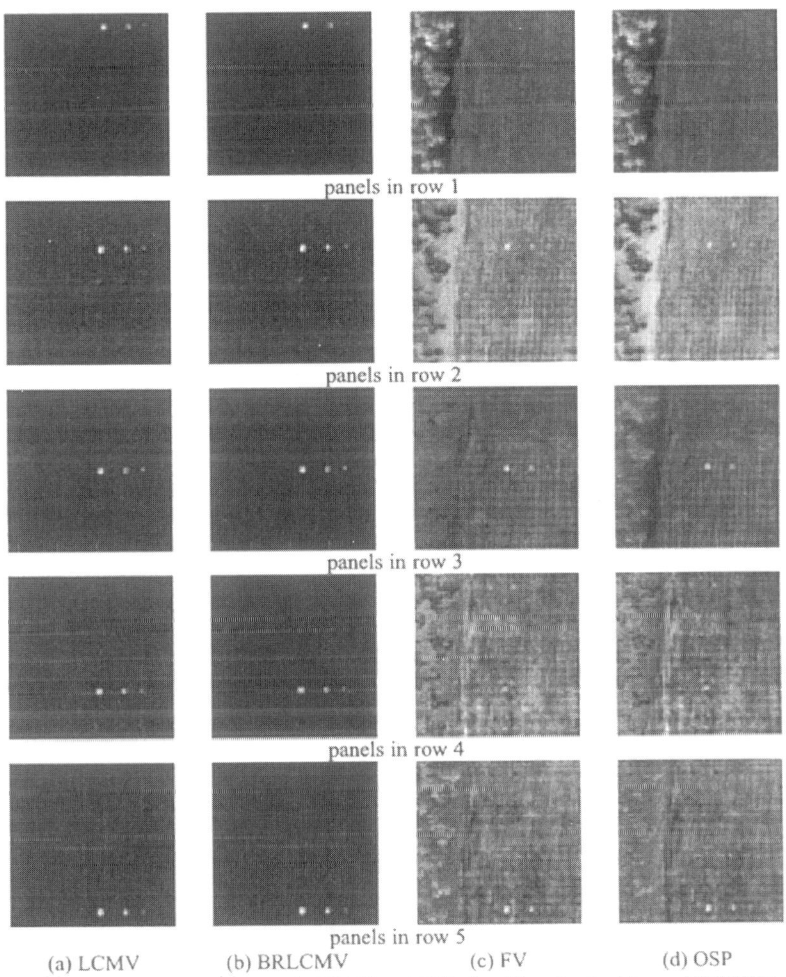

Figure 11.7. Results of the LCMV, BRLCMV, FV and OSP classifiers for the 15-panel HYDICE scene

Based on the ground truth map provided by Fig. 1.7(b) we can actually tally the number of target pixels that are correctly detected and classified by a particular method.

Since the LCMV classifiers only generate the abundance fractions of the specific targets to be classified and not for all the targets, the winner-take-all thresholding approach using the WTAMPCV specified by (9.2) in Chapter 9 is not applicable. In this case, the a%MPCV specified by (9.3) in Chapter 9 was used to segment targets from the image background. Tables 11.1(a-d)-11.2(a-d) tabulate the detection and classification results for the four classifiers, LCMV, BRLCMV, FV and OSP where the 50%MPCV and 25%MPCV were used to produce the following tables.

Table 11.1(a). Detection and classification rates of LCMV using 50% abundance cut-off threshold

	N_R	N_{RD}	R_{RD}	R_C	N_{FA}
P1	3	2	0.6667	0.6667	2
P2	4	3	0.7500	0.7500	2
P3	4	3	0.7500	0.7500	3
P4	4	3	0.7500	0.7500	3
P5	4	3	0.7500	0.7500	2
Sum	19	14	0.7368	0.7368	12

Table 11.1(b). Detection and classification rates of BRLCMV using 50% abundance cut-off threshold

	N_R	N_{RD}	R_{RD}	R_C	N_{FA}
P1	3	2	0.6667	0.6667	2
P2	4	3	0.7500	0.7500	2
P3	4	3	0.7500	0.7500	3
P4	4	3	0.7500	0.7500	3
P5	4	3	0.7500	0.7500	2
Sum	19	14	0.7368	0.7368	12

Table 11.1(c). Detection and classification rates of FV using 50% abundance cut-off threshold

	N_R	N_{RD}	R_{RD}	R_C	N_{FA}
P1	3	0	0.0000	0.0000	239
P2	4	4	1.0000	1.0000	3089
P3	4	4	1.0000	1.0000	8
P4	4	4	1.0000	1.0000	2033
P5	4	4	1.0000	1.0000	1232
Sum	19	16	1.0000	1.0000	6601

Table 11.1(d). Detection and classification rates of OSP using 50% abundance cut-off threshold

	N_R	N_{RD}	R_{RD}	R_C	N_{FA}
P1	3	0	0.0000	0.0000	242
P2	4	4	1.0000	1.0000	3736
P3	4	4	1.0000	1.0000	105
P4	4	4	1.0000	1.0000	1429
P5	4	4	1.0000	1.0000	1209
Sum	19	16	1.0000	1.0000	6721

Table 11.2(a). Detection and classification rates of LCMV using 25% abundance cut-off threshold

	N_R	N_{RD}	R_{RD}	R_C	N_{FA}
P1	3	3	1.0000	1.0000	8
P2	4	4	1.0000	1.0000	10
P3	4	4	1.0000	1.0000	5
P4	4	4	1.0000	1.0000	5
P5	4	4	1.0000	1.0000	4
Sum	19	19	1.0000	1.0000	32

Table 11.2(b). Detection and classification rates of BRLCMV using 25% abundance cut-off threshold

	N_R	N_{RD}	R_{RD}	R_C	N_{FA}
P1	3	3	1.0000	1.0000	8
P2	4	4	1.0000	1.0000	10
P3	4	4	1.0000	1.0000	5
P4	4	4	1.0000	1.0000	5
P5	4	4	1.0000	1.0000	4
Sum	19	19	1.0000	1.0000	32

Table 11.2(c). Detection and classification rates of FV using 25% abundance cut-off threshold

	N_R	N_{RD}	R_{RD}	R_C	N_{FA}
P1	3	3	1.0000	1.0000	3240
P2	4	4	1.0000	1.0000	4011
P3	4	4	1.0000	1.0000	2519
P4	4	4	1.0000	1.0000	3963
P5	4	4	1.0000	1.0000	3950
Sum	19	19	1.0000	1.0000	17683

Table 11.2(d). Detection and classification rates of OSP using 25% abundance cut-off threshold

	N_R	N_{RD}	R_{RD}	R_C	N_{FA}
P1	3	3	1.0000	1.0000	3286
P2	4	4	1.0000	1.0000	4034
P3	4	4	1.0000	1.0000	3150
P4	4	4	1.0000	1.0000	3896
P5	4	4	1.0000	1.0000	3968
Sum	19	19	1.0000	1.0000	18334

In these tables, N is total number of panel pixels, N_R(Pi) is the number of red panel pixels in row i, N_{RD}(Pi) is the number of red panel pixels detected and classified correctly in row i, N_{FA} is the number of false alarm pixels for Pi, R_D(Pi) is the detection rate defined by $R_{RD}(Pi) = N_{RD}(Pi)/N_R(Pi)$ for Pi and $R_C = \sum_{i=1}^{5}(N_R(Pi)/N)R_{RD}(Pi)$ is the classification rate for Pi.

As shown in these tables, the FV and OSP classifiers achieved 100% detection rate and 100% classification rate at 25% while 0% for both detection and classification rates at 50%. On the other hand, both LCMV and BRLCMV classifiers performed identically at 50% and 25%. In both cases, they outperformed the FV and the OSP classifiers. Despite that all the four classifiers achieved 100% detection and classification rates, the number of false alarm pixels generated by the FV and OSP classifiers was significantly greater than (more than 500 times) those produced by the LCMV and BRLCMV classifiers.

Analogous to Fig. 9.5, 3-D ROC curves and 2-D ROC curves can be plotted based on the target hit rate R_H defined by (9.27) versus the false alarm rate via the a%MPCV defined by (9.3) for performance evaluation. In this case, we compare the performance of all the five LCMV classifiers, LCMV, BRLCMV, WTACEM, SCEM, and MTCEM against that of the FV and OSP. Figs. 11.8(a-d) show the 3-D and 2-D ROC curves of the five LCMV-based classifiers, FV and OSP classifiers. As demonstrated in these figures, the ROC curves fall into two categories, one produced by the five LCMV-based classifiers whose ROC curves were very close, and another by the FV and OSP classifiers, which performed very closely but substantially worse than the LCMV-based classifiers. It should be noted that the 3-D ROC curves in Fig. 11.8(a) were generated by the a%MPCV with varying abundance percentage as a thresholding value from 100%

down to 0%. Their corresponding 2-D ROC curves were plotted based on the resulting target hit rate R_H versus the false alarm rate.

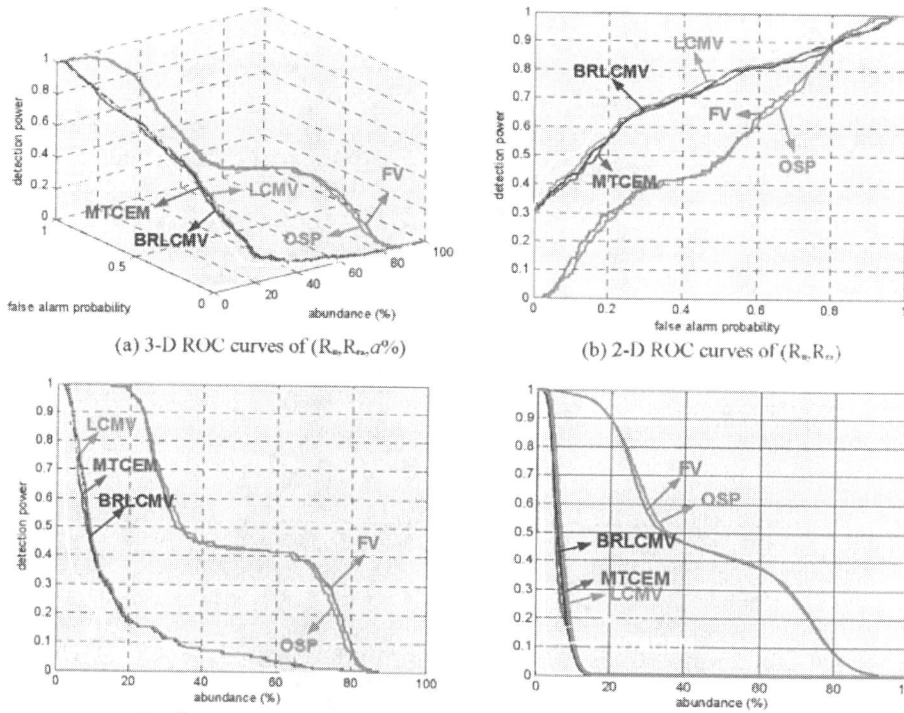

(a) 3-D ROC curves of $(R_H, R_{FA}, a\%)$ (b) 2-D ROC curves of (R_H, R_{FA})

(c) 2-D curves of $(R_H, a\%)$ (d) 2-D curves of $(R_{FA}, a\%)$

Figure 11.8. 3-D ROC and 2-D curves of LCMV, BRLCMV, WTACEM, SCEM, MTCEM, FV and OSP

As we can see from Fig. 11.8, the 3-D ROC curves in Fig. 11.8(a) show the performance of a classifier as a function of three parameters R_H, R_{FA},$a\%$. While the 2-D ROC curves of (R_H, R_{FA}) in Fig. 11.8(b) provide the mean target hit rate of a classifier versus the mean false alarm rate, the 2-D curves of (R_H,$a\%$) and (R_{FA},$a\%$) in Figs. 11.8(c-d) indicate how a threshold value of $a\%$ affects the performance of a classifier. Table 11.3 shows the mean target hit rates by calculating the areas under their 2-D ROC curves in Fig. 11.8(b).

Table 11.3. Detection rates produced by LCMV, BRLCMV, WTACEM, SCEM, MTCEM, FV and OSP

	MTCEM	LCMV	BRLCMV	WTACEM	SCEM	OSP	FV
R_H	0.7334	0.7376	0.7372	0.7805	0.7769	0.4302	0.4578

It is interesting to note that in terms of detecting R and Y target pixels, WTACEM was the best among all the five classifiers, while the OSP classifier was worst. Also, according to Figs. 11.8(c), LCMV, BRLCMV, MTCEM, SCEM classifiers performed very similarly where their R_H dropped rapidly around $a\%$ = 11.8% compared to the WTACEM whose R_H dropped drastically around $a\%$ = 5%. On the other hand, the FV and the OSP classifiers performed very closely where their R_H decreased gradually with two sudden drops occurred around 35% and 65%. This demonstrated that the R_H of the five LCMV-based classifiers was sensitive to small a%'s and but their R_H's dropped below 0.5 after 11.8%. So, the overall performance of these six classifiers was offset by a% greater than 11.8%. However, this is not the case for the FV and the OSP classifiers whose overall performance was significantly offset by a% greater than 65%. Similar phenomena for R_{FA} were also observed in the 2-D curves of (R_{FA},a%) in Fig. 11.8(d). Since R_H and R_{FA} perform against each other, a compromise can be obtained from the 2-D ROC curves of (R_H,R_{FA}) in Fig. 11.8(b).

In order for the five LCMV-classifiers to classify the 15-panel HYDICE in a single image, five colors, red, green, blue, yellow and magenta were assigned to highlight the detected panels in rows 1-5 respectively. Fig. 11.9 shows that classification results produced by these five classifiers. As we can see, there is no visible difference among all the five color images.

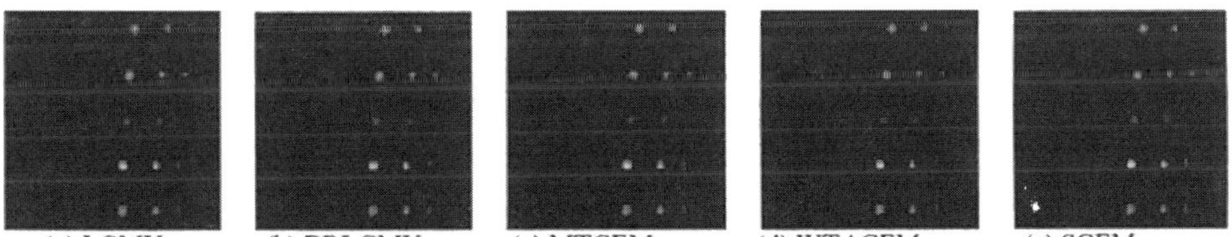

(a) LCMV (b) BRLCMV (c) MTCEM (d) WTACEM (e) SCEM

Figure 11.9. Classification of LCMV, BRLCMV, MTCEM, WTACEM and SCEM using a color assignment

11.8 REAL-TIME IMPLEMENTATION FOR LCMV CLASSIFIERS

As described in Chapter 4, the sample spectral correlation matrix $\mathbf{R}_{L\times L}$ allows us to implement the LCMV classifier in real time via a QR-decomposition. Despite that the real time processing implemented in Chapter 4 was carried out on a line-by-line basis, it can be also implemented in a pixel-by-pixel manner where the causal information is updated pixel-by-pixel rather than line-by-line. So, as a complement to Chapter 4, we will implement the following experiments for the TCIM classifier in real time on a pixel-by-pixel basis. The hyperspectral images to be used for experiments are those in Figs. 1.6 and 1.7.

Example 11.4.1 (LCVF AVIRIS Image)

The AVIRIS image in Fig. 1.6(a) contains five signatures of interest in these images are identified, "vegetation", "cinders", "shade", "playa (dry lakebed)" and "rhyolite". So, it requires five colors for target classification. In order to achieve the best possible visual discrimination, green, white, blue, olive and red were selected for vegetation, cinders, shade, playa and rhyolite respectively for visual discrimination. The choice of these colors is our preference, but can be arbitrary. To implement these five colors for the TCIM classifier, the required number of rows is 5. Since there are five targets, 5 column vectors are required. To implement the TCIM classifier, the constraint matrix $\mathbf{C}$ has size of $k = p = 5$ and was chosen to be the 5×5 identity matrix

$$\mathbf{C}^T = \left[\mathbf{c}_1\ \mathbf{c}_2\ \mathbf{c}_3\ \mathbf{c}_4\ \mathbf{c}_5\right]^T = \begin{bmatrix} 1 & 0 & 0 & 0 & 0 \\ 0 & 1 & 0 & 0 & 0 \\ 0 & 0 & 1 & 0 & 0 \\ 0 & 0 & 0 & 1 & 0 \\ 0 & 0 & 0 & 0 & 1 \\ \text{green} & \text{white} & \text{blue} & \text{olive} & \text{red} \end{bmatrix}. \qquad (11.21)$$

Each column vector $\mathbf{c}_i = (c_{i1}, c_{i2}, c_{i3}, c_{i4}, c_{i5})^T$ with $c_{ii} = 1$ and $c_{ij} = 0$ for $i \neq j$ is used to constrain one single specific target while nulling out all the other targets. As an example, we can set $c_{11} = 1$ in $\mathbf{c}_1$ to constrain $\mathbf{d}$ = [vegetation] in (11.21) to classify vegetation and $c_{1j} = 0$ for $2 \leq j \leq 5$ in $\mathbf{c}_1$ to null out $\mathbf{U}$ = [cinders,shade,playa,rhyolite] so that vegetation can be detected and classified by green while other targets will be eliminated at the same time. The real-time processing was carried out on a pixel-by-pixel basis in a top-to-bottom and left-to-right fashion.

Fig 11.10 shows a sequence of images generated by a pixel-by-pixel real time process where the five targets were detected in images labeled by (b-f). It started with random colors with no target pixels detected in Fig. 11.10(a) until at pixel 188 in line 5 where the first vegetation pixel was detected and was highlighted by green in Fig. 11.10(b). This was followed by detecting the first shade pixel with blue at pixel 89 in line 6 in Fig. 11.10(c), the first cinders pixel with white at pixel 98 in line 10 in Fig. 11.10(d), the first playa pixel with olive at pixel 187 in line 85 in Fig. 11.10(e) and the first rhyolite pixel with red at pixel 60 in line 110 in Fig. 11.10(f). Figs. 11.10(g) and 11.10(h) show the color and gray scale images after the processing was completed. The color map in Fig. 11.10(g) offers an advantage that the gray scale image cannot provide, visual discrimination of different detected targets in one single image. Interestingly, there was an anomaly detected at pixel 182 in line 84 in Fig. 11.10(f) and marked by a white circle at the top edge of the lake despite that we did not have knowledge about the anomaly. This anomaly was detected before the first playa pixel was detected in Fig. 11.10(e). It a two-pixel anomaly and detected in both color and gray scale images in Figs. 11.10(g-h), but is invisible in Fig. 3.1 by visual inspection. It was also detected in Fig. 6.1(c) by the RXD described in Chapter 6. Its detection was missed in Harsanyi and Chang (1994) because the used OSP classifier was supervised. As we examined its color closely, it seemed to be a mixture of white and olive. This indicated that the anomaly could be mixed by cinders and playa. Obviously, such information cannot be provided by the gray scale image in Fig. 11.10(h).

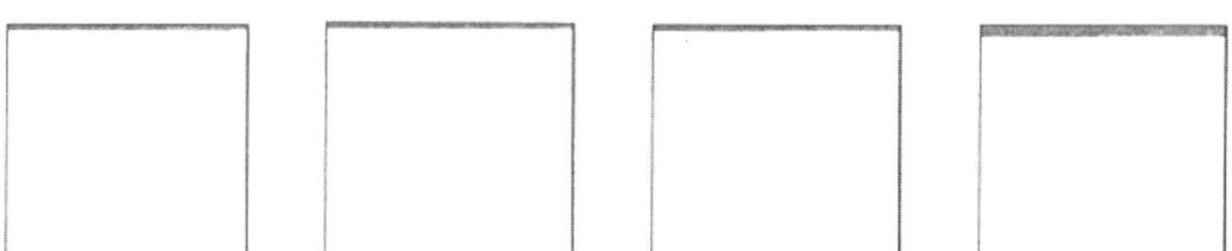

no target detected (line 4 at pixel 186) (a) | 1st detected vegetation pixel (line 5 at pixel 188) (b) | 1st detected shade pixel (line 6 at pixel 89) (c) | 1st detected cinders pixel (line 10 at pixel 98) (d)

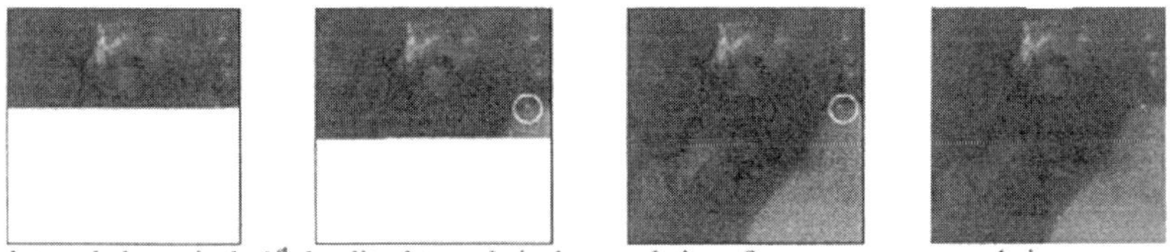

1st detected playa pixel (line 85 at pixel 187) (e) | 1st rhyolite detected pixel (line 110 at pixel 60) (f) | completion of process color map (g) | gray scale image (h)

Figure 11.10. Pixel-by-pixel real time processing of the LCVF AVIRIS image

Example 11.4.2 (15-Panel HYDICE scene)

As shown in Fig. 1.7, there are 15 panels in the image scene where the panels in the same row made from the same material. So, there are five distinct materials, which require five different colors for panel classification. Following a similar experiment conducted in Example 11.4.1, red, green, white, yellow and magenta were selected to specify panels from row 1 to row 5 respectively to achieve the best possible visual discrimination among these 15 panels. To implement these five colors for the TCIM classifier, the required number of rows is 5. On the other hand, to classify the 15 panels into 5 distinct material classes, the required number of columns must be 5. This results in a constraint matrix **C** with size of $k = m = 5$ specified by the 5×5 identity matrix

$$\mathbf{C}^T = \left[\mathbf{c}_1^T\,\mathbf{c}_2^T\,\mathbf{c}_3^T\,\mathbf{c}_4^T\,\mathbf{c}_5^T\right] = \begin{bmatrix} 1 & 0 & 0 & 0 & 0 \\ 0 & 1 & 0 & 0 & 0 \\ 0 & 0 & 1 & 0 & 0 \\ 0 & 0 & 0 & 1 & 0 \\ 0 & 0 & 0 & 0 & 1 \\ \underbrace{}_{\text{red}} & \underbrace{}_{\text{green}} & \underbrace{}_{\text{white}} & \underbrace{}_{\text{yellow}} & \underbrace{}_{\text{magenta}} \end{bmatrix}. \tag{11.22}$$

The constraint vectors, $\mathbf{c}_1, \mathbf{c}_2, \mathbf{c}_3, \mathbf{c}_4, \mathbf{c}_5$ in **C** are used to detect and classify the 15 panels. Each column vector $\mathbf{c}_i = (c_{i1}, c_{i2}, c_{i3}, c_{i4}, c_{i5})^T$ with $c_{ii} = 1$ and $c_{ij} = 0$ for $i \neq j$ corresponds to one constraint vector in (11.1) with $\mathbf{d}$ = P*j* and $\mathbf{U} = [\mathrm{P}i]_{i=1, i\neq j}^{5}$. For example, in order to detect and classify the panels in row 1, we set $c_{11} = 1$ in $\mathbf{c}_1$ to constrain $\mathbf{d}$ = [P1] and $c_{1j} = 0$ for $2 \leq j \leq 5$ in $\mathbf{c}_1$ to null out $\mathbf{U}$ = [P2 P3 P4 P5] so that panels in row 1 can be detected and classified by red while panels in other rows will be eliminated at the same time. The real-time processing was carried out on a pixel-by-pixel basis in a top-to-bottom and left-to-right fashion. Fig. 11.11(a)-11.11(o) shows a sequence of images resulting from implementing the TCIM classifier in a pixel-by-pixel real time process with the constraint matrix **C** specified by (11.21).

In the beginning of data processing, no panels were detected. So, random colors were shown in the image in Fig. 11.11(a). As soon as the process detected the first target, i.e., the panel p_{11} in row 1, the detected p_{11} pixel was immediately changed to red while the background was suddenly changed to dark where the red color was chosen to specify the panels in row 1. Then the process continued to detect the panels in row 1 with red, e.g.

p_{21} in Fig. 11.11(c) until the process detected the first panel pixel, p_{21} in row 2 in Fig. 11.11 (d) where the pre-assigned green color was used to highlight the panel p_{21} and discriminate it from the red pixels that were panel pixels in row 1. Unfortunately, it missed the detection of the panel pixel, p_{13} due to its very small amount of abundance. Figs. 11.11(e-o) how the real time process detected target panel pixel by pixel, p_{22}, p_{23} with green, p_{31}, p_{32}, p_{33} with white, p_{41}, p_{42}, p_{43}, with yellow, p_{51}, p_{52}, p_{53} with magenta in sequence. Finally, Fig. 11.11(p) shows a complete detection. At the time when the last pixel in the image was completed, the classification process was also accomplished and all the 15 panels were successfully classified by five colors, red for the panels in row 1, green for the panels in row 2, white for the panels in row 3, yellow for the panels in row 4 and magenta for the panels in row 5. In order to demonstrate how important to use a color map to visualize the classification result in Fig. 11.11(p), Fig. 11.11(q) shows only the gray scale image with all the detected 15 panels. As we can from Fig. 11.11(g), there is no way to discriminate the panels of one row from the panels of another row.

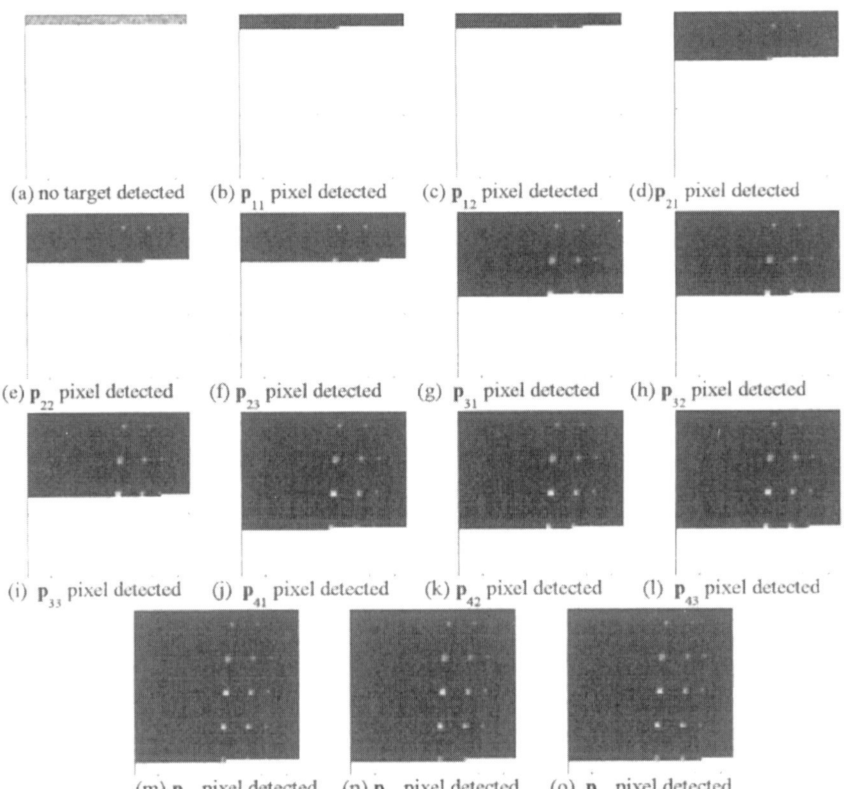

(a) no target detected (b) p_{11} pixel detected (c) p_{12} pixel detected (d) p_{21} pixel detected

(e) p_{22} pixel detected (f) p_{23} pixel detected (g) p_{31} pixel detected (h) p_{32} pixel detected

(i) p_{33} pixel detected (j) p_{41} pixel detected (k) p_{42} pixel detected (l) p_{43} pixel detected

(m) p_{51} pixel detected (n) p_{52} pixel detected (o) p_{53} pixel detected

(p) completion of process (color map)

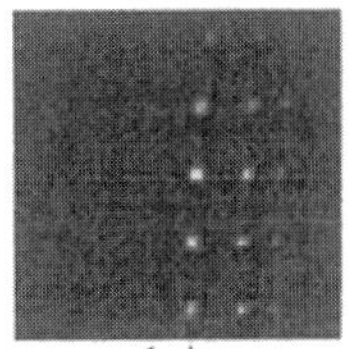
(q) gray scale image

Figure 11.11. Real-time pixel-by-pixel implementation of the 15-panel HYDICE scene

Several experiments were also conducted by including some partial knowledge in the TCIM-based target classifier, for instance, grass and tree in the image scene. These two natural background signatures can be used as undesired targets for background annihilation. The results are not included here since there was not much improvement. This is because there are large fields of trees and grass in the image scene. Two background signatures were not sufficient to represent the image background.

11.9 CONCLUSIONS

This chapter presents several versions of LCMV classifiers, which can detect and classify multiple targets of distinct types in a single image by a single operation. More importantly, the LCMV classifiers can be implemented by a real-time process on a pixel-by-pixel basis. It expands the capability of the LCMV detectors described in Chapter 4 to LCMV multiple-target classifiers. The LCMV multiple-target classifiers use multiple constraints to accommodate similar target signatures to ease the information sensitivity as well as to recover missing target pixels. In addition, the LCMV multiple-target classifiers also take advantage of multiple constraints as a color assignment mechanism so that all the detected targets can be displayed in one single image with different colors. Although the LCMV classifiers studied in this chapter is supervised which require target knowledge, they can be further extended to unsupervised classifiers by including unsupervised algorithms in Chapter 5 or anomaly detectors described in Chapter 6. This will be discussed in Chapters 13 and 14. Finally, Table 11.4 summarizes assumptions and constraints that are imposed in each of the methods presented in this paper where the CEM has three versions, WTACEM, SCEM, MTCEM when it is implemented as a classifier.

Table 11.4. Assumptions/constraints imposed by various TSCMPC methods

	linear mixture model	desired targets (**D**)	undesired targets (**U**)	filter vector **w**	R_{LxL}	background
LCMV	no/no	yes/yes	yes/yes	no/no	yes/no	no/no
BRLCMV	no/no	yes/yes	yes/yes	no/no	yes/no	no/yes
WTACEM	no/no	yes/yes	no/no	no/no	yes/no	no/no
SCEM	no/no	yes/yes	no/no	no/no	yes/no	no/no
MTCEM	no/no	yes/yes	no/no	no/no	yes/no	no/no
TCIMF	no/no	yes/yes	yes/yes	no/no	yes/no	no/no
FV	yes/yes	yes/yes	no/no	no/yes	no/no	no/no
OSP	yes/no	yes/no	no/no	no/no	no/no	no/no

For example, a "yes/no" at the row of the OSP crossed with the column of the linear mixture model in Table 11.4 means that the OSP assumes a linear mixture model but does not impose constraints on this model.

12

TARGET SIGNATURE-CONSTRAINED MIXED PIXEL CLASSIFICATION (TSCMPC): LINEARLY CONSTRAINED DISCRIMINANT ANALYSIS (LCDA)

Fisher's Linear Discriminant Analysis (LDA) is a widely used technique for pattern classification and was discussed in Chapter 9. It uses Fisher's ratio, a ratio of between-class scatter matrix to within-class scatter matrix as an optimal criterion to derive a set of feature vectors by which high dimensional data can be projected onto a low-dimensional feature space in the sense of maximizing class separability. This chapter presents an approach derived from Fisher's LDA, referred to as linear constrained distance-based discriminant analysis (LCDA), that uses a similar criterion to Fisher's ratio for classification. It maximizes the ratio of inter-distance between classes to intra-distance within classes while imposing a constraint that all the class means must be aligned along predetermined directions. Interestingly, the LCDA classifier operates at the same operational form as does OSP in Chapter 8, but achieves better classification resulting from the use of target signature constraints. Because of that, LCDA can be viewed as a constrained version of OSP.

12.1 INTRODUCTION

Many mixed pixel classification (MPC) methods have been proposed in the past such as linear unmixing methods described in Chapter 8. One of principal differences between pure and mixed pixel classification is that the former is a class-labeling process, whereas the latter is actually signature abundance estimation. In Chapter 9 we have seen that MPC generally performed better than the pure pixel classification (PPC) where the estimated target signature abundance fractions were used for classification. Additionally, as shown in Chapter 9, if MPC was converted to PPC, Fisher's linear discriminant analysis (LDA) was among the best. This suggested that applying pure pixel-based LDA directly to mixed pixel classification problems may not be an effective way to best utilize Fisher's LDA. In magnetic resonance imaging (MRI) applications (Soltanian-Zadeh *et al.*, 1996), Soltanian-Zadeh *et al.* recently developed a constrained criterion to characterize brain issues for 3-D feature representation. They used the ratio of the inter-set distance (IED) to intra-set distance (IAD) (Tou and Gonzalez, 1974) as an optimal criterion while imposing a constraint that each class center must be aligned along predeterimined directions. In order to arrive at an analytical solution, Soltanian-Zadeh *et al.* further made an assumption of white Gaussian noise so that IED can be further simplified to a constant. As a result, maximizing the ratio of IED to IAD is equivalent to minimizing IAD. However, the white Gaussian noise assumption may not be valid in hyperspectral images since it has demonstrated (Chang et al., 1998, Du and Chang, 1999) that unknown interferers in hyperspectral imagery such as background signatures, clutters were more severe and dominant than noise and they are generally non-Gaussian and non-stationary. This chapter presents a linear constrained distance-based discriminant analysis (LCDA) for hyperspectral image classification. It implements a criterion similar to Fisher's ratio along with a constraint that the vector directions (or spectral shapes) of targets of interest must be aligned with a set of specific directions. The idea is derived from modifying Soltanian-Zadeh *et al.*'s approach, but does not require the Gaussian assumption. As a matter of fact, the white Gaussian assumption is not necessary and can be taken care of by a whitening process. This is because minimizing IAD is equivalent to minimizing the trace of the data sample covariance matrix $\mathbf{K}$ and a whitening process can be applied to decorrelate $\mathbf{K}$ without making white Gaussian assumption. Furthermore, after the whitening processing is completed, LCDA can be carried out by OSP in Chapter 8. As a result of using direction constraints, LCDA performs significantly better than OSP. From this point of view, LCDA can be thought of as a constrained version of OSP.

Recently, a similar approach, called filter vectors (FV), was also proposed by Bowles et al (1995) and discussed in Chapter 11 (Section 11.3). It assumed that there was a set of targets of interest present in an image scene. These targets were then used to form a linear mixture model described by (3.1). A set of filter vectors was obtained by solving a constrained optimization problem with a constraint imposed on the targets of interest in such a way that their target signatures are orthogonal each other. Finally, each of the resulting filter vectors was used to classify a specific target. Interestingly, if a whitening process is included in LCDA prior to classification, Bowles et al's FV approach can be considered as a special case of LCDA with no need of using the linear mixture model.

Of particular interest is a comparative analysis between LCDA and LCMV, both of which impose a constraint on target signature constraint to improve classification performance. The advantage of LCMV over LCDA is that LCMV only requires knowledge about targets of interest including desired and undesired targets. By contrast, LCDA requires a complete knowledge of target signatures of interest, but in this case, it performs more effectively than does LCMV, specifically for the case that two targets have very similar signatures.

It should be noted that the idea of using signal direction constraints is not new and has been applied to various applications, Minimum Variance Distortionless Response (MVDR) beamformer in array processing (Van Veen and Buckley, 1988, Haykin, 1996), chemical remote sensing (Althouse and Chang, 1991) and LCMV in Chapters 4 and 11.

12.2 LCDA

In Fisher's discriminant analysis, the discriminant vectors generally impose no constraints on their directions. However, in many practical applications, there may exist some partial knowledge that can be useful to constrain desired features along certain directions to minimize the effects of undesired features. As discussed in Chapters 4 and 11, LCMV constrains a set of desired target signatures while minimizing its output energy. In MRI (Soltanian-Zadeh *et al.*, 1996), Soltanian-Zadeh *et al.* constrained normal tissues along prespecified target positions so as to achieve a better clustering for visualization. In this section, we follow a similar idea to derive a linear constrained discriminant analysis (LCDA) for hyperspectral image classification.

Let ϕ denote a linear transformation to project a high dimensional data space into a low dimensional space. Suppose that there are c classes of interest. The ratio of the average between-class distance to average within-class distance via the ϕ is given by

$$J(\phi) = \frac{\{2/[c(c-1)]\}\left[\sum_{i=1}^{c}\sum_{j=i+1}^{c}\left\|\phi(\mu_i) - \phi(\mu_j)\right\|^2\right]}{[1/(cN)]\left[\sum_{i=1}^{c}\sum_{j=1}^{N_i}\left\|\phi(\mathbf{x}_j^i) - \phi(\mu_i)\right\|^2\right]} \tag{12.1}$$

where the global mean μ and local mean μ_i of the i-th class were defined in the Section 9.2.3. Suppose that there are p classes of interest with $p \le c$, denoted by $\{C_i\}_{i=1}^{p}$. We can maximize (12.1) over ϕ subject to a constraint that the desired class means $\{\mu_i\}_{i=1}^{p}$ must be aligned along with the prespecified target directions $\{\mathbf{m}_i\}_{i=1}^{p}$. In other words, what we are interested here is to find an optimal linear transformation ϕ^*, which maximizes

$$J(\phi) \text{ subject to } \mathbf{m}_i = \phi(\mu_i) \text{ for all } i \text{ with } 1 \le i \le p \le c. \tag{12.2}$$

Here the target directions $\{\mathbf{m}_i\}_{i=1}^{p}$ can be interpreted as spectral shapes of target signatures as we did for LCMV in Chapters 4 and 11. For example, $\mathbf{m}_i^T\mathbf{m}_j = \delta_{ij}$ implies that the directions of target signatures, $\mathbf{m}_i$ and $\mathbf{m}_j$ are orthogonal. It also implies that the spectral shapes of target signatures are mutually orthogonal. (12.2) outlines a general constrained optimization problem, which can be solved numerically. The linear transformation ϕ in (12.2) can be expressed a matrix transform via a weight matrix $\mathbf{W}_{d\times p} = [\mathbf{w}_1\,\mathbf{w}_2\cdots\mathbf{w}_p]_{d\times p}$ by

$$\mathbf{y} = \phi(\mathbf{x}) = \mathbf{W}_{d\times p}^T\mathbf{x}. \tag{12.3}$$

Now, assume that $p = c$ and $\mathbf{m}_i$ in (12.2) is a $p\times 1$ unit column vector with one in the i-th component and zeros in all other components that is specified by $\mathbf{m}_j = (0\cdots 0\underset{j}{1}0\cdots 0)^T$, $\mathbf{m}_i$ is a vector along the i-th coordinate in the space R^p. If we further assume that $\{\mathbf{w}_i\}_{i=1}^{p}$ in $\mathbf{W}_{d\times p}$ represented by be the linear transformation ϕ also linearly independent, then (12.2) becomes

$$J(\mathbf{W}_{d\times p}) \text{ subject to } \mathbf{w}_i^T \mu_j = \delta_{ij} \text{ for } 1 \le i, j \le p \tag{12.4}$$

where δ_{ij} is Kronecker's notation given by $\delta_{ij} = 1$ if $i = j$ and $\delta_{ij} = 0$ if $i \neq j$.

The numerator and denominator of $J(\phi)$ in (12.1) can be further reduced to

$$\begin{aligned} &\{2/[p(p-1)]\}\left[\Sigma_{i=1}^{p}\Sigma_{j=i+1}^{p}\left\|\phi(\mu_i)-\phi(\mu_j)\right\|^2\right] \\ &= \{2/[p(p-1)]\}\{\Sigma_{i=1}^{p}\Sigma_{j=i+1}^{p}[2]\} = 2 \end{aligned} \tag{12.5}$$

and

$$\begin{aligned} &[1/(pN)]\Sigma_{i=1}^{p}\left[\Sigma_{j=1}^{N_i}\left\|\phi(\mathbf{x}_j^i)-\phi(\mu_i)\right\|^2\right] \\ &= [1/(pN)]trace\left(\mathbf{W}^T\left\{\Sigma_{i=1}^{p}\left[\Sigma_{j=1}^{N_i}\left(\mathbf{x}_j^i-\mu_i\right)\left(\mathbf{x}_j^i-\mu_i\right)^T\right]\right\}\mathbf{W}\right) \end{aligned} \tag{12.6}$$

Since the numerator in (12.1) becomes a constant given by (12.5), maximizing (12.1) is reduced to minimizing (12.6). More specifically, maximizing (12.4) is reduced to

$$\min_{\mathbf{W}} trace\left(\mathbf{W}^T\left\{\Sigma_{i=1}^{p}\left[\Sigma_{j=1}^{N_i}\left(\mathbf{x}_j^i-\mu_i\right)\left(\mathbf{x}_j^i-\mu_i\right)^T\right]\right\}\mathbf{W}\right) \text{ subject to } \mathbf{w}_i^T\mu_j = \delta_{ij}. \tag{12.7}$$

It should be noted that the term in the brackets in (12.7) turns out to be the within-class scatter matrix $\mathbf{S}_W$ in Section 9.2.3. According to Fisher's LDA minimizing $\mathbf{S}_W$ is equivalent to minimizing the total scatter matrix $\mathbf{S}_T$. If we let $\mathbf{K}$ denote the sample covariance matrix of all training samples, then $\mathbf{S}_T$ and $\mathbf{K}$ are related by the following equation

$$\mathbf{S}_T = N \cdot \mathbf{K}. \tag{12.8}$$

Since N is the total number of samples and is a constant, substituting (12.8) into (12.7) results in the following equivalent problem by

$$\min_{\mathbf{W}} trace\left(\mathbf{W}^T\mathbf{K}\mathbf{W}\right) \text{ subject to } \mathbf{w}_i^T\mu_j = \delta_{ij}. \tag{12.9}$$

Equation (12.9) is now reduced to a simple problem, that is, how to find a matrix $\mathbf{A}$, which decorrelates and whitens the covariance matrix $\mathbf{K}$ into an identity matrix. Let $\mathbf{A}$ be such a matrix. (12.9) can be whitened and further reduced to a simple optimization problem given by

$$\min_{\mathbf{W}}\left\{\Sigma_{i=1}^{p}\mathbf{w}_i^T\mathbf{w}_i\right\} = \min_{\mathbf{W}}\left\{\Sigma_{i=1}^{p}\|\mathbf{w}_i\|^2\right\} \text{ subject to } \mathbf{w}_i^T\hat{\mu}_j = \delta_{ij}. \tag{12.10}$$

where $\{\hat{\mu}_i\}_{i=1}^{p}$ are the vectors resulting from the whitening process using **A**, i.e., $\hat{\mu}_i = \mathbf{A}^T\mu_i$. Since $\|\mathbf{w}_i\|^2$ is nonnegative for each $1 \le i \le p$, solving (12.10) is equivalent to solving

$$\min_{\mathbf{w}_i} \left(\mathbf{w}_i^T\mathbf{w}_i\right)^{-1/2} \text{ subject to } \mathbf{w}_i^T\hat{\mu}_j = \delta_{ij} \text{ for each } i \text{ with } 1 \le i \le p. \tag{12.11}$$

The solution $\mathbf{w}_i^*$ to (12.11) can then be obtained analytically by

$$\mathbf{w}_i^* = \hat{\mu}_i^T P_{\mathbf{U}}^{\perp} \tag{12.12}$$

where

$$P_{\mathbf{U}}^{\perp} = \mathbf{I} - \mathbf{U}(\mathbf{U}^T\mathbf{U})^{-1}\mathbf{U}^T \tag{12.13}$$

and **U** is the space linearly spanned by all cluster centers but $\hat{\mu}_i$, i.e. $\mathbf{U} = \left\langle \{\hat{\mu}_j\}_{j=1, j\neq i}^{p} \right\rangle$. Surprisingly, the solution specified by (12.12) turns out to be OSP in Chapter 8. This implies that the classifier $\mathbf{w}_i^* = \hat{\mu}_i^T P_{\mathbf{U}}^{\perp}$ derived by LCDA is essentially a constrained version of OSP.

A comment on LCDA is noteworthy. Despite the fact that two criteria used in LDA and LCDA look similar where both calculate a ratio of a measure of between-class to a measure of within-class, there are differences between LDA and LCDA. The LDA maximizes the between-class scatter matrix while LCDA minimizing the intra-distance within classes. Another difference is that the number of the discriminant vectors resulting from the LDA is one less than the total number of classes of interest, p, whereas the number of projection vectors generated by LCDA is the same number of classes of interest, p. Most significantly, LCDA can be regarded as a variant of OSP, but LDA cannot.

12.3 WHITENING PROCESS FOR LCDA

In order to reduce the optimization problem given by (12.9) to the one specified by (12.10), Soltanian-Zadeh *et al.* (1996) made the white Gaussian noise assumption for MRI to arrive at (12.10). As a matter of fact, this assumption can be removed by finding a whitening matrix **A** for (12.9) to derive (12.10).

Assume that $\left\{\lambda_i\right\}_{i=1}^{d}$ are the eigenvalues of the sample covariance matrix **K** or the total scatter matrix $\mathbf{S}_{\mathrm{T}}$ and $\left\{\mathbf{v}_i\right\}_{i=1}^{d}$ are their corresponding normalized unit eigenvectors. Since $\mathbf{S}_{\mathrm{T}}$ or **K** is nonnegative definite, all eigenvalues are nonnegative. There exists a unitary matrix **Q** such that Σ can be expressed by

$$\mathbf{Q}^T\mathbf{K}\mathbf{Q} = \Lambda \tag{12.14}$$

where $\mathbf{Q} = [\mathbf{v}_1\, \mathbf{v}_2 \cdots \mathbf{v}_d]$ is an eigenvector matrix made up of all eigenvectors $\{\mathbf{v}_i\}_{i=1}^{d}$ and $\Lambda = diag\{\lambda_i\}_{i=1}^{d}$ is a diagonal matrix with $\{\lambda_i\}_{i=1}^{d}$ as diagonal elements.

If we let $\Lambda^{-1/2} = diag\{\sqrt{\lambda_i}\}_{i=1}^{d}$, multiplying both sides of (12.14) by $\Lambda^{-1/2}$ results in

$$\Lambda^{-1/2}\mathbf{Q}^T\mathbf{K}\mathbf{Q}\Lambda^{-1/2} = \mathbf{I} \tag{12.15}$$

From (12.15), we obtain the desired matrix $\mathbf{A}$ for (12.9), which is given by

$$\mathbf{A} = \mathbf{Q}\Lambda^{-1/2} \tag{12.16}$$

so that $\mathbf{A}^T\mathbf{K}\mathbf{A} = \mathbf{I}$.

Using (12.12) and (12.16) the linearly constrained Euclidean distance-based discriminant analysis optimization problem specified by (12.2) can be solved as follows:

LCDA Algorithm

1. Find the cluster centers $\{\mu_i\}_{i=1}^{p}$ of p-classes and the sample covariance matrix $\mathbf{K}$.
2. Find the eigenvalues $\{\lambda_i\}_{i=1}^{d}$ and their corresponding eigenvectors $\{\mathbf{v}_i\}_{i=1}^{d}$ of $\mathbf{K}$ to form the unitary matrix $\mathbf{Q} = [\mathbf{v}_1\, \mathbf{v}_2 \cdots \mathbf{v}_d]$ and the diagonal matrix $\Lambda = diag\{\lambda_i\}_{i=1}^{d}$.
3. Form the desired matrix $\mathbf{A} = \mathbf{Q}\Lambda^{-1/2}$ using (12.16).
4. Find the $\mathbf{A}$-transformed cluster means $\hat{\mu}_i = \mathbf{A}^T\mu_i$ for each i with $1 \le i \le p$.
5. Find $\mathbf{w}_i^* = \hat{\mu}_i^T P_U^{\perp}$ for each i with $1 \le i \le p$ with $P_U^{\perp}$ given by (12.13). This step is the classification step where $\mathbf{w}_i^*$ is used to classify data samples into the i-th class.

12.4 BOWLES ET AL.'S FILTER VECTORS (FV) ALGORITHM

In Section 11.3, we have shown that Bowles et al.'s FV algorithm (Bowles et al., 1995) was a special case of the BRLCMV classifier. In this section, we will further show that Bowles et al.'s FV algorithm can be also interpreted as a special case of LCDA.

First of all, recalling (3.1), a linear mixture of p target signatures $\mathbf{m}_1, \mathbf{m}_2, \cdots, \mathbf{m}_p$ in an image pixel vector $\mathbf{r}$ can be expressed by

$$\mathbf{r} = \mathbf{M}\alpha + \mathbf{n} \tag{3.1}$$

The first condition imposed on (3.1) by Bowles et al.'s filter vectors, $\{\mathbf{w}_i\}_{i=1}^{p}$ is (11.5) which says that a filter vector, $\mathbf{w}_i$ must be orthogonal to all the signatures $\{\mathbf{m}_j\}_{j=1, j\neq i}^{p}$ except the one, $\mathbf{m}_i$ that it used for training. This condition is exactly the same constraint imposed by LCDA, (12.4). The second condition proposed by Bowles et al.'s filter vectors is (11.6), which states that the sum of all the components in each filter vector must be zero. This implies that mean of $\mathbf{w}_i$ must be zero. The Bowles et al. third condition, (11.7) that minimizes the vector length of each of filter vectors is identical to the objective function in (12.11). Despite that there was no explicit rationale provided in Bowles' et al. (1995) we can derive that the need of these conditions actually arises from the following facts.

Let y_i denote the output of the filter vector $\mathbf{w}_i$ operating on $\mathbf{r}$. Then

$$\begin{aligned} y_i &= \mathbf{w}_i \cdot \mathbf{r} = \mathbf{w}_i^T \mathbf{r} = \mathbf{w}_i^T (\mathbf{M}\alpha + \mathbf{n}) = \mathbf{w}_i^T \left(\sum_{j=1}^{p} \mathbf{m}_j \alpha_j + \mathbf{n} \right) \\ &= \sum_{j=1}^{p} \left(\mathbf{w}_i^T \mathbf{m}_j \alpha_j \right) + \mathbf{w}_i^T \mathbf{n} \\ &= \sum_{j=1}^{p} \left(\delta_{ij} \alpha_j \right) + \mathbf{w}_i^T \mathbf{n} = \alpha_i + \mathbf{w}_i^T \mathbf{n} \end{aligned} \tag{12.17}$$

and its output energy is given by

$$\begin{aligned} y_i^2 &= E\left[\left(\mathbf{w}_i^T \mathbf{r} \right) \left(\mathbf{w}_i^T \mathbf{r} \right)^T \right] = E\left[\left(\alpha_i + \mathbf{w}_i^T \mathbf{n} \right) \left(\alpha_i + \mathbf{w}_i^T \mathbf{n} \right)^T \right] \\ &= \alpha_i^2 + 2\alpha_i E\left[\mathbf{w}_i^T \mathbf{n} \right] + \mathbf{w}_i^T E\left[\mathbf{n}\mathbf{n}^T \right] \mathbf{w}_i \end{aligned} \tag{12.18}$$

Equation (12.17) implies that in order for the filter output y_i to accurately approximate α_i, it requires that $\mathbf{w}_i \cdot \mathbf{n} = \mathbf{w}_i^T \mathbf{n}$ be zero. On the other hand, (12.18) indicates that in order for output y_i to accurately approximate α_i, we must minimize $2\alpha_i E\left[\mathbf{w}_i^T \mathbf{n} \right] + \mathbf{w}_i^T E\left[\mathbf{n}\mathbf{n}^T \right] \mathbf{w}_i$. One way to achieve these two goals is to make the assumption that the noise in model (3.1) is a zero-mean random vector process, i.e.,

$$E[\mathbf{n}] = 0 \tag{12.19}$$

so that $E\left[\mathbf{w}_i^T \mathbf{n} \right] = \sum_{l=1}^{L} w_{il} E[n_l] = 0$. By taking expectation of y_i and using (12.19), (12.17) becomes

$$y_i = \mathbf{w}_i^T \mathbf{r} = \alpha_i \,. \tag{12.20}$$

Assuming the noise variance given by σ^2 and taking expectation of y_i^2, (12.18) is reduced to

$$y_i^2 = \alpha_i^2 + \sigma^2 \mathbf{w}_i^T \mathbf{w}_i \tag{12.21}$$

which turns out to be exactly the same second condition given in Pesses (1999). Both (12.19) and (12.20) suggest that in order for the filter output $y_i = \mathbf{w}_i^T \mathbf{r}$ to accurately represent the abundance fraction of $\mathbf{m}_i$, α_i, we must minimize $\mathbf{w}_i^T \mathbf{w}_i$ in (12.21) with respect to $\mathbf{w}_i$ for each $1 \le i \le p$. As discussed in Section 11.3, the objective of Bowles et al.'s second condition, i.e., $\mathbf{w}_i \cdot \mathbf{1}_{L\times 1} = \mathbf{w}_i^T \mathbf{1}_{L\times 1} = 0$, is to remove the effects caused by image background, not make $\mathbf{w}_i \cdot \mathbf{n} = \mathbf{w}_i^T\mathbf{n} \approx 0$ as proposed by (12.19). However, as shown in (12.20), if the noise is assumed to be a zero-mean white random vector process specified by (12.19), $y_i = \mathbf{w}_i^T\mathbf{r} = \alpha_i$. Therefore, if we replace Bowles et al.'s second condition, $\mathbf{w}_i \cdot \mathbf{1}_{L\times 1} = \mathbf{w}_i^T\mathbf{1}_{L\times 1} = 0$ given by (11.6) with (12.19) which can be easily satisfied by a whitening process described in Section 12.3, Bowles et al's filter vectors become the ones that minimize (12.21). This is equivalent to solving exactly the same constrained optimization problem for LCDA that is specified by (12.11). With this interpretation, Bowles et al.'s filter vectors method essentially performs as a different version of LCDA with zero-mean noise process.

12.5 COMPUTER SIMULATIONS AND HYPERSPECTRAL IMAGE EXPERIMENTS

In order to evaluate the performance of LCDA, we conduct real hyperspectral image experiments in comparison with the results produced by OSP, LDA and LCMV.

Example 12.1 (Computer Simulations)

The following computer simulations are a follow-up example of Section 11.6 considered in Chapter 11. The same 250 simulated mixed pixels were also used in this example. Fig. 12.1 is the result detection and classification results produced by the LCDA, the FV and the OSP classifiers (from left to right) with the target signature matrix **M** = [blackbrush, creosote leaves, sagebrush]. It should be noted that the figures in Fig. 12.1 used different scales to show their ability in classification. Apparently, LCDA performed significantly better than did FV and OSP where the LCDA classifier correctly detected and classified most target pixels while the FV and the OSP classifiers failed to detect all target pixels. The reason for this is because the three target signatures, blackbrush, creosote leaves and sagebrush have very similar signatures according to Tables 2.3-2.6. In this case, LCDA shows its strength in classification by constraining their target signature directions. It is worth noting that none of these three classifiers detected correct amounts of target abundance fractions. If we further compare the results in Fig. 12.1 to the results in Fig. 11.1, LCMV and BRLCMV are the best classifiers among these five classifiers.

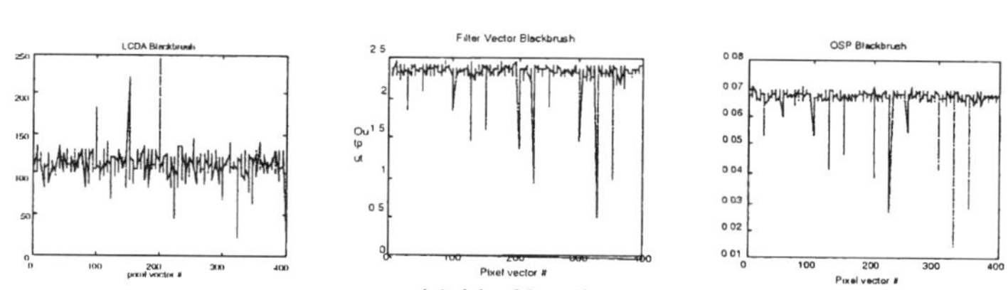

(a) blackbrush

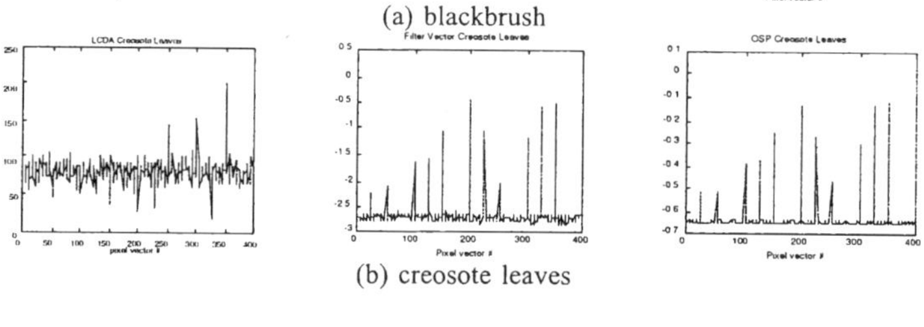

(b) creosote leaves

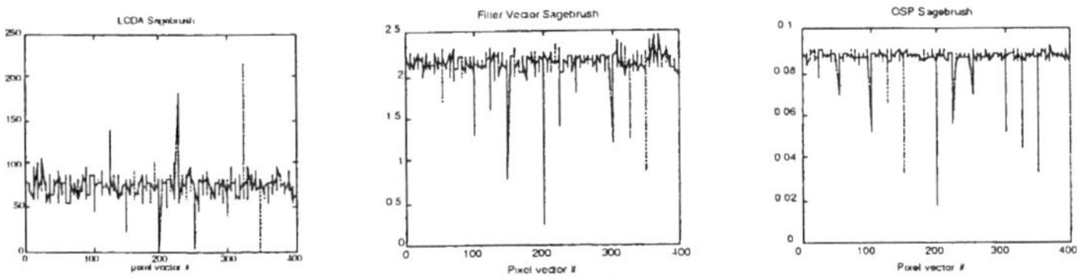

(c) sagebrush

Figure 12.1. Detection and classification results of the LCDA, FV and OSP classifiers (from left to right) with **M** = [blackbrush, creosote leaves, sagebrush].

Similar computer simulations to that in Fig. 11.2 were also conducted using the target signature matrix **M** = [blackbrush, creosote leaves, sagebrush, dry grass, red soil] where the red soil and dry grass were simulated as background signatures. The classification results of the LCDA, the FV and the OSP classifiers are shown in Fig. 12.2. It is interesting to note that the performance of all the three classifiers was improved, specifically, the FV and the OSP classifiers. The LCDA classifier still performed slightly better than the FV and the OSP classifier. This experiment shows that when background signatures are distinct from the target signatures, they actually may help the classifier improve their performance.

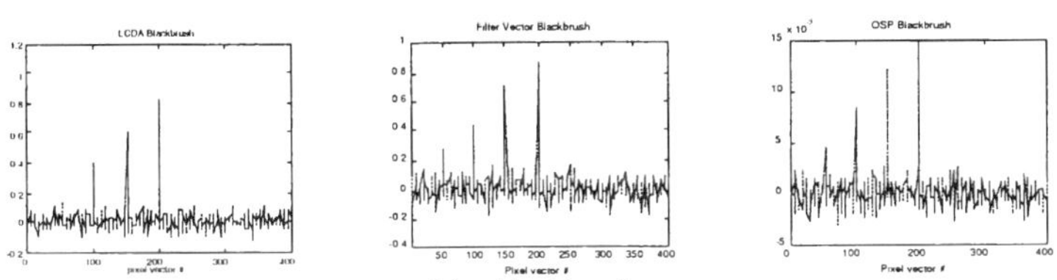

(a) blackbrush

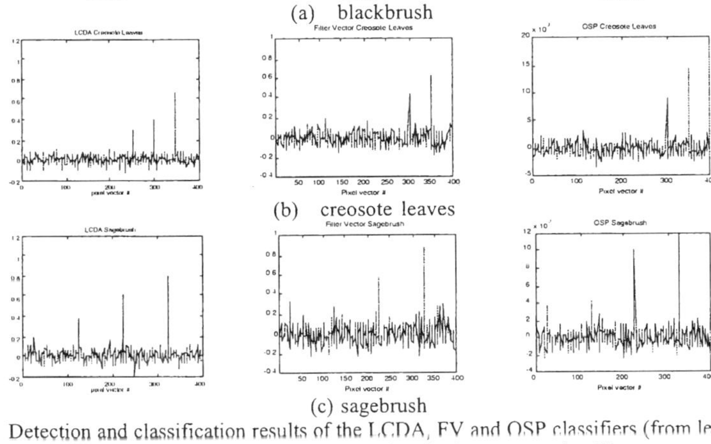

(b) creosote leaves

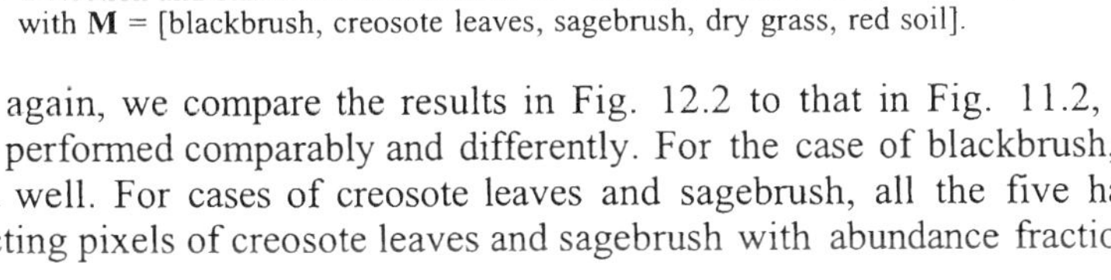

(c) sagebrush

Figure 12.2. Detection and classification results of the LCDA, FV and OSP classifiers (from left to right) with **M** = [blackbrush, creosote leaves, sagebrush, dry grass, red soil].

Once again, we compare the results in Fig. 12.2 to that in Fig. 11.2, all the five classifiers performed comparably and differently. For the case of blackbrush, all the five performed well. For cases of creosote leaves and sagebrush, all the five had difficulty with detecting pixels of creosote leaves and sagebrush with abundance fractions 20% and 40%.

Example 12.2 (LCVF AVIRIS Image)

In this example, the AVIRIS data in Fig. 1.6(a) were used for experiments. In this case, $p = 5$ and $d = 224$, which is the total number of bands. LCDA imposed a constraint that forces these five target signatures along five orthogonal directions. Fig. 12.3 shows results with the images in the first, second, third and fourth columns were produced by the LCDA, the LDA, the FV and the OSP classifiers respectively.

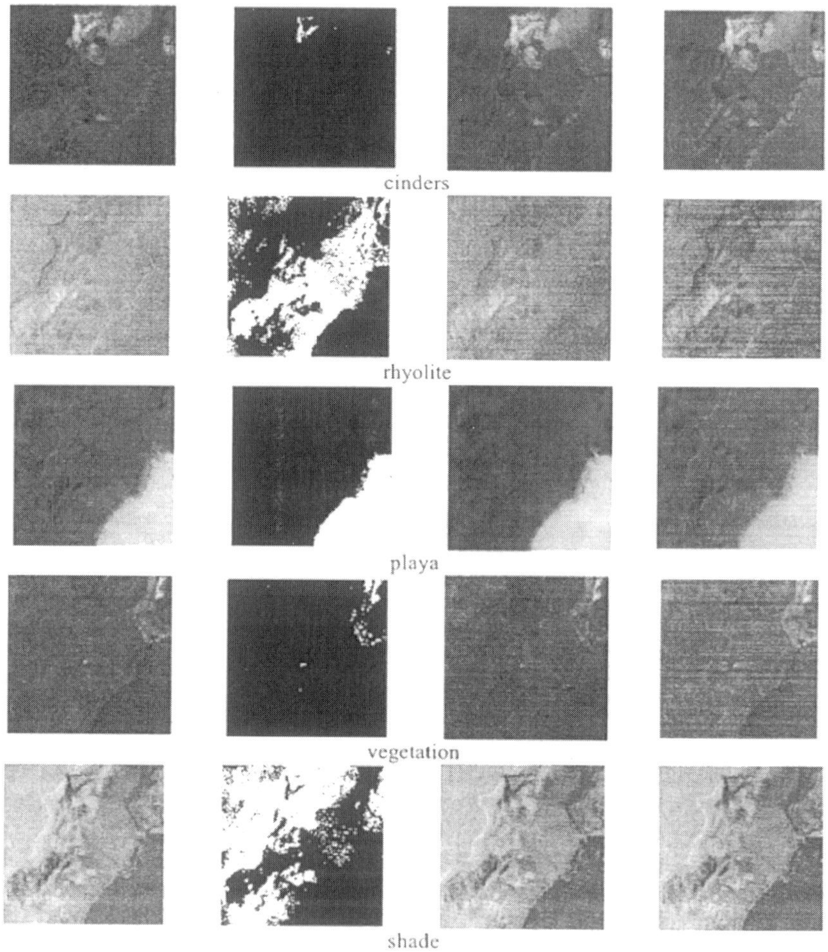

Figure 12.3. Classification results produced by 1st column: LCDA; 2nd column: LDA; 3rd column: FV; 4th column: OSP LCDA, LDA, FV and OSP

As we can see from Fig. 12.3, the results produced by the LCDA performed better than the LDA, FV and OSP classifiers in detection and classification of all five targets, specifically, in classifying cinders, vegetation and shade. It should be noted that LDA performed as a pure pixel classifier rather than a mixed pixel classifier. As a result, the images produced by LDA are binary as opposed to gray scale abundance fractional images produced by the LCDA, the FV and the OSP classifiers. Therefore, it is not surprising to see that LDA produced the worst results.

Example 12.3 (15-Panel HYDICE scene)

As another example, we conducted experiments for LCDA using the 15-panel HYDICE scene in Fig. 1.7(a) where the target spectral signature of the panels in row i is represented by Pi shown in Fig. 1.8. The results are shown in Fig. 12.4 where the results of the LDA, FV and OSP classifiers are also included for comparison.

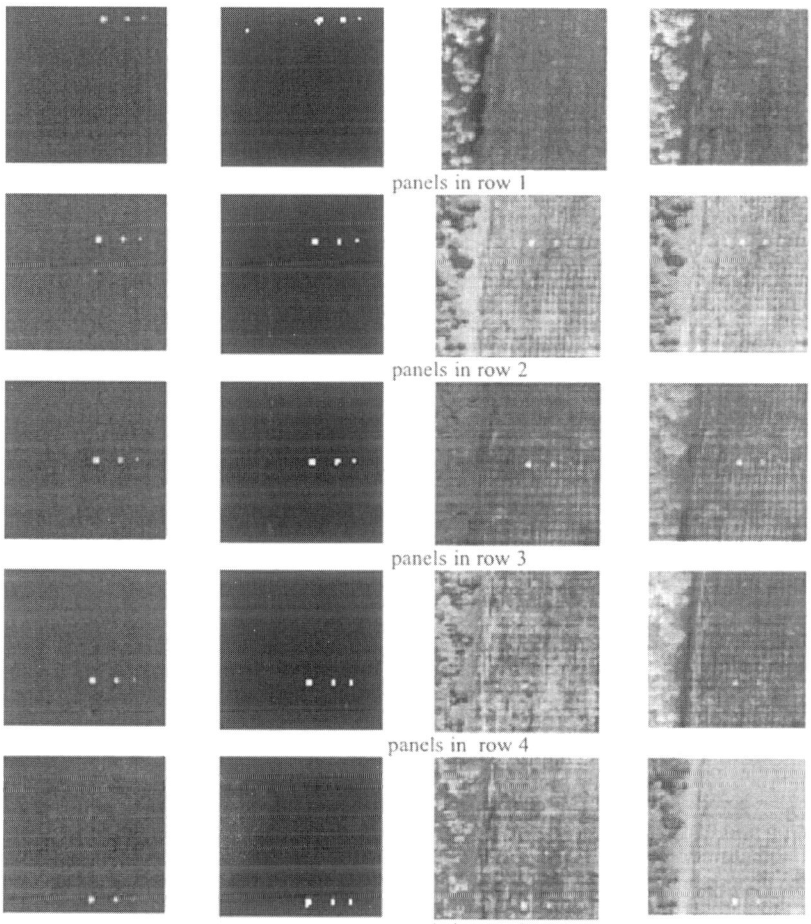

Figure 12.4. Classification results produced by LCDA, LDA, FV and OSP classifiers

Interestingly, LCDA and LDA performed very similarly and both outperformed the FV and the OSP classifiers. Like Tables 10.1-10.3 we can also calculate the detection and classification rates for LCDA by tallying the number of target pixels that were correctly detected and classified. Tables 12.1-12.4 were detection results obtained by the LDA and the LCDA classifiers using the 50%MPCV, 25%MPCV and 20%MPCV respectively.

Table 12.1. Detection and classification rates of LDA

	N_R	N_{RD}	R_D	R_C	N_{FA}
P1	3	2	1.0000	1.0000	9
P2	4	3	1.0000	1.0000	3
P3	4	3	1.0000	1.0000	4
P4	4	3	1.0000	1.0000	4
P5	4	4	1.0000	1.0000	5
Total	19	14	0.7368	0.7368	25

Table 12.2. Detection and classification rates of LCDA using 50% abundance as the cut-off threshold

	N_R	N_{RD}	R_D	R_C	N_{FA}
P1	3	2	0.6667	0.6667	0
P2	4	3	0.7500	0.7500	0
P3	4	3	0.7500	0.7500	0
P4	4	3	0.7500	0.7500	0
P5	4	4	1.0000	1.0000	0
Total	19	14	0.7368	0.7368	0

Table 12.3. Detection and classification rates of LCDA using 25% abundance as the cut-off threshold

	N_R	N_{RD}	R_D	R_C	N_{FA}
P1	3	2	0.6667	0.6667	0
P2	4	4	1.0000	1.0000	0
P3	4	4	1.0000	1.0000	0
P4	4	4	1.0000	1.0000	0
P5	4	3	0.7500	0.7500	0
Total	19	17	0.8947	0.8947	0

Table 12.4. Detection and classification rates of LCDA using 20% abundance as the cut-off threshold

	N_R	N_{RD}	R_D	R_C	N_{FA}
P1	3	3	1.0000	1.0000	0
P2	4	4	1.0000	1.0000	0
P3	4	4	1.0000	1.0000	0
P4	4	4	1.0000	1.0000	0
P5	4	4	1.0000	1.0000	0
Total	19	19	1.0000	1.0000	0

As shown in these tables, the LDA classifier performed very well and achieved 100% detection and classification rates. However, it also produced many false alarm pixels, 25 pixels in total. On the other hand, the performance of the LCDA classifier was determined by the cut-off threshold value. The detection and classification rates were improved as the cut-off threshold was reduced from 25% to 20% while the number of false alarm pixels, N_{FA} remaining zero. In this case, the LCDA classifier performed better than the LDA classifier. Compared to Tables 4.1-4.3 and 1.1-11.3, the LCDA classifier also performed as well as did TCIMF in detection of the 15 panels, but did better than TCIMF with no false alarm pixels.

Analogous to Fig. 11.8, 3-D ROC curves and 2-D ROC curves can be plotted based on the target hit rate R_H defined by (9.27) versus the false alarm rate via the *a*%MPCV defined by (9.3) for performance evaluation. Fig. 12.5 shows the 3-D and 2-D ROC curves generated by LCDA, FV and OSP. As we can see, LCDA performed significantly better than did FV and OSP.

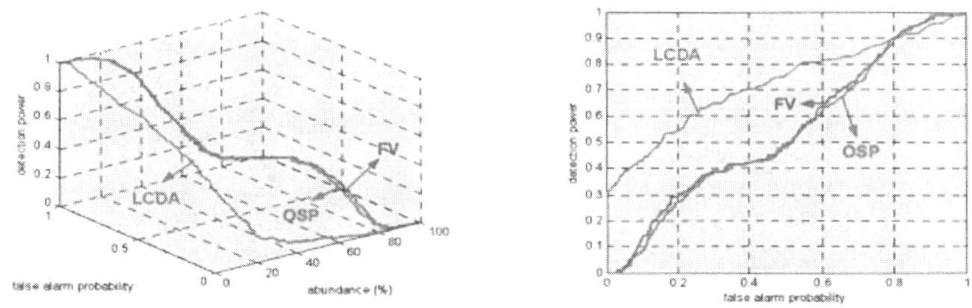

(a) 3-D ROC curves of $(R_D, R_F, a\%)$ (b) 2-D ROC curves of (R_D, R_F)

Figure 12.5. ROC curves of LCDA, FV and OSP

Table 12.5 tabulates the target hit rates by calculating the areas under their 2-D ROC curves in Fig. 12.5(b).

Table 12.5. Detection rates produced by LCDA, FV and OSP

	LCDA	OSP	FV
R_H	0.7548	0.4302	0.4578

12.6 CONCLUSIONS

Fisher's linear discriminant analysis (LDA) has been well accepted as a major technique in pattern classification. It can be also applied to hyperspectral image classification. This chapter presents a similar but different approach to LDA, called LCDA, which replaces Fisher's ratio with the ratio of inter-distance to intra-distance as a criterion for optimality. The advantage of LCDA over LDA is that it constrains the class means along the desired orthogonal directions. Consequently, all the classes of interest are forced to separate along with prescribed directions. By means of this direction constraint LCDA can detect and classify similar targets. Interestingly, it also demonstrates that the Bowles et al.'s filter-vectors approach that was shown to be a special case of the BRLCMV classifier in Chapter 11 can be also interpreted as a special case of LCDA. In analogy with Table 11.4, Table 12.6 summarizes assumptions and constraints that are imposed by LCDA in comparison with FV and OSP.

Table 12.6. Assumptions/constraints imposed by LCDA, FV and OSP

	linear mixture model	desired targets (**D**)	undesired targets (**U**)	filter vector **w**	R_{LxL}	background
LCDA	no/no	yes/yes	no/no	no/no	yes/no	no/no
FV	yes/yes	yes/yes	no/no	no/yes	no/no	no/no
OSP	yes/no	yes/no	no/no	no/no	no/no	no/no

A comparative study among the five classifiers, LCMV, BRLCMV, MTCEM, SCEM, WTACEM discussed in Chapter 11 and the three classifiers described in this chapter, LCDA, FV and OSP was also reported in Chang (2002b) where the eight classifiers were studied in a unified framework.

Finally, it is also worth noting that LCDA can be extended to an unsupervised method for unknown image scenes using unsupervised algorithms in Chapter 5 when no *a priori* signature knowledge is available. The experimental results are very impressive and almost as good as their supervised counterparts.

V

AUTOMATIC MIXED PIXEL CLASSIFICATION (AMPC)

In PART III and PART IV, we considered the problems of unconstrained mixed pixel classification and constrained mixed pixel classification. In PART V, we will further consider the problem of automatic mixed pixel classification (AMPC). Like automatic subpixel detection studied in Chapter 5, there are also two types of AMPC, unsupervised MPC and automatic target detection and classification (ATDC). The former is applied to the case that when there is a lack of target knowledge, the three unsupervised algorithms described in Sections 5.2-5.4 of Chapter 5 can be used to generate the required unsupervised target knowledge to extend the supervised MPC described in Chapters 7-11 to unsupervised MPC. The latter is similar to anomaly detection in Chapter 5 and does not require unsupervised target knowledge for classification. Rather, it detects and classifies anomalies in an unknown image scene. Therefore, technically speaking, unsupervised MPC and ATDC are considered to be different approaches. In Chapter 13, we first investigate the unconstrained MPC. It derives an unsupervised OSP approach to unsupervised MPC. This method can be implemented in two different applications. One is the desired target detection and classification algorithm (DTDCA) that can be used to detect and classify specific targets provided by prior target knowledge. Another is called automatic target detection and classification algorithm (ATDCA) that can be used to detect and classify unknown targets. In Chapter 14, another type of AMPC, ATDC is considered. It extends anomaly detection to anomaly classification. Although a simple approach combining a classifier with an anomaly detector may be desirable, two issues need to be addressed before doing so. Since an anomaly detector does not discriminate anomalies it detects, we need a measure to discern the detected anomalies. Second, it requires target information for a classifier to be effective. In order to resolve these dilemmas, four target discrimination measures are introduced in this chapter. These measures cluster the detected anomalies into different targets classes in an unsupervised manner. Then the means of each clustered target class are calculated to generate the desired target information for classification. By far, all the techniques presented in PART II-PART V assume that the target abundance fractions are unknown constant, i.e., nonrandom parameters. Apparently, this does not have to be the case in general, specifically for hyperspectral imagery, which may uncover many unknown random signal sources. In Chapter 15, an approach to linear spectral random mixture analysis (LSRMA) is proposed where targets of interest are modeled as random signal sources. It is based on the concept of independent component analysis (ICA) which has shown great success in blind source separation in signal process. This chapter provides a promising application of the ICA to hyperspectral image analysis. Chapter 16 presents a similar idea used in the LSRMA but a different approach, called projection pursuit (PP), which can be considered a general version of the LSRMA. It designs a projection index to explore interesting stuctures of image data. When the projection index is the data variance, the PP becomes the well-known principal components analysis (PCA). On the other hand, if the projection index is a measure of statistical independence among projection images, the PP is reduced to the LSRMA where each projection image corresponds to an independent component. One of most challenging problems encountered in AMPC is the estimation of the intrinsic dimensionality of the data, which is generally unknown. The same problem was also encountered in automatic subpixel target detection. Chapter 17 is devoted to this topic where three Neyman-Pearson detection theory-based eigenvalue-thresholding techniques are developed for estimation of the intrinsic dimensionality.

13

AUTOMATIC MIXED PIXEL CLASSIFICATION (AMPC): UNSUPERVISED MIXED PIXEL CLASSIFICATION

The automatic mixed pixel classification (AMPC) considered in this chapter is fully computer automated and can be implemented to automatically detect and classify targets with no human intervention. Like the automatic subpixel detection discussed in Chapters 5-6 AMPC can be also categorized into unsupervised mixed pixel classification and anomaly classification. The former classifies mixed pixels in an unsupervised manner, where the required unsupervised target knowledge is the *a posteriori* target information generated directly from the image data as noted in Chapter 5. By contrast, the latter extends anomaly detection to anomaly classification, in which case the detected anomalies can be classified with no need of unsupervised target knowledge. Depending upon availability of *a priori* target knowledge two versions of unsupervised MPC, referred to as desired target detection and classification algorithm (DTDCA) and automatic target detection and classification algorithm (ATDCA), are presented in this chapter. The DTDCA is applied to a situation that there is knowledge about specific targets to be classified, whereas ATCDA can be used to classify targets of interest present in an unknown image scene without *a priori* target knowledge. As a consequence, they result in different applications.

13.1 INTRODUCTION

Automatic mixed pixel classification (AMPC) is part of automatic target recognition. One type of AMPC is considered in this chapter, which is unsupervised mixed pixel classification (unsupervised MPC). The unsupervised MPC arises from a need, where in many practical applications obtaining complete target knowledge is difficult, if not impossible. As shown in Chapter 8, when the complete prior target knowledge is available, the OSP classifier was the optimal linear Bayes classifier in the least-squares sense. However, when there is a lack of prior target knowledge, the OSP classifier can only produce sub-optimal solutions. The unsupervised MPC can improve the performance by taking advantage of additional *a posteriori* target information that is directly obtained from the image data by an unsupervised means. Such unsupervised target information helps the supervised MPC reduce interfering effects caused by unknown signal sources in the image data. So, in this chapter, the unsupervised MPC is considered. Of particular interest is an extension of OSP to unsupervised OSP. Two unsupervised OSP classifiers can be derived in accordance with different applications. The first unsupervised OSP classifier is referred to as the desired target detection and classification algorithm (DTDCA), which only requires the prior knowledge of specific targets of interest. The problems of this type occur in applications such as reconnaissance, rescue and search where specific targets are known, but image scene are generally unknown. The second unsupervised OSP classifier is called the automatic target detection and classification algorithm (ATDCA), which is developed to search and classify unknown targets in an unknown environment. It is applied to the case that no prior target knowledge is available. Since there are no specific targets of interest, all the targets detected in the image data are considered to be interesting and must be classified. Interestingly, both DTDCA and ATDCA make use of the same unsupervised target generation process (UTGP) described in Section 5.3 to obtain the required unsupervised target information, despite that the initial targets used in their algorithms are different. When the prior knowledge of specific targets is provided, these targets are used as initial targets to initialize DTDCA, which then classifies these particular targets. When no such prior target knowledge is available, the initial targets must be obtained directly from the image data to initialize ATDCA. There is a significant difference between DTDCA and ATDCA. DTDCA considers the targets generated by UTGP as undesired targets and eliminate these UTGP-generated targets to improve its performance. On the contrary, ATDCA does not have any prior target knowledge. As a result, the targets found by UTGP are considered to be interesting targets, all of which must be classified. Therefore, ATDCA can be used for anomaly detection, while DTDCA cannot.

13.2 UNSUPERVISED MPC

The mixed pixel classifiers discussed in Chapters 8-12 required either partial or full prior target knowledge. They can be viewed as partially or completely supervised classifiers. A completely supervised optimal classifier is a Bayes classifier that requires the complete knowledge of the image data. In reality, this is impossible in many applications, particularly for hyperspectral image analysis. On the other hand, a completely unsupervised classifier such as ISODATA, which is widely used in pattern classification (Duda and Hart, 1973), generally performs poorly as was shown in Chapter 9 or in Chang and Ren (2000). This is because it is a pattern classification technique, which is not designed for target classification. A good example is CEM in Chapter 4, which applies partial knowledge to target detection, while suppressing energies resulting from unknown signal sources. It is further improved by TCIMF, which includes unsupervised target knowledge that provides information of unknown signal sources generated by an unsupervised algorithm. As a result, TCIMF eliminates these targets instead of minimizing the energies of these targets as is done in CEM. A similar improvement can be expected for unsupervised MPC. In the following sections, we consider two unsupervised MPC methods.

13.3 DESIRED TARGET DETECTION AND CLASSIFICATION

In many applications, there always exists some partial *a priori* knowledge, which we should be able to take advantage of. For example, in reconnaissance applications, we may be only interested in some specific targets present in an unknown image scene, which may include unknown signal sources that have interfering effects on the desired targets. In order to reduce such interference, we can find these potential interferers and remove them before detection takes place. Since there is no knowledge about such interferers, it must be obtained directly from the image scene by an unsupervised method. The algorithm, referred to as desired target detection and classification algorithm (DTDCA) is developed for this purpose. It uses the unsupervised algorithm, UTGP described in Section 5.2 to generate a collection of interfering signal sources that will be eliminated by the follow-up OSP-based classifier.

Desired Target Detection and Classification Algorithm (DTDCA)

1) Select $\mathbf{t}_0$ to be the target of interest.
2) Use $\mathbf{t}_0$ as the initial target in UTGP in Section 5.3 to generate a set of targets denoted by $\mathbf{U}_k$.
3) Use $\delta_{OSP}(\mathbf{r})$ specified by (3.9) to classify $\mathbf{t}_0$.

It should be noted that when we classify target $\mathbf{t}_0$, any target other than $\mathbf{t}_0$ will be considered to as an interferer or undesired target to target $\mathbf{t}_0$, no matter what this target is, a man-made object or natural background signature. In this case, the generated target set $\mathbf{U}_k$ will be the same $\mathbf{U}$ used in (3.2). Next, we apply the OSP projector $P_{\text{OSP}} = \mathbf{t}_0^T P_{\mathbf{U}_k}^{\perp}$ to all image pixel vectors $\mathbf{r}$ for classification. The resulting image will show only target $\mathbf{t}_0$ with all the target signatures in $\mathbf{U}_k$ being nulled out. In general, if there is a set of desired targets of interest, $\mathbf{D}$, $\mathbf{t}_0$ in step 1) can be replaced with those targets in $\mathbf{D}$.

In the following example, we will show how detection performance can be improved by incorporating unsupervised target information.

Example 13.1 (LCVF AVIRIS Image)

In this example, the LCVF AVIRIS scene in Fig. 1.6(a) was used for experiments where playa, cinders, vegetation and rhyolite were targets of interest and each of these four target signatures was used as the desired target signature $\mathbf{t}_0$ for DTDCA. Since the only assumed target knowledge was $\mathbf{t}_0$ and no other target information was available, the DTDCA must generate additional target information in such a way that this information could be used to improve the detection and classification performance. Fig. 13.1 shows the results of DTDCA in detection and classification of playa, cinders, vegetation and rhyolite. In each of four cases 12 unknown targets were generated by UTGP in step 2) of DTDCA. These UTGP-generated targets were considered to be undesired targets and joined with $\mathbf{t}_0$ to form the target signature matrix $\mathbf{M}$ for the OSP classifier.

The images labeled by (a) in Fig. 13.1 were produced by the OSP classifier with $\mathbf{M} = [\mathbf{t}_0\,\mathbf{U}_1]$ with $\mathbf{U}_1 = \mathbf{t}_1$ where $\mathbf{t}_1$ was the first unknown target generated by UTGP and the η underneath the images was the OPCI defined in (5.3). Then the images labeled by (b) in Fig. 13.1 were produced by the OSP classifier with $\mathbf{M} = [\mathbf{t}_0\,\mathbf{U}_2]$ with $\mathbf{U}_2 = [\mathbf{t}_1\,\mathbf{t}_2]$ where $\mathbf{t}_2$ was the second unknown target generated by UTGP and the η underneath the images was the OPCI defined in (5.4). The performance of DTDCA was improved by eliminating one more undesired target $\mathbf{t}_2$. The UTGP was continued until the 12-th target was generated where $\mathbf{M} = [\mathbf{t}_0 \, \mathbf{U}_{12}]$ with $\mathbf{U}_{12} = [\mathbf{t}_1 \, \mathbf{t}_2 \cdots \mathbf{t}_{12}]$. As we can from Fig. 13.1, the improvement of DTDCA was not visible after first few undesired targets were eliminated. This is largely due to the fact that the image has low spatial resolution (20 m) and target signatures may be mixed with background signatures. As will be shown in Example 13.2, this will be changed when a HYDICE image is used.

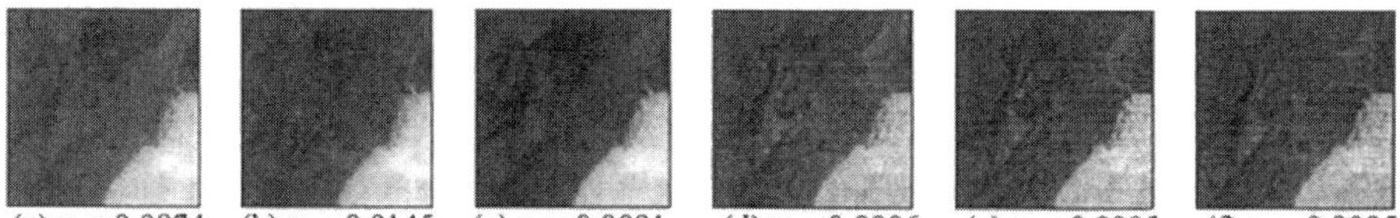

(a) $\eta = 0.0874$ (b) $\eta = 0.0145$ (c) $\eta = 0.0091$ (d) $\eta = 0.0006$ (e) $\eta = 0.0005$ (f) $\eta = 0.0005$

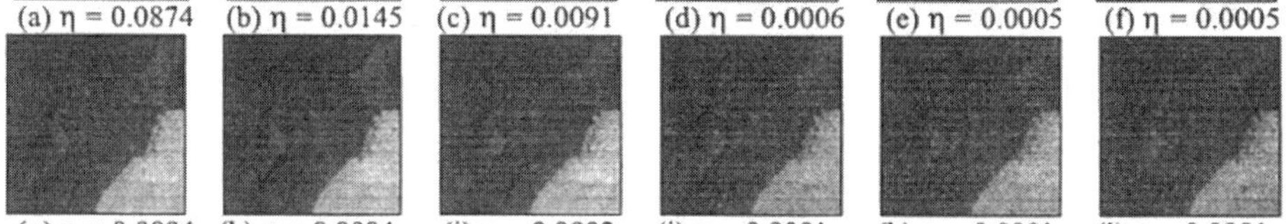

(g) $\eta = 0.0004$ (h) $\eta = 0.0004$ (i) $\eta = 0.0002$ (j) $\eta = 0.0001$ (k) $\eta = 0.0001$ (l) $\eta = 0.0001$

detection of playa by DTDCA

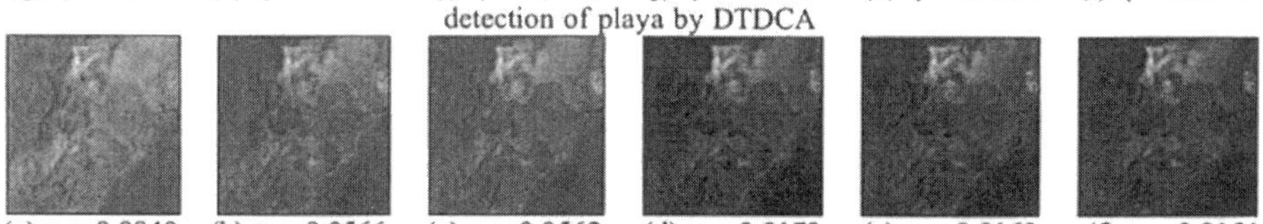

(a) $\eta = 0.0840$ (b) $\eta = 0.0566$ (c) $\eta = 0.0562$ (d) $\eta = 0.0173$ (e) $\eta = 0.0169$ (f) $\eta = 0.0151$

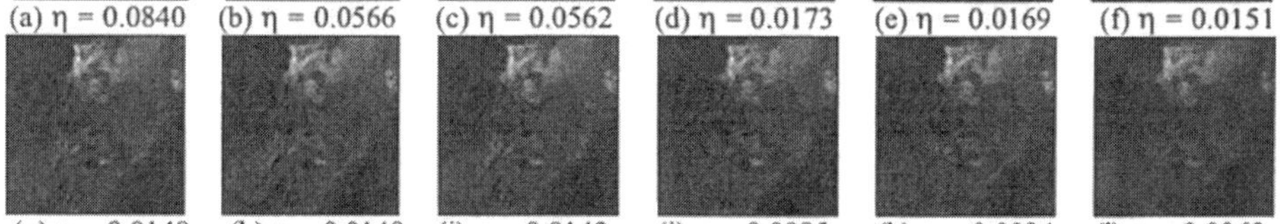

(g) $\eta = 0.0148$ (h) $\eta = 0.0148$ (i) $\eta = 0.0143$ (j) $\eta = 0.0085$ (k) $\eta = 0.0084$ (l) $\eta = 0.0062$

detection of cinders by DTDCA

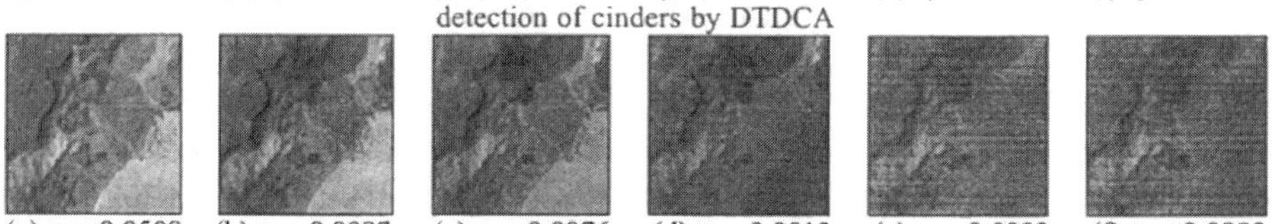

(a) $\eta = 0.0508$ (b) $\eta = 0.0097$ (c) $\eta = 0.0076$ (d) $\eta = 0.0013$ (e) $\eta = 0.0003$ (f) $\eta = 0.0003$

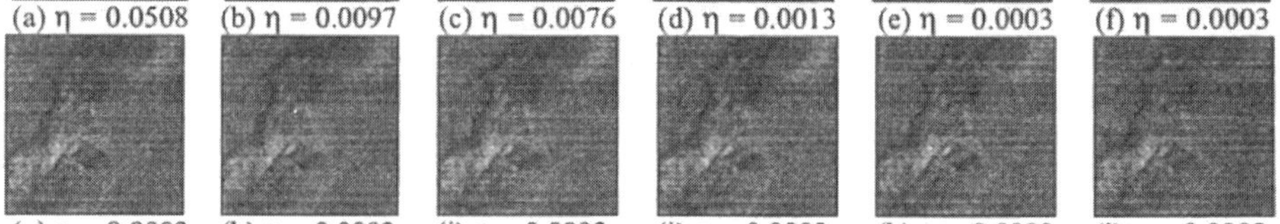

(g) $\eta = 0.0003$ (h) $\eta = 0.0003$ (i) $\eta = 0.0003$ (j) $\eta = 0.0003$ (k) $\eta = 0.0002$ (l) $\eta = 0.0002$

detection of rhyolite by DTDCA

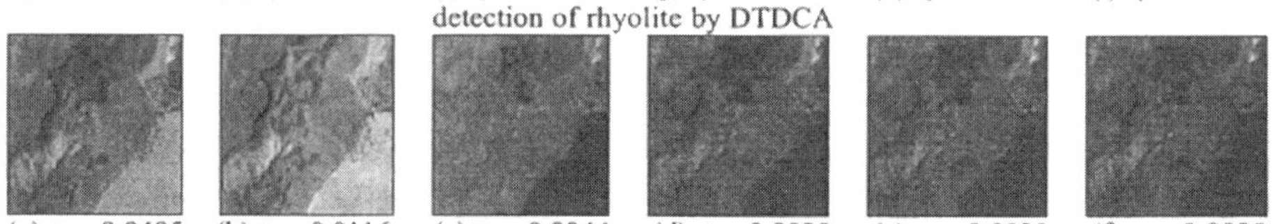

(a) $\eta = 0.0495$ (b) $\eta = 0.0116$ (c) $\eta = 0.0044$ (d) $\eta = 0.0030$ (e) $\eta = 0.0029$ (f) $\eta = 0.0029$

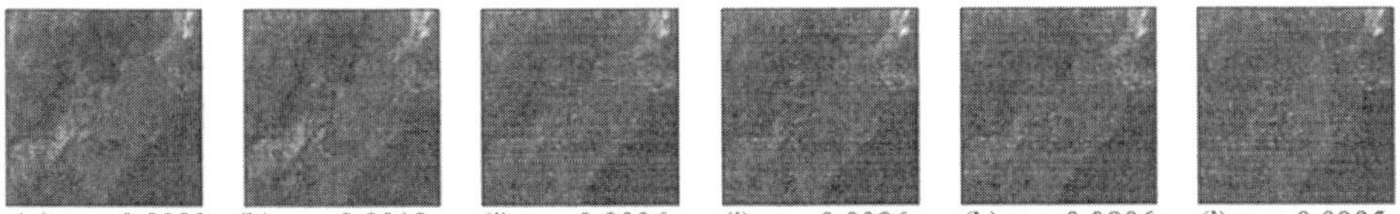

(g) $\eta = 0.0023$ (h) $\eta = 0.0012$ (i) $\eta = 0.0006$ (j) $\eta = 0.0006$ (k) $\eta = 0.0006$ (l) $\eta = 0.0005$

detection of vegetation by DTDCA

Figure 13.1. Results of DTDCA in detection and classification of playa, cinders, vegetation and rhyolite

In the above experiment, DTDCA was used to detect and classify the four desired targets, which were assumed to be resident in the image scene. However, in some applications, the desired target may not be present in the image to be processed. Fig. 13.2 shows how DTDCA works for such cases.

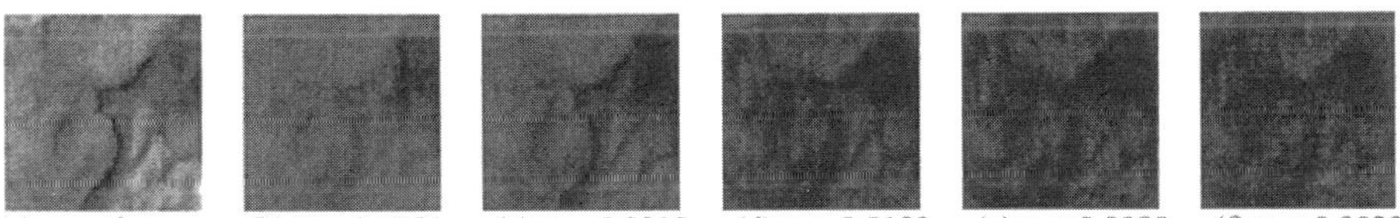

(a) another image scene (b) $\eta = 0.0874$ (c) $\eta = 0.0312$ (d) $\eta = 0.0122$ (e) $\eta = 0.0098$ (f) $\eta = 0.0023$

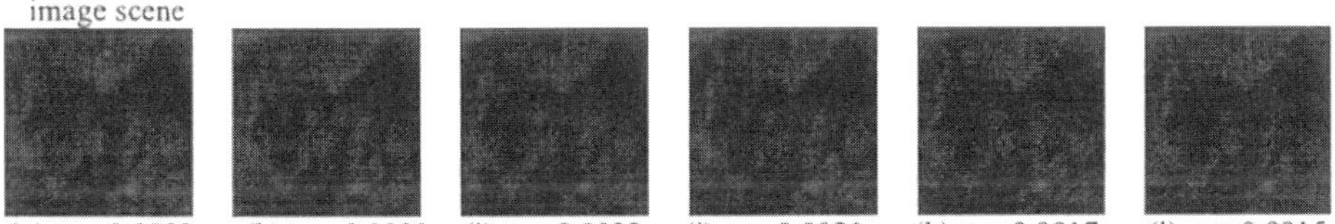

(g) $\eta = 0.0023$ (h) $\eta = 0.0023$ (i) $\eta = 0.0022$ (j) $\eta = 0.0021$ (k) $\eta = 0.0017$ (l) $\eta = 0.0015$

Figure 13.2. No playa was detected by DTDCA

The image used in Fig. 13.2 was shown in Fig. 13.2(a), which was extracted from Fig. 3.1(a) and did not have the playa in the scene. In this case, DTDCA used the playa as the desired target, $\mathbf{t}_0$ and made an attempt to detect and classify the playa in the scene. As we can see from the images, DTDCA may show something going on in the beginning (see Fig. 13.(b-e)). But as more unknown targets were eliminated, the OSP-classified images became darker and showed no indication of the desired target present in the image scene. At this point, we can conclude that no desired target was found and detected in the image scene. It should be noted that DTDCA can be very useful in reconnaissance applications when one searches for a known target in an unknown environment.

Example 13.2 (15-Panel HYDICE scene)

In Example 13.1, the AVIRIS experiments did not show clear advantages of DTDCA. In this example, we will use the 15-panel HYDICE image in Fig. 1.7(a) to show the effectiveness of DTDCA, which was not demonstrated in Example 13.1. In analogy with Fig. 13.1, Fig. 13.3 shows the results of DTDCA in detection and classification of 15 panels where 24 unknown targets were generated by UTGP. The desired target signatures used in the DTDCA were $\{Pi\}_{i=1}^{5}$ shown in Fig. 1.8. Since the panels in the same row are made from the same material, they were all classified into one class as shown in Fig. 13.3. Unlike the AVIRIS images in Fig. 13.1, the HYDICE images in Fig. 13.3 clearly showed the effectiveness of DTDCA. According to spectral analysis in Chapter 2, the spectral signatures of the panels in rows 2 and 3 are very similar, so are the spectral signatures of the panels in rows 4 and 5. It is generally difficult to classify them into different classes. DTDCA seemed to have the same difficulty in the beginning of the process where the panels in both rows were shown in the same image (see detection of panels in row 3 and detection of panels in row 5). However, as more unknown targets were generated and eliminated, the panels in the desired row were enhanced while the panels in the other row beginning to vanish. Eventually, only the panels in the desired row were detected and correctly classified in a single image.

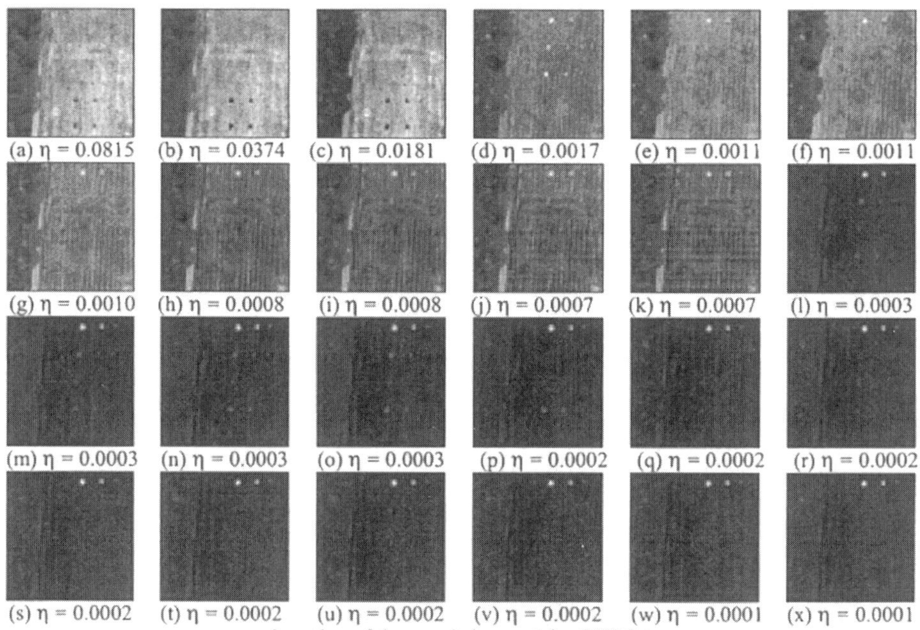

detection of the panels in row 1 by DTDCA

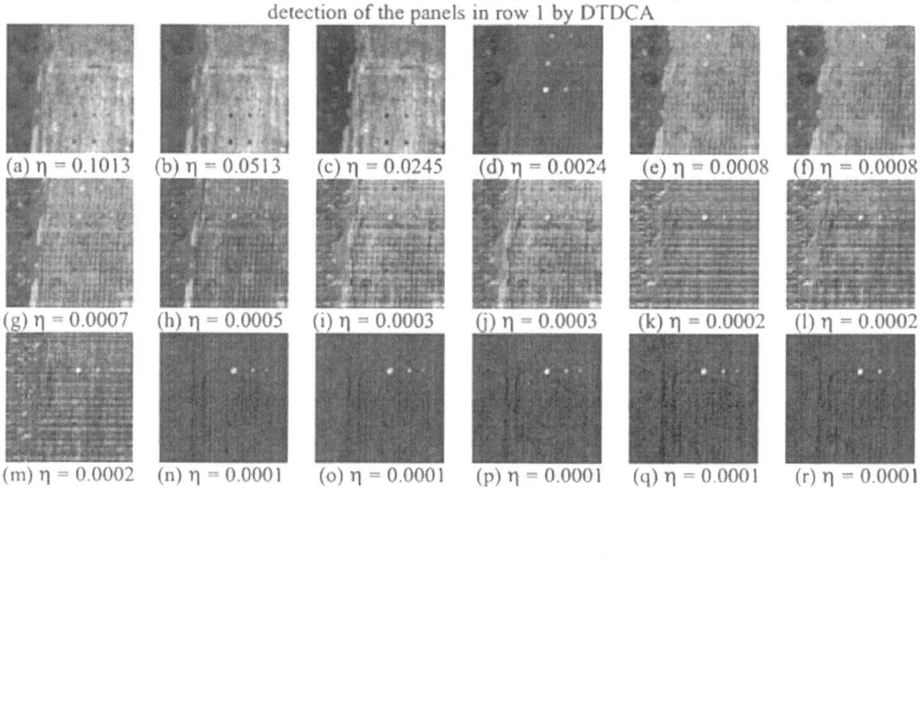

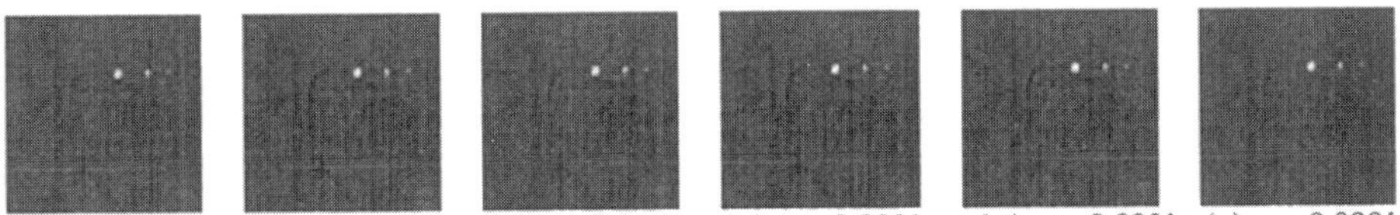

(s) η = 0.0001 (t) η = 0.0001 (u) η = 0.0001 (v) η = 0.0001 (w) η = 0.0001 (x) η = 0.0001

detection of the panels in row 2 by DTDCA

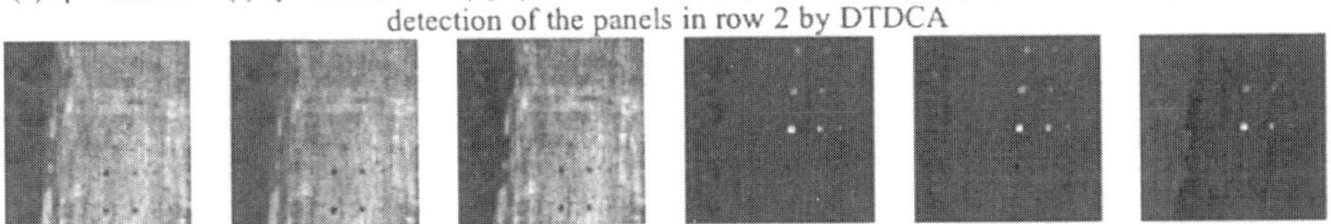

(a) η = 0.1005 (b) η = 0.0563 (c) η = 0.0293 (d) η = 0.0054 (e) η = 0.0050 (f) η = 0.0029

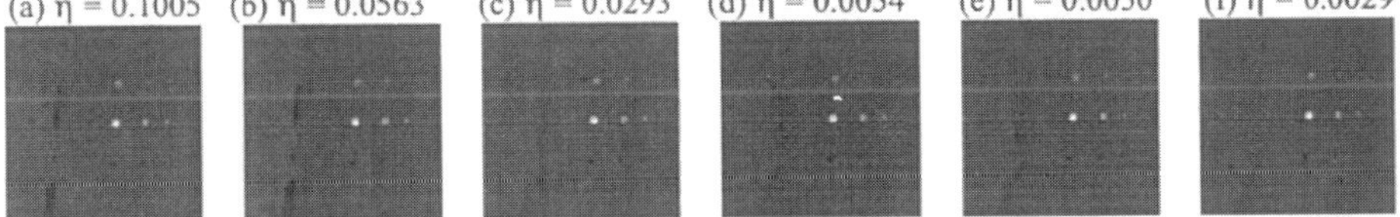

(g) η = 0.0028 (h) η = 0.0026 (i) η = 0.0022 (j) η = 0.0018 (k) η = 0.0018 (l) η = 0.0017

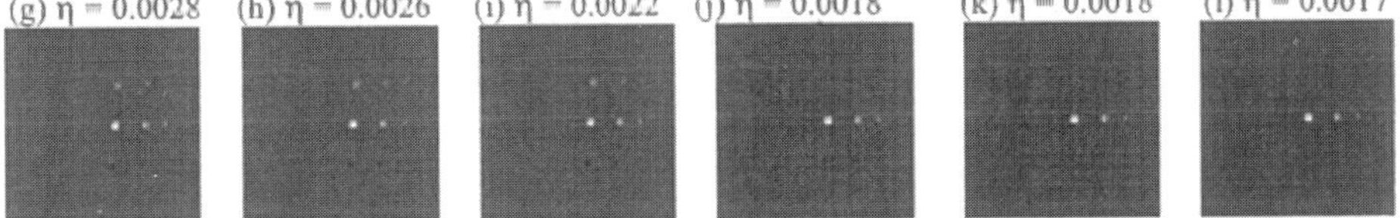

(m) η = 0.0016 (n) η = 0.0016 (o) η = 0.0015 (p) η = 0.0006 (q) η = 0.0006 (r) η = 0.0005

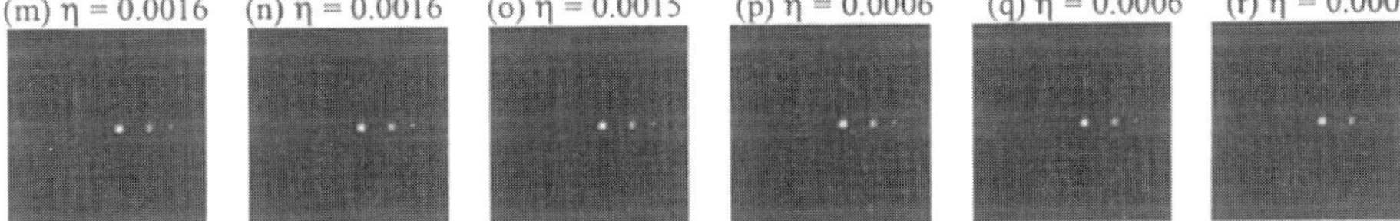

(s) η = 0.0005 (t) η = 0.0005 (u) η = 0.0005 (v) η = 0.0005 (w) η = 0.0005 (x) η = 0.0005

detection of the panels in row 3 by DTDCA

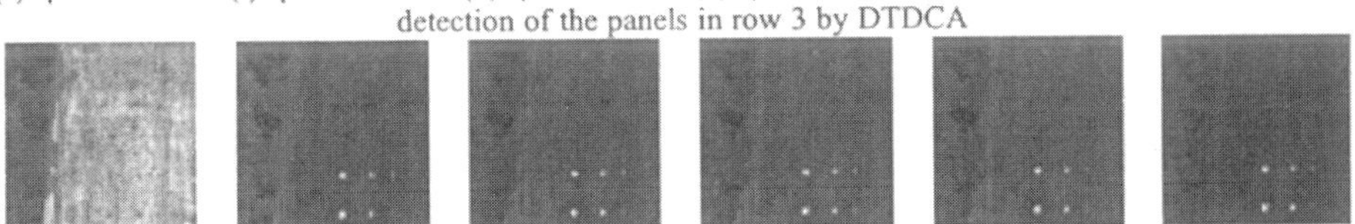

(a) η = 0.0609 (b) η = 0.0142 (c) η = 0.0132 (d) η = 0.0132 (e) η = 0.0132 (f) η = 0.0054

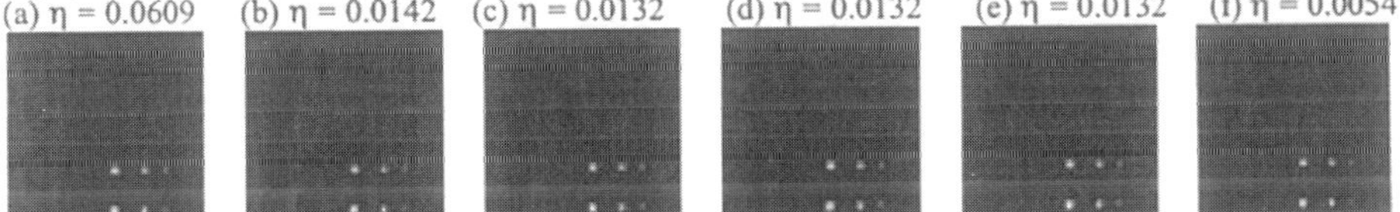

(g) η = 0.0046 (h) η = 0.0037 (i) η = 0.0036 (j) η = 0.0034 (k) η = 0.0031 (l) η = 0.0031

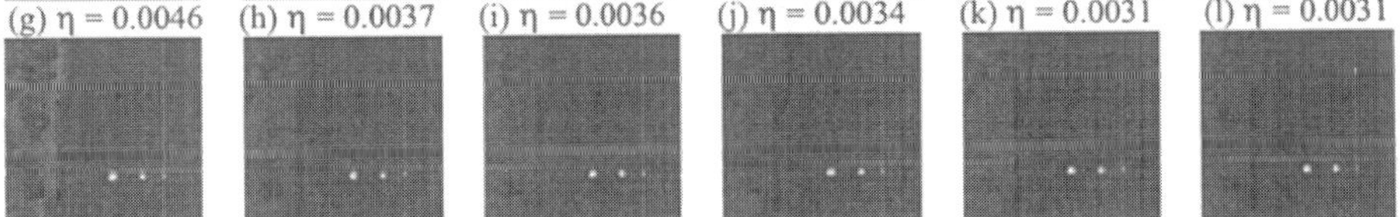

(m) η = 0.0001 (n) η = 0.0001 (o) η = 0.0001 (p) η = 0.0001 (q) η = 0.0001 (r) η = 0.0001

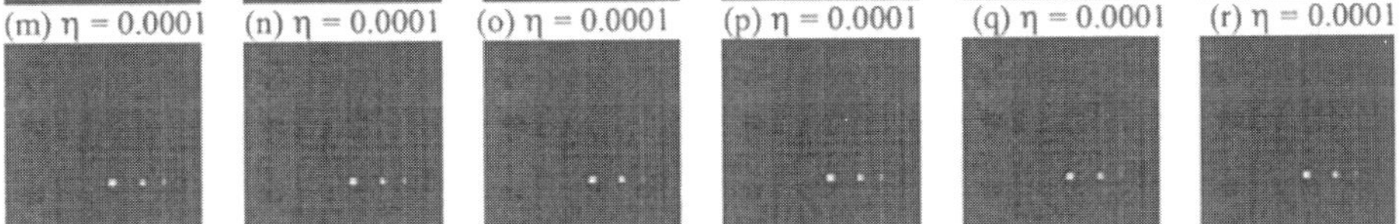

(s) η = 0.0001 (t) η = 0.0001 (u) η = 0.0001 (v) η = 0.0001 (w) η = 0.0001 (x) η = 0.0001

detection of the panels in row 4 by DTDCA

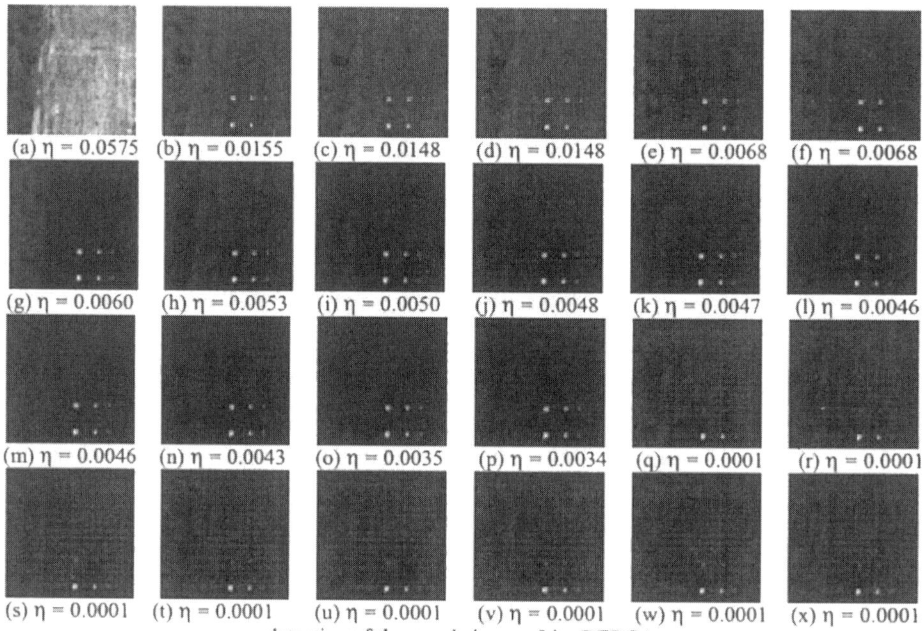

detection of the panels in row 5 by DTDCA

Figure 13.3. Results of DTDCA in detection and classification of 15 panels

In order to further show the utility of DTDCA in searching for a known target not present in an image scene, we selected a HYDICE scene shown in Fig. 13.4(a) in which no P1 signature was present. Fig. 13.4 (b-l) demonstrates a step-by-step process of DTDCA where the images became completely dark after the first four unknown targets were generated and eliminated. This implies that no desired target was detected in the image scene.

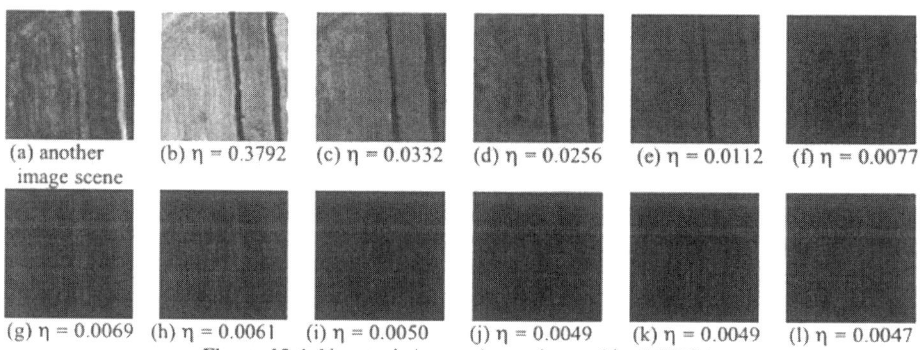

Figure 13.4. No panels in row 1 was detected by DTDCA

To conclude this section, one comment is noteworthy. As mentioned in Section 5.3, there are two ways to terminate UTGP. One is to set a threshold for OPCI. Another is to predetermine the number of unknown targets to generate. Which one is better is determined by applications. In examples considered in this section, we chose the latter which preset 12 targets for Example 13.1 and 24 targets for Example 13.2.

13.4 AUTOMATIC TARGET DETECTION AND CLASSIFICATION

In DTDCA, we assumed that there were specific targets of interest. DTDCA used these targets as desired targets and nulled out all other unknown targets that are generated by UTGP prior to target detection. In this section, we assume that such target knowledge is not available. In this case, no specific targets are of particular interest. The algorithm designed for this purpose is referred to as automatic target detection and classification algorithm (ATDCA). It is very similar to DTDCA but differs in two aspects. First of all, there is no desired target to serve as an initial target $\mathbf{t}_0$ that can be used to initialize the algorithm. It must be generated directly from the image scene. Second, since we do not have any target knowledge, we need to classify all the targets that are generated by the UTGP. As a result, each target will be classified in a separate image individually.

Automatic Target Detection and Classification Algorithm (ATDCA)

1) Select $\mathbf{t}_0 = \arg\{\max_{\mathbf{r}}[\mathbf{r}^T\mathbf{r}]\}$.
2) Use $\mathbf{t}_0$ as the initial target in step 1 of UTGP
3) Follow steps 2-8 outlined in UTGP to generate $\mathbf{U}_k$.
4) Apply $\delta_{\text{OSP}}(\mathbf{r})$ specified by (3.9) to classify $\mathbf{t}_0$ and all targets in $\mathbf{U}_k$ individually. More specifically, assume $\mathbf{T}_k = \{\mathbf{t}_0\} \cup \mathbf{U}_k$ in model (3.1). Apply the OSP classifier $q_{OSP} = \mathbf{t}_j^T P_{\overline{\mathbf{U}}_j}^{\perp}$ to classify all individual targets $\mathbf{t}_j \in \mathbf{T}_k$ with $\overline{\mathbf{U}}_j = \left[\mathbf{t}_0\, \mathbf{t}_1\, \mathbf{t}_2 \cdots \mathbf{t}_{j-1}\, \mathbf{t}_{j+1} \cdots \mathbf{t}_k\right]$. Since there are $k+1$ targets in $\mathbf{T}_k$, $k+1$ images will be generated by UTGP, each of which shows only one particular target in $\mathbf{T}_k$ classified by the OSP classifier.

In order to compare to DTDCA, we conduct the same experiments used for Examples 13.1 and 13.2 and demonstrate the difference between DTDCA and ATDCA.

Example 13.3 (LCVF AVIRIS scene)

With an assumption that no prior target knowledge about the image is available, the initial target $\mathbf{t}_0$ must be obtained directly from the scene by the UTGP, which was a playa signature. Fig. 13.5(a) shows the classification result of $\mathbf{t}_0$. This is followed by the classification result of the first target, $\mathbf{t}_1$ generated by the UTGP, which was a cinder pixel. Then the second target, $\mathbf{t}_2$ found by the UTGP was a vegetation pixel and its classification result is shown in Fig. 13.5(c). The process was continued until a stopping criterion was met. In this case, the same criterion (i.e. 12 targets) that was used to terminate DTDCA was also set to terminate ATDCA. As we examine the images in Fig. 13.5(a-l), playa, rhyolite, shade and anomaly were classified in Figs. 13.5(a,h,j), 13.5(i), 13.5(d,l) and Fig. 13.5 (e,f) respectively. Interestingly, ATDCA detected a single two-pixel wide anomaly, which cannot be detected by DTDCA since the anomaly was not visible in the image scene and can not be identified by visual inspection.

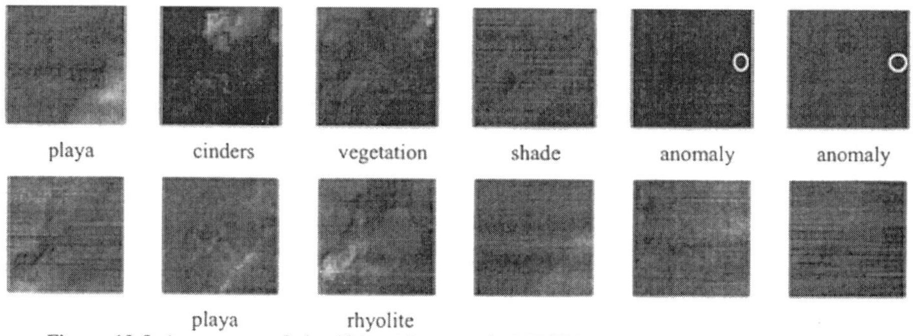

Figure 13.5. A sequence of classifying 12 targets in LCVF image scene detected by ATDCA

It is worth noting that since the playa covers a large area and its spectral signatures varies. As a result, the playa was detected in three different separate images in Fig. 13.5(a,h,j). Similarly, it also found that shade and anomaly were classified in two separate images, Fig. 13.5(d,l) and Fig. 13.5 (e,f) respectively. The same phenomenon was also observed in Chang and Heinz (2000).

Due to a lack of target knowledge, ATDCA cannot compete against DTDCA, which uses the desired target knowledge. It should be also noted that the image sequences shown in Figs. 13.1-13.4 were the progressive improvement of DTDCA in detection of the desired targets, whereas the image sequence in Fig. 13.5 showed the classification results of specific targets found by the UTGP in sequence.

Example 13.4 (15-Panel HYDICE scene)

In this example, we once again used the 15-panel HYDICE scene for experiments to better illustrate the difference between DTDCA and ATDCA. Fig. 13.6(a-l) shows the classification results of 12 targets found by the UTGP in sequence. The initial target $\mathbf{t}_0$ was found at the left upper corner and is a strong interferer, which was also detected in Fig. 6.5(a,e). Then the first target $\mathbf{t}_1$ was found to be a background signature, which was a grass pixel. It was then followed by the second found target $\mathbf{t}_2$ which was the first panel pixel in row 5. As mentioned previously, the spectral signatures of panels in row 4 and 5 are very close. This resulted in classification in Fig. 13.6(c), which classified all panels in these two rows in one image. In this case, we could not discriminate panels in row 5 from those in row 4. A similar result was also obtained in Fig. 13.6(e) where the fourth UTGP-generated target $\mathbf{t}_4$ was the first panel pixel in row 3. The classification could not make distinction between the panels in row 2 and the panels in row 3. Since the panels in row 1 have very distinct spectral signatures from those in other rows, they were classified in Fig. 13.6(f) in a separate image where the first panel pixel in row 1 was detected as the fifth target, $\mathbf{t}_5$. Compared to the results obtained in Fig. 13.3 by DTDCA, DTDCA outperformed ATDCA in classification of the 15 panels. In particular, all panels in different rows were successfully separated and correctly classified by the DTDCA. This is because that the target signatures used by DTDCA were the average signatures obtained in Fig. 1.8, whereas the target knowledge used in ATDCA was single-pixel target information which may not well represent the spectral signatures of panels in the same row. Nevertheless, this problem can be mitigated by replacing a single target pixel with the average of a group of similar target pixels found by a similar hyperspectral measure described in Chapter 2.

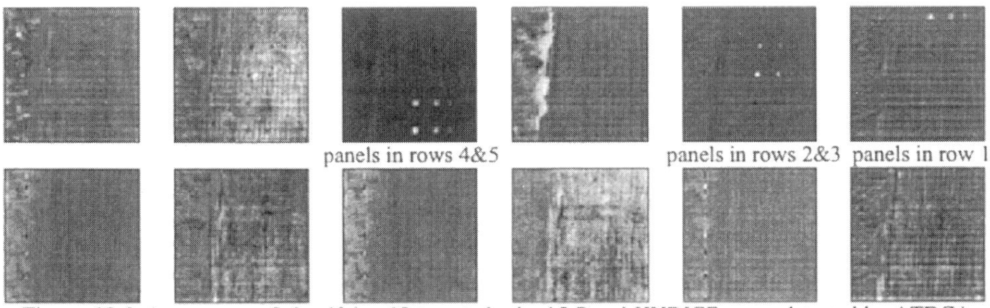

Figure 13.6. A sequence of classifying 12 targets in the 15-Panel HYDICE scene detected by ATDCA

13.5 CONCLUSIONS

This chapter presents two unsupervised MPC methods, DTDCA and ATDCA. Despite that they both look alike, their applications are completely different. DTDCA can be used for cases that some partial prior knowledge is available, such as reconnaissance applications. On the other hand, the ATDCA works for completely unknown data where no *a priori* information is given. In this case, *a posteriori* information is obtained directly from the data for target detection and classification. Applications of this type include surveillance or monitoring of unknown targets.

14

AUTOMATIC MIXED PIXEL CLASSIFICATIO (AMPC): ANOMALY CLASSIFICATION

In Chapter 13, one type of AMPC, the unsupervised MPC, was considered. It made use of an unsupervised algorithm to generate necessary unsupervised target information required for unsupervised MPC. In this chapter, another type of AMPC, automatic target detection and classification (ATDC) is investigated, which does not require any prior target knowledge. It extends anomaly detection to anomaly classification. A natural approach is to combine an anomaly detector with a classifier to classify the detected anomalies. Unfortunately, it is not that simple. Two issues are needed to address before doing so. Since detected anomalies do not necessarily belong to the same target class, a mechanism is required to discern among these targets. Besides, it also needs the knowledge of target classes for classification. In order to cope with these two problems, an automatic thresholding method and four target discrimination measures are introduced in this chapter. The proposed automatic thresholding method is developed to segment anomalies from image background before target discrimination takes place. It can be implemented automatically. The target discrimination measures are designed based on two criteria, the Mahalanobis distance and matched-filter based distance, which can be used to cluster the detected anomalies into different target classes in an unsupervised manner. The mean of each clustered target class is calculated to generate the required target information for that particular target class to be used for classification. Coupled with the automatic thresholding and a target discrimination measure, anomaly detection can be extended to anomaly classification.

14.1 INTRODUCTION

Anomaly detection requires no information at all to detect targets. Accordingly, it cannot discriminate the targets that it detects one another. To make anomaly detection anomaly classification, the detection must be implemented in conjunction with an effective criterion to classify detected targets. For clarity, we make a subtle difference between "target classification" and "target discrimination". Target discrimination allows one to discern among a set of targets of interest, denoted by **T** one from another. The targets in **T** could belong to a same class but may have different spectral features. Or they could be targets of distinct types that belong to different classes. In both cases, target discrimination does not necessarily classify the targets in **T**. On the other hand, target classification clusters or groups targets of interest, **T** in accordance with a certain set of rules, such as minimum distance in Section 8.3, nearest neighboring rule in Section 5.2. Such a clustering process can be carried out by either a supervised or an unsupervised means. The information generated by each cluster is then used for classification. As a result, a classifier may uncover targets in image data, which are not in **T**, but may possess similar spatial or spectral properties to that of some targets in **T**. More specifically, target classification takes advantage of the knowledge of **T** to produce *a posteriori* information which can be used for classification, whereas target discrimination is performed only within a specific set of targets, **T**. For anomaly classification, the set of targets of interest, **T** is produced by anomaly detection. The targets in the **T** are then discriminated by a target discrimination measure to form a set of distinct target classes. The mean of each distinct target class is subsequently calculated and used as the target knowledge for classification. Using these generated class means a classifier can classify targets in the entire image data regardless of whether or not these targets are in **T**. Therefore, generally speaking, anomaly classification can be accomplished by three stages, anomaly detection in the first stage, followed by target discrimination in the second stage and finally, concluded by classification in the third stage. Among these three sequential stage processes anomaly detection and classification have been already discussed in Chapter 6 and Chapters 8-10 respectively. The target discrimination is the only stage process yet to be developed, which is the focus of this chapter.

In general, any hyperspectral measure presented in Chapter 2, such as spectral angle mapper (SAM), spectral information divergence (SID) can be used for target discrimination. Unfortunately, these measures are developed on a single pixel basis and do not take advantage of spectral correlation among pixel vectors as does anomaly detection proposed in Section 6. In order to take into account such sample spectral correlation, the Mahalanobis distance specified by (9.8) and Bhattacharyya distance specified by (9.9) can be included to derive second-order statistics hyperspectral measures. Several of such measures discussed in Chang (1998b) shed light on this approach. Following a similar idea, we derived four target discrimination measures from two criteria, Mahalanobis distance and matched filter-based distance. Interestingly, all of these four target discriminations are also very closely to CEM in Chapter 4 and the anomaly detector, the RXD in Chapter 6. The experimental results also show that these four measures perform very closely and nearly the same.

14.2 TARGET DISCRIMINATION MEASURES

As recalled in Chapter 9, the Mahalanobis distance defined by (9.9) is derived from a Gaussian kernel which measures the distance between a target signature $\mathbf{x}$ and the Gaussian mean $\mathbf{m}_j$ with inclusion of the covariance matrix Σ_j^{-1} to take care of sample correlation. The Bhattacharyya distance defined by (9.10) classifies two target signatures, $\mathbf{m}_i$ and $\mathbf{m}_j$ using their respective class sample covariance matrices, Σ_i and Σ_j to account for second-order statistics. If we assume that $\Sigma_i = \Sigma_j = \Sigma$, the Bhattacharyya distance is reduced to

$$\left(\mathbf{m}_i - \mathbf{m}_j\right)^T \Sigma^{-1}\left(\mathbf{m}_i - \mathbf{m}_j\right). \tag{14.1}$$

As noted, (14.1) has the identical structure to that of RXD, $(\mathbf{x}-\mu)^T\mathbf{K}_{L\times L}^{-1}(\mathbf{x}-\mu)$ with $\mathbf{x}=\mathbf{m}_i$, $\mu=\mathbf{m}_j$ and $\mathbf{K}_{L\times L}=\Sigma$, specified by (6.1). If we replace $\mathbf{m}_i$ and $\mathbf{m}_j$ with two target pixels $\mathbf{t}_i$ and $\mathbf{t}_j$, and $\mathbf{K}_{L\times L}=\Sigma$ in (14.1), the resulting distance

$$(\mathbf{t}_i-\mathbf{t}_j)^T\mathbf{K}_{L\times L}^{-1}(\mathbf{t}_i-\mathbf{t}_j), \tag{14.2}$$

referred to as covariance-Mahalanobis distance (CMD), can be used to measure the discrepancy between two target pixels $\mathbf{t}_i$ and $\mathbf{t}_j$, both of which share the same sample covariance matrix, $\mathbf{K}_{L\times L}$ that is formed by the image data. According to (14.2), the smaller the value is, the harder the discrimination between the two targets is. However, we should bear in our mind that the Mahalanobis distance is effective when the number of samples that form the sample covariance matrix is sufficiently large. It should be noted that for simplicity we have used target pixels and target signatures interchangeably.

The CMD specified by (14.2) is not the only distance measure that can be used for target discrimination. An alternative distance can be also derived by the concept of matched filter as follows

$$(\mathbf{t}_i-\mu)^T\mathbf{K}_{L\times L}^{-1}(\mathbf{t}_j-\mu) \tag{14.3}$$

which is referred to as covariance-matched filter based distance (CMFD). It is worth noting that a major difference between the Mahalanobis distance specified by (9.9) and CMFD is that the former uses the same variable $\mathbf{x}$ as opposed to $\mathbf{t}_i$ and $\mathbf{t}_j$ used in the latter.

Additionally, if the $\mathbf{K}_{L\times L}^{-1}$ in (14.2)-(14.3) is replaced by the inverse of sample correlation matrix, $\mathbf{R}_{L\times L}^{-1}$, two more alternative measures can be modified from (14.2-14.3) by

$$(\mathbf{t}_i-\mathbf{t}_j)^T\mathbf{R}_{L\times L}^{-1}(\mathbf{t}_i-\mathbf{t}_j), \tag{14.4}$$

referred to as correlation-Mahalanobis distance (RMD) and

$$\mathbf{t}_i^T\mathbf{R}_{L\times L}^{-1}\mathbf{t}_j, \tag{14.5}$$

referred to as correlation-matched filter based distance (RMFD).

Here, (14.4) is an alternative form of the Mahalanobis distance. But, we should pay a particular attention to (14.5). It is very close to the form of CEM specified by (4.8) where $\mathbf{t}_i$ is designated as the desired target signature $\mathbf{d}$ and $\mathbf{t}_j$ can be viewed as the image pixel vector $\mathbf{r}$ to match the desired target $\mathbf{t}_i$. So, the larger the matched value is, the more likely the two targets belong to the same class. It basically performs a similar task as does a matched filter. This is the reason why the target discrimination measures specified by (14.3) and (14.5) are referred to as matched-filter distance measures.

14.3 ANOMALY CLASSIFICATION

The purpose of a target discrimination measure is to differentiate the targets detected by an anomaly detector and separate them into different classes. These distinct target classes will be used to generate target information required for a classifier. In other words, these target classes can be used as training sample classes to generate *a posteriori* information needed for classification. Basically, an anomaly classifier is made up of three operators in sequence, an anomaly detector, followed by a target discriminator and concluded by a classifier. The algorithm to implement an anomaly classifier can be summarized as follows.

Anomaly Classification Algorithm

1) Apply an anomaly detector to detect potential anomalous targets, $\{\mathbf{t}_j\}_{j=1}^{n}$.
2) Use one of the target discrimination measures specified by (14.2)-(14.5) to group the targets detected in step 1) into separate target classes, $\{\omega_j\}_{j=1}^{p}$.
3) Find the mean of each target class, $\{\bar{\mathbf{t}}_j\}_{j=1}^{p}$.
4) Apply a classifier to classify targets in image data using $\{\bar{\mathbf{t}}_j\}_{j=1}^{p}$ as the desired target information. The classifier used in this step can be any classifier, such as the OSP classifier specified by (3.9) or LCMV classifier specified by (11.2). In this step, the classification is done based on the target information provided in step (3). As a result, the targets that match $\{\bar{\mathbf{t}}_j\}_{j=1}^{p}$ will be detected and also classified. These classified targets may also include targets, which were missed by the anomaly detector in step 1).

14.4 AUTOMATIC THRESHOLDING METHOD

In view of the fact that the images generated by RXD are generally gray scale, the detection is usually carried out by visual inspection. However, in order to avoid such human interpretation and to make an objective assessment, we need to develop a computer-automated thresholding method that converts a gray scale image to a binary image where the detected targets can be extracted from the image background.

Recalling (14.2), RXD operates a form, which allows one to detect anomalies in a large background by finding high peaks of gray levels in homogeneous regions. Therefore, the larger the gray values of the pixels are, the more likely the pixels are anomalous pixels. This suggests that the gray level values of anomalies should behave as outliers and fall in the right tail of the image distribution. For a given value α, we define a rejection region, denoted by $R(\alpha)=\{\mathbf{r}|\delta_{RXD}(\mathbf{r})<\alpha\}$, by a set made up of all the image pixels in the RXD-detected image whose gray level values less than the α. We use the histogram of the RXD-detected image to define the rejection probability, $P(\alpha)$, as

$$P(a)=\Pr(R(a)). \tag{14.6}$$

Then a threshold α_0 used for anomaly detection can be determined by setting a confidence coefficient γ such that $P(\alpha_0) = \gamma$. If $\delta_{RXD}(\mathbf{r}) > \alpha_0$, **r** will be detected as an anomaly. For example, Fig. 14.1(a) plots the gray level values of Fig. 6.5(a) which was produced by RXD for the 15 panel HYDICE scene and Fig. 14.1(b) shows its corresponding histogram. If the confidence coefficient was set by $\gamma = 0.99$, the corresponding threshold α_0 was found to be 484.75 shown in Fig. 14.1(c).

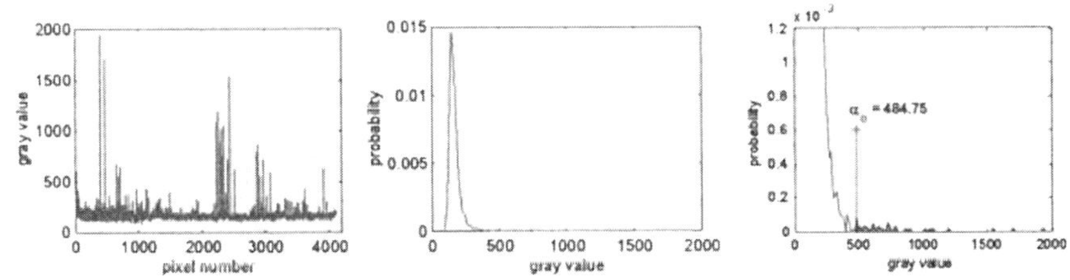

(a) (b) (c)

Figure 14.1. (a) Plot of the gray values produced by the RXD generated image; (b) histogram of (a); (c) An enlargement of right tail in (b).

Using this α_0 to threshold the RXD-detected gray scale image in Fig. 6.5(a) resulted in a binary-thresholded image shown in Fig. 14.2 where there were 39 target pixels, $\{\mathbf{t}_j\}_{j=1}^{39}$ segmented by $\alpha_0 = 484.75$.

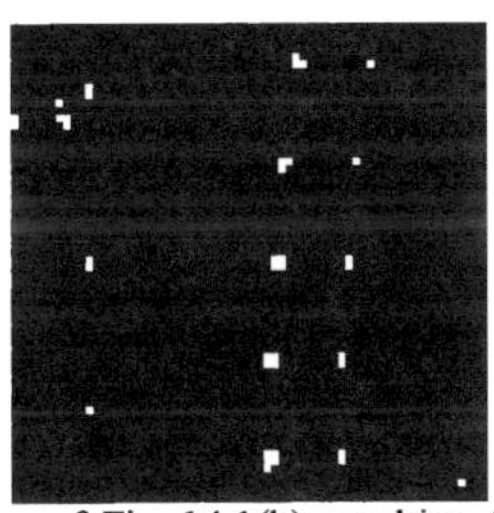

Figure 14.2. A binary-thresholded image of Fig. 14.1(b) resulting from the automatic thresholding method with the confidence coefficient γ = 0.99.

It should be noted that the confidence coefficient of $\gamma = 0.99$ was selected empirically and can be adjusted. When the confidence coefficient γ is close to 1, only a few targets will be detected as anomalies. If the confidence coefficient γ is set too low ($\gamma << 1$), many interferers and background signatures will be extracted as anomalies. Fortunately, with the proposed thresholding technique the sensitivity of selecting an appropriate threshold value can be reduced because it is based on the detected images rather than the original image.

According to Fig. 14.2, 39 target pixels, $\{\mathbf{t}_j\}_{j=1}^{39}$ were detected and were labeled by the order that they were detected from top to bottom and left to right. To further classify these 39 detected targets into separate target classes, the four target discrimination measures, CMD, CMFD, RMD and RMFD specified by (14.2)-(14.5) can be used for this purpose. Since they all produced the same results, only results of RMFD is presented here and shown in Fig. 14.3(a-k), where $\mathbf{t}_i$ was designated as a seed pixel while $\mathbf{t}_j$ running through all the 39 target pixels. For example, in Fig. 14.3(a), $\mathbf{t}_3$ was used as a seed pixel for $\mathbf{t}_i$ and $\mathbf{t}_j$ was chosen to be any target $\mathbf{t}$ from $\{\mathbf{t}_j\}_{j=1}^{39}$. As shown in the plot, the peak values of $\mathbf{t}_i^T\mathbf{R}_{L\times L}^{-1}\mathbf{t}_j$ produced by the RMFM were $\mathbf{t}_1,\mathbf{t}_2,\mathbf{t}_4$. They were all clustered together. So, $\mathbf{t}_3$ and $\mathbf{t}_1,\mathbf{t}_2,\mathbf{t}_4$ were considered to be in the same class. It was clearly shown in Fig. 14.3(a-j) that the 39 detected targets were further clustered into 10 target classes, denoted by $\{\mathbf{t}_1,\mathbf{t}_2,\mathbf{t}_3,\mathbf{t}_4\}$, $\{\mathbf{t}_5,\mathbf{t}_6,\mathbf{t}_{17},\mathbf{t}_{21}\}$, $\{\mathbf{t}_7,\mathbf{t}_9,\mathbf{t}_{10},\mathbf{t}_{12}\}$, $\{\mathbf{t}_8,\mathbf{t}_{11}\}$, $\{\mathbf{t}_{13},\mathbf{t}_{14},\mathbf{t}_{15},\mathbf{t}_{16}\}$, $\{\mathbf{t}_{18},\mathbf{t}_{19},\mathbf{t}_{20},\mathbf{t}_{22},\mathbf{t}_{23},\mathbf{t}_{24}\}$, $\{\mathbf{t}_{25},\mathbf{t}_{26},\mathbf{t}_{27},\mathbf{t}_{28},\mathbf{t}_{29},\mathbf{t}_{30}\}$, $\{\mathbf{t}_{31}\}$, $\{\mathbf{t}_{32},\mathbf{t}_{33},\mathbf{t}_{34},\mathbf{t}_{35},\mathbf{t}_{36},\mathbf{t}_{37},\mathbf{t}_{38}\}$ and $\{\mathbf{t}_{39}\}$ where 15 panels are classified in images in Fig. 14.3(a,e,f,g,h,i).

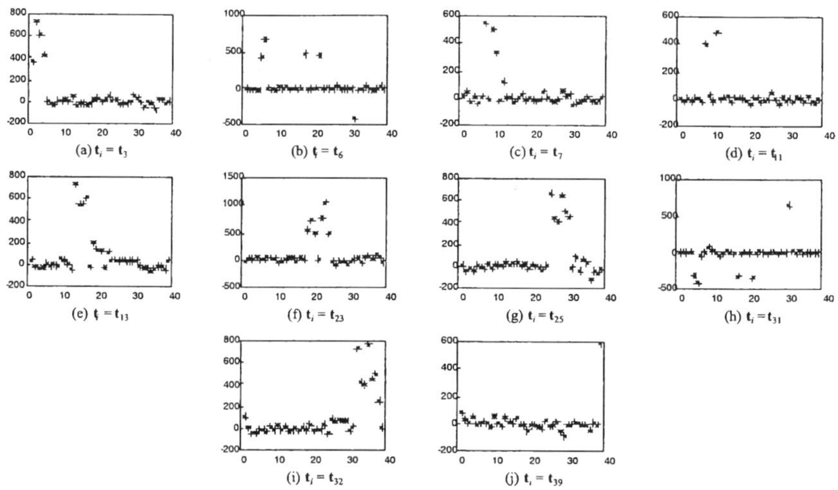

Figure 14.3. Plot generated by RMFD, $\mathbf{t}_i^T\mathbf{R}_{L\times L}^{-1}\mathbf{t}_j$

After target discrimination, we need to find what these targets were. According to the ground truth map provided by Fig. 1.7(b), each of four panels of size $3\text{m}\times 3\text{m}$, p_{21}, p_{31}, p_{41}, p_{51} contains two R pixels except that the panel p_{11} of size $3\text{m}\times 3\text{m}$ has only one R pixel. There is only one R pixel in each of panels of size $2\text{m}\times 2\text{m}$ and $1\text{m}\times 1\text{m}$. Based on this information the 39 detected target pixels can be identified into 10 classes as follows.

(1) Panels in row 1: $\mathbf{t}_2$ = R pixel of p_{11}, $\mathbf{t}_1,\mathbf{t}_3$ = Y pixels of p_{11}, $\mathbf{t}_4$ = R pixel of p_{12}

(2) $\{\mathbf{t}_5,\mathbf{t}_6,\mathbf{t}_{17},\mathbf{t}_{21}\}$ = anomalies located in the forest

(3) $\{\mathbf{t}_7,\mathbf{t}_9,\mathbf{t}_{10},\mathbf{t}_{12}\}$ = anomalies located in the forest

(4) $\{\mathbf{t}_8,\mathbf{t}_{11}\}$ = anomalies located in the forest

(5) Panels in row 2: $\mathbf{t}_{13},\mathbf{t}_{16}$ = R pixels of p_{21}, $\mathbf{t}_{14}$ = Y pixel of p_{21}, $\mathbf{t}_{15}$ = R pixel of p_{22}

(6) Panels in row 3: $\mathbf{t}_{22}, \mathbf{t}_{23}$ = R pixels of p_{31}, $\mathbf{t}_{18}, \mathbf{t}_{19}$ = Y pixels of p_{31}, $\mathbf{t}_{24}$ = R pixel of p_{32}, $\mathbf{t}_{20}$ = Y pixel of p_{32}

(7) Panels in row 4: $\mathbf{t}_{28}, \mathbf{t}_{29}$ = R pixels of p_{41}, $\mathbf{t}_{25}, \mathbf{t}_{26}$ = Y pixels of p_{41}, $\mathbf{t}_{30}$ = R pixel of p_{42}, $\mathbf{t}_{27}$ = Y pixel of p_{42}

(8) $\mathbf{t}_{31}$ = anomaly in the forest

(9) Panels in row 5: $\mathbf{t}_{32}, \mathbf{t}_{35}$, = R pixels of p_{51}, $\mathbf{t}_{27}$, $\mathbf{t}_{33}, \mathbf{t}_{36}, \mathbf{t}_{38}$ = Y pixels of p_{51}, $\mathbf{t}_{37}$ = R pixel of p_{52}, $\mathbf{t}_{34}$ = Y pixel of p_{52}

(10) $\mathbf{t}_{39}$ = anomaly in grass pixel

Fig. 14.4 further shows the correlation among the 39 anomalous target pixels produced by RMFD where strong correlation occurs along the diagonal.

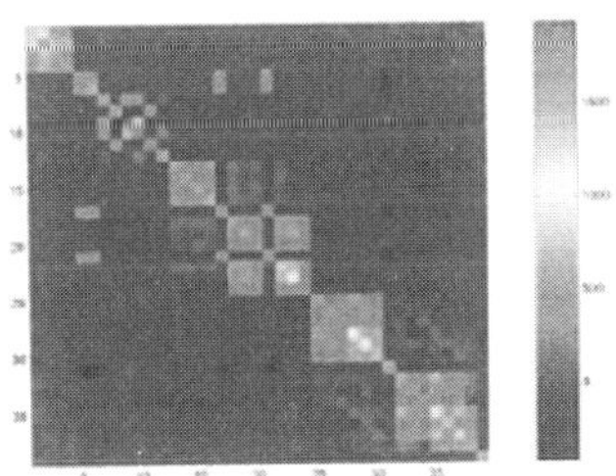

Figure 14.4. Correlation among the 39 detected anomaly pixels produced by RMFD

From Fig. 14.4, it shows that the target classes, $\{\mathbf{t}_1, \mathbf{t}_2, \mathbf{t}_3, \mathbf{t}_4\}$ and $\{\mathbf{t}_{13}, \mathbf{t}_{14}, \mathbf{t}_{15}, \mathbf{t}_{16}\}$ clearly formed two individual and separate blocks. Since each of $\{\mathbf{t}_{31}\}$ and $\{\mathbf{t}_{39}\}$ only correlates itself and does not correlate with other target pixels, there are two blocks formed by a single green pixel along the diagonal. In particular, $\mathbf{t}_{31}$ separated the target class $\{\mathbf{t}_{25}, \mathbf{t}_{26}, \mathbf{t}_{27}, \mathbf{t}_{28}, \mathbf{t}_{29}, \mathbf{t}_{30}\}$ from the target class $\{\mathbf{t}_{32}, \mathbf{t}_{33}, \mathbf{t}_{34}, \mathbf{t}_{35}, \mathbf{t}_{36}, \mathbf{t}_{37}, \mathbf{t}_{38}\}$ into two distinct blocks. Similarly, $\mathbf{t}_{21}$ also broke up the middle block of the target class $\{\mathbf{t}_{18}, \mathbf{t}_{19}, \mathbf{t}_{20}, \mathbf{t}_{22}, \mathbf{t}_{23}, \mathbf{t}_{24}\}$ into two blocks where $\mathbf{t}_{17}$ and $\mathbf{t}_{21}$ form four bright green pixels around the middle block. The block highlighted by orange and red colors is formed by $\{\mathbf{t}_9, \mathbf{t}_{10}\}$ and shows strong correlation between these two anomalous pixels. As a matter of fact, these two pixels are strong interferers and always extracted by automatic and unsupervised detection. There are also three anomaly classes resulting from the forest, $\{\mathbf{t}_5, \mathbf{t}_6, \mathbf{t}_{17}, \mathbf{t}_{21}\}$, $\{\mathbf{t}_8, \mathbf{t}_{11}\}$ and $\{\mathbf{t}_7, \mathbf{t}_9, \mathbf{t}_{10}, \mathbf{t}_{12}\}$. For example, $\{\mathbf{t}_5, \mathbf{t}_6, \mathbf{t}_{17}, \mathbf{t}_{21}\}$ which was made up of three small blocks formed by $\{\mathbf{t}_5, \mathbf{t}_6\}$, $\{\mathbf{t}_{17}\}$ and $\{\mathbf{t}_{21}\}$.

Fig. 14.5 shows the target discrimination image resulting from RMFD where 10 target classes are highlighted by different colors.

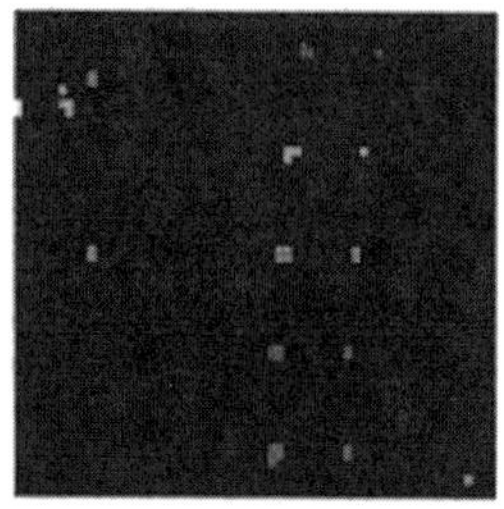

Figure 14.5. 10-class discrimination image of the 39 detected pixels

Using the discrimination results provided by Fig. 14.5 the mean of each target class was calculated to represent the target information for this particular class. These 10 pieces of target information were available for CEM detection. The detection results are given in Fig. 14.6 where the pixels of size $1\text{m} \times 1\text{m}$ in rows 1, 2, 3 and 4 were extracted in Fig. 14.6(a,e,f,g), which were missed by RXD.

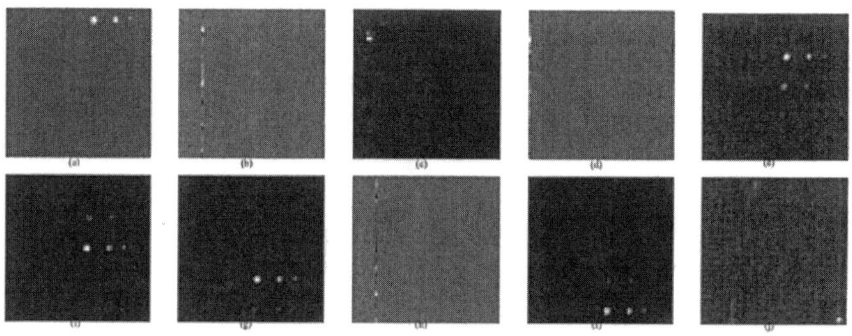

Figure 14.6. Detection of CEM using the target information derived from the discrimination results

Finally, the LCMV classifier was used to classy all the 10 target classes in a single image shown in Fig. 14.7.

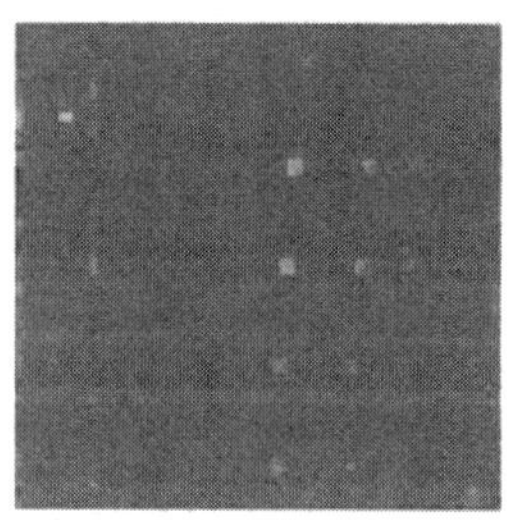

Figure 14.7. Classification by LCMV using the target information derived from the discrimination results

It should be also noted that RXD failed to detect all the five $1\text{m} \times 1\text{m}$ panels p_{i3}, $1 \leq i \leq 5$ due to small amounts of abundance present in these five panel pixels and their size smaller than the 1.5 spatial resolution. However, these missing panels can be extracted if we used one of the four target discrimination measures to generate the necessary target information for each target class. As shown in Fig. 14.6(a,e,f,g) and Fig. 14.7, it was indeed the case. The panels p_{13}, p_{23}, p_{33} and p_{43} which were not picked up by RXD were actually detected 14.6(a,e,f,g) by the CEM detector and classified by

LCMV classifier. This advantage results from the use of target information generated by a discrimination measure.

As a final comment, in order for the four target discrimination measures specified by (14.2)-(14.5) to work effectively, the number of samples that form the sample covariance matrix must be sufficiently large to avoid singularity problems resulting from ill-rank of the sample covariance matrix. In this case, the number of data samples must be greater than or equal to the number of spectral bands. Since the size of the remotely sensed imagery is generally larger than the total number of spectral bands used for acquisition, this requirement is usually satisfied. For the real-time implementation of CRXD, which uses the sample correlation matrix, the real-time processing does not take place until the number of samples is greater than the number spectral bands. In other words, there is no real-time processing in the first few lines before we collect enough data samples to form the non-singular sample correlation matrix.

14.5 ANALYSIS ON TARGET CORRELATION USING TARGET DISCRIMINATION MEASURES

As shown in Fig. 14.4, the color map produced by target correlation could be very useful in target classification since it described the degree of one target correlated with another. Using such a color map to display correlation for visualization was used in Lee and Landgrebe (1993). In this section, a further analysis is conducted for four target discrimination measures on target correlation using the 19 red pixels provided by the ground truth map in Fig. 1.7(b). Fig. 14.8(a-d) shows the color visualization of the correlation among 19 R pixels (i.e., target center pixels) generated by CMD, RMD, CMFD and RMFD, respectively, where vertical bars next to figures are provided to show the degree of correlation using a range of colors ranging from dark red to dark blue. A red color indicates high correlation while a blue color shows less correlation.

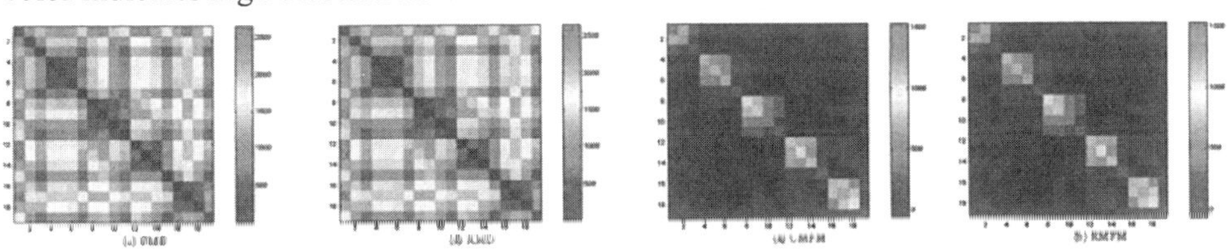

(a) $\left(\mathbf{t}_i - \mathbf{t}_i\right)^T \mathbf{K}^{-1}_{L\times L}\left(\mathbf{t}_i - \mathbf{t}_j\right)$ (b) $\left(\mathbf{t}_i - \mathbf{t}_j\right)^T \mathbf{R}^{-1}_{L\times L}\left(\mathbf{t}_i - \mathbf{t}_j\right)$ (c) $\left(\mathbf{t}_i - \mu\right)\mathbf{K}^{-1}_{L\times L}\left(\mathbf{t}_j - \mu\right)$ (d) $\mathbf{t}_i^T \mathbf{R}^{-1}_{L\times L}\mathbf{t}_j$

Figure 14.8. Correlation among 19 R pixels produced by CMFD, RMFD, CMD and RMD

According to the ground truth, these 19 red pixels are painted by five different materials. This fact is reflected in Fig. 14.8(a-d) where five small square blocks along the diagonal are clearly shown in Fig. 14.8(a-d). Fig. 14.8(a-b) shows the correlation maps produced by the Mahalanobis distance-based distance measures (i.e. CMD and RMD) where a low value implies a high correlation as shown by blue color. By contrast, Fig. 14.7(c-d) shows the correlation maps using the matched filter-based distance measures (i.e. CMFD and RMFD) where a higher value represents a higher correlation as shown by colors other than blue. Furthermore, because the five $1\text{m} \times 1\text{m}$ panels p_{i3}, $1 \le i \le 5$ which were not detected by RXD contained very small amount of abundance, their correlation was weak. This was evidenced in the Fig. 14.7(c-d) where there was one pixel apart between two consecutive blocks.

Table 14.1. Values of CMFD between the 19 R panel pixels

		Panel 1			Panel 2				Panel 3				Panel 4				Panel 5			
		P_{11}	P_{12}	P_{13}	P_{21}	P_{21}	P_{22}	P_{23}	P_{31}	P_{31}	P_{32}	P_{33}	P_{41}	P_{41}	P_{42}	P_{43}	P_{51}	P_{51}	P_{52}	P_{53}
P1	P_{11}	1534	671	249	-15	-37	0	-44	-15	16	30	-1	33	-23	53	-16	2	37	100	-16
	P_{12}	671	594	63	-31	-36	-11	-25	40	51	27	36	-9	-20	53	1	-40	-16	86	10
	P_{13}	249	63	244	29	43	2	45	-18	-37	-24	13	37	19	3	40	-18	-12	-68	-16
P2	P_{21}	-15	-31	29	737	613	548	141	105	34	39	35	36	39	34	-8	-26	-30	-13	3
	P_{21}	-37	-36	43	613	739	533	189	132	37	42	26	19	31	9	-19	-17	-5	-49	20
	P_{22}	0	-11	2	548	533	718	129	101	-8	17	-29	47	35	-14	-40	2	-17	4	-2
	P_{23}	-44	-25	45	141	189	129	294	76	21	28	18	-19	48	10	4	-58	-51	-29	-14
P3	P_{31}	-15	40	-18	105	132	101	76	888	771	432	283	-60	-1	-26	4	-13	-23	4	2
	P_{31}	16	51	-37	34	37	-8	21	771	1048	498	301	-21	-72	-11	-13	36	73	86	8
	P_{32}	30	27	-24	39	42	17	28	432	498	489	172	13	-14	34	-17	-50	-27	29	-14
	P_{33}	-1	37	13	35	26	-29	18	283	301	172	301	-1	-7	12	-8	-15	-18	-21	6
P4	P_{41}	33	-9	37	36	19	47	-19	-60	-21	13	-1	1079	730	706	164	71	148	103	-25
	P_{41}	-23	-20	19	39	31	35	48	-1	-72	-14	-7	730	1007	643	134	68	117	102	32
	P_{42}	53	53	3	34	9	-14	10	-26	-11	34	12	706	643	789	123	-19	99	192	19
	P_{43}	-16	1	40	-8	-19	-40	4	4	-13	-17	-8	164	134	123	287	44	2	17	120
P5	P_{51}	2	-40	-18	-26	-17	2	-58	-13	36	-50	-15	71	68	-19	44	732	767	505	123
	P_{51}	37	-16	-12	-30	-5	-17	-51	-23	73	-27	-18	148	117	99	2	767	1192	774	130
	P_{52}	100	86	-68	-13	-49	4	-29	4	86	29	-21	103	102	192	17	505	774	857	115
	P_{53}	-16	10	-16	3	20	-2	-14	2	8	-14	6	-25	32	19	120	123	130	115	241

Table 14.2. Values of RMFD between the 19 R panel pixels

		Panel 1			Panel 2				Panel 3				Panel 4				Panel 5			
		P_{11}	P_{12}	P_{13}	P_{21}	P_{21}	P_{22}	P_{23}	P_{31}	P_{31}	P_{32}	P_{33}	P_{41}	P_{41}	P_{42}	P_{43}	P_{51}	P_{51}	P_{52}	P_{53}
	P_{11}	1534	672	250	-16	-39	3	-43	-13	20	32	0	33	-24	54	-16	3	37	101	-14
P1	P_{12}	672	595	64	-29	-34	-10	-24	41	51	27	37	-8	-19	54	2	-39	-15	87	11
	P_{13}	250	64	245	30	44	3	46	-17	-37	-23	14	38	21	4	41	-17	-11	-67	-15
	P_{21}	-16	-29	30	734	609	551	143	108	40	42	35	35	37	35	-9	-26	-30	-12	6
P2	P_{21}	-39	-34	44	609	734	537	191	136	44	46	26	19	28	9	-20	-17	-6	-48	23
	P_{22}	3	-10	3	551	537	718	129	101	-10	16	-28	49	37	-13	-38	3	-15	5	-2
	P_{23}	-43	-24	46	143	191	129	295	77	22	28	19	-18	49	11	6	-57	-49	-28	-13
	P_{31}	-13	41	-17	108	136	101	77	887	769	431	285	-58	2	-25	6	-12	-21	5	1
P3	P_{31}	20	51	-37	40	44	-10	22	769	1043	496	303	-18	-67	-10	-9	38	75	87	7
	P_{32}	32	27	-23	42	46	16	28	431	496	489	174	14	-11	35	-15	-48	-25	30	-14
	P_{33}	0	37	14	35	26	-28	19	285	303	174	301	0	-7	12	-8	-14	-18	-20	8
	P_{41}	33	-8	38	35	19	49	-18	-58	-18	14	0	1076	730	707	165	72	148	104	-23
P4	P_{41}	-24	-19	21	37	28	37	49	2	-67	-11	-7	730	1006	644	133	68	117	103	34
	P_{42}	54	54	4	35	9	-13	11	-25	-10	35	12	707	644	790	124	-18	100	193	20
	P_{43}	-16	2	41	-9	-20	-38	6	6	-9	-15	-8	165	133	124	288	44	2	18	122
	P_{51}	3	-39	-17	-26	-17	3	-57	-12	38	-48	-14	72	68	-18	44	733	768	506	124
P5	P_{51}	37	-15	-11	-30	-6	-15	-49	-21	75	-25	-18	148	117	100	2	768	1192	775	132
	P_{52}	101	87	-67	-12	-48	5	-28	5	87	30	-20	104	103	193	18	506	775	858	116
	P_{53}	-14	11	-15	6	23	-2	-13	1	7	-14	8	-23	34	20	122	124	132	116	241

Table 14.3. Values of CMD between the 19 R panel pixels

		Panel 1			Panel 2				Panel 3				Panel 4				Panel 5			
		P_{11}	P_{12}	P_{13}	P_{21}	P_{21}	P_{22}	P_{23}	P_{31}	P_{31}	P_{32}	P_{33}	P_{41}	P_{41}	P_{42}	P_{43}	P_{51}	P_{51}	P_{52}	P_{53}
	P_{11}	0	786	1280	2302	2348	2252	1917	2453	2550	1964	1837	2547	2588	2217	1854	2261	2653	2192	1807
P1	P_{12}	786	0	711	1392	1405	1334	937	1401	1540	1030	822	1691	1641	1277	879	1405	1818	1279	815
	P_{13}	1280	711	0	923	897	957	449	1168	1366	781	519	1249	1212	1026	451	1011	1459	1237	516
	P_{21}	2302	1392	923	0	250	358	749	1415	1717	1149	967	1744	1665	1457	1041	1521	1988	1620	972
P2	P_{21}	2348	1405	897	250	0	392	655	1363	1714	1145	987	1780	1685	1511	1064	1505	1941	1695	941
	P_{22}	2252	1334	957	358	392	0	755	1404	1781	1174	1078	1703	1656	1534	1085	1447	1944	1568	963
	P_{23}	1917	937	449	749	655	755	0	1030	1299	728	558	1410	1205	1063	573	1141	1587	1210	562
	P_{31}	2453	1401	1168	1414	1363	1404	1030	0	394	514	622	2086	1897	1729	1168	1646	2127	1737	1125
P3	P_{31}	2550	1540	1366	1717	1714	1781	1299	394	0	540	747	2168	2199	1859	1360	1707	2095	1733	1273
	P_{32}	1964	1030	781	1149	1145	1174	728	514	540	0	446	1543	1525	1210	810	1321	1735	1289	758
	P_{33}	1837	822	519	967	987	1078	558	622	747	446	0	1380	1322	1067	604	1062	1529	1200	528
	P_{41}	2547	1691	1249	1744	1780	1703	1410	2086	2168	1543	1380	0	625	456	1038	1668	1975	1729	1369
P4	P_{41}	2588	1641	1212	1665	1685	1656	1205	1897	2199	1525	1322	625	0	510	1027	1603	1966	1660	1184
	P_{42}	2217	1277	1026	1457	1511	1534	1063	1729	1859	1210	1067	456	510	0	830	1559	1782	1262	992
	P_{43}	1854	879	451	1041	1064	1085	573	1168	1360	810	604	1038	1027	830	0	932	1475	1112	288
	P_{51}	2261	1405	1011	1521	1505	1447	1141	1646	1707	1321	1062	1668	1603	1559	932	0	390	579	727
P5	P_{51}	2653	1818	1459	1988	1941	1944	1587	2127	2095	1735	1529	1975	1966	1782	1475	390	0	501	1173
	P_{52}	2192	1279	1237	1620	1695	1568	1210	1737	1733	1289	1200	1729	1660	1262	1112	579	501	0	869
	P_{53}	1807	815	516	972	941	963	562	1125	1273	758	528	1369	1184	992	288	727	1173	869	0

Table 14.4. Values of RMD between the 19 R panel pixels

		Panel 1			Panel 2				Panel 3				Panel 4				Panel 5			
		P_{11}	P_{12}	P_{13}	P_{21}	P_{21}	P_{22}	P_{23}	P_{31}	P_{31}	P_{32}	P_{33}	P_{41}	P_{41}	P_{42}	P_{43}	P_{51}	P_{51}	P_{52}	P_{53}
P1	P_{11}	0	784	1278	2301	2345	2247	1915	2448	2537	1959	1836	2547	2587	2216	1854	2261	2653	2191	1803
	P_{12}	784	0	711	1387	1397	1333	937	1399	1535	1028	822	1690	1638	1276	878	1405	1817	1279	814
	P_{13}	1278	711	0	918	890	956	449	1167	1361	779	518	1248	1209	1026	450	1010	1458	1237	516
P2	P_{21}	2301	1387	918	0	250	348	743	1404	1969	1138	965	1742	1665	1454	1040	1518	1986	1616	963
	P_{21}	2345	1397	890	250	0	379	646	1349	1688	1130	984	1776	1684	1505	1062	1500	1938	1689	928
	P_{22}	2247	1333	956	348	379	0	754	1404	1779	1174	1074	1699	1649	1533	1081	1444	1940	1566	963
	P_{23}	1915	937	449	743	646	754	0	1029	1294	727	557	1410	1202	1063	571	1141	1586	1210	561
P3	P_{31}	2448	1399	1167	1404	1349	1404	1029	0	393	514	618	2083	1890	1726	1163	1644	2122	1735	1125
	P_{31}	2537	1535	1361	1696	1688	1779	1294	393	0	539	737	2158	2184	1852	1349	1699	2084	1726	1270
	P_{32}	1959	1028	779	1138	1130	1174	727	514	539	0	442	1539	1517	1208	805	1318	1730	1287	758
	P_{33}	1836	822	518	965	984	1074	557	618	737	442	0	1380	1321	1066	604	1062	1529	1200	526
P4	P_{41}	2547	1690	1248	1742	1776	1699	1410	2083	2158	1539	1380	0	624	456	1037	1668	1975	1729	1366
	P_{41}	2587	1638	1209	1665	1684	1649	1202	1890	2184	1517	1321	624	0	508	1027	1602	1965	1658	1178
	P_{42}	2216	1276	1026	1454	1505	1533	1063	1726	1852	1208	1066	456	508	0	829	1559	1782	1262	991
	P_{43}	1854	878	450	1040	1062	1081	571	1163	1349	805	604	1037	1027	829	0	932	1475	1111	285
P5	P_{51}	2261	1405	1010	1518	1500	1444	1141	1644	1699	1318	1062	1668	1602	1559	932	0	389	579	725
	P_{51}	2653	1817	1458	1986	1938	1940	1586	2122	2084	1730	1529	1975	1965	1782	1475	389	0	501	1170
	P_{52}	2191	1279	1237	1616	1689	1566	1210	1735	1726	1287	1200	1729	1658	1262	1111	579	501	0	868
	P_{53}	1803	814	516	963	928	963	561	1125	1270	758	526	1366	1178	991	285	725	1170	868	0

If we compare Fig. 14.7(c-d) against Fig. 14.7(a-b), it is obvious that the former characterizes correlation better than does the latter except for the panels, p_{i3}, $1 \leq i \leq 5$. In other words, the matched filter-based distance measures performed a little bit better than the Mahalanobis distance-based distance measures in terms of capturing correlation among most panel pixels.

In order to further analyze quantification performance, we calculate the correlation values for these 19 panel pixels using all the four target discrimination measures and tabulate their results in Tables 14.1-14.4. As shown in Tables 14.1-14.4, the five small blocks showing strong correlation in Fig. 14.7(a-d) are also shown in the tables. Like Fig. 14.7(a-d), the values in Tables 14.1-14.2 also characterize panel correlation better than those in Tables 14.3-14.4.

14.6 ON-LINE IMPLEMENTATION

In order to demonstrate an on-line processing for anomaly classification, we will implement CRXD-LCMV hybrid classifier using RMFD specified by (14.5) as the target discrimination measure along with the automatic thresholding described in Section 14.4. Since there was no appreciable difference between RMD specified by (14.3) and RMFD as shown in Chiang and Chang (2001) and Chiang (2001), we chose RMFD due to the fact that RMFD can be easily implemented by a QR-decomposition. The whole processing consists of three stages, the CRXD implementation in step 1), a target-clustering process using RMFD and a real time processing of LCMV classification developed in Chang et al. (2001). Like RXD which can be implemented in real time, the CRXD-LCMV anomaly detection and classification can be also implemented on-line processing with negligible time lag. Since it requires two pass processes, a time delay is inevitable. Nevertheless it can be minimized. A similar phenomenon was also discussed in Section 10.5 where near real-time process was used to make distinction from the on-line process used here.

The first process is carried out in the same manner as does CRXD to detect anomalies in real time. The second pass of discrimination and classification process takes place only a few lines delay (or a set of pixels delay if it is implemented on a pixel-by-pixel basis) right after CRXD was executed. The algorithm takes advantage of time lag of a few lines to generate sufficient target information for the follow-up classification. Since the classifier used in anomaly classification is the LCMV classifier that can be also implemented in real time as described in Chang et al. (2001), the proposed two-pass anomaly detection and classification can be essentially implemented in a timely manner with a few lines delay between two passes. Such real time processing with a few lines time delay is referred to as on-line process.

It should be noted that instead of the sample covariance matrix the sample correlation matrix was used in CRXD and the target discrimination measure (i.e., either RMD or RMFD) to achieve real-time capability. As demonstrated in Chapter 11, a real-time process can be implemented by either a line-by-line or pixel-by-pixel fashion. However, a difficulty arises in the process of target discrimination, which must be done right after detection and prior to classification. In addition, the target information generated by discrimination also needs to be updated either every line or every pixel depending upon which real-time process is used. In order to alleviate this problem, we build a look-up table and update it while the detection process in process. The look-up table contains target templates produced by an anomaly detector and a target discrimination measure. Each template represents one distinct type of targets and will be used as the desired target information for classification. Initially, there is no target template in the look-up table. As soon as the anomaly detector detects its first anomaly, it will be included as the first target template in the table. When a second anomaly is detected, the target discrimination measure then compares the new detected anomalous target to the first target template. If the detected target is close and similar to the target template, the first target template will be updated and replaced by a new target template which is obtained by averaging the new detected target and the first target template, and this new target will be then discarded. Otherwise, the second detected anomaly will be included in the table as the second target template. This look-up table keeps track of new target templates and is updated by comparing the next detected anomaly against the existing target templates. If there is a match between the newly detected anomaly and a target template in the table, the matched target template will be updated by the average of the matched target template and the new target, and the new target will be eliminated. If there is no match, the new target will be used to create a new target template by its own and will be included in the table. This particular on-line anomaly classifier is referred to as the RXD-LCMV hybrid classifier. Although there may be a small time lag of implementing the target discrimination between the target detection and classification, the three operations can be actually processed in parallel. Even in the case that the detection, discrimination and classification are implemented sequentially by computer, the whole process can be nearly simultaneously if the size of an image scene to be process is not very large. For the 15-panel HYDICE scene with size of $64 \times 64 \times 169$ in Fig. 1.7(a), there was no appreciable time delay in real-time implementation. Fig. 14.9 shows an on-line target discrimination result produced by the CRXD using the RMFD as the target discrimination measure where the line number underneath each image indicates the result was obtained by using the causal information up to that particular line. As we can see, the results in Fig. 14.9 are completely different from that in Fig. 6.5(a). Other than the 2-pixel anomaly detected in Fig. 6.5(a) many additional anomalies were picked up and discriminated by RMFM. This is because the process of the target discrimination occurred immediately right after the detection process. For example, cinders, shade and vegetation were detected in Fig. 14.9(a-b), then followed by playa detected in after line 80 in Fig. 14.9(c-g). The rhyolite was extracted after line 120 in Fig. 14.9(e-g).

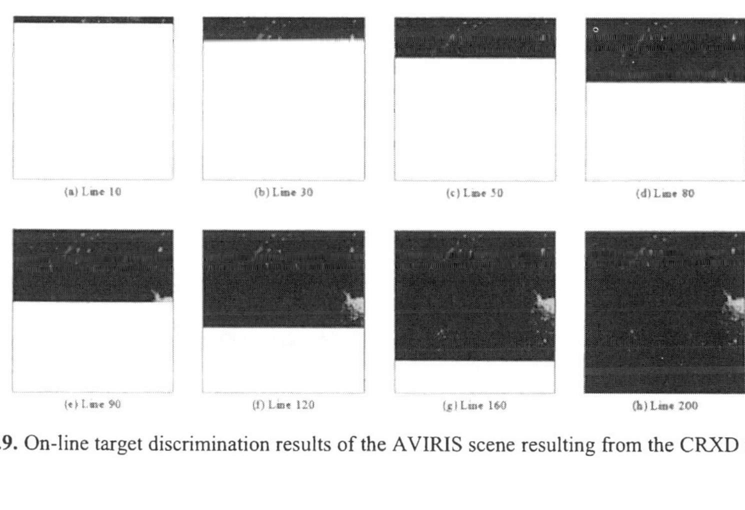

Figure 14.9. On-line target discrimination results of the AVIRIS scene resulting from the CRXD using RMFD

When the process was completed, the CRXD-LCMV classifier actually detected and classified all the five major substances in the image scene, cinders, playa, rhyolite, shade and vegetation shown Fig. 14.10. Unfortunately, it missed the anomaly because it was averaged out by target discrimination.

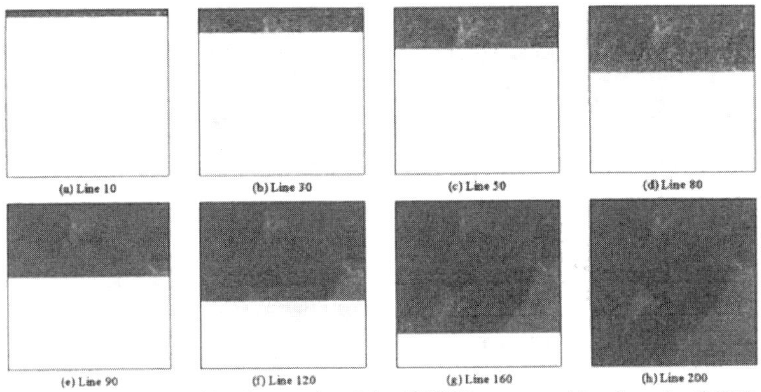

Figure 14.10. On-line target classification results of the AVIRIS scene resulting from the CRXD using RMFM in conjunction with the LCMV classifier

Like the CRXD-LCMV classifier, CRXD also detected many anomalies as well in Fig. 14.11(a-h). Unfortunately, all these detected anomalies suddenly vanished as soon as a two-pixel anomaly detected at the upper edge of the lake in Fig. 14.11(i). The same image almost remained unchanged until the whole process completed.

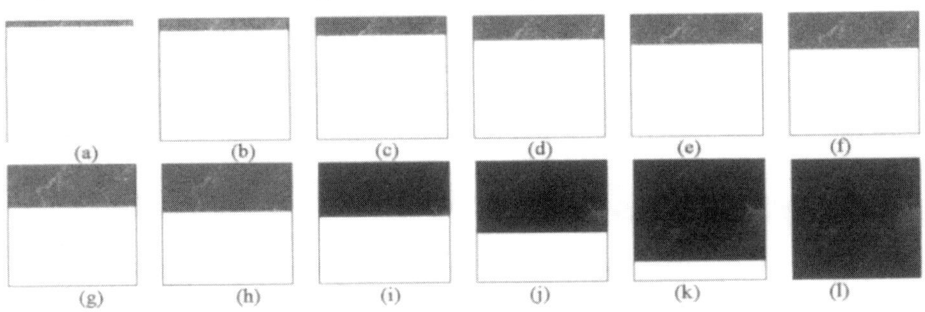

Figure 14.11. On-line target detection results of the AVIRIS scene resulting from the CRXD without target discrimination

A problem arises in RXD is that RXD could not discriminate among the anomalies it detected and all the detected anomalies are shown in images according to their gray scales. As a consequence, a weak anomaly of one type may be dominated by a strong anomaly of another type. This explains why there was only the 2-pixel anomaly detected by the RXD in Fig. 6.5(a) because its abundance was so strong and eventually dominated other anomalies which may belong to different types of targets, but had relatively low abundance fractions. It did not imply that they were not detected by RXD. It is simply too difficult to detect their presence by visual inspection since their gray level values were scaled down by the strong 2-pixel anomaly. However, if a target discrimination is applied right after RXD, targets of different types can be separated and would not be dominated by targets of another types. In this case, different classes of targets can be

detected as shown in Fig. 14.9. This experiment provides an interesting example. That is, when target discrimination is implemented in conjunction with CRXD, the resulting CRXD-LCMV becomes a classifier rather than an anomaly detector.
A similar experiment was also conducted using the 15-panel HYDICE image scene. Fig. 14.12 (a-g) shows the results by the line-by-line on-line process of the CRXD-LCMV classifier.

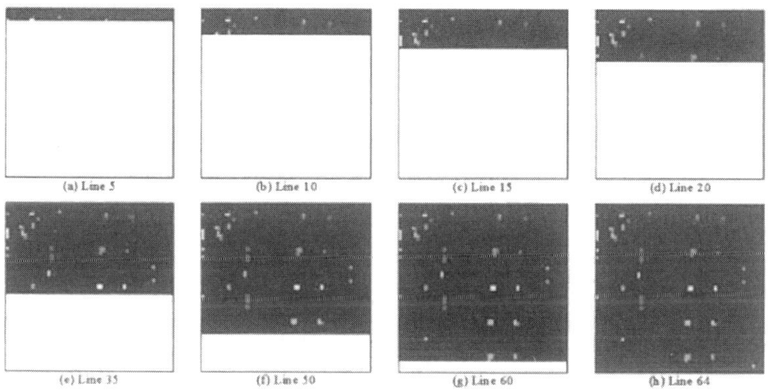

Figure 14.12. On-line target discrimination results of the 15-panel HYDICE scene resulting from the CRXD using RMFD

As shown in Fig. 14.12(a), an interferer and the first panel pixel in row 1 were detected and discriminated by yellow and green colors respectively. Then two more intereferers were picked up in Fig. 14.12(b) highlighted by red. This was followed by Fig. 14.12(c-g) where panels in rows 2-5 were extracted respectively and also discriminated by different colors. Fig. 14.12(h) shows the results of the complete on-line target discrimination process. Fig. 14.13 (a-h) shows the results of the on-line CRXD-LCMV classification right after the on-line target discrimination process.

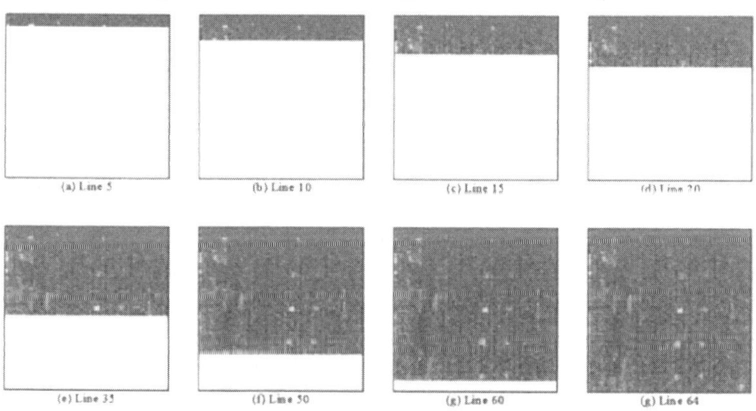

Figure 14.13. On-line target classification results of the 15-panel HYDICE scene resulting from the CRXD using RMFM in conjunction with the LCMV classifier

It used the target information generated from the on-line target discrimination process to classify different types of targets. If we compare Fig. 14.10 to Fig. 14.9 and Fig. 14.13 to Fig. 14.12, the images in Figs. 14.9 and 14.12 were more clean than those in Figs. 14.10 and 14.13. This is because the target discrimination was performed on only the pixels detected by RXD while the target classification was carried out for all image pixel vectors. Fig. 14.14 also shows the results produced by CRXD without target discrimination.

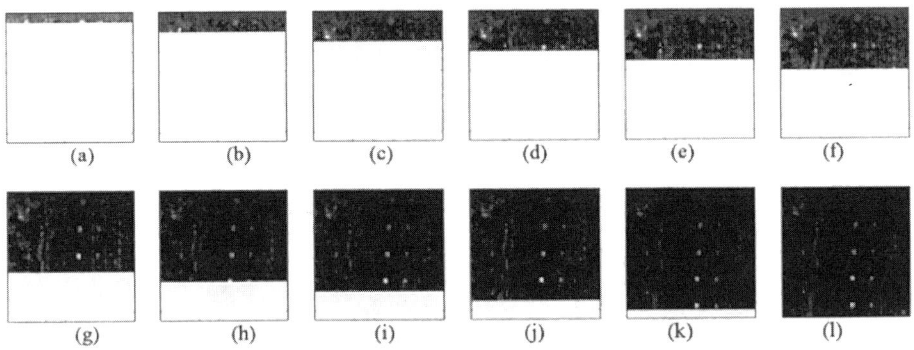

Figure 14.14. On-line target detection results of the HYDICE scene resulting from CRXD without target discrimination

As we can see from Fig. 14.14 CRXD detected some panels in rows 1-3 in Fig. 14.14(a-f). However, as soon as the panels in row 4 were detected in Fig. 14.14(g), the previous detected target pixels suddenly became dim as shown in Fig. 14.14 (h-l). This phenomenon remained until the end of the process. As we compare the final result of CRXD in Fig. 14.14(l) to that generated by the CRXD-LCMV in Fig. 14.13(g), CRXD-LCMV using target discrimination performed significantly better than CRXD with no target discrimination included in target detection and classification. In particular, the former extracted a few panel pixels in row 2 and one panel pixel in row 1 which were missed by the latter in the first column of Fig. 14.10(g). So, this experiment further shows that the RXD-LCMV classifier using target discrimination actually performed and operated as a classifier.

14.7 CONCLUSIONS

This chapter extends the anomaly detection in Chapter 6 to anomaly classification. Such extension is challenging because an anomaly detector does not necessarily discriminate among the targets it detected. It requires a discrimination measure to differentiate the detected targets one another. To meet this need, four target discrimination measures are presented which can be categorized into two classes, the Mahalanobis distance-based measures and matched filter-based target discrimination measures. The former measures the distance between two target pixel vectors using the data covariance/correlation matrix to take care of sample spectral correlation. The latter measures the distance between two target pixel vectors based on the degree to which both pixel vectors match each other. Interestingly, experimental results show that these four target discrimination measures perform very similarly. In order to further implement anomaly classification in real time as the way we did for anomaly detection in Chapter 6, a hybrid of RXD and LCMV, called RXD-LCMV anomaly classifier, is also developed for real-time implementation. It combines CRXD with an LCMV classifier by incorporating a target discrimination measure prior to target classification. The experiments demonstrate that the RXD-LCMV anomaly classifier performs very differently from the anomaly detector, RXD that does not use target discrimination. This shows that in order for an anomaly detector to classify its detected anomalous targets, target discrimination is a necessary and critical step to success in anomaly classification.

15

AUTOMATIC MIXED PIXEL CLASSIFICATION (AMPC): LINEAR SPECTRAL RANDOM MIXTURE ANALYSIS (LSRMA)

Independent component analysis (ICA) has shown much success in blind source separation and channel equalization. Its applications to remotely sensed images are investigated in recent years. Linear spectral mixture analysis (LSMA) has been widely used for subpixel detection and mixed pixel classification in remote sensing image processing and already studied in Chapters 3, 8-10, 13. It models the spectral signature of an image pixel vector as a linear mixture of spectral signatures of targets present in the image data and the target abundance fractions are assumed to be *unknown*, but *nonrandom constants*. This chapter combines these two approaches to one, called ICA-based linear spectral random mixture analysis (LSRMA), which describes an image pixel vector as a random process resulting from a random composition of multiple spectral signatures of distinct targets in the image data. It differs from LSMA in that the abundance fractions of the target signatures in LSRMA are considered to be *unknown random independent signal sources*. Two major advantages can be benefited from LSRMA. First, LSRMA does not require any prior knowledge of the targets used in the linear mixture model. Second and most importantly, LSRMA models each of target signatures as an independent random signal source so that the spectral variability of target signatures can be captured more effectively in a stochastic manner. The only required knowledge for LSRMA is the number of signal sources, p assumed to be present in the image data. This number is estimated by Neyman-Pearson detection theory-based eigen-thresholding methods to be investigated in Chapter 17.

15.1 INTRODUCTION

The linear spectral mixture analysis (LSMA) assumes that the spectral signature of an image pixel vector is a linear mixture of the spectral signatures of target signatures resident in image data. Two restrictions generally limit the utility of LSMA. One is that the complete target knowledge must be given *a priori*. In many practical applications, obtaining such *a priori* information is usually very difficult. To relax this requirement, several unsupervised methods discussed in Chapter 5 can be used to generate necessary target information directly from the image data. The second restriction is that the abundance fractions of targets must be constrained to achieve better estimation. When the target abundance fractions are considered as unknown nonrandom constants, they can be estimated by statistical methods such as least-squares estimation. However, due to many unknown factors such as atmospheric effects, nonstationary image background, etc., the spectral signatures of targets may vary pixel by pixel. Therefore, it is more realistic to assume that the abundance fractions of targets in a pixel vector are random quantities rather than unknown constants so that the spectral variability of a target can be described and captured by a random process. In this case, it is more appropriate to assume that an image pixel vector is a random linear mixture where the abundance fraction of each target signature is considered as a random signal source.

Independent component analysis (ICA) has been studied extensively in blind source separation and other applications (Hyvarinen and Oja, 2000; Comon, 1994; Karhunen et al., 1997; Oja et al., 1995; Bell and Sejnowski, 1995; Lee, 1998; Girolami, 2000). The idea of ICA offers a feasible approach to solving the random abundance mixture problem described above. Applications of the ICA to remotely sensed image analysis have been recently investigated (Bayliss et al., 1997; Zhang and Chen, 2002; Szu, 1999; Tu, 2000; Chiang et al., 2000). It differs from the principal components analysis (PCA) in many aspects. The PCA decorrelates the sample data covariance matrix in the sense that the data set is decomposed into a set of uncorrelated and orthogonal components where each component is oriented by an eigenvector. With such eigen-decomposition PCA can achieve data compaction, dimensionality reduction by preserving most significant information in principal components in terms of data variance. So, PCA is generally used for information preservation, but not for target detection and image classification. In contrast to PCA, ICA looks for components, which are *statistically independent* rather than *uncorrelated*; thus, it requires statistics of orders higher than variance and covariance. In addition, the ICA-generated components are not necessarily geometrically orthogonal. The ICA utilizes a linear model to describe a mixture of a set of unknown random signal sources, then unmixes these signal sources in separate components so as to achieve signal detection and classification. If we further assume that the abundance fraction of each target signature in LSMA is an unknown and independent random signal source, the source mixing model considered in ICA can be readily applied to LSMA. In this case, ICA can be used to solve for random abundance fractions of target signatures. Such an ICA-based LSMA can be viewed as a random version of LSMA and will be referred to as linear spectral random mixture analysis (LSRMA).

In order for LSRMA to be effective, several assumptions must be made. One is that the source components must mutually statistically independent. This implies that the target signatures present in the image data must be spectrally distinct. A second assumption is that at most one source component is allowed to be Gaussian. This is because a sum of Gaussian processes is also Gaussian and ICA cannot separate Gaussian processes using a linear mixture model. In remotely sensed imagery, the number of target pixel vectors, such as small man-made targets, anomalies or rare minerals is relatively small compared to the entire image. Due to a large number of background pixels, we may assume that image background pixel vectors are Gaussian-distributed and small target pixel vectors are non-Gaussian signal sources. If we further assume that the noise is additive white Gaussian, then the background pixel vectors and the white Gaussian noise pixel vectors will be classified in a single class that is assigned to a same Gaussian independent component since ICA cannot separate two Gaussian sources. Since the targets of interest are considered as non-Gaussian signal sources in LSRMA, they will be separated by ICA in different components. As will be shown in experiments, it is indeed the case.

The ICA proposed in LSRMA is slightly different from the commonly used ICA in two ways. First, the separating matrix (also referred to as unmixing matrix) **W** derived from ICA is usually assumed to be a square matrix of full rank and an orthogonal matrix. In this case, the number of signal sources, say p, must be equal to data dimensionality, L. As demonstrated in our experiments, this assumption may not be valid for hyperspectral images where the number of signal sources is generally much smaller than the number of bands, i.e. $p < L$. This results in that the matrix **W** is not of full rank and the learning rule derived from inverting **W** may become unstable and may not converge. The problem of this type is called under-complete ICA and has been investigated in Amari (1999). Second, the mixing matrix **M** is often assumed to be orthogonal. In remote sensing image classification, the matrix **M** made up of the spectral signatures of targets of interest, which are distinct but not necessarily orthogonal. With the orthogonality assumption for **W** the learning algorithm designed for ICA is forced to look for an orthogonal decomposition of signal sources. The resulting independent components may be able to unmix targets with signatures spectrally orthogonal to each other, but may not effectively unmix targets with similar spectral signatures. So, instead of imposing the constraint that the covariance matrix of the separating matrix **W** must be an identity matrix as is done in the standard ICA, our proposed ICA approach is derived by using the constraint that the covariance matrix of unmixed abundance fractions of target signatures must be an identity matrix. This advantage enables us to design a learning algorithm to converge to independent components that can separate spectrally similar targets. A similar approach was also investigated in Cardoso (1996). The experiments show that such learning algorithm is very useful and suitable to hyperspectral image analysis because many targets extracted from a hyperspectral image do have similar spectral signatures, which may not be orthogonal.

Like LSMA, LSRMA also requires knowing the number of independent sources, p, in the image data. In order to estimate the p, three Neyman-Pearson detection theory-based eigen-thresholding methods recently developed for estimation of the intrinsic dimensionality can be used to estimate p (Chang and Du, 1999; Harsanyi et al., 1994) and also discussed in Chapter 17. It converts the determination of the number of signal sources to a binary composite hypothesis testing problem where the number of times a Neyman-Pearson test fails is the number of unknown signal sources present in the image data. It turns out that this approach provides a good estimate of the p.

15.2 INDEPENDENT COMPONENT ANALYSIS (ICA)

Following the same notations used in Section 3.2 and (3.1) a linear spectral mixture model is described by

$$\mathbf{r} = \mathbf{M}\alpha + \mathbf{n} \tag{15.1}$$

where **r** be an $L \times 1$ column pixel vector, **n** is noise, **M** is an $L \times p$ target signature matrix, denoted by $\left[\mathbf{m}_1\ \mathbf{m}_2 \cdots \mathbf{m}_p\right]$ with $\mathbf{m}_j$ being an $L \times p$ column vector represented by the signature of the j-th target and p is the total number of targets assumed to be in the image. Suppose that $\alpha = \left(\alpha_1, \alpha_2, \cdots, \alpha_p\right)^T$ is a $p \times 1$ abundance column vector associated with **r** where α_j denotes the abundance fraction of $\mathbf{m}_j$ and **r** will be used to represent either the pixel vector **r** or its spectral signature. A spectral linear unmixing method estimates the unknown abundance fractions, $\alpha_1, \alpha_2, \cdots, \alpha_p$ via an inverse of a linear mixture model described by (15.1). One requirement of LSMA is that the signature matrix **M** must be known *a priori*. Many approaches have been proposed in the past to obtain **M** directly from the image data in an unsupervised means. Here, we investigate a rather different and interesting approach, independent component analysis (ICA) which is also based on model (15.1), but does not require the prior knowledge of **M**. Unlike previous techniques discussed so far in this book, it assumes that the p abundance fractions, $\alpha_1, \alpha_2, \cdots, \alpha_p$ are unknown random signal sources rather than unknown nonrandom parameters. In this case, three additional assumptions must be made on the abundance vector $\alpha = \left(\alpha_1, \alpha_2, \cdots, \alpha_p\right)^T$.

(i) The p target signatures, $\mathbf{m}_1, \mathbf{m}_2, \cdots, \mathbf{m}_p$ in **M** must be spectrally distinct.
(ii) The p abundance fractions, $\alpha_1, \alpha_2, \cdots, \alpha_p$ are mutually statistically independent random sources.
(iii) Each of the p abundance fractions, $\alpha_1, \alpha_2, \cdots, \alpha_p$ must be a zero-mean random source and at most one source is Gaussian.

In remotely sensed imagery, the first assumption simply says that there are p spectrally distinct targets in a scene and each column vector in **M** represents one distinct target in the scene. The second assumption implies that the compositions of distinct targets in a pixel **r** are random quantities, one independent of another. The third assumption suggests that if targets of interest are not Gaussian sources, ICA can detect and classify them from the image background in separate independent components. This is based on the fact that due to a very large number of background pixels in the image the natural background pixels can be considered as a Gaussian-like source while targets can be viewed as non-Gaussian signal sources. Since ICA cannot separate Gaussian processes using a linear mixture model, ICA classifies the background area and additive Gaussian noise into one single independent component, whereas different targets are classified into other separate independent components. Accordingly, automatic target detection and classification can be accomplished by LSRMA. It should be noted that except for the three above assumptions, no prior information is assumed about model (15.1).

15.3 ICA-BASED LSRMA

In order to implement ICA using model (15.1), the mixing matrix and the unknown signal sources used in the blind source separation are replaced with the target signature matrix **M** and the p random abundance fractions denoted by $\alpha_1, \alpha_2, \cdots, \alpha_p$ respectively. With this interpretation ICA finds a $p \times L$ separating matrix **W** and applies it to an image pixel **r** to unmix the p abundance fractions, $\alpha_1, \alpha_2, \cdots, \alpha_p$. More precisely, ICA solves an inverse problem of model (15.1) for the $p \times L$ separating matrix **W** via the following equation

$$\hat{\alpha}(\mathbf{r}) = \mathbf{W}\mathbf{r}, \tag{15.2}$$

where $\hat{\alpha}(\mathbf{r}) = \left(\hat{\alpha}_1(\mathbf{r}), \cdots, \hat{\alpha}_p(\mathbf{r})\right)^T$ is the estimate of abundance fractional vector $\left(\alpha_1, \alpha_2, \cdots, \alpha_p\right)^T$ derived from $\mathbf{r}$. Since changing order of components in $\hat{\alpha}(\mathbf{r})$ does not affect their statistical independence, the estimate of the i-th abundance fraction α_i may appear as any component $\hat{\alpha}_j(\mathbf{r})$ of $\hat{\alpha}(\mathbf{r})$. Further, because multiplying random variables by non-zero scalar factors does not affect the statistical independence, it is also impossible to determine the true amounts of the abundance fractions, $\alpha_1, \alpha_2, \cdots, \alpha_p$, from model (15.1) without additional assumptions. Unless we are interested in target quantification the order and the true abundance fractions are generally not crucial in detection and classification of targets. In this case, we can normalize each abundance fraction source to unit variance so that the covariance matrix of the abundance fraction sources becomes the identity matrix. This can be simply done by a sphering (whitening) process studied in Section 12.3 of Chapter 12.

15.3.1 Relative Entropy-Based Measure for ICA

In order to use ICA, a criterion is required to measure the statistical independence among the estimated abundance fractions $\hat{\alpha}_1(\mathbf{r}), \hat{\alpha}_2(\mathbf{r}), \cdots, \hat{\alpha}_p(\mathbf{r})$. According to information theory (Cover and Thomas, 1991), relative entropy or Kullback-Leibler information distance function is an appropriate measure. Let $p(\hat{\alpha}(\mathbf{r})) = p(\hat{\alpha}_1(\mathbf{r}), \hat{\alpha}_2(\mathbf{r}), \cdots, \hat{\alpha}_p(\mathbf{r}))$ be the joint probability density function (pdf) of the estimated random abundance vector $\hat{\alpha}(\mathbf{r})$ obtained from (15.2) and $p(\hat{\alpha}_j)$ be the marginal pdf of the j-th abundance fraction $\hat{\alpha}_j$ for $1 \le j \le p$ given by (15.2). Since $\alpha_1, \alpha_2, \cdots, \alpha_p$ in model (15.1) are assumed to be independent, $p(\alpha) = \prod_{j=1}^{p} p(\alpha_j)$. If we assume that α is a random source vector and $\hat{\alpha}(\mathbf{r})$ is the estimate of α from the observation vector $\mathbf{r}$, then the entropy of α relative to $\hat{\alpha}(\mathbf{r})$ is defined by

$$\begin{aligned} D\left(p(\hat{\alpha}(\mathbf{r})) \,\|\, p(\alpha) = \textstyle\prod_{j=1}^{p} p(\alpha_j)\right) &= \Sigma\, p(\hat{\alpha}(\mathbf{r})) \log\left(p(\hat{\alpha}(\mathbf{r})) / p(\alpha)\right) \\ &= \Sigma\, p(\hat{\alpha}(\mathbf{r})) \log\left(p(\hat{\alpha}(\mathbf{r})) / \textstyle\prod_{j=1}^{p} p(\alpha_j)\right) \quad (15.3) \\ &= -H(\hat{\alpha}(\mathbf{r})) - \Sigma\, p(\hat{\alpha}(\mathbf{r}))\left[\textstyle\sum_{j=1}^{p} \log p(\alpha_j)\right] \end{aligned}$$

where $H(\hat{\alpha}(\mathbf{r}))$ is the entropy of the estimated abundance vector $\hat{\alpha}(\mathbf{r})$ as defined in (2.35) in Chapter 2. Minimizing (15.3) over $\hat{\alpha}(\mathbf{r})$ through $\mathbf{W}$ in (15.2) is equivalent to minimizing the discrepancy between pdfs $p(\alpha)$ and $p(\hat{\alpha}(\mathbf{r}))$. That is, the smaller the $D\left(p(\hat{\alpha}(\mathbf{r})) \| p(\alpha)\right)$ is, the less the discrepancy between $p(\alpha)$ and $p(\hat{\alpha}(\mathbf{r}))$ is; thus, the more likely to be independent the $\hat{\alpha}_1(\mathbf{r}), \hat{\alpha}_2(\mathbf{r}), \cdots, \hat{\alpha}_p(\mathbf{r})$ are. Because the pdf of α is generally unknown and needs to be estimated, the $p(\alpha) = \prod_{j=1}^{p} p(\alpha_j)$ in (15.3) is generally replaced by its estimate $p(\hat{\alpha}(\mathbf{r})) = \prod_{j=1}^{p} p(\hat{\alpha}_j(\mathbf{r}))$. Substituting this estimate into (15.3) results in

$$\begin{aligned} D\left(p(\hat{\alpha}(\mathbf{r})) \,\|\, p(\hat{\alpha}(\mathbf{r})) = \Pi_{j=1}^{p} p(\hat{\alpha}_j(\mathbf{r}))\right) &= E\left[-\Sigma_{j=1}^{p} \log\left(\hat{\alpha}_j(\mathbf{r})\right)\right] - H(\hat{\alpha}(\mathbf{r})) \\ &= \Sigma_{j=1}^{p} H(\hat{\alpha}_j(\mathbf{r})) - H(\hat{\alpha}(\mathbf{r})) \end{aligned} \tag{15.4}$$

where E[.] is the expectation with respect to $p(\hat{\alpha}(\mathbf{r}))$ and $H(\hat{\alpha}_j(\mathbf{r}))$ is the entropy of the j-th estimated source $\hat{\alpha}_j(\mathbf{r})$. Unfortunately, even in this case, finding the pdf of $\hat{\alpha}(\mathbf{r})$ is also difficult in practice. In order to mitigate this dilemma, Comon introduced an alternative criterion (Common, 1994) that approximates the $D\left(p(\hat{\alpha}(\mathbf{r}))\middle\| \Pi_{j=1}^{p} p(\hat{\alpha}_j(\mathbf{r}))\right)$ given in (15.4). Instead of minimizing $D\left(p(\hat{\alpha}(\mathbf{r}))\middle\| \Pi_{j=1}^{p} p(\hat{\alpha}_j(\mathbf{r}))\right)$, Comon suggested to maximize the higher order statistics of the data, called contrast function given by

$$\psi(\mathbf{W}) = \Sigma_{j=1}^{p} 4\kappa_{jjj}^2 + \kappa_{jjjj}^2 + 7\kappa_{jjj}^4 - 6\kappa_{jjj}^2 \kappa_{jjjj} \tag{15.5}$$

where κ_{jjj} is the third order standardized cumulant of the j-th $\hat{\alpha}_j(\mathbf{r})$ for $1 \le j \le p$ representing " skewness" of $p(\hat{\alpha}_j(\mathbf{r}))$ and κ_{jjjj} is the fourth standardized cumulant of the j-th $\hat{\alpha}_j(\mathbf{r})$ for $1 \le j \le p$ representing " kurtosis" of $p(\hat{\alpha}_j(\mathbf{r}))$. If the skewness of $p(\hat{\alpha}_j(\mathbf{r}))$ is sufficiently large, (15.5) can be further approximated by $7\Sigma_{j=1}^{p} \kappa_{jjj}^4$. On the other hand, If the kurtosis of $p(\hat{\alpha}_j(\mathbf{r}))$ is sufficiently large, a good approximation of (15.5) is $\Sigma_{j=1}^{p} \kappa_{jjjj}^2$. In either case, (15.5) is reduced to a much simpler criterion.

15.3.2 Learning Algorithm to Find Separating Matrix W

Since the second-order statistics can be taken care of by decorrelation, the data vectors $\mathbf{r}$ are first pre-whitened prior to separation. In this case, data are completely characterized by statistics with orders higher than 2. So, for simplicity we assume that the data vectors have been pre-whitened in this section. In order to derive a learning algorithm, we impose a constraint that the covariance matrix of the estimated abundance vector $\hat{\alpha}(\mathbf{r})$ in (15.2) must be an identity matrix. To further simplify notations, we denote $\hat{\alpha}(\mathbf{r})$ by $\mathbf{y}$ with $y_j = \hat{\alpha}_j(\mathbf{r})$. The learning algorithm to be developed must solve the following constrained optimization problem.

$$\text{maximize } \psi(\mathbf{W}) = \Sigma_{j=1}^{p} E\left[y_j^m\right] \text{ over } \mathbf{W} \text{ for } m \ge 3 \text{ subject to } E\left[\mathbf{y}\mathbf{y}^T\right] = \mathbf{I} \tag{15.6}$$

where $\mathbf{I}$ is the $p \times p$ identity matrix. For such constrained problems, we use exterior penalty methods discussed in Cichocki and Unbehauen (1993) to eliminate some or all of the constraints. The idea is to add to the objective function specified by (15.6) so-called penalty function terms, which assign a higher cost to infeasible points. In our case, the penalty function terms imposed on (15.6) are defined by

$$U(h_{ij}(\mathbf{y})) = \begin{cases} 0; & \text{if } h_{ij}(\mathbf{y}) = 0 \\ > 0; & \text{if } h_{ij}(\mathbf{y}) \neq 0 \end{cases} \tag{15.7}$$

where constraints are $h_{ij}(\mathbf{y})$ defined by $h_{ij}(\mathbf{y}) = E\left[y_i y_j\right] - \delta_{ij}$. Now using the penalty function terms specified by (15.6) and (15.7) we can define a penalty function $P(\mathbf{y})$ by

$$P(\mathbf{y}) = \Sigma_{i,j=1}^{p} \lambda_{ij} U(h_{ij}(\mathbf{y})) = (1/2)\Sigma_{i,j=1}^{p} \lambda_{ij} \left(E[y_i y_j] - \delta_{ij}\right)^2 \tag{15.8}$$

where $\lambda_{ij} > 0$ is called a penalty parameter or penalty multiplier (typically all $\lambda_{ij} = \lambda$). Thus, maximizing the constrained problem described by (15.5) is equivalent to maximizing the following function,

$$J(\mathbf{W}) = \psi(\mathbf{W}) - P(\mathbf{y}) = \psi(\mathbf{W}) - (\lambda / 2)\Sigma_{i,j=1}^{p}\left(E[y_i y_j] - \delta_{ij}\right)^2 \tag{15.9}$$

which subtracts the penalty function given by (15.8) from the objective function in (15.5). In order to find the separating matrix $\mathbf{W} = \left[w_{st}\right]_{p \times l}$, we calculate gradient of the function $J(\mathbf{W})$ by differentiating $\psi(\mathbf{W})$ and $P(\mathbf{y})$ with respect to w_{st} respectively:

$$\frac{\partial \psi(\mathbf{W})}{\partial w_{st}} = \frac{\partial}{\partial w_{st}}\left(\Sigma_{j=1}^{p} E[y_j^m]^2\right) = 2mE[y_s^m]E[y_s^{m-1} r_t] \tag{15.10}$$

$$\begin{aligned} \frac{\partial P(\mathbf{y})}{\partial w_{st}} &= \frac{\lambda}{2} \frac{\partial\left(\Sigma_{j=1}^{p}\left(E[y_i y_s] - \delta_{is}\right)^2\right)}{\partial w_{st}} \\ &= 2\lambda \Sigma_{i=1}^{p}\left(E[y_i y_s] - \delta_{is}\right)E[y_i y_t] \end{aligned} \tag{15.11}$$

(15.10) and (15.11) can be expressed in terms of matrix forms as follows.

$$\nabla_{\mathbf{W}} \psi(\mathbf{W}) = 2m\Lambda E[g(\mathbf{y})\mathbf{r}^T] \tag{15.12}$$

$$\nabla_{\mathbf{W}} P(\mathbf{y}) = 2\lambda\left(E[\mathbf{y}\mathbf{y}^T] - \mathbf{I}\right)\mathbf{y}\mathbf{r}^T. \tag{15.13}$$

where $g(\mathbf{y}) = \left(y_1^{m-1}, y_2^{m-1}, \cdots, y_p^{m-1}\right)^T$ and $\Lambda = diag\left\{E[y_i^m]\right\}$ is the diagonal matrix with the i-th diagonal element given by $E[y_i^m]$. From Eqs. (15.12) and (15.13), a learning algorithm to generate the desired separating matrix W can be designed by

$$\mathbf{W}_{k+1} = \mathbf{W}_k + \mu E[g(\mathbf{y})\mathbf{r}^T] - \eta\left(E[\mathbf{y}\mathbf{y}^T] - \mathbf{I}\right)\mathbf{y}\mathbf{r}^T. \tag{15.14}$$

where μ and η are learning parameters in (15.12) and (15.13) respectively. It was found empirically that μ controls the convergence speed and is generally less than 1 while η controls the constraint and is generally greater than μ.

15.4 EXPERIMENTS

Since there is no prior knowledge about the signal sources assumed in ICA, the number of targets, p, resident in an image scene must be estimated from the image data. For multispectral imagery, L is not large and the p can be estimated by exhausting all the components. However, it may also have a drawback. In some occasions, $p > L$ which results in an over-complete ICA (Lee et al., 1999). In hyperspectral imagery, the number of bands is usually greater than hundreds. In this case, p is much smaller than L, $p << L$ and selecting an appropriate p for LSRMA becomes crucial. In the past, many criteria such as an information theoretic criterion (AIC) (Akaike, 1974), minimum description length (MDL) (Schwarz, 1977; Rissanen, 1978; Rissanen, 1983), network information criterion (NIC) (Murata, et al., 1994), transformed Gershgorin radii (TGR) (Wu et al., 1995) have been proposed for estimating the number of signals in sensor array processing. However, these criteria require the prior knowledge of likelihood functions, generally of Gaussian functional forms. In addition, in all these criteria, the noise assumed to be independent identically distributed, an assumption generally not true in remotely sensed images. Recently, the TGR was modified by a noise adjusted transformed Gershgorin disk (NATGD) to estimate p (Tu, 2000) which was determined by a graphical plot. However, Chiang et al. (2000) showed by experiments that the p was underestimated by the NATGD. Since the p is closely related to the virtual dimensionality (VD) in Chapter 17, the p is estimated by VD + 1 due to the fact that at least one component is required to accommodate the noise in LSRMA. The issue of the VD estimation will be postponed to Chapter 17.

In the following sections, two sets of real hyperspectral image data, AVIRIS in Fig. 1.6(a) and 15-panel HYDICE scene in Fig. 1.7(a) were used for experiments to demonstrate the performance of LSRMA with skewness ($m = 3$ in (15.6)) and kurtosis ($m = 4$ in (15.6)) used as optimal criteria.

15.4.1 AVIRIS Image Experiments

It was demonstrated in Harsanyi and Chang (1994) that there were five targets of interest, cinders, rhyolite, playa (dry lake), vegetation and shade. So, at least five components, $p \geq 5$, are needed to classify and separate these five targets. However, this was obtained by visual inspection supported by ground truth. If we assume that no prior knowledge is available about the image scene, the number of targets in an image scene, p must be estimated from the data. As shown in Chapter 6, there was a single anomaly of two-pixel size located at the top edge of the lake shown inside a circle marked in Fig. 6.1(b). This anomaly cannot be seen or detected visually from the scene. In order for LSRMA to detect this anomaly along with the five targets, the p must be greater than 5 so that the cinders, rhyolite, playa dry lake, vegetation, shade and the anomaly can be detected and classified in 6 separate components. If we use the three methods, the HFC, modified HFC and NSP methods proposed in Chapter 17, the VD was estimated to be 4, 5 and 8 respectively. So, in this case the corresponding $p = \text{VD} + 1$ will be 5, 6 and 9 with one extra component included to accommodate the noise. In Tu (2000) the same identical image scene in Fig. 1.6(a) was considered and the p was estimated by the proposed NATGD to be 4. Unfortunately, with $p = 4$ the shade and the anomaly were not detected. In order to demonstrate the impact of different values of p on the performance, we conducted experiments with $p = 4$ estimated by NATGD, $p = 5$ by HFC, $p = 6$ by modified HFC and $p = 9$ by NSP respectively. The learning parameters μ and η used in (15.14) for the following experiments were empirically set to $\mu = 0.5$ and $\eta = 1$.

Example 15.1 ($p = 4$)

Fig. 15.1(a-d) shows the results generated by LSRMA using skewness as a criterion. Only the cinders, vegetation and center of playa were classified in Fig. 15.1(a-c) and the noise component was shown in the 4-th component. It missed the detection of the rhyolite, shade and part of playa.

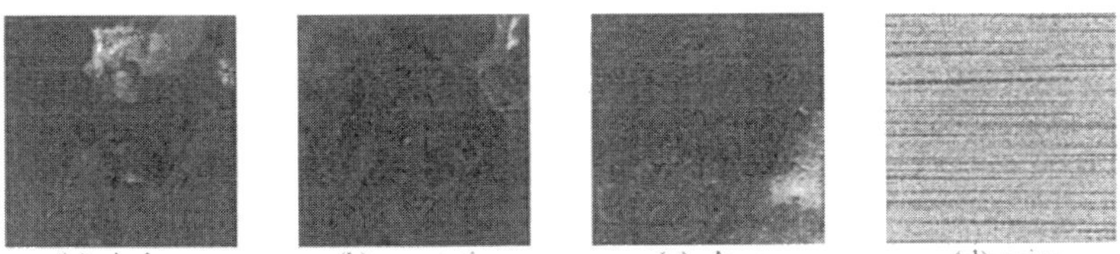

(a) cinders (b) vegetation (c) playa (d) noise

Figure 15.1. Results produced by LSRMA using skewness for $p = 4$

Fig. 15.2 (a-d) shows the results produced by LSRMA using kurtosis where only the anomaly, the vegetation and the bottom part of playa were classified in Fig. 15.2(a-c) and the noise was represented by the 4-th component. It also missed the detection of the cinders, rhyolite, shade and a large portion of playa. Interestingly, Fig. 15.2(a) detected the anomaly observed in Fig. 6.1(b). This experiment demonstrated that the skewness and the kurtosis performed very differently, but kurtosis is more effective to detect small targets.

(a) anomaly (b) vegetation (c) playa (d) noise

Figure 15.2. Results produced by LSRMA using kurtosis for $p = 4$

Example 15.2 ($p = 5$)

In this example, we conducted similar experiments to Example 15.1 with $p = 5$ where Figs. 15.3 and 15.4 were results obtained for skewness and kurtosis respectively. Comparing Fig. 15.3 to Fig. 15.1, little has been changed in results for skewness except that the 5-th component showed nothing but noise.

(a) cinders (b) vegetation (c) playa (d) noise (e) noise

Figure 15.3. Results produced by LSRMA using skewness for $p = 5$

However, when the kurtosis was used, there was a drastic change between Figs. 15.2 and 15.4. Both detected the anomaly in their first components shown in Fig. 15.2(a) and Fig. 15.4(a). However, the cinders and shade were extracted in Fig. 15.4(c-d) respectively which were missed in Fig. 15.2. Unfortunately, it still also missed the detection of the rhyolite and the playa.

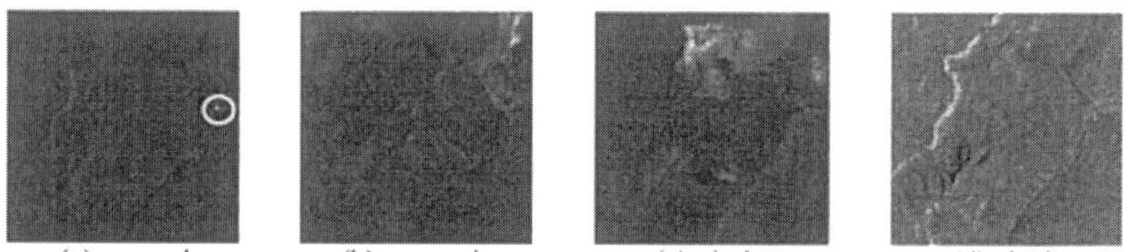

(a) anomaly (b) vegetation (c) cinders (d) shade

Figure 15.4. Results produced by LSRMA using kurtosis for $p = 5$

Example 15.3 ($p = 6$)

Figs. 15.5-15.6 show the results produced by LSRMA using skewness and kurtosis for $p = 6$. Both criteria did not perform well either. In Fig. 15.5, the skewness missed the anomaly, vegetation, rhyolite and a large portion of dry lake. On the other hand, the kurtosis did detect the anomaly, but also missed vegetation and part of dry lake.

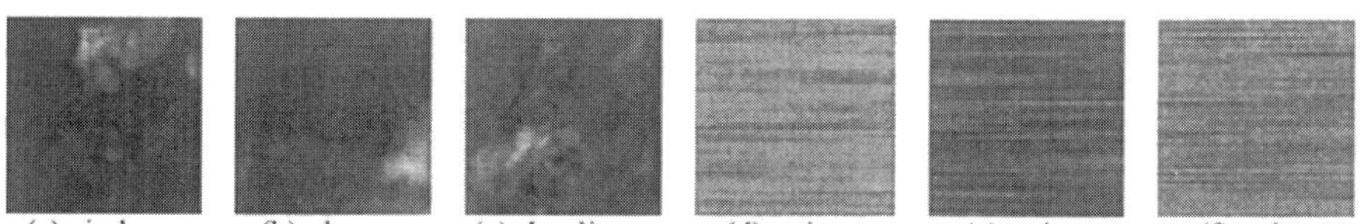

(a) cinders (b) playa (c) rhyolite (d) noise (e) noise (f) noise

Figure 15.5. Results produced by LSRMA using skewness with $p = 6$

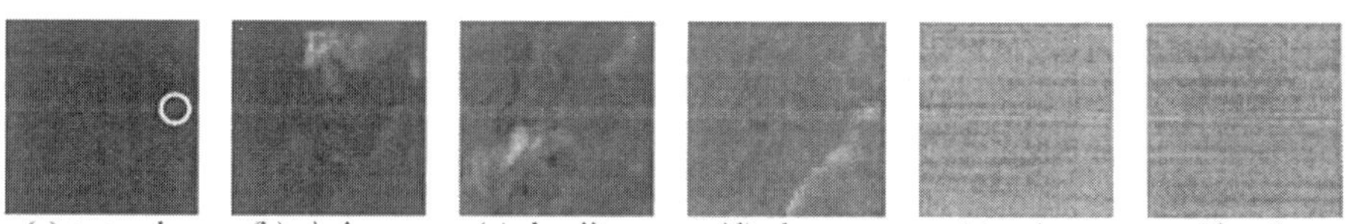

(a) anomaly (b) cinders (c) rhyolite (d) playa (e) noise (f) noise

Figure 15.6. Results produced by the LSRMA using kurtosis with $p = 6$

Example 15.4 ($p = 9$)

Similar experiments were also conducted for $p = 9$. Figs. 15.7 and 15.8 show the results produced by LSRMA using skewness and kurtosis respectively where the skewness and kurtosis demonstrated their different strengths in target extraction. The cinders, the vegetation, the playa, the rhyolite and the shade were detected by skewness in the first 6 components shown by Fig. 15.7(a-f). Interestingly, the playa was detected in two separate components in Figs 15.7(c) and 15.7(e) and the anomaly was barely detected along with the shade in Fig. 15.7(f).

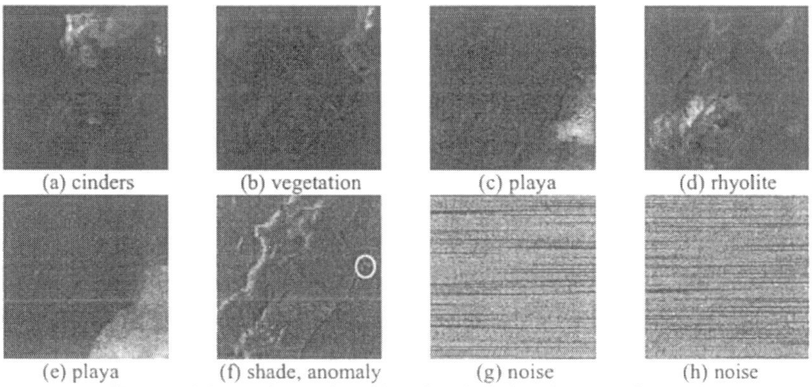

Figure 15.7. Results produced by LSRMA using skewness for $p = 9$

By contrast, kurtosis extracted the anomaly, vegetation and cinders in the first 3 components shown by Fig. 15.8(a-c) while the playa was extracted in the next three separate components in Fig. 15.8(d-f). Unfortunately, the rhyolite and the shade, which were extracted in Figs. 15.7(c) and 15.7(f) by the skewness were not detected by the kurtosis. Since the playa covers a very large area of the image scene, the dynamic range of its spectral variability is relatively large. Therefore, the playa was classified in three separate components with the center detected in Figs. 15.7(c) and 15.8(d), the edge detected in Fig. 15.8(e), and the bottom detected in Figs. 15.7(e) and 15.8(f). Interestingly, there was an interferer picked up in Fig. 15.7(h), which was not shown in all the previous images.

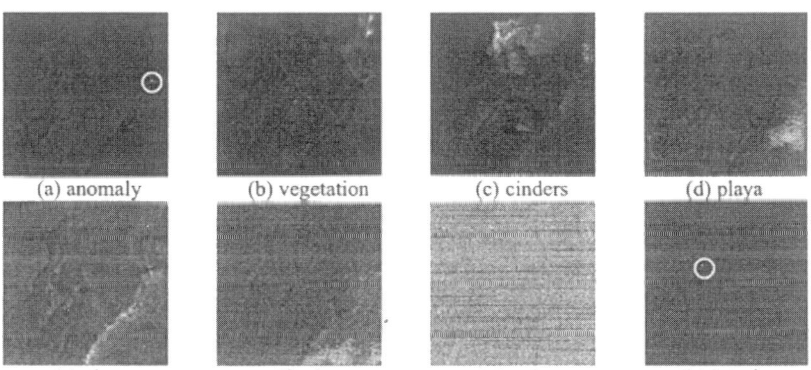

Figure 15.8. Results produced by the LSRMA using kurtosis for $p = 9$

This phenomenon was also observed in Chang and Heinz (2000) and Heinz and Chang (2001). Such subtle spectral variations were overlooked in Harsanyi and Chang (1994) because the playa signature was obtained by averaging a large dry lake area by visual inspection. As a consequence, the entire playa was extracted, and the anomaly was averaged out and could not be detected. A similar problem was also found in Tu (2000) where the lake was detected as an entity, and both the shade and anomaly were not detected. This resulted from the fact that the estimated p was too small.

To conclude the AVIRIS experiments, various values of p were also tested and evaluated by skewness and kurtosis. The results for $p = 4,5,6,7,8$ were not as good as that obtained by $p = 9$. When the p increases from 10 to 12, the performance of both criteria was not improved significantly except that different portions of the lake and Gaussian sources were detected in the increased components. In order to see whether or not the performance of LSRMA can be improved with $p = L$, Figs. 15.9 and 15.10 show the result of LSRMA for $p = L = 158$ using skewness and kurtosis respectively where playa and vegetation were detected and classified in three and two separate components in Fig. 15.9(b,f,g) and Fig. 15.9(c,e) by skewness respectively.

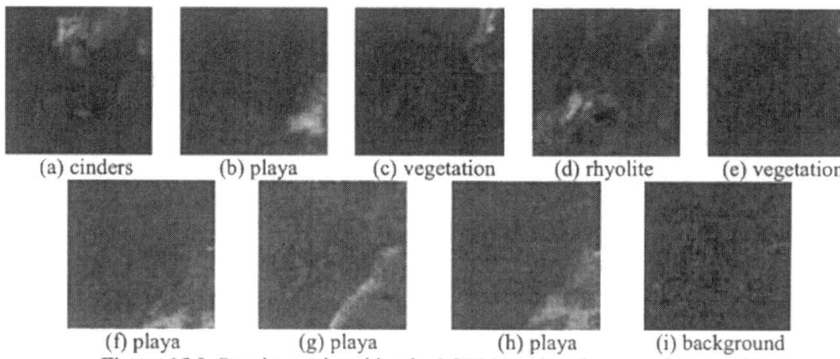

Figure 15.9. Results produced by the LSRMA using skewness for $p = 158$

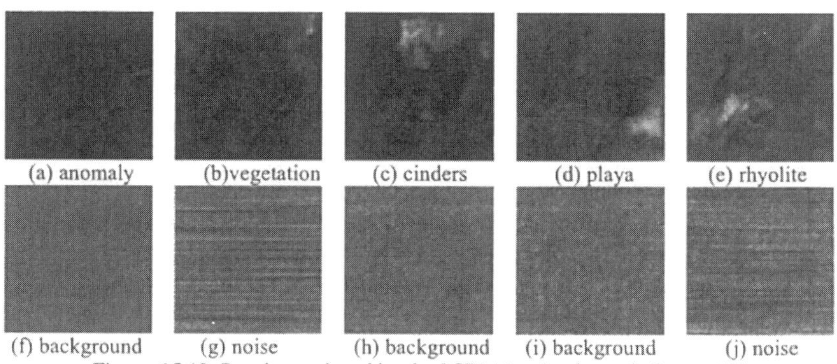

Figure 15.10. Results produced by the LSRMA using kurtosis for $p = 158$

As we can see from Figs. 15.9 and 15.10, increasing p did not necessarily improve detection and classification performance. In Fig. 15.9, skewness missed the detection of the anomaly and the shade compared to Fig. 15.10 where the kurtosis missed the detection of the shade as well as a large portion of the lake.

The above experiments demonstrated that skewness may be a good criterion for classification of large areas while the kurtosis may be effective in extracting small targets or insignificant targets.

In general, the number of components produced by LSRMA should be equal to p where each component accommodates one signal source. Additionally, a different p should also generate a different set of component images. However, it is interesting to note that according to our experiments, if the p exceeds a certain number, the difference (e.g. Euclidean distance) between two consecutive projection vectors generated by the LSRMA is very small. In this case, LSRMA will stop generating additional components and the LSRMA will be called to converge to this number, denoted by $N_{LSRMA}(p)$ which is usually much smaller than the p. Table 15.1 tabulates $N_{LSRMA}(p)$ generated by LSRMA using various p.

Table 15.1. $N_{LSRMA}(p)$ produced by different values of p

	$p = 4$	$p = 5$	$p = 6$	$p = 9$	$p = 122$	$p = 151$	$p = 158$
sknewness	4	5	6	4	122	151	9
kurtosis	4	5	6	139	18	151	10

For instance, for $p = 158$ we were supposed to have 159 components. However, after the 9th (skewness) or 10th (kurtosis) components was generated, the subsequent components were very similar to 9th (skewness) or 10th (kurtosis) components within very small least squares errors. So, $N_{LSRMA}(p)$ for skewness and kurtosis are 9 and 10 respectively. It should be also noted that if p is less than $N_{LSRMA}(p)$, a different value of p will generate a complete different set of component images. This suggests that finding an appropriate $N_{LSRMA}(p)$ for LSRMA could be very challenging. The VD proposed in Chapter 17 may be a good estimate for N_{LSRMA}.

In all the experiments conducted in this chapter, the difference between two projection vectors is measured by Euclidean distance and the error tolerance ε was set to 0.1. It turns out that the error threshold was rather robust where N_{LSRMA} was the same for ε ranging from 0.1 to 0.001.

15.4.2 HYDICE Image Experiments

Since the HYDICE scene in Fig. 1.7(a) provides a ground truth map in Fig. 1.7(b), we can take advantage of target center pixels to simulate synthetic images to evaluate the performance of LSRMA.

Example 15.4 (Computer Simulations)

In this experiment, we simulated a scene with size of 50×50 that is similar to Fig. 1.7(a). It consists of 25 single-pixel panels simulated by P1, P2, P3, P4 and P5 in Fig. 1.8. These 25 simulated pixels are shown in Fig. 15.11(a) and arranged in 5 rows with 5 pixels in each row

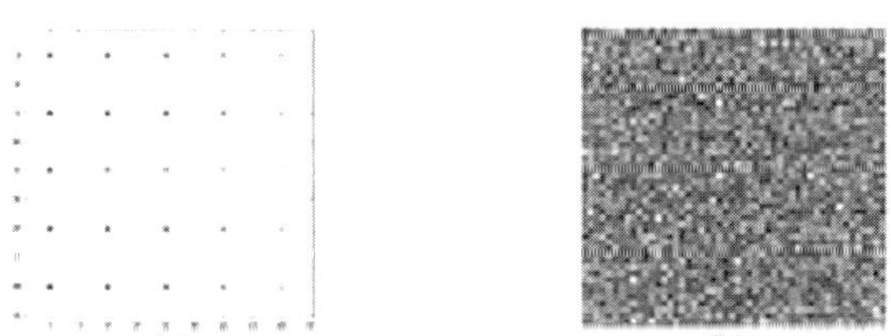

(a) (b)

Figure 15.11. A synthetic HYDICE scene with 25 simulated pixels

The 5 pixels in the same row were simulated by the same signature with different abundance fractions, 1.0, 0.8, 0.6, 0.4, 0.2 respectively starting from the first column to the 5-th column. In this case, the mixing matrix $\mathbf{M}$ is formed by $\mathbf{M} = [\text{P1 P2 P3 P4 P5}]$ and the abundance vector α is assigned by $\alpha = (1.0, 0.8, 0.6, 0.4, 0.2)^T$.

In addition to these 25 simulated pixels, we also added zero-mean Gaussian noise with different variances to achieve different levels of signal-to-noise ratio (SNR). Fig. 15.11(b) shows the resulting image with a zero-mean Gaussian noise added to the synthetic image in Fig. 15.11(a).

In the following simulations, 6 scenarios with SNR = 30dB, 25dB, 20dB, 15dB, 10dB, 5dB were conducted to evaluate the performance of LSRMA with using skewness and kurtosis. In the skewness case, $N_{LSRMA}(5)$ = 5 for SNR = 30dB, 25dB, 20dB, 15dB, 10dB, and all 25 panel pixels were detected and accurately classified into their own classes in these five components. For the case of SNR = 5dB, $N_{LSRMA}(5)$ = 4 and the panel pixels in rows 2 and 3 were detected but forced to be classified into one class. This is because that the spectra of P2 and P3 are very similar as demonstrated previously. For comparison, only results for SNR = 5dB and 10dB are shown in Figs. 15.12 and 15.13 respectively where the amounts of detected abundance fractions are also plotted for reference.

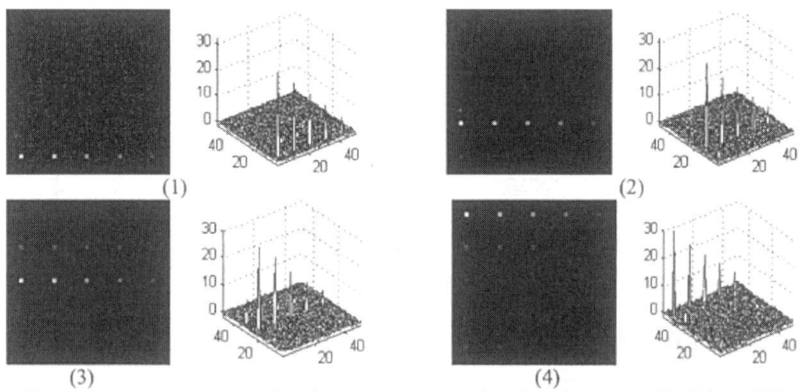

Figure 15.12. Detection results of 25 simulated panels using skewness with SNR = 5dB

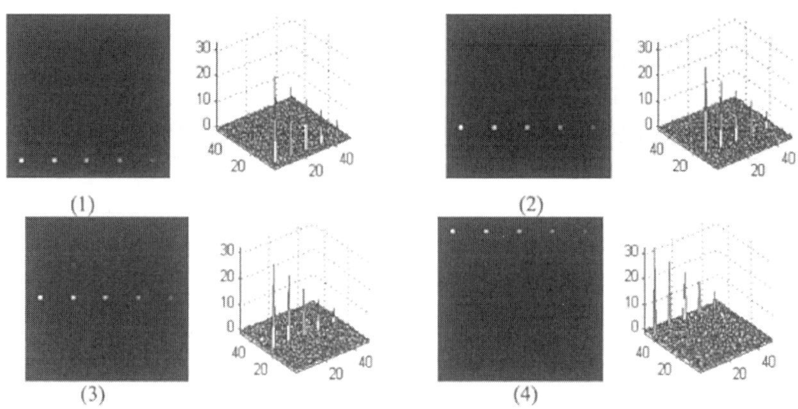

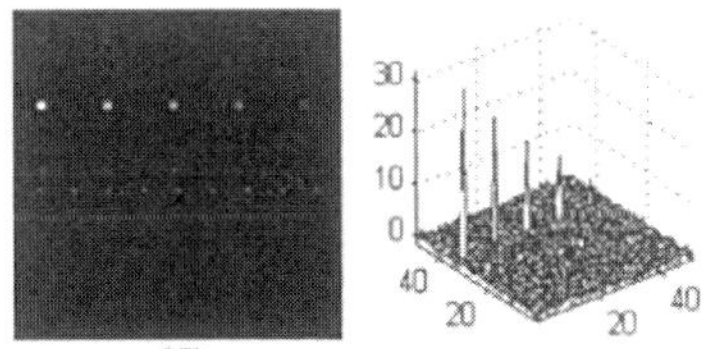

(5)

Figure 15.13. Detection results of 25 simulated panels using skewness with SNR = 10dB

Unlike skewness, LSRMA using kurtosis produced a very large number of components for SNR = 30dB, 25dB, 20dB, 15dB, 10dB, 5dB, but all the 25 panel pixels were detected and accurately classified into their own classes in the first five components. Since the results for all cases were similar except different amounts of detected abundance fractions, only results for the case of SNR = 5 dB is shown in Fig. 15.14.

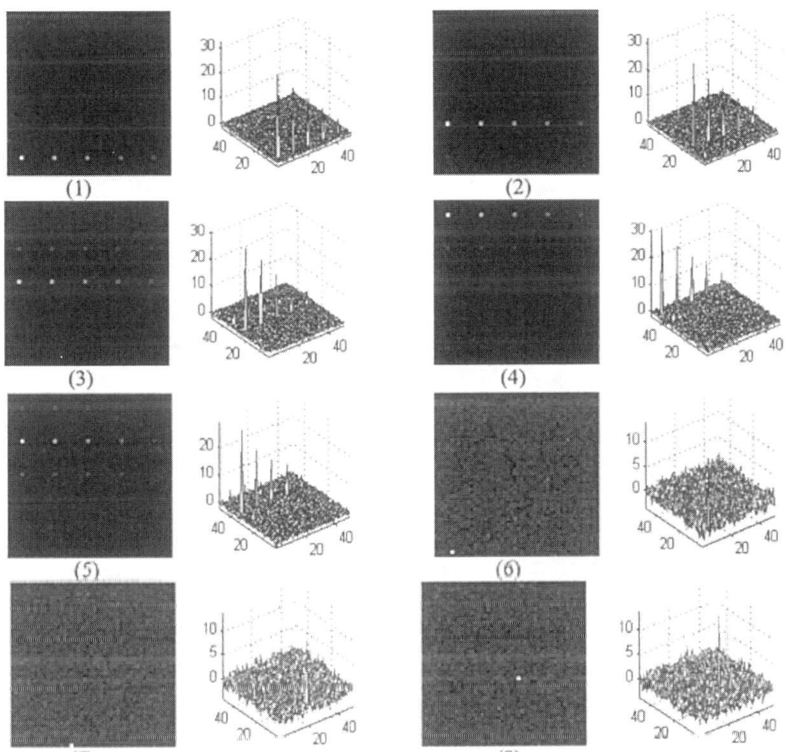

Figure 15.14. Detection results of 25 simulated panels using kurtosis with SNR = 5

It is worth noting that there were also interferers and background signatures detected with relatively small abundance fractions in the components that are beyond the first five components. Furthermore, as SNR increased, so was $N_{LSRMA}(p)$, but the amount of detected abundance fractions were more accurate and close to the true abundance fractions. This also resulted in that more interferers were detected, each of which was classified into a separate and individual component. According to the above computer simulations, kurtosis was very effective but also sensitive to target detection. In addition, the performance of both criteria was also proportional to the strength of SNR.

Example 15.5 (15-panel HYDICE scene)

In this example, we used the 15-panel HYDICE scene in Fig. 1.7(a) to conduct experiments for LSRMA using skewness and kurtosis. The number of mixing targets was estimated to be p = 14, 16, 21, 125 and 86 using p = VD + 1 where the VD was obtained by Table 17.3 in Chapter 17. Interestingly, in these experiments, the interferer on the left corner and the 15 panels were all picked up and classified in the first 6 components and their results were all very similar. The only difference is that a different p produced a different number of components for LSRMA. The components beyond the first six components generally extracted natural background signatures such as grass, tree or road and noise. Table 15.2 tabulates their corresponding values of $N_{LSRMA}(p)$ generated by LSRMA. It is worth noting from Table 15.2 that a higher value of p does not necessarily imply a larger $N_{LSRMA}(p)$.

Table 15.2. $N_{LSRMA}(p)$ produced by different values of p

	p = 14	p = 16	p = 21	p = 86	p = 125	p = 169
skewness	9	8	17	12	6	6
kurtosis	8	6	9	11	13	12

Experiments for p ranging from 6 to 30 were also conducted for comparison. Figs. 15.15-15.18 show both skewness and kurtosis detected the panels in row 2 effectively with p = 14 and 16. We also conducted experiments with p = 15 for both the skewness and kurtosis. In this case, both of them did not detect panels in row 2. However, after p reached a certain number, the results were similar and did not differ very much in terms of detecting 15 panels. For the 15-panel scene, we have found this number is 17 for skewness and 18 for kurtosis, in which case the performance of LSRMA was very robust and stable. Apparently, the VD in Chapter 17 provides a good estimate of this number as shown in Tables 15.1 and 15.2.

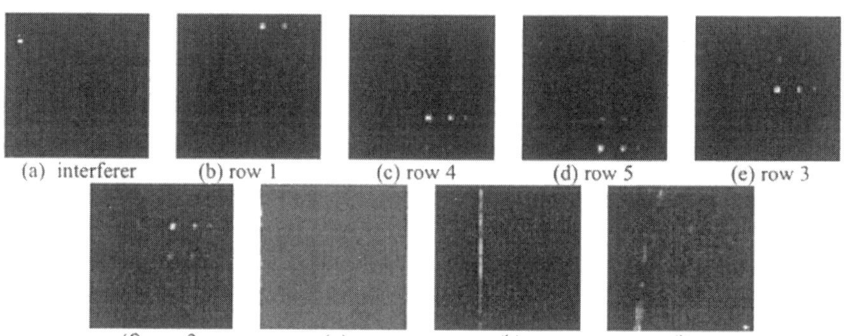

(a) interferer (b) row 1 (c) row 4 (d) row 5 (e) row 3

(f) row 2 (g) (h) (i)

Figure 15.15. Results of LSRMA using skewness with p = 14 (9 components)

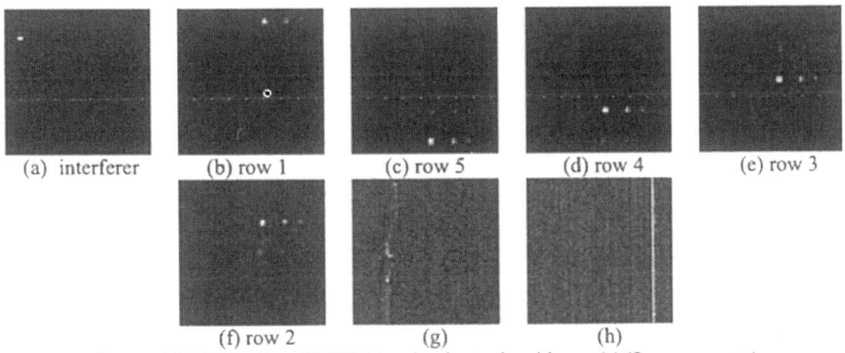

Figure 15.16. Results of LSRMA using kurtosis with $p = 14$ (8 components)

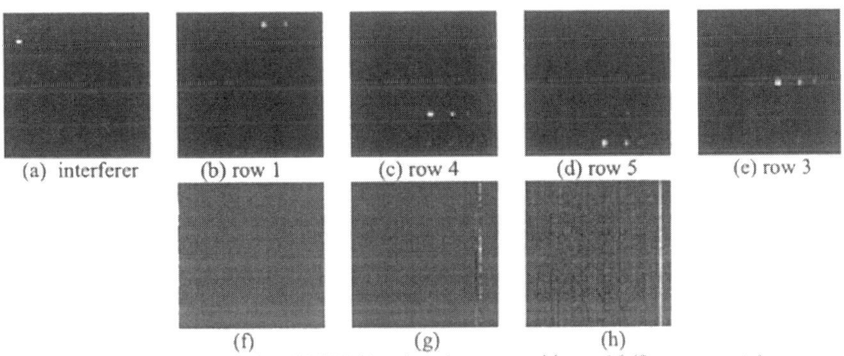

Figure 15.17. Results of LSRMA using skewness with $p = 16$ (8 components)

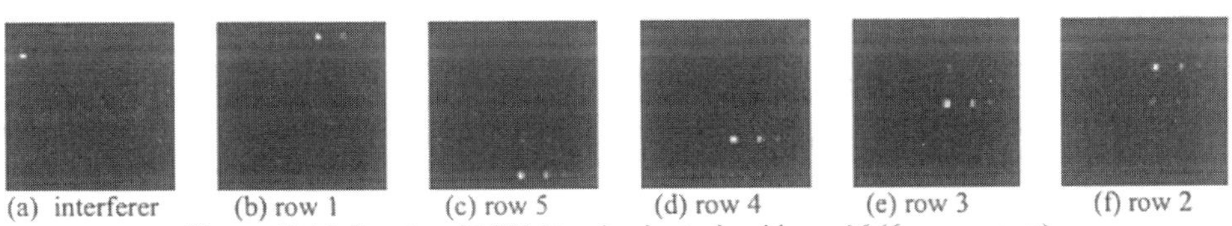

Figure 15.18. Results of LSRMA using kurtosis with $p = 16$ (6 components)

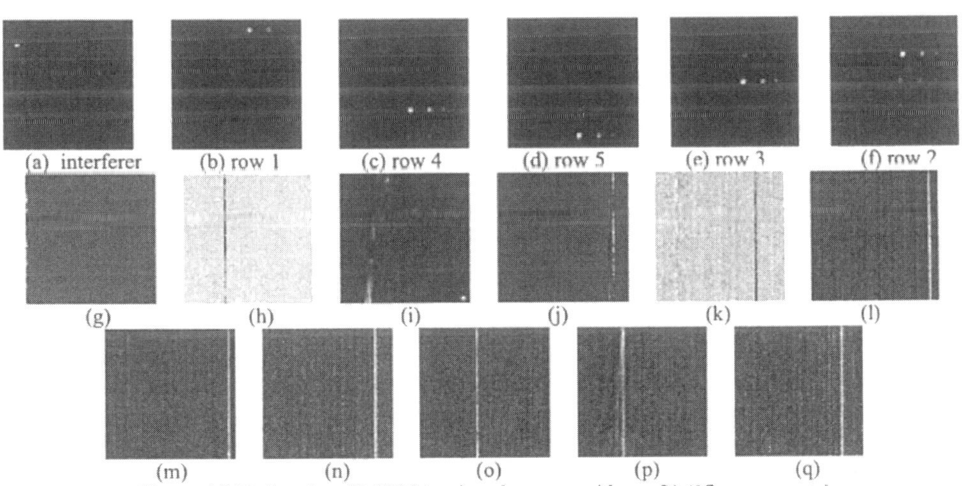

Figure 15.19. Results of LSRMA using skewness with $p = 21$ (17 components)

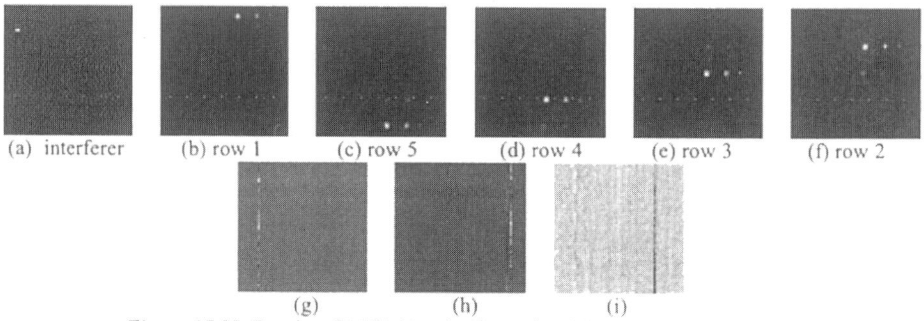

Figure 15.20. Results of LSRMA using kurtosis with $p = 21$ (9 components)

As demonstrated in all the figures, LSRMA was able to detect all the 15 panels in the second through sixth components when $p = 21$. The strong interferer located in the left upper corner in the forest of Fig. 1.7(a) was also detected by both skewness and kurtosis in their first components. By visual inspection of the scene in Fig. 1.7(a) there is no way to identify this interferer. The detection of this unidentified interferer in Fig. 15.15(a)-15.20(a) demonstrated that LSRMA can be used as an anomaly detector to detect unknown targets. However, as also shown, it was unable to discriminate panels in row 2 from those in row 3. The same phenomenon was also witnessed in Fig. 15.12(a) by the computer simulation using skewness with SNR = 5dB. In order to make comparison, we also conducted similar experiments for $p = 169$. The results for skewness and kurtosis are shown in Figs. 15.21 and 15.22 respectively where the interferer and all the 15 panels were detected and classified into 6 separate components. Like previous experiments, both skewness and kurtosis have the same difficulty with discriminating panels in row 2 from those in row 3.

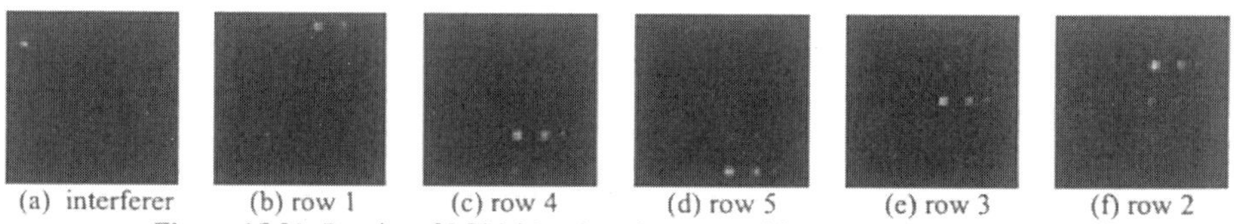

Figure 15.21. Results of LSRMA using skewness with $p = 169$ (6 components)

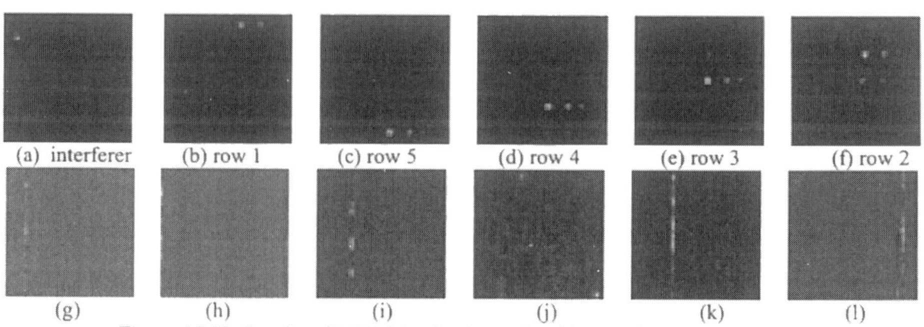

Figure 15.22. Results of LSRMA using kurtosis with $p = 169$ (12 components)

It should be noted that $N_{LSRMA}(169)$ generated by skewness was 6 and 12 by kurtosis. Like the AVIRIS image experiments, this also implies that when p is greater than the number of targets (18 in this example), a large value of p does not necessarily generate a large $N_{LSRMA}(p)$.

15.5. 3-D ROC ANALYSIS FOR LSRMA

Since LSRMA is unsupervised, an automatic thresholding method is required to segment targets from background. In this section, we use the thresholding technique developed in Section 14.4 for this purpose.

Let $\delta_{\mathrm{LSRMA}}(\mathbf{r})$ be the abundance fraction of an image pixel $\mathbf{r}$ resulting from LSRMA, which represents the gray level value of $\mathbf{r}$. Due to the fact that $\delta_{\mathrm{LSRMA}}(\mathbf{r})$ cannot detect true amount of target abundance fraction, its value is generally a real number and does not necessarily lie in the range [0,1]. For any given value α, we use the rejection probability, $P(\alpha) = \Pr(R(\alpha))$ defined by (14.6) along with a prescribed confidence coefficient γ specified by

$$P(\alpha_0) = \gamma. \tag{15.15}$$

to determine a desired threshold value α_0. More precisely, assume that the confidence coefficient is set to γ with $P(\alpha_0) = \gamma$. If $\delta_{\mathrm{LSRMA}}(\mathbf{r}) > \alpha_0$, $\mathbf{r}$ will be detected as a target pixel and a background pixel, otherwise. It should be noted that the confidence coefficient can be adjusted and is determined by target size. For small targets such as panels in Fig. 1.7(a), a panel size of 3m $\times$ 3m with pixel resolution 1.5m would have at most 4 pixels. So, the ratio of a panel with 4 pixels to the entire image size with $64 \times 64 = 4096$ pixels is no more than 0.001. In this case, a reasonable estimate of the confidence coefficient γ would be approximately $\gamma = 1 - 0.001 = 0.999$. As an example, Tables 15.3(a-c) tabulate the values of threshold α_0 yielded by the confidence coefficient γ = 0.999, 0.998, 0.997 where Pi represents detection of panels in row i.

Table 15.3(a). Values of threshold α_0 yielded by confidence coefficient γ = 0.999 via (15.5)

	P1	P2	P3	P4	P5
skewness	15.3502	17.7059	17.9903	19.7057	17.6416
kurtosis	14.7830	17.2765	17.4983	19.1237	16.6457

Table 15.3(b). Values of threshold α_0 yielded by confidence coefficient γ = 0.998 via (15.5)

	P1	P2	P3	P4	P5
skewness	10.8525	7.2989	6.8064	5.2319	6.4973
kurtosis	10.3464	6.7873	6.7814	5.3131	6.1286

Table 15.3(c). Values of threshold α_0 yielded by confidence coefficient γ = 0.997 via (15.5)

	P1	P2	P3	P4	P5
skewness	4.9019	5.7652	4.0263	3.6379	4.3647
kurtosis	4.7596	5.2888	4.4243	3.6922	4.1389

Tables 15.4(a-b)-15.6(a-b) tally the number of panel R and Y pixels detected in Figs. 15.19 and 15.20 with γ = 0.999, 0.998, 0.997 respectively where N_R, N_{RD}, N_Y, N_{YD} N_{R+Y}, $N_{(R+Y)D}$, N_{FA} and N_M were defined in Section 9.3. As we can see from the above tables, skewness and kurtosis performed very closely and their results were nearly the same. When γ = 0.999, both skewness and kurtosis achieved 0% false alarm rate, but they also missed 6 R pixels. On the other hand, when γ was decreased to 0.997, both skewness and kurtosis detected all the 19 R pixels at the expense of falsely detecting 9 and 7 pixels respectively.

Table 15.4(a). Tally of panel R-pixel detection by LSRMA using skewness with $\gamma = 0.999$

T	N_{R+Y}	N_R	N_Y	N_{RD}	N_{YD}	$N_{(R+Y)D}$	N_M	N_{FA}
P1	50	3	47	2	3	5	45	0
P2	45	4	41	3	1	4	41	0
P3	41	4	37	2	2	4	37	0
P4	44	4	40	3	2	5	39	0
P5	43	4	39	3	1	4	39	0
Total	223	19	204	13	9	22	201	0

Table 15.4(b). Tally of panel R-pixel detection by LSRMA using kurtosis with $\gamma = 0.999$

T	N_{R+Y}	N_R	N_Y	N_{RD}	N_{YD}	$N_{(R+Y)D}$	N_M	N_{FA}
P1	50	3	47	2	2	4	46	0
P2	45	4	41	3	1	4	41	0
P3	41	4	37	2	2	4	37	0
P4	44	4	40	3	2	5	39	0
P5	43	4	39	3	2	5	38	0
Total	223	19	204	13	9	22	201	0

Table 15.5(a). Tally of panel R-pixel detection by LSRMA using skewness with $\gamma = 0.998$

T	N_{R+Y}	N_R	N_Y	N_{RD}	N_{YD}	$N_{(R+Y)D}$	N_M	N_{FA}
P1	50	3	47	2	6	8	42	0
P2	45	4	41	3	3	6	39	2
P3	41	4	37	4	4	8	33	0
P4	44	4	40	4	5	9	35	0
P5	43	4	39	3	6	9	34	0
Total	223	19	204	16	24	40	183	2

Table 15.5(b). Tally of panel R-pixel detection by LSRMA using kurtosis with $\gamma = 0.998$

T	N_{R+Y}	N_R	N_Y	N_{RD}	N_{YD}	$N_{(R+Y)D}$	N_M	N_{FA}
P1	50	3	47	2	6	8	42	0
P2	45	4	41	4	3	7	38	1
P3	41	4	37	4	4	8	33	0
P4	44	4	40	4	5	9	35	0
P5	43	4	39	3	5	8	35	0
Total	223	19	204	17	23	40	183	1

Table 15.6(a). Tally of panel R-pixel detection by LSRMA using skewness with $\gamma = 0.997$

T	N_{R+Y}	N_R	N_Y	N_{RD}	N_{YD}	$N_{(R+Y)D}$	N_M	N_{FA}
P1	50	3	47	3	10	13	37	0
P2	45	4	41	4	6	10	35	3
P3	41	4	37	4	5	9	32	4
P4	44	4	40	4	7	11	33	1
P5	43	4	39	4	8	12	31	1
Total	223	19	204	19	36	55	168	9

Table 15.6(b). Tally of panel R-pixel detection by LSRMA using kurtosis with $\gamma = 0.997$

T	N_{R+Y}	N_R	N_Y	N_{RD}	N_{YD}	$N_{(R+Y)D}$	N_M	N_{FA}
P1	50	3	47	3	9	12	38	0
P2	45	4	41	4	6	10	35	2
P3	41	4	37	4	5	9	32	3
P4	44	4	40	4	8	12	32	1
P5	43	4	39	4	7	11	32	1
Total	223	19	204	19	35	54	169	7

If we further used $\gamma = 0.999$, 0.998 and 0.997 to threshold images in Fig. 15.19(b-f) and Figs. 15.20(b-f), Figs. 15.23(a-b)-15.25(a-b) show their corresponding binary images produced by threshold values of α_0 given in Tables 15.3(a-c).

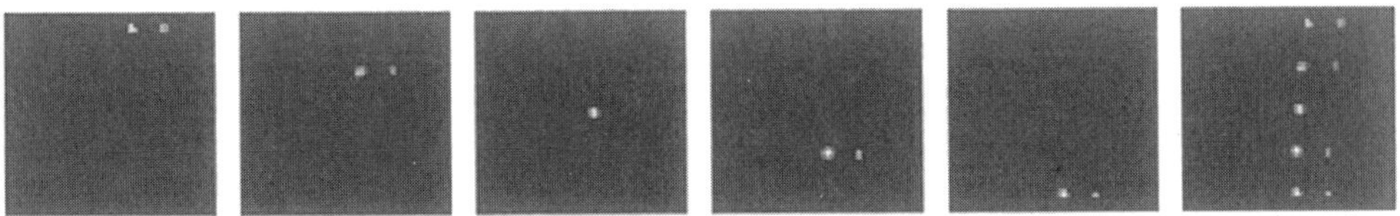

panels in row 1 panels in row 2 panels in row 3 panels in row 4 panels in row 5 total panels

Figure 15.23(a). Target detection resulting from LSRMA using skewness with $\gamma = 0.999$

panels in row 1 panels in row 2 panels in row 3 panels in row 4 panels in row 5 total panels

Figure 15.23(b). Target detection resulting from LSRMA using kurtosis with $\gamma = 0.999$

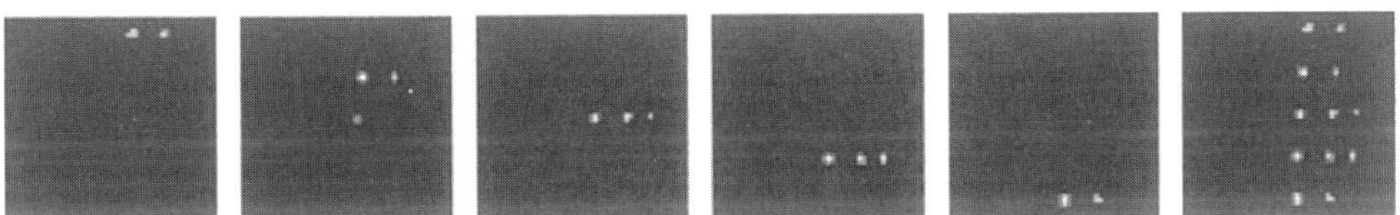

panels in row 1 panels in row 2 panels in row 3 panels in row 4 panels in row 5 total panels

Figure 15.24(a). Target detection resulting from LSRMA using skewness with $\gamma = 0.998$

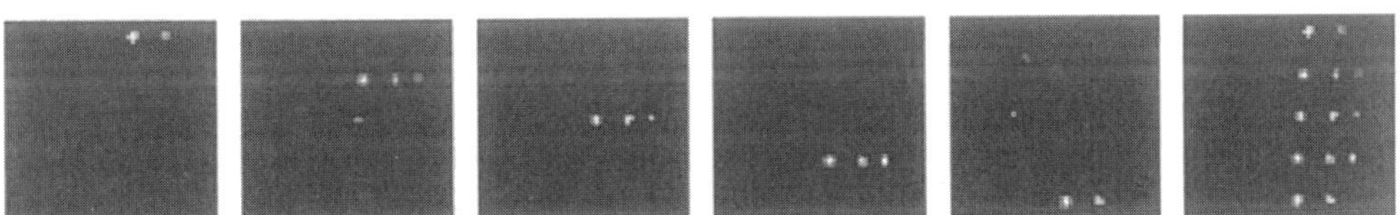

panels in row 1 panels in row 2 panels in row 3 panels in row 4 panels in row 5 total panels

Figure 15.24(b). Target detection resulting from LSRMA using kurtosis with $\gamma = 0.998$

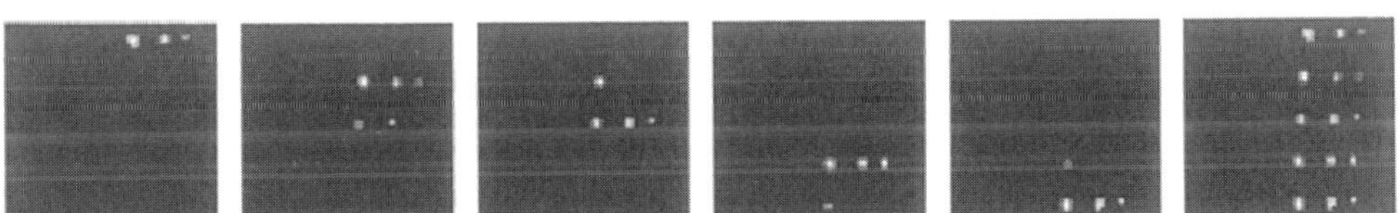

panels in row 1 panels in row 2 panels in row 3 panels in row 4 panels in row 5 total panels

Figure 15.25(a). Target detection resulting from LSRMA using skewness with $\gamma = 0.997$

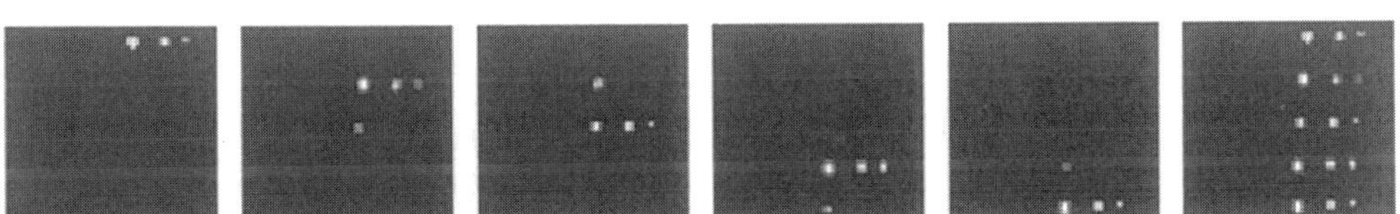

panels in row 1 panels in row 2 panels in row 3 panels in row 4 panels in row 5 total panels

Figure 15.25(b). Target detection resulting from LSRMA using kurtosis with $\gamma = 0.997$

As we can see from Fig. 15.23(a-b) the 5 R panel pixels (i.e., p_{13}, p_{23}, p_{33}, p_{43}, p_{53}, one for each row in Fig. 1.7(b)) in the third column were not detected with $\gamma = 0.999$. When $\gamma = 0.998$ was used, p_{23}, p_{33}, p_{43} were extracted in Fig. 15.24(a-b) while some false alarm pixels were also shown up. As γ was further deceased to 0.997, the 5 R panel pixels, p_{13}, p_{23}, p_{33}, p_{43}, p_{53}, were effectively detected in Fig. 15.25(a-b) by both skewness and kurtosis at the expense of more false alarm pixels. It should be noted that these panel pixels have size of $1\text{m} \times 1\text{m}$, which is smaller than the 1.5m pixel resolution. As a result, they cannot be seen by visual inspection from Fig. 1.7(a). This example further demonstrates ability of LSRMA in subpixel detection. As also noted in Fig. 15.25(a-b) and Tables 15.6(a)-15.6(b), LSRMA also falsely detected a panel pixel in row 3 when it intended to detect panels in row 3, and vice versa because the panel signatures in rows 2 and 3 are very close. Similarly, it was also true for detection panels in rows 4 and 5. These phenomena occurred due to the lack of prior target knowledge.

In order to evaluate the performance of LSRMA, OSP and CEM were used for comparative analysis. Such selection was made based on two reasons. One is that both OSP and CEM have shown success in target detection and classification in Chapters 3, 4, 8 and 9. Another is from the level of used target knowledge. OSP requires complete target knowledge opposed to CEM that only needs the knowledge of the desired target of interest. Compared to OSP and CEM, LSRMA does not require target knowledge *a priori*. As a matter of fact, if the number of signal sources, p can be estimated reliably, LSRMA does not need any information at all.

Tables 15.6 and 15.7 also tally the results produced by OSP and CEM using the same confidence coefficients $\gamma = 0.999, 0.998, 0.997$ where OSP used the five panel signatures, P1, P2, P3, P4, P5 in Fig. 1.8 as its complete target knowledge and CEM only used Pi as its desired target knowledge to detect panels in row i.

Table 15.7(a). Tally of panel R-pixel detection by OSP with $\gamma = 0.999$

T	N_{R+Y}	N_R	N_Y	N_{RD}	N_{YD}	$N_{(R+Y)D}$	N_M	N_{FA}
P1	50	3	47	0	0	0	50	5
P2	45	4	41	0	0	0	45	4
P3	41	4	37	0	0	0	41	5
P4	44	4	40	0	0	0	44	4
P5	43	4	39	0	0	0	43	4
Total	223	19	204	0	0	0	223	22

Table 15.7(b). Tally of panel R-pixel detection by OSP with $\gamma = 0.998$

T	N_{R+Y}	N_R	N_Y	N_{RD}	N_{YD}	$N_{(R+Y)D}$	N_M	N_{FA}
P1	50	3	47	0	0	0	50	8
P2	45	4	41	0	0	0	45	8
P3	41	4	37	0	0	0	41	8
P4	44	4	40	0	0	0	44	9
P5	43	4	39	1	0	1	42	8
Total	223	19	204	1	0	1	222	41

Table 15.7(c). Tally of panel R-pixel detection by OSP with $\gamma = 0.997$

T	N_{R+Y}	N_R	N_Y	N_{RD}	N_{YD}	$N_{(R+Y)D}$	N_M	N_{FA}
P1	50	3	47	0	0	0	50	13
P2	45	4	41	0	0	0	45	15
P3	41	4	37	0	0	0	41	12
P4	44	4	40	0	0	0	44	14
P5	43	4	39	1	0	1	42	12
Total	223	19	204	1	0	1	222	66

Table 15.8(a). Tally of panel R-pixel detection by CEM with $\gamma = 0.999$

T	N_{R+Y}	N_R	N_Y	N_{RD}	N_{YD}	$N_{(R+Y)D}$	N_M	N_{FA}
P1	50	3	47	2	3	5	45	0
P2	45	4	41	3	2	5	40	0
P3	41	4	37	3	2	5	36	0
P4	44	4	40	3	2	5	39	0
P5	43	4	39	3	1	4	39	0
Total	223	19	204	14	10	24	199	0

Table 15.8(b). Tally of panel R-pixel detection by CEM with $\gamma = 0.998$

T	N_{R+Y}	N_R	N_Y	N_{RD}	N_{YD}	$N_{(R+Y)D}$	N_M	N_{FA}
P1	50	3	47	2	7	9	41	0
P2	45	4	41	4	4	8	37	1
P3	41	4	37	4	4	8	33	0
P4	44	4	40	4	5	9	35	0
P5	43	4	39	4	4	8	35	0
Total	223	19	204	18	24	42	181	1

Table 15.8(c). Tally of panel R-pixel detection by CEM with $\gamma = 0.997$

T	N_{R+Y}	N_R	N_Y	N_{RD}	N_{YD}	$N_{(R+Y)D}$	N_M	N_{FA}
P1	50	3	47	3	9	12	38	0
P2	45	4	41	4	7	11	34	2
P3	41	4	37	4	7	11	30	1
P4	44	4	40	4	7	11	33	2
P5	43	4	39	4	9	13	30	0
Total	223	19	204	19	39	58	165	5

Interestingly, if we compare Tables 15.4-15.6 to Tables 15.8(a)-15.8(c), LSRMA performed as well as did CEM and their results were nearly the same even if LSRMA did not assume any target knowledge. On the other hand, Tables 15.7(a-c) show that OSP performed extremely poor. This is because the target knowledge used in OSP did not well represent the image scene. It was made up of only five panel signatures {P1,P2,P3,P4,P5} and did not include background signatures such as the large grass field, the forest on the left edge and the road on the right edge of the scene. Fig. 15.26(a-c) show the detection results of OSP with γ = 0.999, 0.998, 0.997 where the binary images were obtained by thresholding the OSP-generated gray scale images using the values of α_0 that were determined by γ via (15.15).

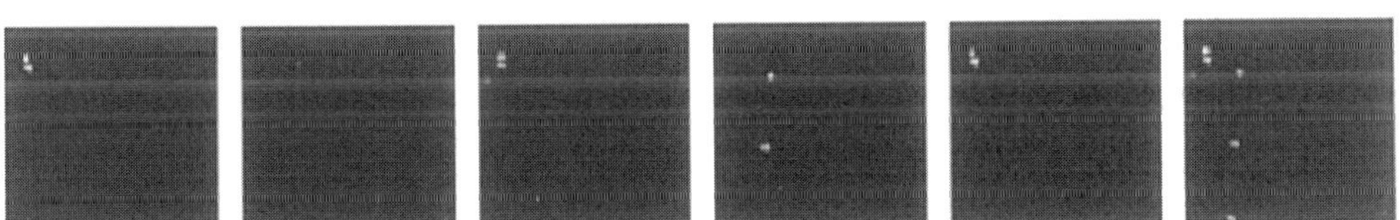

panels in row 1 panels in row 2 panels in row 3 panels in row 4 panels in row 5 total panels

Figure 15.26(a). Target detection resulting from OSP using confidence coefficient $\gamma = 0.999$

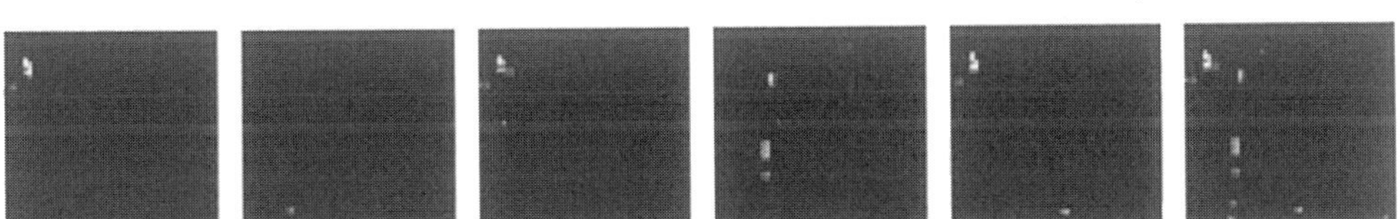

panels in row 1 panels in row 2 panels in row 3 panels in row 4 panels in row 5 total panels

Figure 15.26(b). Target detection resulting from OSP using confidence coefficient $\gamma = 0.998$

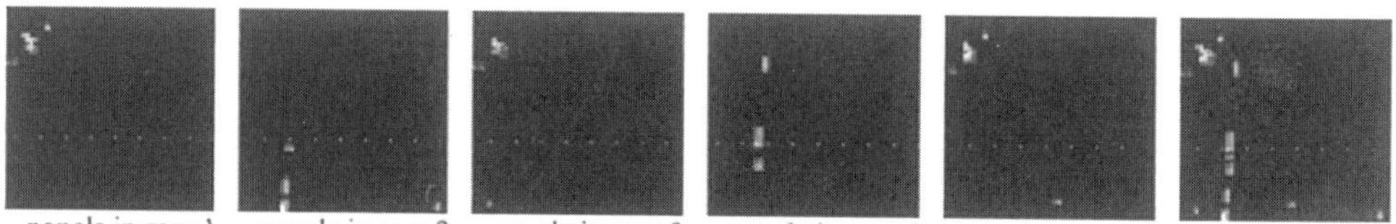

panels in row 1 panels in row 2 panels in row 3 panels in row 4 panels in row 5 total panels

Figure 15.26(c). Target detection resulting from OSP using confidence coefficient $\gamma = 0.997$

As we can see from Tables 15.7(a)-15.7(c) and the above detection images, OSP failed to detect most of panel pixels. No R pixel was detected when $\gamma = 0.999$ and only one R pixel, p_{51} in row 5 was detected by $\gamma = 0.998$, 0.997. By contrast, CEM performed very well. Fig. 15.27(a-c) shows its detection results with $\gamma = 0.999$, 0.998, 0.997 where the binary images were obtained by thresholding the CEM-generated gray scale images using the values of α_0 that were determined by γ via (15.15).

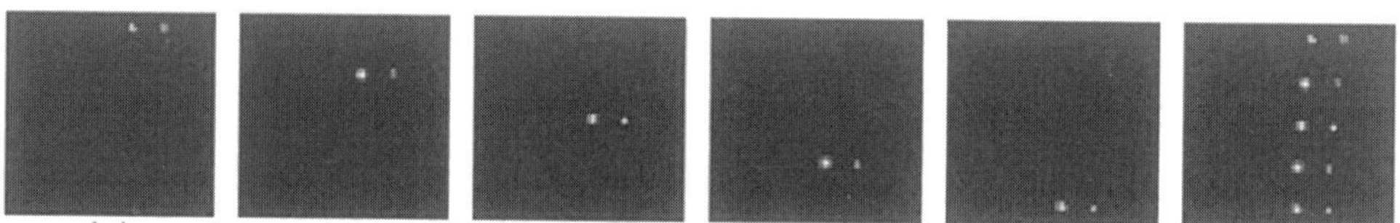

panels in row 1 panels in row 2 panels in row 3 panels in row 4 panels in row 5 total panels

Figure 15.27(a). Target detection resulting from CEM using confidence coefficient $\gamma = 0.999$

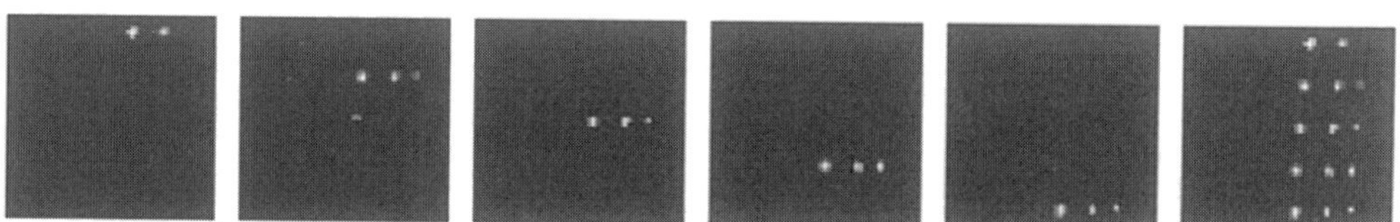

panels in row 1 panels in row 2 panels in row 3 panels in row 4 panels in row 5 total panels

Figure 15.27(b). Target detection resulting from CEM using confidence coefficient $\gamma = 0.998$

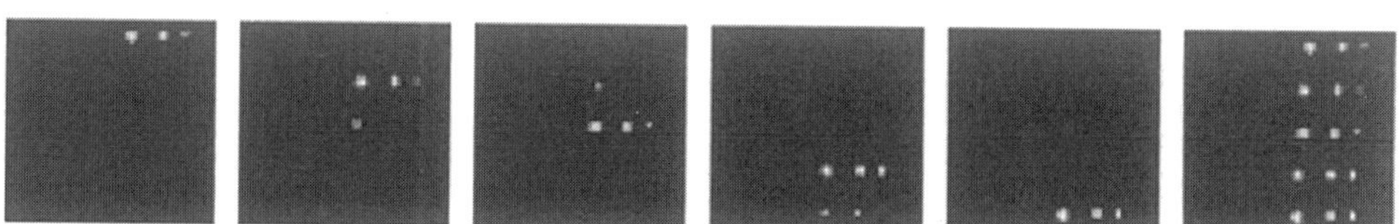

panels in row 1 panels in row 2 panels in row 3 panels in row 4 panels in row 5 total panels

Figure 15.27(c). Target detection resulting from CEM using confidence coefficient $\gamma = 0.997$

If we compare Fig. 15.27(a,b) to Figs. 15.23(a,b)-15.25(a,b), the performance of CEM is very close to that of LSRMA except that CEM had fewer false alarm pixels. These experiments further demonstrate that with no complete target knowledge which should also include background signatures OSP simply cannot compete with CEM and LSRMA.

In analogy with Figs. 9.5, 11.8 and 12.5 we can also plot for LSRMA a 3-D ROC curve and 2-D ROC curves based on three parameters, R_H, R_F and *a*% via *a*%MPCV defined in (9.3). However, if we replace *a*% with the confidence coefficient γ as another parameter, we can derive a new a mixed-to-pure converter from *a*%MPCV defined in (9.3), referred to as, γMPCV which can be defined by

$$\chi_{\gamma\mathrm{MPCV}}(\mathbf{r}) = \mathbf{Mu}_j = \mathbf{m}_j \text{ if } \hat{\alpha}_j(\mathbf{r}) \geq \alpha_0. \tag{15.16}$$

where α_0 is determined by γ through $P(\alpha_0) = \gamma$ in (15.15).

By means of (15.16) we can also plot 3-D ROC curves and 2-D ROC curves via γMPCV for LSRMA, OSP and CEM based on (R_H,R_{FA},γ). In this case, two types of 3-D ROC curves can be generated and plotted by *a*%MPCV and γMPCV for comparative analysis. One is generated by *a*%MPCV based on (R_H,R_{FA},*a*%) as we have seen in Figs. 9.5 and 12.5. Another is produced by γMPCV based on (R_H,R_{FA},γ) where the confidence coefficient γ is determined by the proposed automatic thresholding method via (15.15). Fig. 15.28(a-b) plot these two types of 3-D ROC curves of (R_H,R_{FA},*a*%) and (R_H,R_{FA},γ) respectively, which were generated by LSRMA using skewness, LSRMA using kurtosis, OSP and CEM for target hit of the 15 panels in Fig. 1.7(a) where a target is hit if either R and Y pixel is detected.

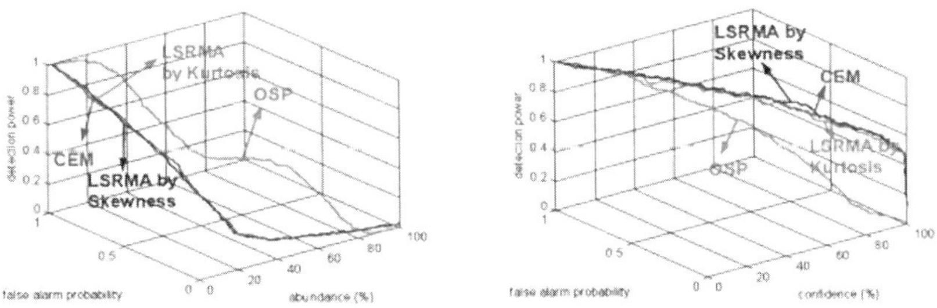

(a) 3-D ROC curves of (R_H,R_{FA},*a*%) (b) 3-D ROC curves of (R_H,R_{FA},γ)

Figure 15.28. 3-D ROC curves generated by *a*%MPCV and γMPCV

Fig. 15.29(a-b) plots their corresponding 2-D ROC curves of (R_H,R_{FA}) produced by Fig. 15.28(a-b) respectively.

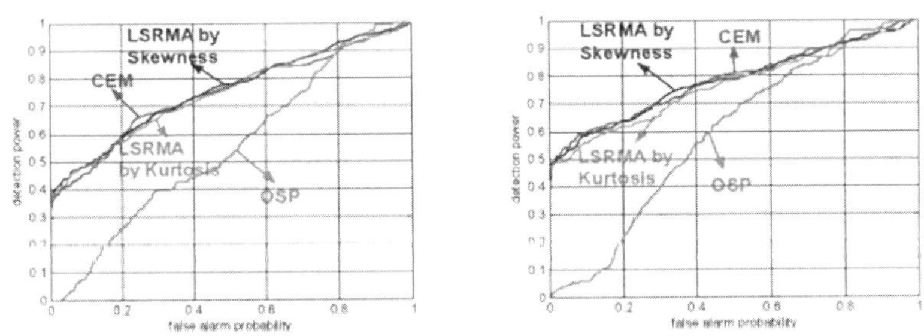

(a) abundance *a*% (b) confidence coefficient γ

Figure 15.29. 2-D ROC curves of (R_H,R_{FA})

As shown in these figures, LSRMA using skewness and kurtosis performed very closely to CEM and clearly outperformed OSP. In order to quantitatively evaluate their target hit performance, the areas under their corresponding ROC curves were calculated and tabulated in Table 15.8.

Table 15.8. Target hit rates calculated from Fig. 15.29

	LSRMA (skewness)	LSRMA (kurtosis)	OSP	CEM
(R_H,R_{FA},*a*%)	0.77798	0.76008	0.59133	0.78376
(R_H,R_{FA},γ)	0.77826	0.76031	0.59171	0.78356

Interestingly, a%MPCV and γMPCV yielded nearly the same target hit rates in Table 15.8 despite that the shapes of their corresponding 2-D ROC curves of (R_H,R_{FA}) in Fig. 15.29(a-b) were different, particularly those generated by OSP. This is because the target hit rate is calculated based on overall performance not on a specific percentage or a particular level of confidence coefficient. To further see their relative difference, we also plot 2-D curves of (R_H, γ) and (R_H,a%) in Fig. 15.30(a-b) for comparison. As we can see from Fig. 15.30(a), R_H is dropped rapidly after a% is greater than 15%. By contrast, R_H is gradually decreased as γ increases.

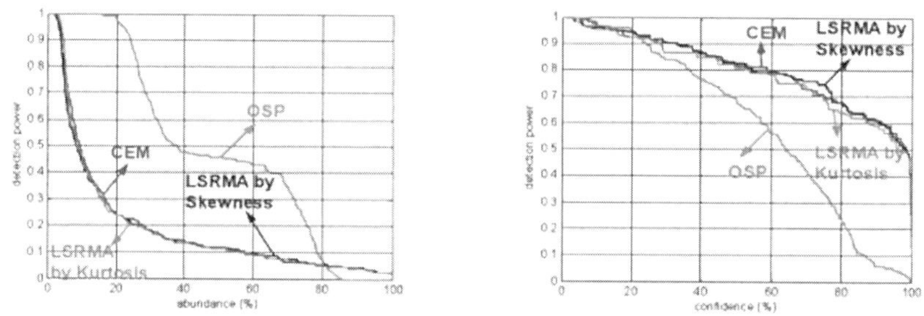

(a) 3-D curves of (R_H,a%) (b) 2-D curves of (R_H,γ)

Figure 15.30. 2-D curves of LSRMA using skewness and kurtosis, OSP and CEM

A similar observation can be also made on the 2-D curves of (R_{FA},γ) and (R_{FA},a%) plotted in Figs. 15.31(a-b) with sudden drops of R_{FA} during [5%,15%] for LSRMA and CEM.

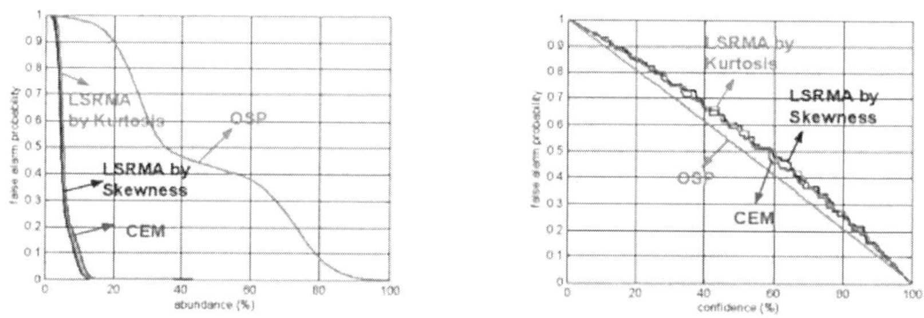

(a) 2-D curves of (R_{FA},a%) (b) 2-D curves of (R_{FA},γ)

Figure 15.31. 2-D ROC curves of LSRMA using skewness and kurtosis, OSP and CEM

The 2-D curves in Figs. 15.30-15.31 suggest that a%MPCV is sensitive to selection of a%. In particular, the high performance of a%MPCV resulting from small abundance percentages (<15%) was offset by poor performance resulting from higher abundance percentages (>15%). Compared to a%MPCV, the performance of γMPCV did not have such drastic changes as γ increased. Nevertheless, their overall performance is very close as shown in Table 15.8.

15.6 CONCLUSIONS

This chapter presents an ICA-based linear spectral random mixture analysis approach, linear spectral random mixture analysis (LSRMA) to hyperspectral target detection and classification. It is different from the commonly used ICA approach in that the learning algorithm is derived from the orthogonality constraint imposed on the abundance vector rather than the separating matrix **W**. In addition, the separating matrix **W** is not necessarily of full rank. Neither is the mixing matrix **M** orthogonal. These advantages are very useful in hyperspectral image classification. Since hyperspectral sensors can uncover targets with subtle differences, their spectral signatures are generally similar to some degree and may not be orthogonal. The new designed learning algorithm is able to converge to non-orthogonal independent components. In addition, two types of 3-D ROC curves generated by a%MPCV and γMPCV are also introduced for comparative study and analysis. According to the conducted experiments, skewness and kurtosis used in LSRMA have demonstrated different strengths in classification. Because skewness measures asymmetry of a distribution, it generally can detect changes in large areas. On the other hand, kurtosis measures the flatness of a distribution, thus it can detect small targets. That explains why kurtosis worked effectively for 1.5m HYDICE data, while skewness performed well for 20-m AVIRIS.

As a concluding remark, LSRMA can be considered as a special case of projection pursuit (PP) which will be studied in Chapter 16 where PP designs a projection index (PI) to look for projections of interestingness. The two criteria, relative entropy specified by (15.3) and statistical moments specified by (15.6) (particularly skewness and kurtosis) used for LSRMA can be viewed as particular projection indices. Interestingly, developing a PP approach for unsupervised hyperspectral image analysis using the concept of the relative entropy as a PI was explored in Ifarraguerri and Chang (2000). On the other hand, a recent PP approach using skewness and kurtosis as a PI for unsupervised target detection and classification was also investigated in Chiang, Chang and Ginsberg (2001). These ideas are identical to that of LSRMA. As a general case, if the relative entropy specified by (15.3) is used as the PI, which measures statistical independence among the components, the LSRMA actually becomes a special case of PP. More details can be found in Chapter 16.

16

AUTOMATIC MIXED PIXEL CLASSIFICATION (AMPC): PROJECTION PURSUIT

In Chapter 15, LSRMA made use of skewness and kurtosis to detect and classify potential targets automatically without prior target knowledge. These two criteria were a simplification of the contrast function in (15.5) derived by Comon. In this chapter, we present a similar approach developed by Chiang, Chang and Ginsberg (2001), called projection pursuit (PP), also used skewness and kurtosis as a base to design various project indices for automatic target detection and classification. Both LSRMA and PP model small targets as anomalies with an understanding that their sizes are relatively small in the spatial image scene compared to the image background. With this assumption, such small anomalies can be detected by finding outliers of the background distribution. The criteria, skewness and kurtosis are well suited for this purpose. Unlike LSRMA that requires a linear mixture model, the proposed PP projects a high dimensional data set into a low dimensional data space while retaining desired target information. It utilizes skewness and kurtosis as projection indices to explore projections of interestingness which are those caused by small man-made targets in a large unknown background. To find optimal solutions for projection indices, an evolutionary algorithm is developed to prevent the solutions from being trapped in local optima. Finally, target detection and classification is achieved by projecting the image data into separate projection images. In order to segment potential targets from the image background in these projection images, a zero-detection process is developed to threshold the projection images into a sequence of binary images, each of which detects a particular type of targets. This approach is different from LSRMA where the number of independent components is determined by an estimate using VD in Chapter 17 or is selected *a priori*.

16.1 INTRODUCTION

Principal components analysis (PCA) is a versatile technique which has been used for a wide range of applications. It is based on the concept that the interesting projections that orient principal components are determined by data variance. However, in some applications the data variance may not be the only projection we are interested in. In target detection, one of most interesting targets is an anomaly or a small man-made target. In general, such a target cannot be extracted from a large scene by PCA because the variance it contributes to the data is very small. Projection pursuit (PP) provides a solution to this problem. The PP was first developed by Friedman and Tukey (1974) as a technique for exploratory analysis of multivariate data and has been studied extensively since then (Friedman, 1974; Huber, 1985; Jones and Sibson, 1987; Bullock et al., 1994). Unlike most of developed target detection algorithms that require statistical models such as linear mixture, PP is a linear mapping that searches for interesting low-dimensional projections from a high-dimensional data space via a projection index (PI). The PI is a measure used to explore projections of interestingness. In particular, it can be designed to characterize nonlinear structures in projected distributions. For example, if the desired direction of a PI is one pointing to data variance, PP is reduced to PCA. On the other hand, if we would like the desired components to be statistically independent, ICA used to develop LSRMA in Chapter 15 becomes a special case of PP. So, PP is a power technique in many signal/image processing applications. Using PP for hyperspectral image classification has been studied previously by Jimenez and Landgrebe (1999) who designed a PI based on Bhattacharyya's distance to reduce the dimensionality of feature space. Ifarraguerri and Chang (2000) also used the information divergence (relative entropy) as a PI to look for interesting projections that deviates from Gaussian distributions. A similar idea was further explored in Chiang, Chang and Ginsberg (2001), where they modeled and considered small targets as anomalies that caused outliers of the background distribution.

In military applications, man-made target detection in hyperspectral imagery is of major interest since their size is generally small and sometimes even smaller than the ground sampling distance (GSD). Such target detection must rely on subpixel spectral detection, not pixel-based spatial detection. From this point of view, an interesting structure of an image scene is the one caused by man-made targets in a large unknown background. So, a small man-made target can be viewed as an anomaly in an image scene due to the fact that its size is relatively small in the spatial sense compared to the image background and its spectral properties are distinct from that of its surrounding pixels. As a result, detecting such small targets in an unknown image scene can be reduced to finding the outliers or deviations from the background distribution. It is known that *skewness* defined by normalized third moment of the sample distribution measures the asymmetry of the distribution and *kurtosis* defined by normalized fourth moment of the sample distribution measures the flatness of the distribution. They both are susceptible to outliers. So, using skewness and kurtosis as a base to design a PI may be an effective means for target detection. However, it may also occur that a small region or set of background pixels can be also detected as anomalies. This results from no availability of prior target knowledge.

Once a PI is determined, finding optimal solutions for the desired PI is crucial. Unfortunately, there are generally no analytic solutions and they must be solved by numerical algorithms. Two major principles, "hill-climbing" and "random move" have widely used to design optimization algorithms. For a relatively smooth PI where the first derivatives exist, "hill-climbing"-based gradient descent methods are usually preferred. However, they may be trapped in local optimal solutions that are close to the initial starting points. In this case, a good guess of initial conditions is a key to success of these methods. On the other hand, a random move-based method such as simulated annealing (SA) can escape from local optima to some degree where its success is determined by initial starting points. In this chapter, we consider a similar random move-based approach to SA, called evolutionary algorithm (EA) (Golderg, 1989; Michalewicz,

1996; Dasgupta and Michalewicz, 1997), which is more likely to lead solutions to global optima.

Prior to implementation of EA, it generally requires to preserve the translation invariance property of PI. This can be accomplished by centering the original data matrix using a linear translation followed by an eigen-decomposition method such as singular value decomposition (SVD) to whiten the centered data matrix. The latter whitening process is referred to as sphering in the PP literature. There are four stage processes implemented in EA, which are population selection, crossover, mutation and termination. Of particular interest are crossover and mutation processes. While the crossover process generates a descendant population by combining pairs of individuals randomly chosen from a parent population, the mutation process performs a local optimization which makes random changes in individual points to search for local optimum points. The proposed EA is a little bit different from the EA referred in the literature in the sense that it includes a zero-detection thresholding to achieve target detection. Since we are interested in small man-made targets, the outliers resulting from skewness and kurtosis are attributed to these targets. In order to extract these targets, a zero detection process is proposed to threshold each projection image so that targets of different types can be segmented by thresholding the projection images into a sequence of binary images. The algorithm implemented by EA in conjunction with PP and a zero-detection thresholding method is referred to as PPEA.

16.2 PROJECTION PURSUIT

The term of "Projection Pursuit (PP)" was first coined by Friedman and Tukey and was used as a technique for exploratory analysis of multivariate data. The idea is to project a high dimensional data set into a low dimensional data space while retaining the information of interest. It utilizes a PI to explore projections of interestingness in data.

Following the approach suggested in Jones and Sibson (1987), we assume that there are N data points $\{\mathbf{x}_n\}_{n=1}^{N}$ each with dimensionality K and $\mathbf{X} = \left[\mathbf{x}_1 \mathbf{x}_2 \cdots \mathbf{x}_N\right]$ is a $K \times N$ data matrix and $\mathbf{a}$ is a K-dimensional column vector which serves as a desired projection. Then $\mathbf{a}^T\mathbf{X}$ represents an N-dimensional row vector that is made up of orthogonal projections of all sample data points mapped onto the direction $\mathbf{a}$ where T is the matrix transpose. Now if we let $H(\cdot)$ is a function measuring the degree of the interestingness of the projection $\mathbf{a}^T\mathbf{X}$ for a fixed data matrix $\mathbf{X}$, a projection index (PI) is a real-valued function of $\mathbf{a}$, $I(\mathbf{a})$: $R^K \to R$ defined by

$$I(\mathbf{a}) = H(\mathbf{a}^T\mathbf{X}). \tag{16.1}$$

The PI can be easily extended to multiple directions, $\{\mathbf{a}_j\}_{j=1}^{J}$. In this case, $\mathbf{A} = \left[\mathbf{a}_1 \mathbf{a}_2 \cdots \mathbf{a}_J\right]$ is a $K \times J$ projection direction matrix and the corresponding projection index is also a real-valued function, $I(\mathbf{A})$: $R^{K \times J} \to R$ is given by

$$I(\mathbf{A}) = H(\mathbf{A}^T\mathbf{X}). \tag{16.2}$$

The choice of the $H(\cdot)$ in (16.1) and (16.2) is application-dependent. Its purpose is to reveal interesting structures within data sets such as clustering. In Chapter 15, LSRMA with skewness, κ_3 and kurtosis, κ_4 as criteria were shown effective in detecting man-made targets. This suggests that using κ_3 and κ_4 as a PI may be appropriate criteria for target detection. The following four projection indices will be used as criteria for target classification.

$$I_{\text{skew}}(\mathbf{a}) = H_1(\mathbf{a}^T\mathbf{X}) = \kappa_3^2 \tag{16.3}$$

$$I_{\text{kurtosis}}(\mathbf{a}) = H_2(\mathbf{a}^T\mathbf{X}) = \kappa_4^2 \tag{16.4}$$

$$I_{mix}(\mathbf{a}) = H_3(\mathbf{a}^T\mathbf{X}) = \kappa_3^2 + (\kappa_4^2 / 12) \tag{16.5}$$

where $I_{mix}(\mathbf{a})$ is a linear mixture of κ_3 and κ_4, and

$$I_{\text{product}}(\mathbf{a}) = H_4(\mathbf{a}^T\mathbf{X}) = \kappa_3^2 \cdot \kappa_4^2 = \left(\kappa_3\kappa_4\right)^2 \tag{16.6}$$

where $I_{\text{product}}(\mathbf{a})$ is a product of κ_3^2 and κ_4^2.

16.3 EVOLUTIONARY ALGORITHM (EA)

Since the four PI's specified by (16.3)-(16.6) are designed to capture the third and fourth moments, the analysis can be largely simplified if the second-order statistics is taken care of by a whitening processes described in Section 12.3. Such whitening process is also referred to as sphering in the PP literature. So, in this section, we assume that the data have been whitened and the resulting whitened data matrix is still represented by $\mathbf{X}$. As indicated in the introduction, finding optimal PI's is not a simple matter. Here we present an optimization algorithm, known as the evolutionary algorithm (EA) developed in Golderg (1989), Michalewicz (1996), Dasgupta and Michalewicz (1997). It can be implemented by four stage processes (population selection, crossover, mutation and termination), each of which is briefly described as follows. For details we refer to Golderg (1989), Michalewicz (1996), Dasgupta and Michalewicz (1997).

1) Initial population:
   Unlike gradient methods and simulated annealing, an EA requires an initial population to start with to perform a multi-directional search for optimal points simultaneously during each iteration. A population is made up of possible projection vectors, each of which can be viewed as individuals. To initialize EA, an initial population can be created randomly or with prior knowledge if there is any. A simple way to form an initial population by including all basis vectors of the space to projected.
2) Population Selection Process
   Once a population is formed. Then each individual in the population will be evaluated by a fitness function specified by the desired projection index, $f(\mathbf{x})$. The

individual which has better fitness will have a more chance to generate new population in the next iteration. The procedure is described as follows.

(a) Calculate $f(\mathbf{v}_i)$ for each individual $\mathbf{v}_i$.

(b) Calculate the probability of selecting $\mathbf{v}_i$ by $p_i = f(\mathbf{v}_i) / \sum_k f(\mathbf{v}_k)$.

(c) Calculate the cumulative probability q_i for $f(\mathbf{v}_i)$ by $q_i = \sum_{j=1}^{i} p_j$.

(d) Generate a random number r in the range $[0,1]$.

(e) Select the i-th individual $\mathbf{v}_i$ such that $q_{i-1} < r \le q_i$

(f) Repeat steps (d) and (e) until a desired population is formed.

It should be noted that the above selection process is stochastic and some individuals may be selected more than once. This is in accordance with the natural principle of "survival of fitness" which says that the best individuals get more descendants while the average stays even and the worst dies off.

3) Crossover Process

The crossover process used in EA is called arithmetical crossover which can be defined as follows. Assume there are two individuals, denoted by two vectors, $\mathbf{v}$ and $\mathbf{w}$. Then the two vectors formed by $\mathbf{v}' = a\mathbf{v} + (1-a)\mathbf{w}$ and $\mathbf{w}' = (1-a)\mathbf{v} + a\mathbf{w}$ are called crossover offsprings of $\mathbf{v}$ and $\mathbf{w}$ with $a \in [0,1]$. Let p_c be the pre-selected crossover probability which will be used to determine the number of individuals needed to produce crossover offsprings. This number is $N_c = \lfloor p_c \cdot N_k \rfloor$ with $p_c \cdot N_k - 1 < \lfloor p_c \cdot N_k \rfloor \le p_c \cdot N_k$ where N_k is the size of population at the k-th iteration. The crossover is performed as follows.

(a) For each individual $\mathbf{v}$ in the current population generate a random number r in the range $[0,1]$. If $r < p_c$, the $\mathbf{v}$ will be selected as a crossover vector. The selection procedure will be continued to form a crossover vector group until the expectation of number of selected crossover vectors is equal to N_c.

(b) For each two individuals $\mathbf{v}$ and $\mathbf{w}$ in the selected crossover group, form their two crossover offsprings $\mathbf{v}' = a\mathbf{v} + (1-a)\mathbf{w}$ and $\mathbf{w}' = (1-a)\mathbf{v} + a\mathbf{w}$ where a is a random number generated from the uniform distribution in the range $[0,1]$. This requires N_c being even. So, if N_c is odd, we need select one more individual randomly to make it even.

Through the crossover process it provides us with a more effective means to find more competitive projections other than the ones already in the population.

4) Mutation Process

Like the mutation defined in a genetic algorithm, the mutation in EA is also performed on a bit-by-bit basis. For each individual $\mathbf{v} = v_1 v_2 \cdots v_n$, each component, say v_i is expressed in a binary expansion with l precisions by $v_i = v_{i1} v_{i2} \cdots v_{il}$. Then $\mathbf{v}$ can be represented by a binary string, called a chromosome, formed by concatenating the l-precision binary expansions of all the n components, $\mathbf{v} = v_1 v_2 \cdots v_n = v_{11} v_{12} \cdots v_{1l} \cdots v_{n1} v_{n2} \cdots v_{nl}$. Now the mutation process will be performed on the chromosome bit by bit. Let p_m be a predetermined probability of mutation. It determines the number of bits needed to be mutated, denoted N_m, which is the expectation of the number of mutated bits is equal to $\lfloor p_m \cdot N_c \cdot n \cdot l \rfloor$.

(a) Encode each individual in the current population into a chromosome.

(b) For each bit in the chromosome, generate a random number r from the uniform distribution in the range [0,1]. If $r < p_m$, mutate the bit by flipping the bit between 0 and 1.

(c) Decode each chromosome into a new individual.

The goal of the mutation is to explore the neighborhood of each individual to find an optimal projection.

5) Termination Process

There are several ways to terminate EA. (1) The algorithm reaches an optimal solution. (2) No significant changes are made. (3) The number of iterations exceeds a predetermined limit.

16.4 THRESHOLDING OF PROJECTION IMAGES USING ZERO-DETECTION

Like PCA, PPEA produces a sequence of projection images, which show the information in decreasing order of magnitudes of projection values produced by a PI. It should be noted that the magnitude referred here is the absolute projection values where the projection values of a PI can be positive or negative. Assume that $\mathbf{a}_j$ is the j-th projection vector that maximizes a PI. Using $\mathbf{a}_j$ we can project the data matrix $\mathbf{X}$ into a projection space generated by $z_n^j = \mathbf{a}_j^T \mathbf{x}_n$ where the data matrix $\mathbf{X}$ is formed by all pixel vectors in the image, i.e. $\mathbf{X} = \left[\mathbf{x}_1 \mathbf{x}_2 \cdots \mathbf{x}_N\right]$. For example, if we assume that the histogram of the projection values $\{z_n^j\}_{n=1}^N$ behaves like a Gaussian distribution, the outliers that creates ripples on either side of tails of the Gaussian distribution can be considered to be caused by small targets. So, the first values in the histogram that occur at zero will be selected as the desired thresholds. Then those projected pixels with gray scales exceed the thresholds will be extracted and considered to be target pixels. As a result, two images can be obtained by such zero-detection thresholding, one for target detection and another for the use of next projection. The one used for target detection is a binary image which is obtained by setting those pixels that exceed the thresholds to 1 and zero, otherwise. This thresholded binary image, denoted by $\mathbf{B}_1$ will be used to tally target pixels detected by this particular projection image. A second image is obtained by assigning 0 to all the pixels in the binary image $\mathbf{B}_1$ while the gray scales of the pixels not in the $\mathbf{B}_1$ remaining unchanged. This resulting image is referred to as the first projection image $\mathbf{X}_1$ and will be used for next projection. The reason of producing such projection image is to prevent the target pixels detected in $\mathbf{B}_1$ from being considered again in the new projection image, $\mathbf{X}_1$. A second projection vector $\mathbf{a}_2$ is then found again by maximizing a PI based on the first projection image, $\mathbf{X}_1$. The projection space is then formed by $\left\{z_k^2\right\}_{k=1}^N$ which are generated by $\mathbf{z}^2 = \mathbf{a}_2^T \mathbf{X}_1$ via $\mathbf{a}_2$. Similarly, a binary image $\mathbf{B}_2$ can be obtained from $\mathbf{z}^2$ by using zero-detection in the same manner that $\mathbf{B}_1$ was generated. In analogy with finding $\mathbf{X}_1$, a second projection image $\mathbf{X}_2$ can be also formed by setting to zero the gray scales of pixels that are in $\mathbf{B}_2$ while the pixels that are not in $\mathbf{B}_2$ remaining unchanged. The same procedure is continued to generate a third projection vector $\mathbf{a}_3$ that maximizes PI based on the second projection image, $\mathbf{X}_2$ to form two projection images $\mathbf{z}^3$ and $\mathbf{X}_3$. A third thresholded binary image $\mathbf{B}_3$ is generated by the same way that $\mathbf{B}_1$ and $\mathbf{B}_2$ were generated for subsequent target detection, etc. until the PPEA converges. Accordingly, a set of the desired targets can be extracted from a sequence of $\{\mathbf{B}_j\}$ that was obtained by applying zero-detection thresholding to the sequence of projection images $\{\mathbf{z}^j\}$.

Using the 15-panel HYDICE scene in Fig. 1.7(a) as an example, Fig. 16.1(a) shows a Gaussian-like histogram of the first projection image formed by $\{z_n^1\}_{n=1}^N$ that were generated by $\mathbf{z}^1 = \mathbf{a}_1^T\mathbf{X} = (z_1^1, z_2^1, \cdots, z_N^1)^T$ via $z_n^1 = \mathbf{a}_1^T\mathbf{x}_n$. The histogram in Fig. 16.1(a) is plotted based on the number of projected pixels versus their projection values generated by $z_n^1 = \mathbf{a}_1^T\mathbf{x}_n$ for $1 \leq n \leq N$. Since the right tail of the histogram does not clearly show ripples, its long and flat right tail is enlarged in Fig. 16.1(b).

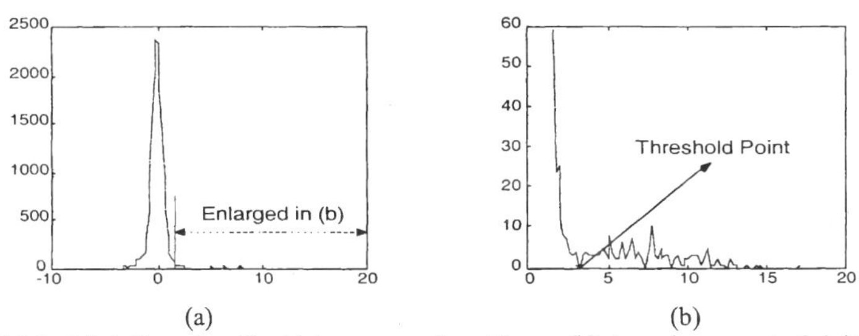

Figure 16.1. (a) A Gaussian-like histogram produced by $\mathbf{z}^1$; (b) An enlargement of right tail in (a)

As we can see, there are many ripples. The first value that detects zero is selected as the desired threshold value as shown by an arrow. It should be noted that the histogram in Fig. 16.1(a) is a Gaussian-like distribution, but it is asymmetric. In this case, the projection index using the skewness was able to detect small targets in Figs. 16.4-16.5 that caused ripples.

16.5 EXPERIMENTS

Two sets of hyperspectral data are used in this section to evaluate the performance of PPEA in target detection.

16.5.1 AVIRIS Data Experiments

The 224-band AVIRIS data in Fig. 1.6(a) were used for the experiments. It has been shown that in general there was no need of using all full bands for image classification (Chang et al., 1999d), only 12 bands were used in the following experiments to ease computation of PPEA. These 12 bands were uniformly selected among 224 bands. Additionally, it was also noted that there was a two-pixel anomaly located at the top edge of the dry lake marked by a white circle in Fig. 6.1(c). This single anomaly of two-pixel size was seen or detected visually in Harsanyi and Chang (1994) because the orthogonal subspace projection (OSP) was supervised. Fig. 16.2 shows the detection results of the first three projection images of all the four PI's.

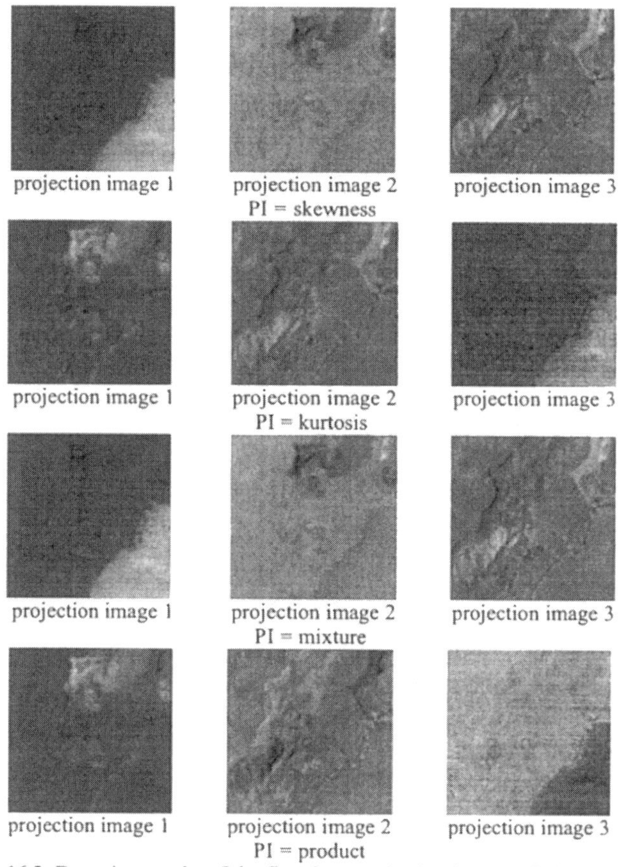

Figure 16.2. Detection results of the first three projection images of all the four PI's

Fig. 16.3 also shows their negative counterparts of Fig. 16.2. The reason that we also included negative projections in Fig. 16.3 is because the projection values are not necessarily positive and they can be negative. The significance of interestingness of a projection is determined by its magnitude not its signs.

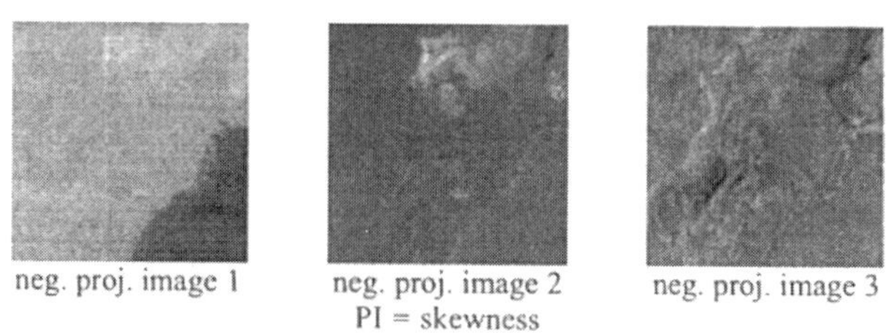

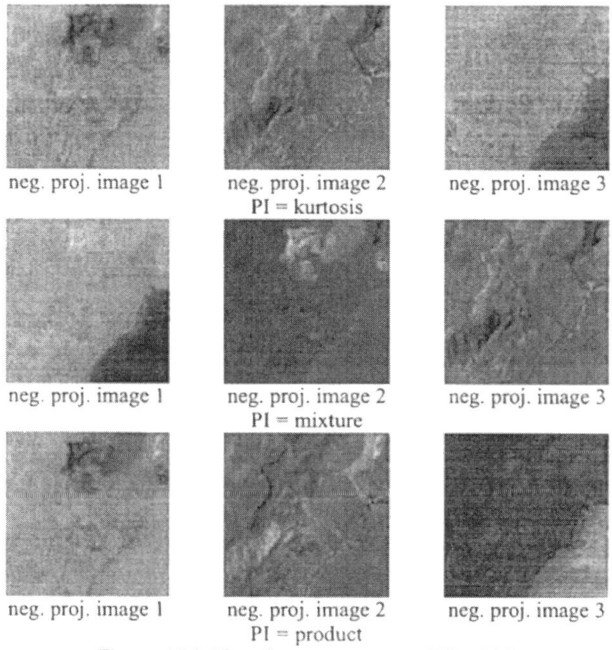

Figure 16.3. Negative counterparts of Fig. 16.2

As shown in Figs. 16.2-16.3, all the six signatures were detected, specifically, the anomaly detected in all the four PI's. Since kurtosis is more interesting than skewness in many other applications, we use kurtosis as an example for illustration. The cinders showed in the first projection positive image while the vegetation was detected in its negative counterpart. In the second projection, the vegetation and the rhyolite were extracted in the positive projection image while the shade and anomaly were detected in its negative counterpart. In third projection, the playa was pulled out from the positive image. It turned out that both positive and negative projection images were very similar. We also noted that if a signature is interesting, it generally shows in more than one projection image such as the vegetation in Figs. 16.2-16.3.

16.5.2 HYDICE Data Experiments

The HYDICE scene in Fig. 1.7(a) was used for the following experiments. Like the AVIRIS data, only 12 bands were selected uniformly for experiments to ease computation of PPEA. As will be shown in the following experiments, 12 bands are sufficiently enough for PPEA to effectively detect target pixels. Since Fig. 1.7(b) provides a ground truth map, we can evaluate the performance of PPEA by actually tallying number of panel pixels detected in the thresholded binary images $\{\mathbf{B}_j\}$ generated by the PPEA.

Following a similar analysis in Chapter 9 and criteria in Section 9.3, R and Y pixels were used to tally results of target detection. Fig. 16.4 shows the detection results based on the first six projection images using the four PI's where projection images were produced by skewness, kurtosis, mixture and product and are labeled by (a), (b), (c) and (d) respectively. As we can see from these images in Fig. 16.4(a-d), all the panels were extracted in the first three projection images.

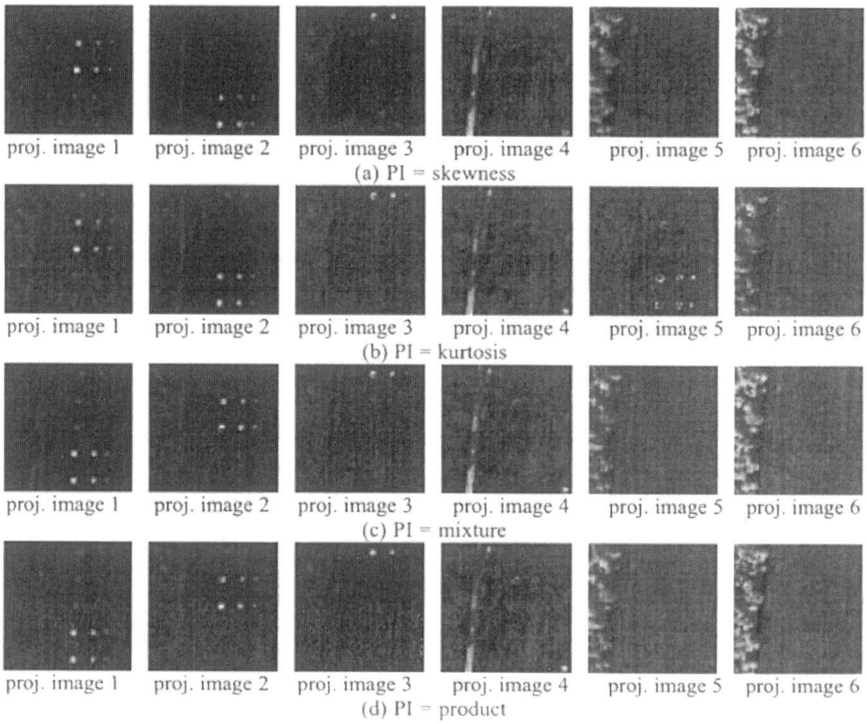

Figure 16.4. Detection results of the first six projection images

Fig. 16.5(a-d) shows the thresholded binary images of the first three projection images in Fig. 16.4(a-d) where all the four PI's detected the panels in the first row in their third projection images. The panels in the second and third rows were detected in one projection image with the panels in the fourth and fifth rows detected in another projection image. The reason that the panels in the second and third rows were detected by the same projection image because their spectra are very similar. It is also true for the panels in the fourth and fifth rows.

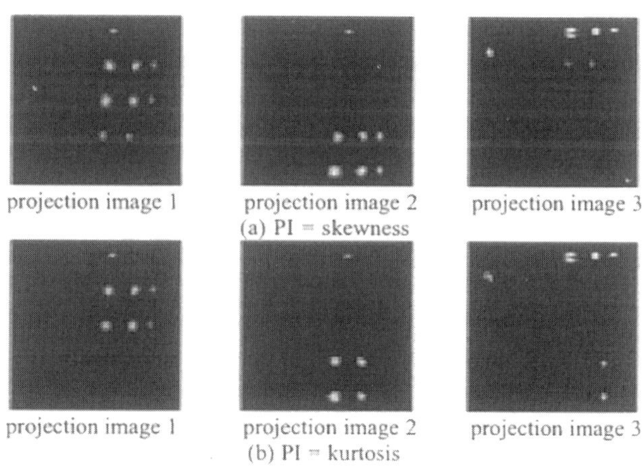

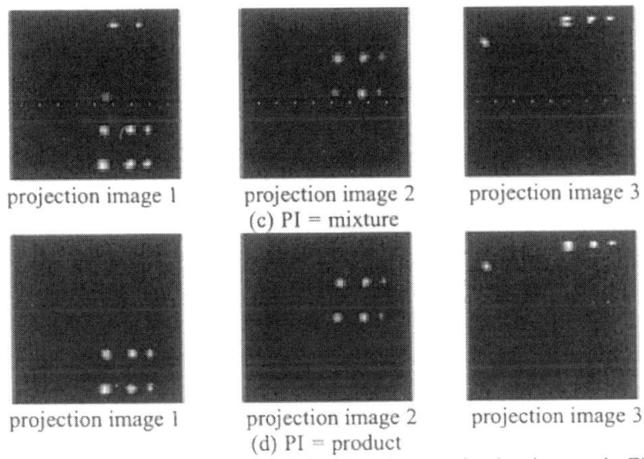

Figure 16.5. Thresholded binary images of the first three projection images in Fig. 16.4(a-d)

As shown in Fig. 16.5(a-d), all the 15 panels were successfully detected and extracted. These detection results are impressive. Since the panels of $1m \times 1m$ in the third column of the scene have spatial resolution smaller than the pixel resolution 1.5 m, they cannot be seen by visual inspection, but they are actually detected and extracted at subpixel level by PPEA. Table 16.1 tabulates the values of PI's of the first six projection images for these four PI's where the first three values of all the four PI's are significantly greater than those of their remaining PI's. This observation is consistent with the detection results in Figs. 16.4-16.5 and provides a clue that only three projections may be sufficiently enough for target detection.

Table 16.1. Values of the first six projections produced by four PI's

	Projection 1	Projection 2	Projection 3	Projection 4	Projection 5	Projection 6
skewness	141.92	198.53	61.53	22.19	7.77	6.14
kurtosis	43666	53827	18126	839	1423	194
mixture	1142	1138	496	26	4.44	1.735
product	12880834	10769387	20991206	22674	1099	245

In order to calculate various detection rates specified by (9.25)-(9.33), Fig. 16.6(a-d) shows the detection results that were obtained by combining the first three thresholded binary projection images in Fig. 16.5(a-d). Using Fig. 16.6(a-d) we can actually tally detected panel pixels.

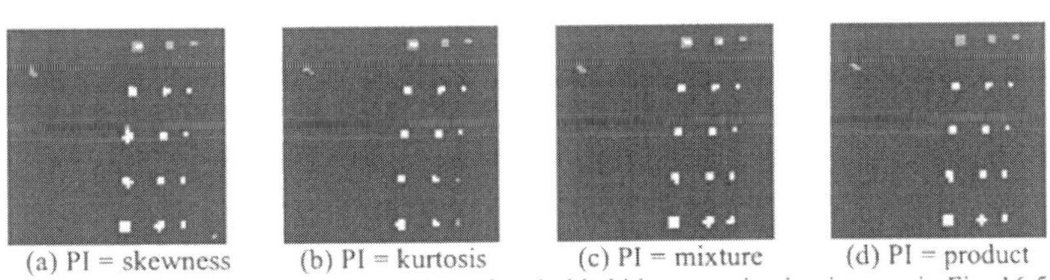

Figure 16.6. Combined image the first three thresholded binary projection images in Fig. 16.5(a-d)

Interestingly, three pixels in the tree line detected in Fig. 16.6(a-b) by skewness and kurtosis, and four pixels detected in Fig. 16.6(c-d) by mixture and product were not panel pixels but interferers. This comes at a price of lacking prior knowledge. It should be noted that for each PI the sets of pixels detected by all the three thresholded binary projection images were disjoint. Therefore, the overall detected pixels by a particular PI is the sum of pixels detected by its first three thresholded binary projection images. All the four PI's produced comparable results as shown in Tables 16.2-16.6.

Table 16.2. Detection rates for PI using skewness

T	N_{R+Y}	N_R	N_Y	$N_{(R+Y)D}$	N_{RD}	N_{YD}	N_{TPM}	R_{RTD}	R_{YTD}	R_{TH}	R_{TPM}
P1	50	3	47	15	3	12	35	1.000	0.255	0.300	0.700
P2	45	4	41	11	4	7	34	1.000	0.171	0.244	0.756
P3	41	4	37	12	4	8	29	1.000	0.216	0.293	0.707
P4	44	4	40	12	4	8	32	1.000	0.200	0.273	0.727
P5	43	4	39	17	4	13	26	1.000	0.333	0.395	0.605

Table 16.3. Detection rates for PI using kurtosis

T	N_{R+Y}	N_R	N_Y	$N_{(R+Y)D}$	N_{RD}	N_{YD}	N_{TPM}	R_{RTD}	R_{YTD}	R_{TH}	R_{TPM}
P1	50	3	47	15	3	12	35	1.000	0.255	0.300	0.700
P2	45	4	41	8	4	4	37	1.000	0.098	0.178	0.822
P3	41	4	37	9	4	5	32	1.000	0.135	0.220	0.781
P4	44	4	40	8	3	5	36	0.750	0.125	0.182	0.818
P5	43	4	39	12	4	8	31	1.000	0.205	0.279	0.721

Table 16.4. Detection rates for PI using mixture

T	N_{R+Y}	N_R	N_Y	$N_{(R+Y)D}$	N_{RD}	N_{YD}	N_{TPM}	R_{RTD}	R_{YTD}	R_{TH}	R_{TPM}
P1	50	3	47	15	3	12	35	1.000	0.255	0.300	0.700
P2	45	4	41	8	4	4	37	1.000	0.098	0.178	0.822
P3	41	4	37	9	4	5	32	1.000	0.135	0.220	0.781
P4	44	4	40	11	4	7	33	1.000	0.175	0.250	0.750
P5	43	4	39	18	4	14	25	1.000	0.359	0.419	0.581

Table 16.5. Detection rates for PI using product

T	N_{R+Y}	N_R	N_Y	$N_{(R+Y)D}$	N_{RD}	N_{YD}	N_{TPM}	R_{RTD}	R_{YTD}	R_{TH}	R_{TPM}
P1	50	3	47	15	3	12	35	1.000	0.255	0.300	0.700
P2	45	4	41	8	4	4	37	1.000	0.098	0.178	0.822
P3	41	4	37	9	4	5	32	1.000	0.135	0.220	0.781
P4	44	4	40	11	4	7	33	1.000	0.175	0.250	0.750
P5	43	4	39	16	4	12	27	1.000	0.308	0.372	0.628

Table 16.6. Overall detection rates for four PI's

	N = 4096		N_{R+Y} = 223		N_R = 19		N_Y = 204			
PI	$N_{(R+Y)D}$	N_{RD}	N_{YD}	N_{TPF}	N_{TPM}	R_{RTD}	R_{YTD}	R_{TH}	R_{TPF}	R_{TPM}
skewness	67	19	48	4	156	1.000	0.235	0.300	0.001	0.700
kurtosis	52	18	34	3	171	0.947	0.167	0.233	0.001	0.767
mixture	61	19	42	4	162	1.000	0.206	0.274	0.001	0.726
product	59	19	40	4	164	1.000	0.196	0.265	0.001	0.735

In Tables 16.2-16.5, no false alarm rate was calculated. Only Table 16.6 includes the overall panel false alarm rate, R_{TPF}. This is because we were interested in detecting the panel pixels in the scene rather than particular panel pixels. From Table 16.6, we can see that the overall R-target pixel detection rate, R_{RTD} are 100% except of $I_{kurtosis}(\mathbf{a}) = \kappa_4^2$ which produced 0.9474, but was very close to 1. On the other hand, all of the four PI's yielded low Y-target pixel detection rates, R_{YTD} and the target hit rates R_{TH}. This shows that the proposed PPEA using four PI's are indeed very effective since a higher R_{RTD}

reduces R_{YTD}. In addition, R_{TPF} is very low and less than or equal to 0.1%. By contrast, the panel miss rate, R_{TPM} is very high and close to 70%-75%. This is due to the fact that R_{TPM} takes into account W panel pixels for target detection. In reality, Y-pixels generally are not panel center pixels, and they may be mixed by background pixels. So, a high R_{TPM} does not necessarily imply low R_{RTD}. This is particularly true when the number of Y-pixels is relatively large compared to that of R-pixels.

In order to further demonstrate the performance of PPEA, PCA was also studied for comparison. It is known that PCA can be considered to be a special case of PP with data variance used as a projection index. The results of the first 6 component images are shown in Fig. 16.7.

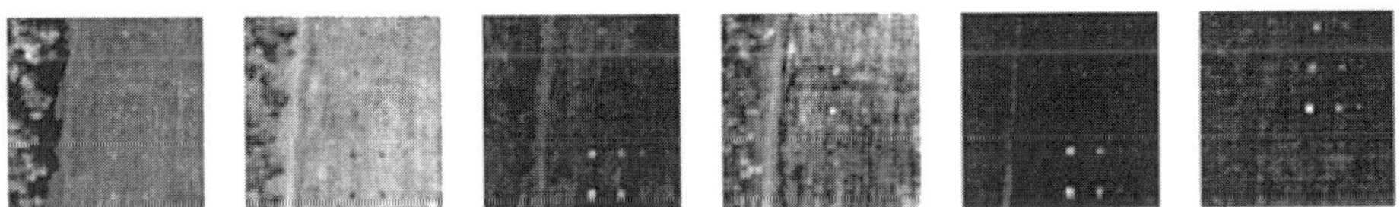

1st component 2nd component 3rd component 4th component 5th component 6th component

Figure 16.7. Results of the first 6 component images produced by PCA

Comparing to Fig. 16.4 where the four proposed PI's extracted all the 15 panels in their first three projection images, PCA picked up these panels in the 4th, 5th and 6th component images. This is because the panels are considered to be small targets in the scene and they do not generate significant information as opposed to that provided by the large grass field and tree line. As a consequence, the grass and tree line were extracted in the first three principal component images in Fig. 16.7. Interestingly, the interferers in the tree line were also shown in the first and third component images. This example shows that PCA can be used for preserving target information but not for the purpose of detecting small targets. Furthermore, in order to account for nonstationarity of the data, the sample correlation matrix instead of sample covariance matrix was used for PCA.

To conclude our experiments, we further plotted in Fig. 16.8(a-d) the learning curves of the first six projections (row 1-row 6 respectively) using PPEA for four PI's specified by (16.3)-(16.6) respectively with the x-axis specified by the number of iterations and the y-axis specified by the values of PI's. In all cases, no more 200 iterations are required in each optimal projection search. In particular, only approximate 150 iterations are needed to achieve the optimal search for the first three projections.

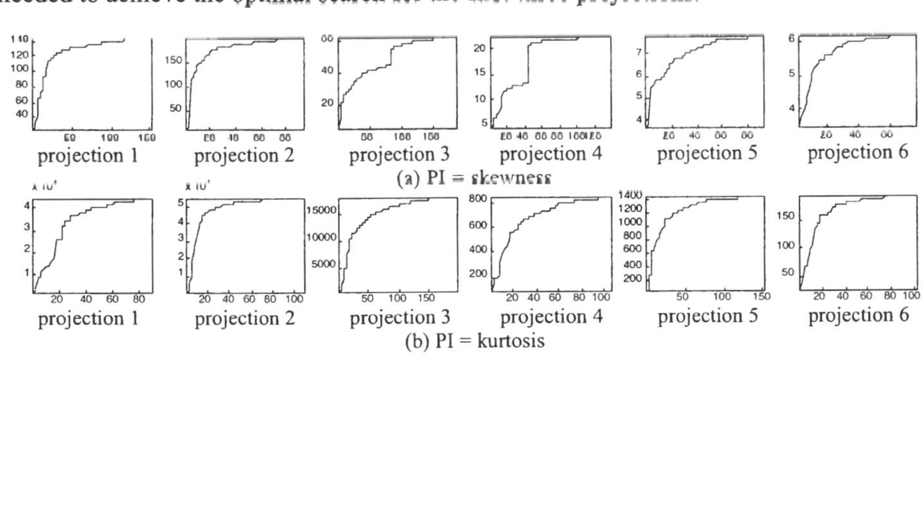

projection 1 projection 2 projection 3 projection 4 projection 5 projection 6

(a) PI = skewness

projection 1 projection 2 projection 3 projection 4 projection 5 projection 6

(b) PI = kurtosis

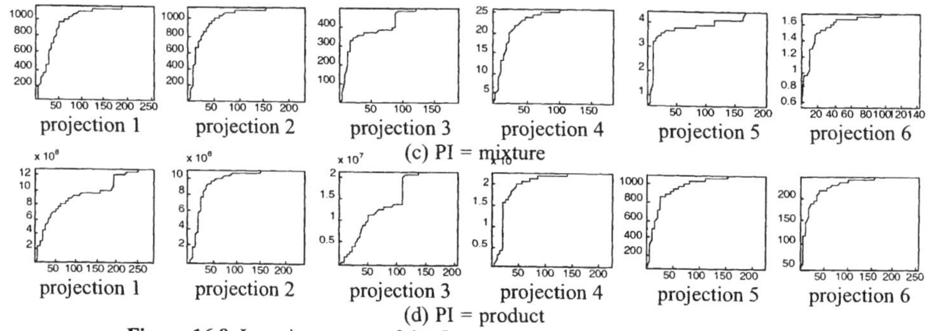

Figure 16.8. Learning curves of the first six projections using PPEA for four PI's

A final comment is noteworthy. Due to the fact that the spatial resolution of HYDICE data is 1.5 m as opposed to 20 m for AVIRIS data, high spatial resolution generally uncovers small targets that may cause the outliers of the background distribution. As expected, PPEA performed better for HYDICE data than for AVIRIS data.

16.6 CONCLUSIONS

Projection pursuit (PP) has received considerable interest in multivariate analysis because it can be used to explore interesting projections by which a high dimensional data set can be projected into a low dimensional space for various applications. This chapter presents a new application of PP in AMPC for hyperspectral imagery. Since in military applications targets of interest are relatively small compared to their surrounding background, these targets can be viewed as pixels that cause outliers of the background distribution. To meet this need, four projection indices are suggested which are based on skewness and kurtosis. A revised PP-based evolutionary algorithm (PPEA) is developed to find the optimal projections. Furthermore, a zero-detection thresholding is included in PPEA to achieve target detection. In order to use PPEA for target identification, a database or spectral library will be required. In this case, spectral criteria introduced in Chapter 2 can be used to measure the spectral similarity between the detected pixels and spectra in database to achieve target identification. Finally, due to the nature of EA the computational complexity of PPEA is generally very expensive. However, this can be compensated for high computer powers.

17

ESTIMATION FOR VIRTUAL DIMENSIONALITY OF HYPERSPECYRAL IMAGERY

Determination of intrinsic dimensionality (ID) for remotely sensed imagery is a challenging problem. According to the definition given in Fukunaga (1990, p. 280), the ID, also referred to effective dimensionality, is the minimum number of parameters required to account for the observed properties of the data. A general approach to ID estimation is principal components analysis (PCA) that makes use of the eigenvalue distribution to determine ID. This approach may be suitable to multispectral imagery since only a small number of bands are used and the resulting ID is expected to be small. However, the PCA method can be difficult to implement if it is applied to hyperspectral imagery. More importantly, it may not be effective even if it is applicable. With very high spectral resolution hyperspectral sensors which can extract many unknown subtle material substances, determining ID of hyperspectral imagery is more problematic than that of multispectral imagery. In order to account for such unknown signal sources, we introduce a new definition, referred to as virtual dimensionality (VD) in this chapter, which is the minimum number of spectrally distinct signal sources that characterize the hyperspectral data from a perspective view of target detection and classification. These signal sources may include unknown interfering sources, which cannot be identified by *a priori* knowledge. With this new definition, three eigen-thresolding based methods are presented to determine VD of hyperspectral imagery. They are all derived from the Neyman-Pearson detection theory. Since an information criterion (AIC) and the minimum description length (MDL) have been commonly used in sensor array processing to estimate the number of signals impinging upon the array, they will be also investigated and evaluated for comparison.

17.1 INTRODUCTION

The true dimensionality of multivariate data is difficult to determine in practice since its intrinsic dimensionality (ID) cannot be simply determined by the dimensionality of a data sample vector, which is the number of components in a data vector, referred to as component dimensionality. In particular, when very high dimensional data are well structured, the data tend to be clouded in a low dimensional space. In this case, ID is expected to be much smaller than the component dimensionality. Several well-known methods have been proposed such as principal components analysis (PCA) (Richards, 1993), factor analysis (Malinowski, 1977a; Malinowski, 1977b) which make use of the eigenvalue distribution to determine ID. Such approaches may be suitable to multispectral imagery where its ID is expected to be small. However, a direct application of these methods to hyperspectral imagery may not be feasible. More importantly, it may not be effective even if it is applicable (Chang and Du, 1999). This is particularly true for hyperspectral imagery where the component dimensionality is specified by hundreds of spectral channels and ID is generally much smaller. Accordingly, determination of the ID is crucial and also critical to success of hyperspectral image analysis, such as linear unmixing methods, which require the knowledge of how many image endmembers in image data. Unfortunately, ID determination can be very tricky and challenging. In order to account for such unknown signal sources and also not confuse with the ID, we introduce a new definition, referred to as virtual dimensionality (VD) in this chapter. It is defined by the minimum number of spectrally distinct signal sources that characterize the hyperspectral data from a perspective view of target detection and classification rather than the image endmembers defined in Schowengerdt (1997) which are idealized, pure signatures. These signal sources may include known and unknown image endmembers, natural signatures, anomalies and interferers (Chang et al. 1998; Chang and Du, 1999). The definition of VD used here has a very narrow implication driven by the need of target detection and classification. Based on this new definition, three methods are presented to estimate VD.

Recently, Harsanyi, Farrand and Chang developed an eigen-thresholding method, referred to as Harsanyi-Farrand-Chang (HFC) to determine the number of spectral endmembers in AVIRIS data (Harsanyi et al., 1997). It was derived from the Neyman-Pearson detection theory (Poor, 1994). Their idea can be briefly described as follows. Let the eigenvalues of the sample correlation matrix and the sample covariance matrix be denoted by correlation-eigenvaules and covariance-eigenvalues respectively. Since the component dimensionality is equal to the total number of eigenvalues, each eigenvalue specifies a component dimension with a certain level of significance provided by that particular component in terms of energy or variance. If there is no signal source contained in a component, its corresponding correlation-eigenvalue and covariance-eigenvalue should reflect only noise energy, in which case, both correlation-eigenvalue and covariance-eigenvalue must be equal. This fact allows us to formulate the difference between the correlation-eigenvalue and the covariance-eigenvalue as a binary composite hypothesis testing problem where the null hypothesis represents the case of the zero-difference while the alternative hypothesis being the case that the difference is greater than zero. When the Neyman-Pearson test is applied to each pair of a correlation-eigenvalue and its corresponding covariance-eigenvalue, the number of times the test fails indicates how many signal sources are present in the image. In other words, a failure of the Neyman-Pearson in a component implies a truth of the alternative hypothesis, which indicates that there is a signal source in this particular component. Using this approach we can estimate VD where the receiver operating characteristics (ROC) analysis can be used to evaluate the effectiveness of the decision. Since the HFC method does not decorrelate the data, an alternative approach is to include a noise whitening process in the HFC method to remove the second-order statistical correlation prior to the use of the HFC method. The resulting method will be referred to as noise-whitened HFC (NWHFC) method. In this case, noise estimation is required for the NWHFC method.

An important fact used by the HFC and NWHFC methods was that the sample size must be sufficiently large so that the covariance between the correlation-eigenvalue and the corresponding covariance-eigenvalue is asymptotically zero. However, this may not be valid for a small sample size. In order to alleviate this problem, a third method, referred to as noise subspace projection (NSP) approach was recently proposed in Du (2000) and Du and Chang (1999). It can be viewed as an extension of HFC via NWHFC. It first estimates the noise covariance matrix of image data, then uses it to whiten the noise variances to unity. The binary composite hypothesis problem used in the HFC method can be then simplified and reduced to that the null hypothesis is specified by the case of unity variance while the alternative hypothesis is the case that variance is greater than unity. As a result, calculating both correlation-eigenvalues and covariance-eigenvalues can be avoided. An immediate benefit from the NSP method is that the sample size is not necessarily large as required by the HFC and NWHFC methods. In order to estimate the noise covariance matrix, three methods will be considered. Two were suggested by Roger and Arnold (1996) and the third was recently developed by Du and Chang (1999) which was used to estimate the number of hidden nodes for a radial basis function neural network.

One advantage resulting from a noise whitening process is to enable us to examine applicability of two well-established information theoretic criteria, an information criterion (AIC), minimum description length (MDL) to VD determination. The AIC and MDL have been widely used in passive sensor array processing to determine the number of signals impinging upon the sensors. However, a crucial requirement for AIC and MDL to be effective is that the noise must be independent identically distributed (i.i.d). It is our hope that a noise whitening process prior to the AIC and MDL may help relax this limitation. Unfortunately, as will be demonstrated by experiments, the noise whitening process is not good enough. In fact, AIC and MDL require more than second-order independency to be effective.

17.2 NEYMAN-PEARSON DETECTION THEORY-BASED EIGEN-THRESHOLDING ANALYSIS (HFC METHOD)

A Neyman-Pearson detection theory-based eigen-threshold method, referred to as HFC method, was previously developed by Harsanyi et al. (1994) to determine the number of endmembers in AVIRIS data. It first calculated the sample correlation matrix, $\mathbf{R}_{L\times L}$ and sample covariance matrix $\mathbf{K}_{L\times L}$, then found the difference between their corresponding eigenvalues. Let $\{\hat{\lambda}_1 \geq \hat{\lambda}_2 \geq \cdots \geq \hat{\lambda}_L\}$ and $\{\lambda_1 \geq \lambda_2 \geq \cdots \geq \lambda_L\}$ be two sets of eigenvalues generated by $\mathbf{R}_{L\times L}$ and $\mathbf{K}_{L\times L}$, called correlation eigenvalues and covariance eigenvalues respectively. Assume that signal sources have positive energies and noise variance in band l is given by σ_l^2. We can expect that

$$\hat{\lambda}_l - \lambda_l > 0 \text{ for } l = 1,2,\cdots,\text{VD}, \tag{17.1}$$

and

$$\hat{\lambda}_l - \lambda_l = 0 \quad \text{for } l = \text{VD}+1,\cdots,L \tag{17.2}$$

where

$$\hat{\lambda}_l = \lambda_l + \sigma_l^2 \text{ for } l = 1,2,\cdots,\text{VD and } \hat{\lambda}_l = \lambda_l = \sigma_l^2 \text{ for } l = \text{VD}+1,\cdots,L. \tag{17.3}$$

In order to determine VD, Harsanyi, Farrand and Chang (1994) formulated the problem outlined by (17.1) and (17.2) as a binary hypothesis problem as follows.

$$\begin{aligned} &H_0: \quad z_l = \hat{\lambda}_l - \lambda_l = 0 \\ &versus \qquad\qquad\qquad\qquad \text{for} \quad l = 1, 2, \cdots, L \\ &H_1: \quad z_l = \hat{\lambda}_l - \lambda_l > 0 \end{aligned} \tag{17.4}$$

where the null hypothesis H_0 and the alternative hypothesis H_1 represent the case that the correlation-eigenvalue is equal to its corresponding covariance eigenvalue and the case that the correlation-eigenvalue is greater than its corresponding covariance eigenvalue respectively. In other words, when H_1 is true (i.e., H_0 fails) for some component l, it implies that there is a signal source contributing the correlation-eigenvalue in this particular component l in addition to noise. As a result, the l-th eigenvalue in $\mathbf{R}_{L\times L}$ is greater than the corresponding l-th eigenvalues in $\mathbf{K}_{L\times L}$.

According to Anderson (1984), we can model each pair of eigenvalues, $\hat{\lambda}_l$ and λ_l in (17.1)-(17.3) under hypotheses H_0 and H_1 as random variables with their asymptotic conditional probability densities given by

$$p_0(z_l) = p(z_l \mid H_0) \cong N(0, \sigma_{z_l}^2) \quad \text{for } l = 1, 2, \cdots, L \tag{17.5}$$

and

$$p_1(z_l) = p(z_l \mid H_1) \cong N(\mu_l, \sigma_{z_l}^2) \text{ for } l = 1, 2, \cdots, L \tag{17.6}$$

respectively where μ_l is an unknown constant and the variance $\sigma_{z_l}^2$ is given by

$$\sigma_{z_l}^2 = Var[\hat{\lambda}_l - \lambda_l] = \sigma_{\hat{\lambda}_l}^2 + \sigma_{\lambda_l}^2 - 2Cov(\hat{\lambda}_l, \lambda_l) \text{ for } l = 1, 2, \cdots, L \tag{17.7}$$

with $\sigma_{\hat{\lambda}_l}^2$ and $\sigma_{\lambda_l}^2$ given by (Anderson,1984)

$$\sigma_{\hat{\lambda}_l}^2 = Var[\hat{\lambda}_l] \cong \frac{2\hat{\lambda}_l^2}{N} \tag{17.8}$$

$$\sigma_{\lambda_l}^2 = Var[\lambda_l] \cong \frac{2\lambda_l^2}{N} \tag{17.9}$$

and N is the total number of samples. Since $Cov(\hat{\lambda}_l, \lambda_l)$ in (17.7) is bounded by Schwarz's inequality

$$Cov(\hat{\lambda}_l, \lambda_l) \leq \sqrt{Var[\hat{\lambda}_l]Var[\lambda_l]} \cong (2/N)\left(\hat{\lambda}_l \cdot \lambda_l\right), \tag{17.10}$$

$Cov(\hat{\lambda}_l, \lambda_l) \to 0$ as $N \to \infty$. This results in

$$\sigma_{z_l}^2 \cong \sigma_{\hat{\lambda}_l}^2 + \sigma_{\lambda_l}^2 \cong \frac{2\hat{\lambda}_l^2}{N} + \frac{2\lambda_l^2}{N} \quad \text{as } N \to \infty. \tag{17.11}$$

(17.11) implies that $\sigma_{z_l}^2$ is asymptotically equal to sum of $\sigma_{\hat{\lambda}_l}^2$ and $\sigma_{\lambda_l}^2$, and can be approximated by $2\left(\hat{\lambda}_l^2 + \lambda_l^2\right)/N$ asymptotically.

From (17.4)-(17.6) and (17.11), we define the false alarm probability and detection probability (detection power) as follows:

$$P_F = \int_{\tau}^{\infty} p_0(z)\,dz \tag{17.12}$$

$$P_D = \int_{\tau}^{\infty} p_1(z)\,dz. \tag{17.13}$$

Using (17.4), (17.12)-(17.13), the maximum detection probability P_D in determining VD for any given false alarm probability P_F is given by the Neyman-Pearson detector, δ_{NP}.

17.3 ESTIMATION OF NOISE COVARIANCE MATRIX

As noted in (17.8), $Cov(\hat{\lambda}_l, \lambda_l)$ may not be zero when the sample size is small. In this case, we consider an alternative approach which only requires computation of $\mathbf{R}_{L\times L}$ or $\mathbf{K}_{L\times L}$, but not both. The trade-off is that we need to know the noise second-order statistics to be used for whitening.

There exist many methods to estimate the noise covariance matrix, such as residual-based estimation (Roger and Arnold, 1996) taking into account intra-band correlation, nearest neighbor difference (NND) (Green et al, 1988; Lee et al. 1990) using inter-band correlation, and linear regression model-based prediction (Roger and Arnold, 1999) taking advantage of intra/inter-band correlation. Since NND has shown to be a poor estimation (Chang and Du, 1999), only residual analysis and linear regression model-based prediction methods developed by Roger and his colleagues (Roger, 1999; Roger and Arnold, 1999) will be discussed in this chapter for noise covariance matrix estimation.

17.3.1 Residual Analysis (Roger, 1996)

Assume that there a hyperspectral image Ω is acquired by L spectral bands. Ω is indeed an image cube made up of L band images, $\{B_l\}_{l=1}^{L}$. As a result, each pixel vector in a L-band multispectral/hyperspectral image cube Ω is denoted by an L-dimensional column vector $\mathbf{x}_{(i,j)} = \left(x_{(i,j)1}, x_{(i,j)2}, \cdots, x_{(i,j)L}\right)^T$ with $x_{(i,j)l}$ being the pixel at the spatial location (i,j) in B_l. Let $\mathbf{K}_{L\times L}$ be the sample covariance matrix given by $\mathbf{K}_{L\times L} = \frac{1}{IJ}\left[\sum_{i=1}^{I}\sum_{j=1}^{J}\left(\mathbf{x}_{(i,j)} - \boldsymbol{\mu}\right)\left(\mathbf{x}_{(i,j)} - \boldsymbol{\mu}\right)^T\right]$ where $\boldsymbol{\mu}$ is the sample mean given by $\boldsymbol{\mu} = (1/IJ)\left[\sum_{i=1}^{I}\sum_{j=1}^{J}\mathbf{x}_{(i,j)}\right]$ and IJ is the total number of pixels in the image. Then $\mathbf{K}_{L\times L}$ can be decomposed as follows.

$$\mathbf{K}_{L\times L} = \mathbf{D}_{\mathbf{K}}\mathbf{R}_{\mathbf{K}}\mathbf{D}_{\mathbf{K}} \tag{17.14}$$

where $\mathbf{D}_{\mathbf{K}}$ is a diagonal matrix given by $\mathbf{D}_{\mathbf{K}} = diag\left\{\sigma_1^2, \sigma_2^2, \cdots, \sigma_L^2\right\}$ with $\left\{\sigma_l^2\right\}_{l=1}^{L}$ being diagonal elements of $\mathbf{K}_{L\times L}$, σ_l^2 is the variance of B_l and

$$\mathbf{R}_K = \begin{bmatrix} 1 & \rho_{12} & \rho_{13} & \cdots & \rho_{1L} \\ \rho_{21} & 1 & \rho_{23} & \cdots & \rho_{2L} \\ \rho_{21} & \rho_{21} & \ddots & \ddots & \vdots \\ \vdots & \vdots & \ddots & \ddots & \rho_{(L-1)L} \\ \rho_{L1} & \rho_{L2} & \cdots & \rho_{L(L-1)} & 1 \end{bmatrix} \tag{17.15}$$

with ρ_{mn} being the correlation coefficient at the (m,n)-th entry of $\mathbf{K}_{L\times L}$ and $m \neq n$.

Similarly, in analogy with decomposition of $\mathbf{K}_{L\times L}$, its inverse $\mathbf{K}_{L\times L}^{-1}$ can be also decomposed as

$$\mathbf{K}_{L\times L}^{-1} = \mathbf{D}_{\mathbf{K}^{-1}}\mathbf{R}_{\mathbf{K}^{-1}}\mathbf{D}_{\mathbf{K}^{-1}} \tag{17.16}$$

where $\mathbf{D}_{\mathbf{K}^{-1}}$ is a diagonal matrix given by $\mathbf{D}_{\mathbf{K}^{-1}} = \left\{\varsigma_1^2, \varsigma_2^2, \cdots, \varsigma_L^2\right\}$ with $\left\{\varsigma_l^2\right\}_{l=1}^{L}$ being the diagonal elements of $\mathbf{K}_{L\times L}^{-1}$ and

$$\mathbf{R}_{\mathbf{K}^{-1}} = \begin{bmatrix} 1 & \xi_{12} & \xi_{13} & \cdots & \xi_{1L} \\ \xi_{21} & 1 & \xi_{23} & \cdots & \xi_{2L} \\ \xi_{21} & \xi_{21} & \ddots & \ddots & \vdots \\ \vdots & \vdots & \ddots & \ddots & \varsigma_{(L-1)L} \\ \xi_{L1} & \xi_{L2} & \cdots & \xi_{L(L-1)} & 1 \end{bmatrix} \tag{17.17}$$

with ξ_{mn} being the correlation coefficient at the (m,n)-th entry of $\mathbf{K}^{-1}$ and $m \neq n$. It turns out that ς_l in $\mathbf{D}_{\mathbf{K}^{-1}}$ can be related to σ_l $\mathbf{D}_{\mathbf{K}}$ by the following relation

$$\varsigma_l = \sigma_l^{-1}\left(1 - r_{L-l}^2\right)^{-1/2} \tag{17.18}$$

where r^2_{L-l} is the multiple correlation coefficients of band B_l on the other $L-1$ bands $\{B_k\}^L_{k=1,k\neq l}$ obtained by using the multiple regression theory. So, ς^2_l is the reciprocal of a good noise variance estimate of band B_l. It should be noted that the $\mathbf{D}_{\mathbf{K}^{-1}}$ in (17.16) is not an inverse of the $\mathbf{D}_{\mathbf{K}}$ in (17.14), nor is $\mathbf{R}_{\mathbf{K}^{-1}}$ in (17.16). The major advantage of using ς_l over σ_l is that as shown in (17.18), ς_l removes its correlation on other ς_l's for $l \neq k$ while σ_l does not.

17.3.2 Inter/Intra-Band Prediction Noise Estimation: Spatial/Spectral Prediction Noise Estimation (Roger and Arnold, 1996)

Let $x_{l,(i,j,k)}$ be the pixel at the spatial location (i,j) in $B_{l,k}$ which is the k-th block of the l-th band B_l. We use a linear regression model to describe the intra/inter-band correlation as followed.

$$\hat{x}_{l,(i,j,k)} = a_1 x_{l-1,(i,j,k)} + a_2 x_{l+1,(i,j,k)} + a_3 x_{l,(i-1,j,k)} + a_0 \tag{17.19}$$

where a_0, a_1, a_2, a_3 are the linear regression coefficients need to be determined.

Now we define the residual error of $x_{l,(i,j,k)}$ estimated by $\hat{x}_{l,(i,j,k)}$ via (17.19) is given by

$$e_{l,(i,j,k)} = x_{l,(i,j,k)} - \hat{x}_{l,(i,j,k)}. \tag{17.20}$$

So, the residual error of $B_{l,k}$ is

$$e^2_{l,k} = \Sigma_{(i,j)} e^2_{l,(i,j,k)}. \tag{17.21}$$

In order to estimate the noise variance of $B_{l,k}$, Roger and Arnold (1996) proposed a method to estimate (17.21) subject the constraint that the mean of the residual errors incurred within the block must be zero, i.e., $\Sigma_{(i,j)} e_{l,(i,j,k)} = 0$. The resulting constrained solution was obtained by

$$\sigma^2_{l,k} = \left(N(B_{l,k}) - 4\right)^{-1} e^2_{l,(i,j,k)}. \tag{17.22}$$

where $N(B_{l,k})$ is the total number of pixels within $B_{l,k}$.

By means of (17.22) we can further estimate the 6variance of B_l by

$$\sigma^2_l = \text{mean of } \left\{\sigma^2_{l,k}\right\}_k \text{ or } \sigma^2_l = \text{median of } \left\{\sigma^2_{l,k}\right\}_k. \tag{17.23}$$

Since the noise is generally random, the band-to-band noise correlation can be assumed to be uncorrelated or weak, the noise covariance matrix $\mathbf{K}_n$ can be estimated by a diagonal matrix with the (l,l) diagonal element given by σ_l^2 in (17.21), that is, $\mathbf{K}_n = diag\{\lambda_1, \lambda_2, \cdots, \lambda_L\}$.

17.4 NOISE ESTIMATION-BASED EIGEN-THRESHOLDING

In many signal detection applications, a common practice is noise whitening which generally improve and enhance signal detectability. In order for a noise whitening process to be effective, the noise second-order statistics must be estimated accurately or at least reliably. The techniques introduced in Section 17.3 have shown success in estimation of noise covariance matrix of AVIRIS data and can be used for this purpose.

17.4.1 Noise-Whitened HFC (NWHFC) Method

The HFC method described in Section 17.2 does not have a noise whitening process. In this case, weak signal sources may be obscured by noise and cannot be detected. As an alternative, the HFC method can be modified by including a noise whitening process as a preprocessing to remove the second-order statistical correlation so that signal sources can be decorrelated to achieve better signal detection. The resulting HFC method will be referred to as noise-whitened HFC (NWHFC) method.

17.4.2 Noise Subspace Projection (NSP)

The effectiveness of the HFC and NWHFC methods relies on (17.8), which guarantees that $Cov(\hat{\lambda}_l, \lambda_l) \to 0$ as $N \to \infty$. However, for the case that the number of samples is not sufficiently large, $Cov(\hat{\lambda}_l, \lambda_l)$ may not be zero. So, in order to resolve this problem, a noise subspace projection (NSP) is developed in this section. It can be viewed as a hybrid of the HFC and NWHFC methods.

If we assume that the number of spectrally distinct signal sources is VD, the data sample covariance matrix, $\mathbf{K}_{L\times L}$ can be then represented by $\mathbf{K}_{L\times L} = \mathbf{K}_s + \mathbf{K}_n$ where $\mathbf{K}_s$ is the covariance matrix of signals present in the image data. Using the eigen-decomposition of $\mathbf{K}_{L\times L}$ (Poor, 1994) we can express $\mathbf{K}_{L\times L}$ as

$$\mathbf{K}_{L\times L} = \sum_{l=1}^{\mathrm{VD}} \left(\lambda_l + \sigma_l^2\right) \mathbf{u}_l \mathbf{u}_l^T + \sum_{l=\mathrm{VD}+1}^{L} \sigma_l^2 \, \mathbf{u}_l \mathbf{u}_l^T \tag{17.24}$$

where $\{\mathbf{u}_l\}_{l=1}^{\mathrm{VD}}$ and $\{\mathbf{u}_l\}_{l=\mathrm{VD}+1}^{L}$ represent two sets of orthonormal vectors to span signal space and noise space respectively and VD is the virtual dimensionality. The λ_l and σ_l^2 in (17.24) correspond to the energies of signals and noise in the l-th component (i.e. l-th band) image respectively. The only problem of (17.24) that remains unsolved is to find VD.

In the HFC method, it used difference between correlation-eigenvalues and covariance-eigenvalues as a criterion to determine VD. In this section, we take advantage of the eigen-decomposition of $\mathbf{K}_{L\times L}$ and premultiply $\mathbf{K}_{L\times L}$ in (17.24) by the diagonal matrix $\mathbf{K}_n^{-1}$. This results in

$$\mathbf{K}_n^{-1}\mathbf{K}_{L\times L} = \Sigma_{l=1}^{\mathrm{VD}}\left(\lambda_l / \sigma_l^2 + 1\right)\mathbf{u}_l\mathbf{u}_l^T + \Sigma_{l=\mathrm{VD}+1}^{L}\mathbf{u}_l\mathbf{u}_l^T \tag{17.25}$$

where the variances of the noise components have been whitened and normalized to unity. If we further assume that the energies of signals is greater than noise energies, that is, $\lambda_l > \sigma_l^2$ for $l = 1,2,\cdots,\mathrm{VD}$, (17.4)-(17.6) become the following binary composite hypothesis testing problem

$$\begin{aligned} &H_0: \quad z_l = \mu_l = 1 \\ &versus \qquad\qquad\qquad\qquad \text{for } l = 1,2,\cdots,L \\ &H_1: \quad z_l = \mu_l = \lambda_l / \sigma_l^2 + 1 > 1 \end{aligned} \tag{17.26}$$

where

$$p_0(z_l) = p(z_l \mid H_0) \cong N(1,\sigma_{z_l}^2) \;\; \text{for } l = 1,2,\cdots,L \tag{17.27}$$

and

$$p_1(z_l) = p(z_l \mid H_1) \cong N(\mu_l,\sigma_{z_l}^2) \;\; \text{for } l = 1,2,\cdots,L \tag{17.28}$$

with $\sigma_{z_l}^2$ given by (17.9), i.e., $\sigma_{z_l}^2 = \sigma_{\lambda_l}^2$.

Now using (17.26)-(17.28) we can derive the Neyman-Pearson detector δ_{NP} for (17.26), which can be used as a decision to determine VD.

17.4.3 AIC AND MDL

In Chapters 3, 8-9 and 15, subpixel detection and mixed pixel classification problems were formulated as linear mixing problems which could be solved by using a linear mixture model specified by (3.1) and (15.2). This approach requires the knowledge of the target signature matrix **M**. Despite that the unsupervised algorithms proposed in Chapter 5 can be used to find target signatures directly from the image data, they still need to know how many target signatures should be generated. This dilemma can be resolved by estimation of the VD. In the passive sensor array processing, a similar problem also arises from estimation of how many signal sources impinging upon the array. This problem is related to how to select an appropriate model for a parameterized family of probability density functions used to best fit the sensor array data. It is known that two commonly used methods, AIC (An Information Theoretic Criterion) suggested by Akaike (1973) and MDL (Minimum Description Length) proposed by Schwartz (1978) and Rissanen (1978) can be used for model selection. They can be derived from the following formulas that were obtained by Wax and Kailath (1985)

$$\begin{aligned} \mathrm{AIC}(p) = &- 2\,\log\left(\Pi_{j=p+1}^{L}\lambda_j^{1/(L-p)} / \left[(1/L-p)\Sigma_{j=p+1}^{L}\lambda_j\right]\right)^{(L-p)N} \\ &+ 2p(2L-p) \end{aligned} \tag{17.29}$$

$$\mathrm{MDL}(p) = - \log\left(\prod_{j=p+1}^{L} \lambda_j^{1/(L-p)} / \left[(1/L-p)\sum_{j=p+1}^{L} \lambda_j\right]\right)^{(L-p)N} + (1/2)p(2L-p)\log N \tag{17.30}$$

where p is the number of free parameters that specifies a family of probability density functions, $\lambda_1 \geq \lambda_2 \geq \cdots \geq \lambda_L$ are eigenvalues generated by the sample covariance matrix $\mathbf{K}_{L\times L}$ given by $\mathbf{K}_{L\times L} = (1/N)\sum_{i=1}^{N} \mathbf{r}_i \mathbf{r}_i^T$.

In Chapter 4 we described that a linear unmixing problem could be formulated as a passive sensor array problem by interpreting signal sources impinging on a sensor array as desired target sources to be detected from a set of co-registered images acquired by a bank of spectral channels. With this interpretation a linear mixing problem can be solved by techniques that solves sensor array problems. In order to take advantage of these two criteria and apply them to the estimation of the VD, we need to examine their underlying assumptions, (1) the noise must be an independent identically distributed (i.i.d.) process and (2) the observation process is a zero-mean Gaussian random process. Part of the first assumption can be taken care of by the whitening technique proposed in Section 17.3, in which case the noise variance will be normalized to unity. Then the resulting sample covariance matrix can be used in the Gaussian process made in the second assumption. Now, the problem of estimating VD can be solved by minimizing (17.29) and (17.30) as follows.

$$\mathrm{VD}_{\mathrm{AIC}} = \arg\left\{\min_p \mathrm{AIC}(p)\right\} \tag{17.31}$$

$$\mathrm{VD}_{\mathrm{MDL}} = \arg\left\{\min_p \mathrm{MDL}(p)\right\}. \tag{17.32}$$

17.5 COMPUTER SIMULATIONS AND HYPERSPECTRAL IMAGE EXPERIMENTS

The experiments conducted in this section consist of computer simulations and real hyperspectral image experiments. The purpose of these experiments is to demonstrate the ability of the methods presented in this chapter and two information theoretic criteria, AIC and MDL in estimation of VD. As will be shown, the AIC and MDL generally do not work effectively if the assumption that the noise is i.i.d. is violated.

17.5.1 Computer Simulations

The data set contains five field reflectance spectra, dry grass, red soil, creosote leaves, blackbrush and sagebrush shown in Fig. 1.5. Using the spectra in Fig. 1.5 each spectral signature was used to 1000 simulated pixels with abundance generated by random numbers in the range (0,1), then they were mixed to form 1000 simulated mixed pixels. Additive Gaussian noise was added to achieve signal-to-noise ratio 25:1 for entire data. In order to simulated an i.i.d. case, the same Gaussian variance was added to each band. Fig. 17.1(a)-(c) shows the log eigen-spectra generated by the HFC, the NWHFC and NSP-based eigen-thresholding methods respectively where the eigenvalues produced by $\mathbf{R}_{L\times L}$ and $\mathbf{K}_{L\times L}$ are denoted by "*"s and "○"s respectively. As noted in Fig. 17.1, when the eigenvalue index increases, the correlation-eigenvalues become to overlay the covariance-eigenvalues. This phenomenon can be explained by the fact that noise is the only signal sources contributed to these eigenvalues. With this interpretation the VD was estimated by the HFC, the NWHFC and NSP-based eigen-thresholding methods to be 5, 4 and 4 respectively using the Neyman-Pearson test with the false alarm probability P_F set to $P_F = 0.001$. The HFC eigen-thresholding method correctly predicted the true data dimensionality, which is 5, while both NWHFC and NSP-based eigen-thresholding methods predicting 4. Since the spectra of creosote leaves and blackbrush shown in Fig. 1.5 are so close, they were considered to be the same signature by the NWHFC and NSP methods.

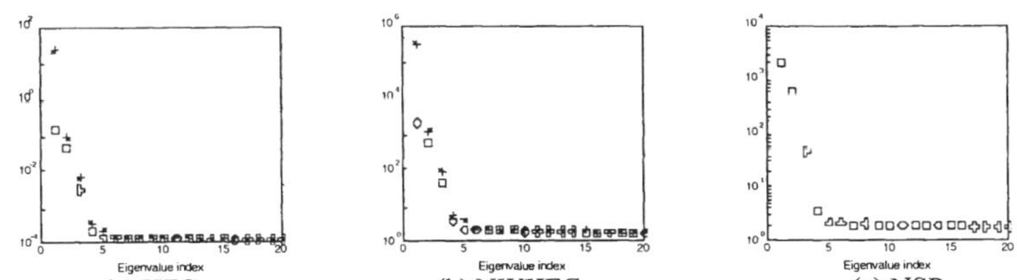

(a) HFC (b) NWHFC (c) NSP

Figure 17.1. Computer simulation with 5 AVIRIS signatures

This is also true for the AIC and MDL where their VD was estimated to be 4 as shown in Fig. 17.2(a-b).

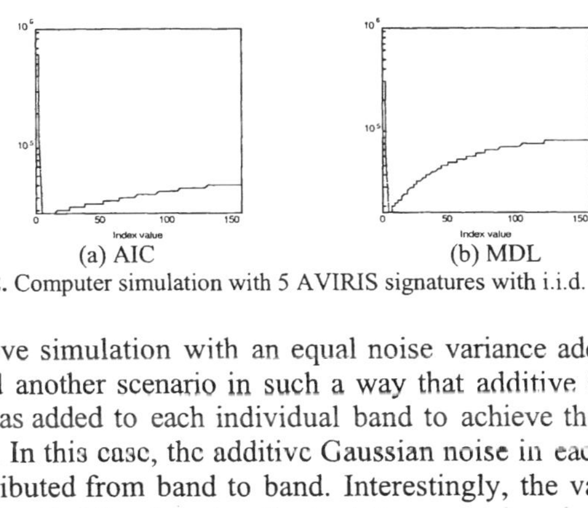

(a) AIC (b) MDL

Figure 17.2. Computer simulation with 5 AVIRIS signatures with i.i.d. Gaussian noise

Unlike the above simulation with an equal noise variance added to each individual band, we simulated another scenario in such a way that additive Gaussian noise with a different variance was added to each individual band to achieve the fixed signal-to-noise ratio (SNR) = 25:1. In this case, the additive Gaussian noise in each band was white but not identically distributed from band to band. Interestingly, the values of VD estimated by the three eigen-thresholding based methods still remained unchanged, but the AIC and MDL completely failed as shown in Fig. 17.3, in which their estimated VD was 139 and 110 respectively. The experiment of the second simulation demonstrates that the i.i.d assumption is crucial to the success of the AIC and MDL, while it has very little effect on the eigen-thresholding based methods.

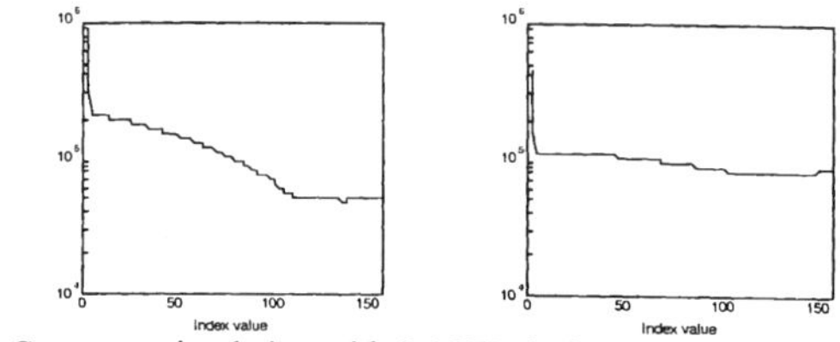

Figure 17.3. Computer simulation with 5 AVIRIS signatures with white Gaussian noise

Table 17.1 summarizes the VD results estimated by the HFC, NWHFC, NSP, AIC and MDL for the two scenarios of simulated Gaussian noise, i.i.d. and non-i.i.d. cases.

Table 17.1. Estimation of VD

	HFC	NWHFC	NSP	AIC	MDL
i.i.d. Gaussian noise (1st scenario)	5	4	4	4	4
non-i.i.d Gaussian noise (2nd scenario)	5	4	4	139	110

Similar experiments were also conducted for various SNRs. It was found that when the SNR is high and greater than 25:1, the VD results remained the same, 5, 4, 4, 4, 4 for the HFC, NWHFC, NSP, AIC and MDL methods respectively. However, if the SNR is lower than 25:1, the VD estimated by the five methods, HFC, NWHFC, NSP and MDL was the same, 3 except for the AIC which yielded 3 for SNR 10:1 and 4 for SNR greater than 10:1. Table 17.2 summarizes the results.

Table 17.2. Estimation of VD for various SNRs

	HFC	NWHFC	NSP	AIC	MDL
SNR = 10:1	3	3	3	3	3
SNR = 15:1	3	3	3	4	3
SNR = 20:1	4	3	3	4	3
SNR = 25:1	5	4	4	4	4
SNR = 30:1	5	4	4	4	4
SNR = 35:1	5	4	4	4	4
SNR = 40:1	5	4	4	4	4

This makes sense. Due to increased noise interference the blackbrush becomes indiscernible from creosote leaves and sagebrush as shown in Chang (2000), Du (2000) and Du and Chang (2000). In this case, the three signatures will be considered to belong to one class and this class along with dry grass and red soil makes up three distinct classes in computer simulations.

17.5.2 AVIRIS and HYDYCE Image Experiments

In analogy with computer simulations, the AVIRIS image in Fig. 1.6 was used for experiments. The log eigen-spectra generated by the HFC, NWHFC and NSP-based eigen-thresholding methods are shown in Fig. 17.4(a-c) respectively with "*"s and "o"s denoted by the eigenvalues produced by $\mathbf{R}_{L\times L}$ and $\mathbf{K}_{L\times L}$. Once again, using the Neyman-Pearson test with the false alarm probability P_F set to $P_F = 0.001$, the VD predicted by the HFC, NWHFC and NSP-based eigen-thresholding methods were 4, 5 and 8 respectively.

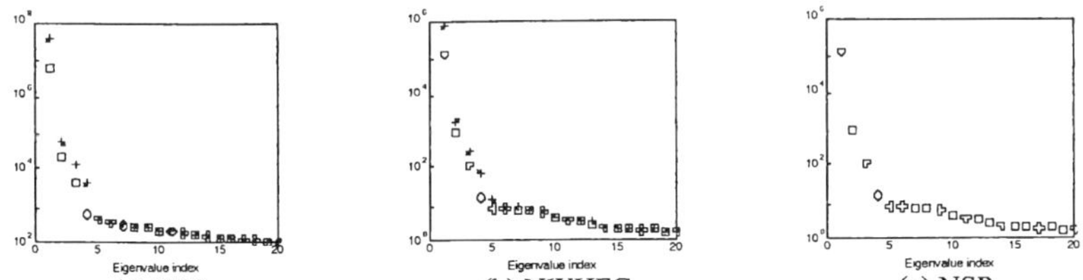

(a) HFC (b) NWHFC (c) NSP

Figure 17.4. AVIRIS image

As shown in Fig. 5.9 the unsupervised classification results produced by the UNCLS-OSP classifier detected at least 5 target signatures, cinders, playa, rhyolite, shade, vegetation in the image scene plus an anomaly at the upper edge of the dry lake. This implies that the VD must be equal or greater than 6. As a matter of fact, due to a wide coverage of the dry lake, the playa signature varied significantly from its top to bottom. Also, according to the results shown in Fig. 15.8, three areas within the dry lake, the edge, center and bottom regions were found to have different playa signatures. If these three signatures are considered to be different signal sources, the VD turns out to be 8 which is exactly the same value estimated by the NSP method. When the AIC and the MDL were further applied to the AVIRIS image scene, the estimated VD was 150 for AIC and 121 for MDL shown in Fig. 17.5. These values are significantly higher than the numbers estimated by the three eigen-thresholding based methods.

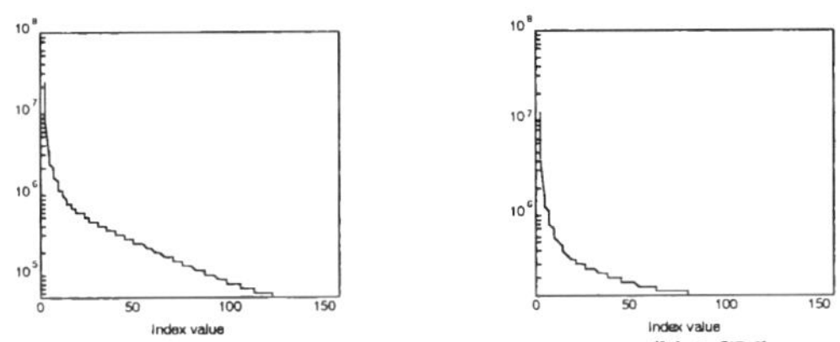

(a) AIC (b) MDL

Figure 17.5. AVIRIS image

As illustrated in computer simulations, due to the fact that the noise in remotely sensed imagery is generally not i.i.d. as usually assumed in array processing, the VD estimated by the AIC and the MDL for the AVRIS image was largely affected by this assumption. Consequently, the AIC and MDL performed poorly in the estimation of VD

Similar experiments were also conducted for the 15-panel HYDICE image in Fig. 1.7 using the HFC, the NWHFC and the NSP-based eigen-thresholding methods. Fig. 17.6(a-c) shows the distributions of the log eigenvalues with "*"s and "∘"s produced by $\mathbf{R}_{L\times L}$ and $\mathbf{K}_{L\times L}$ respectively.

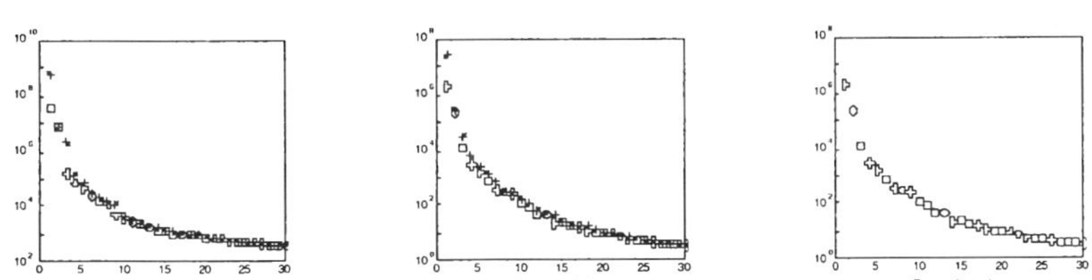

(a) HFC (b) NWHFC (c) NSP

Figure 17.6. 15-panel HYDICE image

Once again, the VD predicted by the HFC, NWHFC and NSP-based eigen-thresholding methods were 13, 15 and 20 with the false alarm probability P_F set to $P_F = 0.001$. As shown in the experiments in Example 15.5 of Section 15.4.2, these values were all adequate because all the 15 panels could be detected and separated in the fist 6 components with p = VD+1. However, in order for N_{LSRMA} to be stable, the N_{LSRMA} obtained for this 15-panel HYDICE scene was shown to be 17 for skewness or 18 for kurtosis which are very close to VD = 20 estimated by the NSP method. If we further examine the unsupervised classification results obtained by the UTGP in Fig. 5.17, the number of target signatures required to detect and separate the 15 panels was shown to be 20. These two experiments demonstrate that the NSP method produces a very good estimate for the VD.

If the AIC and the MDL were used, the estimated VD was 124 and 85 respectively shown in Fig. 17.7. Once again, the AIC and the MDL overestimated the VD. The real hyperspectral image experiments demonstrate that the AIC and the MDL may not be appropriate criteria to be used to estimate the VD of remotely sensed images.

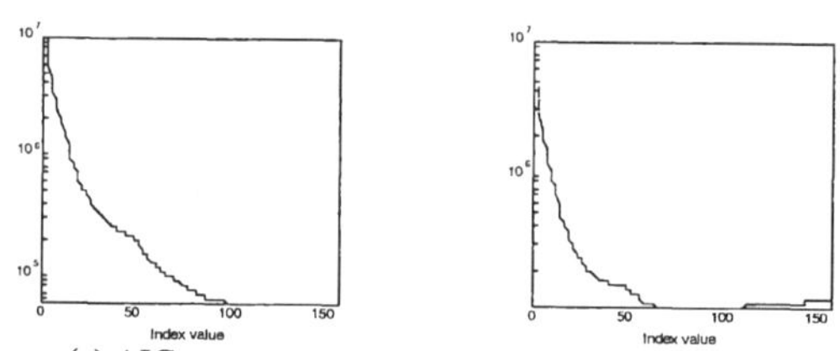

(a) AIC (b) MDL

Figure 17.7. Estimation of VD by (a) AIC and (b) MDL

Table 17.3 summaries VD results obtained by the HFC, the NWHFC, the NSP-based eigen-thresholding methods, the AIC and the MDL for AVIRIS and HYDICE images.

Table 17.3. Estimation of VD for real hyperspectral images

	HFC	NWHFC	NSP	AIC	MDL
AVIRIS image	4	5	8	150	121
15-panel HYDICE image	13	15	20	124	85

As a final comment, the value of VD only provides an estimate of how many spectrally distinct signal sources assumed to be in the image data. It is determined by intrinsic spectral properties not by a specific unsupervised method. For example, when the UNCLS-OSP method was applied to this 15-panel scene, 34 target signatures were required to generate in order to separate the 15 panels into five different target classes shown in Fig. 5.20. However, as shown in Fig. 5.17, only 20 target signatures were needed for UTGP-OSP to correctly classify these 15 panels into their own classes. As we can see from Fig. 5.20, most of 34 target signatures are either natural or background signatures. Some of these signatures may belong to the same signal sources, but were forced to split by the algorithms. So, the number of spectrally distinct signal sources is generally fewer than 34. Figs. 15.19-15.20 in Chapter 15 also confirms this. Interestingly, it was found that if the UTGP-OSP was applied to the AVIRIS scene, 12 target signatures were required to classify all five target signatures, cinders, rhyolite, shade, shade and vegetation shown in Fig. 13.5. This number was higher than 8 that was required for both the UNCLS-OSP and the LSRMA to classify all the five target signatures in Figs. 5.9 and 15.8 respectively. These real AVIRIS and HYDICE image experiments demonstrate a fact that a specific unsupervised algorithm cannot be used to determine the VD. It requires methods, which are particularly designed for this purpose. The three methods, HFC, NWHFC and NSP methods proposed in this paper seem to fit this need.

17.6 CONCLUSIONS

Determination of VD is a difficult and challenging problem. This chapter presents three Neyman-Pearson detection theory-based eigen-thresholding methods. The key idea of these methods is to formulate the VD determination as a Neyman-Pearson detection problem. The first approach was proposed by Harsanyi, Farrand and Chang (1994) which cast the eigenvalue difference between the sample correlation matrix and sample covariance matrix as a binary composite hypothesis testing problem with the Neyman-Pearson detector used as a binary decision maker. An alternative approach is also proposed, noise-whitened the HFC (NWHFC) method which incorporates a whitening process to decorrelate the sample correlation matrix. These two approaches assume that the sample size must be sufficiently large so that the covariance between eigenvalues of the sample correlation matrix and the sample covariance matrix will converge to zero. In order to alleviate this assumption, a third approach, noise subspace projection (NSP) method is further developed which can be derived from NWHFC. It makes use of an NSP-based whitening process prior to the Neyman-Pearson detection. Instead of comparing correlation-eigenvalues and covariance-eigenvalues as done in HFC and NWHFC, the NSP method only needs to deal with only the sample correlation matrix. The AIC and MDL have been widely used in passive sensor array processing to estimate the number of signals impinging on the array. In order to see their utility in estimation of VD for remotely sensed images, they were also included for comparison. For real hyperspectral image experiments, the results showed that all the three eigen-thresholding based methods, the HFC, NWHFC and NSP methods produced very close values of VD while the AIC and MDL over-estimating VD significantly. This is because that the noise in remote sensing images is generally not i.i.d., which is a crucial assumption for AIC and MDL to be effective criteria. As a final comment, the methods proposed in this chapter only provide an estimate of number of spectrally distinct signal sources assumed to be present in image data, which is generally greater than ID.

18

CONCLUSIONS AND FURTHER TECHNIQUES

As a final chapter, we will highlight several features that are considered to be unique in this book and conclude with a number of interesting techniques for which we are unable to cover in this book. The most unique feature is a wide coverage of statistical signal processing techniques that are applied to subpixel detection and mixed pixel classification with various functional taxonomies. On one end, the techniques are categorized according to the level of target knowledge-supervised, unsupervised and automatic. On another end, the techniques are classified according to constraints imposed on targets of interest-target abundance constraint and target signature constraint. A second unique feature is inclusion of sufficient details of mathematical derivations of all the presented techniques that make this book self-contained. In addition, all the relevant references are also included where they are necessary. A third feature is extensive experiments accompanied with comparative study and quantitative analysis to illustrate each described technique. A fourth unique feature is an introduction of a 3-D ROC analysis that can be used to evaluate detection and classification performance. A fifth unique feature is study on issues of sensitivity to level of target knowledge and noise, which play a significant role in design of techniques. Finally, another unique feature is demonstration of ability of the presented techniques in real-time processing through experiments.

18.1 FUNCTIONAL TAXONOMY OF TECHNIQUES

The techniques presented in this book can be categorized according to their functionalities, subpixel detection and mixed pixel classification. In general, it is difficult to draw a line between subpixel detection and mixed pixel classification. A subpixel detection technique can be used for mixed pixel classification, in which case it must be implemented one for each target class. In this case, a stack of images, each of which shows detection results of a different target, can be used for classification. On the other hand, a mixed pixel classification technique can be also used for target detection without classification. Therefore, a logical presentation for this book is to first consider subpixel detection in various aspects, and then expands subpixel detection to mixed pixel classification. The first subpixel detection technique considered is the orthogonal subspace projection (OSP), which is a supervised and unconstrained technique. This technique is then extended to two supervised target-abundance constrained techniques, sum-to-one constrained least squares (SCLS) and non-negatively constrained least squares (NCLS) methods, which impose different constraints on the abundance fractions of targets of interest. Since target abundance-constrained techniques require the complete prior target knowledge of the image data, another type of constrained techniques is developed to reduce the level of required target knowledge. Two supervised target signature-constrained techniques are presented, which only require the knowledge of targets of interest with no need of knowing image background. They are constrained energy minimization (CEM) and target-constrained interference-minimized filter (TCIMF), both of which impose constraints on vector directions of target signatures rather than target abundance fractions. In order to eliminate dependency on target knowledge, automatic subpixel detection is further considered. Two types of automatic subpixel detection are of interest. One is unsupervised subpixel detection, which automatically generates necessary target information directly from the image data by an unsupervised means. Such obtained unsupervised information is referred to as *a posteriori* target information. Three algorithms are introduced, which are unsupervised vector-quantization (UVQ), unsupervised target generation process (UTGP) and unsupervised least squares error-based algorithm. Another is anomaly detection, which does not require any *a priori* or *a posteriori* target information. Two anomaly detectors, RX detector (RXD) and low probability detector (LPD) along with their variants are studied. Since the *a posteriori* target information is obtained from the image data, it may not be accurate. In this case, the sensitivity to target knowledge and noise must be addressed. Unfortunately, this issue has not been investigated in the past. So, as a conclusion of subpixel detection, the sensitivity problem is further considered.

Following an analogous approach to subpixel detection, mixed pixel classification is presented in the same manner. It first begins with supervised and unconstrained mixed pixel classification in which a family of OSP-based classifiers is derived. It extends the OSP detector to an OSP classifier using a least-squares subspace projection approach. Three *a posteriori* OSP classifiers, signature subspace projection (SSP) classifier, target subspace projection (TSP) classifier and oblique subspace projection (OBSP) classifier are developed. A detailed experiment-based comparative study and analysis among these least-squares subspace projection based classifiers is also conducted. In order for mixed pixel classifiers to further perform material quantification, supervised fully-constrained mixed pixel classification is considered which imposes two constraints on target abundance, abundance sum-to-one constraint (ASC) and abundance non-negativity constraint (ANC). Two methods are developed for finding solutions numerically, fully constrained least squares (FCLS) method and modified fully constrained least squares (MFCLS) method, both of which include the SCLS solution as an initial estimate to produce solutions. In analogy with subpixel detection, supervised target signature-constrained mixed pixel classification techniques are also developed to extend CEM and TCIMF subpixel detection techniques to mixed pixel classification, which are linearly constrained minimum variance (LCMV) classifiers. It is known that Fisher's linear discriminant analysis has been widely used as one of most successful pattern classification techniques. This technique is further explored for hyperspectral data exploitation and extended to the linearly constrained discriminant analysis (LCDA) where the directions of the obtained linear discriminants must be aligned with mutual orthogonal directions.

Like subpixel detection, mixed pixel classification can be also carried out without *a priori* target knowledge as unsupervised mixed pixel classification and anomaly classification. Two algorithms are developed for unsupervised mixed pixel classification, which are desired target detection and classification algorithm (DTDCA) for reconnaissance applications, and automatic target detection and classification algorithm (ATDCA) for surveillance applications. Anomaly classification is a hybrid of anomaly detection and classification. It requires an additional measure for target discrimination bridging the gap between detection and classification. In addition to unsupervised mixed pixel classification and anomaly detection, a new set of mixed pixel classification techniques that are not considered in subpixel detection is projection pursuit-based techniques where neither *a priori* nor *a posteriori* target knowledge is required for classification. They use higher orders of statistics to detect and classify interesting targets in the image. Two approaches are developed along this line. One is a Kullback-Leibler information distance-based projection pursuit that use third-order and fourth-order statistics for automatic mixed pixel classification. Another is an independent components analysis-based linear spectral random mixture analysis (LSRMA). It can be considered as a random version of the linear spectral mixture analysis where the targets are assumed to be random signal sources. One major issue for automatic mixed pixel classification is how many targets are present in the image dada. This is closely related to determination of the intrinsic dimensionality (ID) of the image data. So, as a final topic in the book, we introduce a new definition of virtual dimensionality (VD), which is the number of spectrally distinct target signatures resident in the image data. Three eigen-threshold based techniques are also derived to determine appropriate VD for given image data.

18.2 MATHEMATICAL TAXONOMY OF TECHNIQUES

The techniques discussed in this book can be categorized into five groups, (1) linear mixture model-based methods, (2) linearly constrained minimum variance-based methods, (3) maximum likelihood estimation-based methods, (4) discriminant analysis-based methods and (5) projection pursuit-based methods. The first group of techniques consists of OSP and least-squares approaches, which include unconstrained or constrained methods. These are OSP, SSP, TSP, OBSP, SCLS, NCLS, FCLS and MFCLS methods discussed in Chapters 3, 8-10, 13 and 15. The techniques in this group require the complete target knowledge for the linear mixture model, which can be obtained in either a supervised or an unsupervised fashion. They take advantage of spectral information and correlation provided by the target signature matrix to achieve subpixel detection and mixed pixel classification. The second group is discussed in Chapters 4 and 11, and is made of CEM, TCIMF and LCMV methods, which constrain targets of interest while minimizing the energy of unknown signal sources. So, the techniques in this group do not need to know image background. The only required knowledge is the targets of interest, which can be also obtained *a priori* or by an unsupervised means. Since they do not have full knowledge of targets in the image data, they make use of spectral correlation among all the image pixel vectors to approximate the information that could have been provided by undesired target signatures if the complete target knowledge was given as the case for the first group of techniques. The third group contains the Gaussian maximum likelihood (GML) classifier in Chapter 8 and the RXD discussed in Chapter 6 and Chapter 14. It takes advantage of a Gaussian form to derive analytical optimal linear filters. The difference between these two techniques is that the GML classifier only uses the second-order band-to-band spectral correlation within a single image pixel vector, whereas RXD uses the second-order spectral correlation among image pixel vectors, i.e. sample spectral correlation matrix formed by all the image pixel vectors in the image data. Surprisingly, the GML classifier can be shown to be the

OBSP classifier if the noise in the linear mixture model is assumed to be white Gaussian. The techniques in fourth group are the filter-vectors (FV) classifier in Chapters 11-12 and the linearly constrained discriminant analysis (LCDA) classifier in Chapter 12. Interestingly, all the techniques in the first three groups produce linear filters, which operate on a matched filter form using different matched signatures. In particular, the LCDA classifier can be shown to be a constrained OSP classifier. The OSP and the FV classifiers can be derived as a whitened CEM in Chapter 4 (Chang, 2002a) and a whitened BRLCMV classifier in Chapter 11 (Chang, 2002b) respectively, in which case the image data are zero-mean and whitened. Recently, an information-processed matched filter approach has been proposed in Chang (2003) to interpret and illustrate these techniques based on a two-stage filter process, information processing filter followed by a matched filter. A diagram showing a mathematical taxonomy of these techniques is delineated in Fig. 18.1.

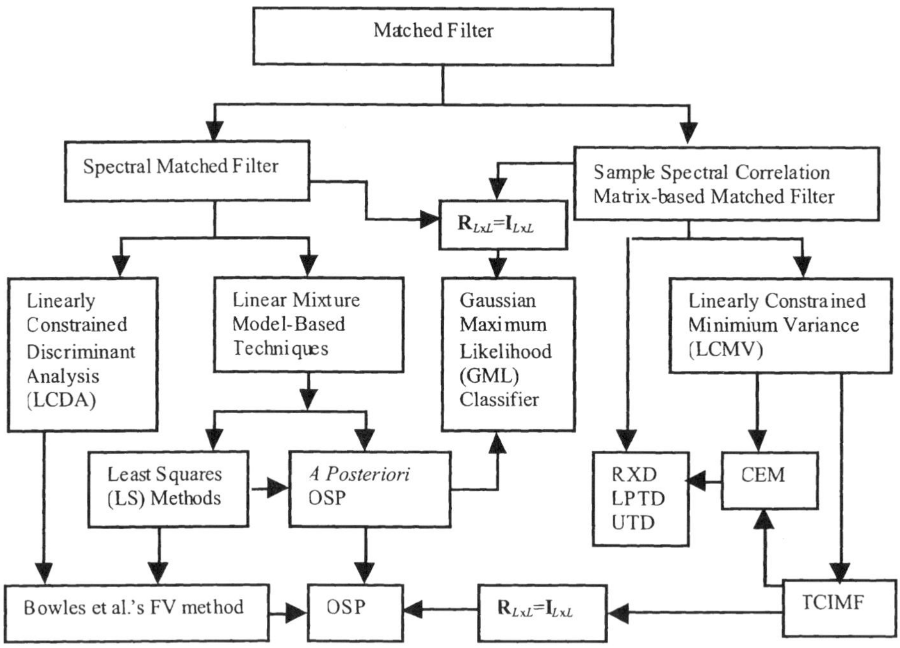

Figure 18.1. Mathematical taxonomy of techniques

In particular, several interesting techniques, OSP, CEM, LCMV, BRLCMV, LCDA, FV and RXD have been studied and analyzed in great depth. Their relationships have been also discussed in various places in the book, specifically, relationship among OSP, CEM, TCIMF in Sections 4.3 and 8.2.4, relationship between CEM and RXD in Section 6.4, relationship between BRLCMV and the FV method in Section 11.3, relationship between LCDA, the FV method in Section 12.4 and relationship between OSP and RXD in Chang (2002a, 2003). Fig. 18.2 summarizes various relationships among them. For example, LCMV is a very general target signature-constrained technique, which includes CEM, TCIMF, RXD, BRLCMV and the FV methods as its special versions, where the FV method can be considered a whitened version of BRLCMV. On one hand, OSP is a versatile technique, which can be extended in various ways. For instance, LCDA and the FV method can be treated as constrained versions of OSP, where OSP can be shown as a whitened version of CEM via TCIMF. In other words, if no sample spectral correlation matrix, $\mathbf{R}_{L\times L}$ is involved in the filter design, OSP turns out to be a linear optimal filter. All other optimal filters can be considered as its variants. On the other hand, if the sample spectral correlation matrix, $\mathbf{R}_{L\times L}$ is used to account for spectral correlation among sample image vectors, LCMV may be used as a universal filter for optimal design which yields different optimal linear filters with specific constraints.

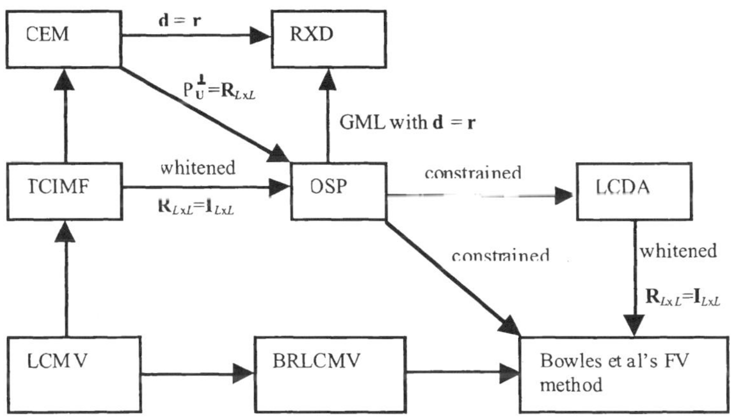

Figure 18.2. Various relationships among OSP, CEM, LCMV, TCIMF, BRLCMV, LCDA, FV and RXD

The techniques in the fifth group discussed in Chapters 15-16 are considered as high-order statistics methods, which go beyond the second-order statistics-based techniques studied in the first four groups. They utilize projection indices measured by statistical independence in Chapter 15, third and fourth moments in Chapter 16 to find interesting projections that can explore potential targets. The techniques in this group are particularly effective in detection of relatively small targets and anomalies. They can be thought of as a generalization of RXD, UTD and LPTD (Chang and Chiang, 2002), which use second-order statistics such as sample covariance/correlation matrix for anomaly detection.

18.3 EXPERIMENTS

A great number of experiments are conducted in this book to demonstrate and illustrate the utility of each presented technique. Two types of hyperspectral image data, AVIRIS and HYDICE are used for experiments. Although both data have the same spectral resolution of 10 nm, their spatial resolutions are different, 20 meters for AVIRIS and 1-4 meters for HYDICE. This difference allows us to examine more closely how spatial resolution affects the performance of spectral analysis-based non-literal techniques. One of nice features of the HYDICE data used in the experiments is that they come with ground truth maps that provide the precise spatial locations of all the targets pixels. By taking advantage of ground truth a set of custom-designed criteria can be developed for comparative analysis and quantitative study. In addition to these two data sets, which are radiance data, a set of AVIRIS laboratory reflectance data is also used for computer simulations. This data set enables us to simulate various scenarios for performance evaluation. With these three data sets we can evaluate the strengths and weaknesses of each of the techniques presented in the book as well as conditions under which how these techniques perform. Such an assessment shall provide a useful guideline for their practical implementation.

18.4 ROC ANALYSIS FOR SUBPIXEL DETECTION AND MIXED PIXEL CLASSIFICATION

Receiver operating characteristic (ROC) analysis has been widely used for signal detection in many applications. An ROC curve is a plot of detection probability (power) versus false alarm probability, which can be used to evaluate the effectiveness of a detector. It is generated by a binary hypothesis testing problem with the null hypothesis and the alternative hypothesis representing noise and signal plus noise respectively. A signal is detected only if the alternative hypothesis is true. In order to calculate detection probability and false alarm probability, it requires knowing probability distributions under each hypothesis. In reality, these two probability distributions must be estimated. If the estimated probability distributions are not accurate, using the resulting ROC curves for analysis will be misleading. In subpixel detection and mixed pixel classification the produced images are gray scale with gray level values representing abundance fraction estimates. The detection and classification is then performed by visual inspection of these abundance fractional images. In order to apply the ROC analysis these abundance fractions images must be appropriately thresholded into binary images. However, such thresholding requires a likelihood ratio test. A general approach is to assume that the noise is Gaussian. Unfortunately, the Gaussian assumption is generally not valid in remote sensing images. Another alternative is to use well-established thresholding techniques available in image processing such as Otsu's method (Otsu, 1979) entropy-based methods (Pal and Pal, 1989) etc. However, the techniques of this type are generally developed for pure-pixel images and may not be applicable to mixed-pixel remote sensing images. This book describes three techniques particularly designed to threshold gray scale abundance fractional images into binary images. Such thresholding is carried out by a process, called mixed-to-pure pixel conversion (MPCV), which converts mixed pixels in abundance fractional images to pure pixels. These three MPCV techniques are abundance percentage cut-off thresholding (a%MPCV), winner-take-all thresholding (WTAMPCV) and probability distribution-based thresholding (γ%MPCV).

Since each particular threshold value produces a pair of false probability rate and detection probability, normalizing the thresholding values to the range of [0,1] and varying the value used for thresholding from 0 to 1 results in a 3-D ROC curve where the three coordinates are specified by detection probability, false alarm probability and abundance fractions. With this interpretation a 3-D ROC analysis can be conducted. In particular, three 2-D curves can be further derived from a 3-D ROC curve and used as additional evaluation tools. One is the standard 2-D ROC curve, which is a plot of detection probability versus false alarm probability. Another is a 2-D curve plotted based on detection probability against abundance fractions. The third one is a 2-D curve, which shows a plot of false alarm probability versus abundance fractions. A combination of these three 2-D curves describes detailed interaction among the three parameters, detection probability, false alarm probability and abundance fractions.

18.5 SENSITIVITY ISSUES

Another unique and important feature in this book is investigation of sensitivity of subpixel detection and mixed pixel classification to level of target knowledge and noise.

18.5.1 Sensitivity to Level of Target Information

In many practical applications, the target knowledge is generally not available and must be obtained directly from the image data. Such generated target knowledge is referred to as *a posteriori* target information and the level of target information is determined by how closely *a posteriori* target information approximates the true target knowledge. However, since the true target knowledge is usually unknown, the techniques designed for subpixel detection and mixed pixel classification must perform robustly and insensitive to the used target information. There are three cases considered for level of target information. One is the most general case that *a posteriori* target information is not precise or accurate where the target information is either estimated or contaminated by unknown signal sources such as natural background. Another case is that the level of target information is provided by partial knowledge *a priori*, not complete prior knowledge. A third case is that the level of target information is deteriorated by interference even the target information is accurate. In this book, numerous experiments are conducted in Chapter 7 to simulate these three cases to produce various scenarios for performance evaluation.

18.5.2 Sensitivity to Noise

Noise sensitivity generally occurs in spectral filters that use the sample covariance/correlation matrix to suppress image background. Since the sample pixel vectors in image data used to form the sample covariance/correlation matrix are real image pixel vectors, noise effects must be factored into filter design. A general approach is to estimate noise covariance matrix, and then pre-white the noise prior to target detection. However, finding a reliable estimate of the noise covariance matrix poses a challenging problem. Another approach is to estimate the true rank of the sample covariance/correlation matrix, and then use the singular value decomposition to eliminate the noise subspace from the original data space. In this case, how to reliably estimate the true rank is the key to success. This is closely related to the problem of determination of the data intrinsic dimensionality. In this book, both issues are addressed in Chapter 7 and Chapter 15. A further study on determination of how many spectrally distinct signatures is also provided in Chapter 17.

18.6 REAL-TIME IMPLEMENTATION

Recent advances in computer technology have greatly improved our ability to display and process enormous amounts of digital data. These advances combined with recent advances in remote sensing sensor technology can now expand hyperspectral data applications. The need of real-time processing algorithms becomes increasingly imperative in applications of law enforcement, military operations, environmental monitoring, disaster and damage control, etc., where real-time processing becomes crucial to resolving critical situations. It is also particularly useful and effective to detect and locate moving targets such as moving vehicles, chemical vapors and plumes. So, another important unique feature in this book is to demonstrate feasibility of real-time implementation of various algorithms. Technically speaking, most of the algorithms presented in this book can be implemented on line or in real time or near real time. Not only can subpixel detection be implemented in real time in Section 4.5 of Chapter 4 (CEM, LCMV) and Chapter 6 (RXD), but also can mixed pixel classification be carried out in a real-time processing as shown in Section 10.5 of Chapter 10 (FCLS), Section 11.5 of Chapter 11 (LCMV) and Section 14.5 of Chapter 14 (RXD-LCMV anomaly classification).

18.7 FURTHER TECHNIQUES

In the past years we have witnessed a surge of fast development in algorithm design for hyperspectral imaging. Due to limited space the techniques presented in this book are mainly focused on linear unmixing methods for target detection and classification. We have not been able to address many interesting but yet important approaches to hyperspectral imaging, which were also developed in the Remote Sensing Signal and Image Processing Laboratory at UMBC. These include generalized orthogonal subspace projection approach (Ren and Chang, 2000); convex cone analysis (Ifarraguerri and Chang, 1999); projection pursuit (Ifarraguerri and Chang, 1999); (interference-annihilated eigen-analysis (Chang and Du. 1999), Kalman filtering-based linear unmixing (Chang and Brumbley 1999a, 1999b); band selection (Chang et al. 1999); linear mixture analysis-based data compression (Du et al., 2000); radial basis function neural network approach (Du and Chang, 1999, Guilfoyle et al., 2001). In what follows, we briefly describe each of these approaches and provide readers with a glimpse of their ideas.

18.7.1 Generalized Orthogonal Subspace Projection

The orthogonal subspace projection (OSP) has been shown to be a versatile technique and have also been successfully applied to a wide range of applications in hyperspectral image analysis. In order for OSP to be effective, the number of spectral bands must be greater than or equal to that of signatures to be detected or classified. This ensures that there are sufficient dimensions for orthogonal projection so that the target signatures resulting from OSP can be accommodated separately. In other words, different target signatures should be projected onto separate dimensions for detection and classification. Such inherent constraint is not an issue for hyperspectral images since they generally have hundreds of spectral bands, which is usually more than the number of signatures resident in images. However, this may not be true for multispectral images where the number of signatures to be classified is likely greater than the number of spectral bands such as 3-band SPOT (Satellite Pour l'Observation de la Terra) images. This phenomenon was first observed in Chang and Brumbley (1999a, 1999b) when OSP was applied to 3-band SPOT data and four signatures were used for classification. It was found that OSP performed poorly in discriminating four signatures using 3-band SPOT data, particularly target signatures with similar spectra. More precisely, if we want to classify target signatures effectively using OSP, each target signature requires a separate dimension for orthogonal projection. If there is a dimension used to accommodate two or more target signatures, it is impossible to discriminate these signatures using a single dimension through orthogonal projection. This constraint, referred to as Band Number Constraint (BNC) in Chang and Brumbley (1999a) points out an inherent limitation on OSP. In order to mitigate this dilemma, two approaches were proposed (Chang and

Brumbley, 1999a, 1999b). One is to replace OSP by a Kalman filtering-based linear unmixing method, which combines a linear unmixing method with a Kalman filter (Chang and Brumbley, 1999a). Since this approach includes an auxiliary state equation in its Kalman filter design, it can be considered as an extension or a generalization of OSP. Another approach is to reduce the number of signatures to be classified in order to satisfy the BNC. But this approach requires that all possible combinations of signatures be examined from which a best classification result could be selected for final classification. Even in doing so, the results may be still not good (Chang and Brumbley, 1999a). As an alternative, Ren and Chang (2000) developed a new approach, called generalized OSP (GOSP). Instead of reducing the number of target signatures, GOSP includes a band generation process, which creates extra new bands from the original spectral bands by nonlinear operations. These new images are generated in a nonlinear manner to capture nonlinear correlation between the original bands so that the classification performance can be further improved. Combining the original bands with these new nonlinearly correlated bands produces a new set of spectral bands so that the BNC can be relaxed to make OSP applicable to multispectral image classification. A similar idea using such band expansion was also applied to CEM, referred to as generalized CEM (GCEM) (Chang et al., 2000).

18.7.2 Convex Cone Analysis

Convex cone analysis was first introduced by Mavrovouniotis *et al.* (1994a, 1994b) for classification of biological samples using time-resolved pyrolysis mass spectrometry at the US Army, Edgewood Research Development Engineering Center (ERDEC). Since then it has progressed to the point where it can produce useful results on many different types of multispectral and hyperspectral images, including chemical plumes. It can be used to uncover constituents in a chemical multi-component admixture. The idea is to model individual component spectral signatures as vertices of a convex cone. The objective of the convex cone is to find a smallest possible convex cone that will embrace all data samples resulting from mixtures of vertices. In Mavrovouniotis *et al.* (1994a, 1994b), Mavrovouniotis *et al.*'s algorithm was improved by Ifarraguerri and Chang for hyperspectral image analysis (Ifarraguerri and Chang, 1999). The evidence provided by these experiments shows that the convex cone analysis can be an effective and useful technique for detection and identification of chemical vapors, plumes and aerosol clouds. For CCA to be successful, several assumptions must be satisfied. One is the assumption that the spectral mixture model is valid for the data cube or some transform thereof. The CCA also requires that there exists significant relative abundance modulation of the various spectral signatures present. In other words, if two or more endmembers are present in the same relative proportion for every pixel in which they are present, they will appear to a convex cone as one endmember with a signature equivalent to the weighted sum of the individual signatures, where the weights are the relative abundances. This is true even if the total abundance of this endmember subset varies across the image. For chemical plumes arising from, e.g. an industrial process, it can be assumed that the various gases are well-mixed, and therefore have constant relative concentration. In this case, CCA alone will not be sufficient to uncover the individual signatures of the various plume constituents. Application of a spectral filter detector such as CEM is not practical here because the number of filters needed, combined with the number of pixels in the scene will typically make the task computationally prohibitive. However, CCA can be combined with analytical spectral analysis techniques, which rely on existing infrared spectral libraries to identify the constituents. In such an approach CCA is used as a way to obtain a "clean" spectrum where the influence of the background clutter has been sharply reduced. This clean spectrum is then automatically analyzed to determine the most likely set of chemical species in the plume. Once the plume constituents have been identified, their relative column contents (concentration-pathlength product) can be obtained by unmixing, using their absorption coefficient spectra. Additionally, if the plume temperature is known or can be estimated, the absolute column contents can be obtained.

18.7.3 Kalman Filter-Based Linear Unmixing

Kalman filtering has been shown to be a very powerful technique in signal processing and control applications. It implements two linear equations, referred to as measurement (output) equation and state equation with the former describing the system output and the latter characterizing the state of the system structure. There are several advantages resulting from the Kalman filtering approach. One is that the filter can be implemented recursively. It updates the equations only based on its immediate past not its entire past history as is done in the Wiener filtering. Since the filter updates its information at every time instant, it can be applied to time-varying systems and nonstationary environments. Another is that due to its recursive structure it can be also implemented in real time. In order to take advantage of the success of the Kalman filtering in signal estimation, Chang and Brumbley (1999a, 1999b) recently investigated an approach, called Kalman filter-based linear unmixing (KFLU), which substituted the linear spectral mixture model specified by (3.1) for the measurement equation in a Kalman filter with the state equation characterized by a Gaussian–Markov process. As is often the case, the remotely sensed imagery is generally affected by many unknown signal sources, a realistic assumption is that the imagery is nonstationary. In this case, the state equation included in KFLU can capture the spectral variability from one pixel to another more effectively. Most of linear unmixing methods such as OSP are spectral processing techniques and developed on a single pixel basis. They discard the spatial correlation among pixels. The KFLU makes use of a state equation to keep track of inter-pixel correlation to improve abundance estimation derived from the measurement equation, that is the linear spectral mixture model. Although the LCMV and RX detector also utilize the sample covariance/correlation matrix, they only capture spectral correlation among sample vectors and do not account for inter-pixel spatial correlation as does KFLU in the state equation. So, they can be considered as special cases of KFLU. Details of the KFLU treatment can be found in Chang and Brumbley (1999a, 1999b). Recently, an unsupervised KFLU was proposed to extend KFLU to the case that the target knowledge could be obtained directly from the image data (Wang and Chang, 2002).

18.7.4 Interference-Annihilated Eigen-Analysis

The goal of Principal Components Analysis (PCA) is to find principal components in accordance with the maximum variance of a data matrix. However, it has been shown recently that such variance-based principal components may not adequately represent image quality. As a result, a modified PCA approach based on maximization of signal-to-noise ratio, called Maximum Noise Fraction (MNF) transformation or Noise-Adjusted Principal Components (NAPC) transform was proposed to arrange principal components in decreasing order of signal-to-noise ration (SNR) rather than variance. One of major disadvantages of this approach is that the noise covariance matrix must be estimated accurately from the data. Another is that the factor of interference is not taken into account in MNF or NAPC where the interfering effect tends to be more serious than noise in hyperspectral images. In order to address these problems, Chang et al. (1998a) and Chang and Du (1999) considered the interference as a separate unknown signal source and extended the standard signal/noise model to a signal/interference/noise model. As a result, more reliable noise estimation can be achieved by eliminating both signals and interference. More interestingly, such interference annihilation can even further improve PCA performance as shown in Chang and Du (1999). A similar idea was also investigated in Thai and Healey (2002) where they considered background as a separate signal source. If we interpret the background sources as part of interference, their signal/background/noise model is reduced to a special case of the signal/interference/noise (SIN) model. A more detailed study on comparison between these two models is given in Du and Chang (2001).

The benefit of considering interference separately has been demonstrated in Chang et al. (1998a) and Chang and Du (1999). Several advantages can be gained by interference annihilation. In hyperspectral images hyperspectral sensors can now extract many unknown signal sources, which cannot be resolved by multispectral imagers. Under this circumstance, signal-to-interference and noise ratio (SINR) criterion is more appropriate than SNR. Alternatively, SNR can be still used as a criterion after interference is annihilated. In either case, separating and eliminating these interferers from noise and signals can certainly improve traditional SNR-based methods. In the traditional signal/noise model, the role of interference is generally overlooked. It is either treated as part of the signal or included in noise. This is primarily due to the fact that it is generally difficult to obtain the knowledge of interference in practice. However, according to experiments conducted for hyperspectral image data, the interference was shown to be an important factor in target detection and classification. Accordingly, estimating the noise covariance matrix without adequately eliminating interference may lead to an inaccurate and biased estimate. In order to find and locate these unknown signal sources, an unsupervised vector quantization algorithm was proposed in Chang et al. (1998c) and an orthogonal projection-based algorithm was also used in Chang and Du (1999). These two algorithms are among the three unsupervised algorithms presented in Chapter 5. As a matter of fact, all the three can be used for the same purpose.

18.7.5 Band Selection

Hyperspectral sensors can image an area with hundreds of spectral channels at different wavelengths for identification of composition of various materials. As a result, each image scene represents an image cube with the third dimension specified by spectral range. Such 3-D representation creates enormous amounts of data for computer processing and data transmission. Band selection for remotely sensed image data is an effective means to mitigate the curse of dimensionality. Taking advantage of spectral correlation to achieve optimal band selection is one of unique features in multispectral/hyperspectral images. Many criteria have been proposed for band selection in the past to find bands that are crucial and significant in terms of information conservation (Masusel et al., 1990; Richards, 1993; Conese and Maselli, 1993). For instance, distance measures (Bhattacharyya distance, Jeffreys-Matusita distance), information theoretic approaches (divergence, transformed divergence, mutual information) and eigen-analysis (PCA) have been applied to multispectral images for optimal band selection. In particular, the use of the divergence measure for band selection has received considerable interest in multispectral imagery. More recently, the divergence was also used as a band selection criterion for hyperspectral pixel classification (Stearns et al., 1993). However, it requires computing divergences for all the possible combinations of band subsets. When the divergence measure is calculated for hyperspectral imagery with more than 200 bands, such a direct calculation becomes formidable. In order to cope with this problem, an alternative approach was reported in (Stearns et al., 1993). More recently, a new band selection method was developed in Chang et al. (1999), which comprises a band prioritization and a band decorrelation. The band prioritization prioritizes all bands in accordance with their contained information to be used for classification. Bands are then selected on the basis of their associated priorities. Since the band prioritization does not consider the spectral correlation, a band decorrelation using the divergence is used to decorrelate prioritized bands.

18.7.6 Linear Mixture Analysis-Based Data Compression

With significantly improved spatial and spectral resolution hyperspectral imagery expands capability of multispectral imagery in many ways, such as subpixel target detection, object discrimination, mixed pixel classification, material quantification, etc. It also presents new challenges to image analysts, particularly, how to effectively deal with its enormous data volume while still achieving their desired goals. One common practice is to compress data prior to image analysis. Two types of data compression can be performed, lossless and lossy according to redundancy removal. More specifically, lossless data compression is generally referred to as data compaction, which eliminates *unnecessary* redundancy without loss of any information. By contrast, lossy data compression removes *unwanted* redundancy or *insignificant* information, which results in entropy reduction. Which compression is appropriate is determined by different applications. For example, in medical imaging, lossless compression is preferred to lossy compression in order to avoid potential lawsuits against doctors. However, in this case, only small compression ratios can be achieved, generally less than 3:1. On the other hand, video processing can be benefited from lossy compression, such as HDTV (High Definition TV). For remotely sensed imagery, both types of compression can be beneficial and have been studied and investigated extensively in the past. If our main interest is applications in target detection and image classification, the accuracy in detection and classification is generally determined by how we can effectively use target features rather than the entire data. As a result, lossless compression does not provide additional advantages over lossy compression in the sense of feature extraction.

Recently, a linear spectral mixture analysis-based orthogonal projection (OP) filter was proposed to compress AVIRIS data (Farison et al., 1997). The proposed OP filter was based on feature extraction and is the same OSP classifier discussed in Chapter 8. This approach directly deals with the abundance fractional images, which may be more effective than with the whole stack of images in a hyperspectral image cube. Unfortunately, the OSP filter approach in Farison et al. (1997) is an unconstrained linear unmixing method and can not accurately estimate the abundance fractional images to represent desired features as shown in Chapter 9. In order to overcome this problem, a UFCLS-based hyperspectral compression approach was developed in Du et al. (2000) where UFCLS has been discussed in Chapter 10. It differs from the approach in Farison et al. (1997) in three important aspects. It is a fully constrained LSMA-based compression method as opposed to the unconstrained OP filter-based method used in Farison et al. (1997). Additionally, compared to the OP filter-based method, the UFCLS-based approach is completely unsupervised. Most importantly, it generates a set of target fractional abundance images in an unsupervised manner and compresses image data by encoding these target fractional abundance images. Since only targets along with their corresponding abundance fractions are required for encoding, a high compression ratio can be achieved. More study on this topic is currently under investigation.

18.7.7 Radial Basis Function Neural Network Approach

Historically, linear spectral mixture analysis (LSMA) has been widely used to describe mixed spectra collected by hyperspectral imagers or laboratory based spectrometers. However, on some occasions, a non-linear mixture model may better describe the mixture spectra for certain endmember distributions (Guilfoyle et al., 2001, 2002). It has been shown by Mustard and Pieters (1989) that the use of a non-linear mixture model based on Hapke's bi-directional spectroscopy theory (1981), referred to as intimate mixture model, improves abundance estimates in intimate mixtures of soils. In order to evaluate both of these mixing models, a testbed set is developed in which both models can be easily applied and verified. A linear mixture model is often used to describe those situations in which the endmember components are distributed in block-like areas within the field-of-view of the instrument. These situations may occur when the spectrometer field-of-view passes over discrete regions such as fields, lakes, rivers and forests. In these cases, one would expect the resultant reflectance spectrum to be a linear combination of endmembers present in the region, for example, a river water component and perhaps a soil component (for the river bank). The linear model assumes that source radiation is singly reflected from the endmember substances and then collected by the imaging spectrometer.

Non-linear mixtures occur in situations where endmember components are randomly distributed throughout the field-of-view of the instrument. Such situations may occur, for example, when viewing striated soils, in areas where multiple rock types are all visible on the region's surface, or in identifying trees in a forest (assuming reflectance spectra differences exist). In these cases, the resultant mixture reflectance spectrum may best be described by assuming that source radiation is multiply scattered by the randomly distributed endmembers before being collected by the imaging spectrometer. Recently Guilfoyle, et al. (2001, 2002) conducted experiments in such a manner that linear and non-linear mixtures have been created with colored sand according to the above descriptions. The resultant mixed spectra are analyzed by a radial basis function neural network (RBFNN) developed by Leung and Lo (1995). The results show that the non-linear model does match the randomly distributed mixture spectra, while the linear model better matches the discrete region mixture (Mustard and Pieters, 1989).

18.8 APPLICATIONS TO MAGNETIC RESONANCE IMAGING

As a final section of this chapter, we conclude this book by a discussion of applications of hyperspectral imaging techniques to magnetic resonance imaging (MRI). Nuclear magnetic resonance (NMR) has recently developed as a versatile technique in many fields such as chemistry, physics, engineering because its signals provide rich information about material structures that involve the nature of a population of atoms, the structure of their environment, and the way in which the atoms interact with environment. When NMR is applied to human anatomy, NMR signals can be used to measure the nuclear spin density, the interactions of the nuclei with their surrounding molecular environment and those between close nuclei, respectively. It produces a sequence of multiple spectral images of tissues with a variety of contrasts using three magnetic resonance (MR) parameters, spin-lattice (T1), spin-spin (T2) and dual echo-echo proton density (PD). By appropriately choosing pulse sequence parameters, echo time (TE) and repetition time (TR) a sequence of images of specific anatomic area can be generated by pixel intensities that represent characteristics of different types of tissues throughout the sequence. As a result, magnetic resonance imaging (MRI) becomes a more useful image modality than X-ray computerized tomography (X-CT) when it comes to analysis of soft tissues and organs since the information about T1 and T2 offers a more precise picture of tissue functionality than that produced by X-CT. Interestingly, such obtained MR images can be viewed as image sequences acquired by different pulse sequence parameters remotely. So, an MR image pixel can be considered as a multispectral image pixel vector, of which each component represents an image pixel acquired by a particular band that is specified by a pulse sequence parameter. With this interpretation Harsanyi and Chang applied the eigen-image technique developed for MRI (Miller et al., 1992) to derive their OSP technique (Harsanyi and Chang, 1994) for hyperspectral image classification. Most recently, Du and Chang (2000) further extended an MR image classification technique developed by Soltanian-Zadeh et al. (1996) to LCDA for constrained mixed pixel classification. Despite great success in applying MRI to hyperspectral imaging, the opposite direction seems not explored and yet to be investigated. In a series of recent papers by Wang et al. (Wang, et al., 2002a, 2002b, 2002c) and Wang's Ph.D. dissertation (Wang, 2002) applications of hyperspectral imaging techniques to MRI have been studied. They first applied the OSP discussed in Chapters 3 and 8 to MR image classification (Wang et al., 2002a) and later extended it to an unsupervised (UOSP) (Wang et al., 2002b) where the DTDCA described in Section 13.3 of Chapter 13 was used as an unsupervised algorithm to generate unknown objects in MR images. More details about applications of OSP and UOSP can be found in Wang (2002). In order to minimize effects caused by unknown brain tissues in the image background, Wang et al. (2003) further applied CEM to detection of spectral features in MR images and has shown that CEM could be a promising technique for MRI when the knowledge of MR image background is not available. Specifically, as shown in Wang et al. (2002c) the 3-D ROC analysis proposed in Section 15.5 of Chapter 15 provided an effective tool for performance evaluation. The work in Wang et al. (2002a, 2002b, 2002c) demonstrates that the potential of hyperspectral imaging techniques presented in this book is not necessarily limited to applications of remote sensing image processing. Other applications such as MRI can be also benefited from these techniques as well.

GLOSSARY

AASC: Absolute Abundance Sum-to-one Constraint, Chapters 10
AIC: An Information Criterion, Chapter 17
AMPC: Automatic Mixed Pixel Classification, Chapters 13-16
ANC: Abundance Nonnegativity Constraint, Chapters 3, 10
a%MPCV: *a*% Mixed-to-Pure Conversion (Converter), Chapters 9, 11, 12, 15
ASC: Abundance Sum-to-one Constraint, Chapters 3, 10
ASD: Automatic Subpixel Detection, Chapters 5-6
ATDCA: Automatic Target Detection and Classification Algorithm, Chapter 13
AVIRIS: Airborne Visible/Infrared Imaging Spectrometer, Chapter 1
BRLCMV: Background Removed Linearly Constrained Minimum Variance, Chapter 11
CBD: City Block Distance, Chapter 2
CCA: Convex Cone Analysis, Chapter 18
CEM: Constrained Energy Minimization, Chapters 4, 7, 11
CMD: Covariance-based Mahalanobis Distance, Chapter 14
CMFD: Covariance-based Matched Filter Distance, Chapter 14
CRXD: Causal RXD, Chapter 6
DTDCA: Desired Target Detection and Classification Algorithm, Chapter 13
EA: Evolutionary Algorithm, Chapter 16
ED: Euclidean Distance, Chapters 2, 9
FCLS: Fully Constrained Least-Squares method, Chapter 10
FIR: Finite Impulse Response
FNNLS: Fast NNLS method, Chapter 3
FNNLSb: A second version of FNNLS, Chapter 3
FV: Filter Vectors Approach developed by Bowels et al., Chapters 11, 12
γ: Confidence coefficient, Chapters 15-16
γMPCV: γMixed-to-Pure Conversion (Converter), Chapters 15-16
GCEM: Generalized Constrained Energy Minimization, Section 18.7.1
GML: Gaussian maximum likelihood, Chapter 3, 8
GOSP: Generalized Orthogonal Subspace Projection, Section 18.7.1
GSD: Ground Sampling Distance
HFC: Harsanyi-Farrand-Chang, Chapter 17
HMM: Hidden Markov Model, Chapter 2
HMMID: Hidden Markov Model-Based Information Divergence, Chapter 2
HYDICE: HYperspectral Digital Imagery Collection Experiment, Chapter 1
IAD: IntrA-Distance, Chapter 12
ICA: Independent Component Analysis, Chapter 15
ID: Intrinsic Dimensionality, Chapter 17
IED: IntEr-Distance, Chapter 12

JMD: Jeffries-Matusita Distance, Chapter 2
LBG: Linde-Buzo-Gray
LCDA: Linearly Constrained Discriminant Analysis, Chapter 12
LCMV: Linearly Constrained Minimum Variance, Chapters 4, 11
LCVF: Lunar Crater Volcanic Field, Chapter 1
LDA: Linear Discriminant Analysis, Chapter 9
LPD: Low Probability Detection/Detector, Chapter 6
LPTD: Low Probability Target Detector, Chapter 6
LSE: Least-Squares Error, Chapters 5, 10
LSMA: Linear Spectral Mixture Analysis, Chapters 3, 8
LSRMA: Linear Spectral Random Mixture Analysis, Chapter 15
MD: Mahalanobis Distance, Chapter 2
MFCLS: Modified Fully Constrained Least-Squares method, Chapter 10
MLC: Maximum Likelihood Classifier, Chapters 3, 8
MNF: Maximum Noise Fraction, Chapter 18
MPC: Mixed Pixel Classification, Chapters 8-16
MPCV: Mixed-to-Pure pixel ConVerter (Conversion), Chapter 9
MRI: Magnetic Resonance Imaging, Section 18.8
MRXD: Modified RX Detector, Chapter 6
MTCEM: Multiple-Target Constrained Energy Minimization, Chapter 11
MVDR: Minimum Variance Distortionless Response, Chapter 4
NAPC: Noise Adjusted Principal Component, Chapter 18
NCLS: Nonnegativity Constrained Least-Squares method, Chapters 3, 5, 10
NNLS: NonNegative Least-Squares method, Chapter 3
NNNLS: Normalized NonNegative Least-Squares method, Chapter 10
NNR: Nearest Neighbor Rule, Chapter 5
NRXD: Normalized RX Detector, Chapter 6
NSCLS: Normalized Sum-to-one Constrained Least-Squares, Chapter 10
NSP: Noise Subspace Projection, Chapter 17
NWHFC: Noise-Whitened Harsanyi-Farrand-Chang, Chapter 17
OBSP: OBlique Subspace Projection, Chapter 8
OPCI: Orthogonal Projection Correlation Index, Chapter 13
OPD: Orthogonal Projection Distance, Chapter 2
OSP: Orthogonal Subspace Projection, Chapter 3
PCA: Principal Components Analysis, Chapter 6
PCLS: Partially Constrained Least-Squares method(s), Chapter 3
PP: Projection Pursuit
PPC: Pure Pixel Classification, Chapter 9
PPEA: Projection Pursuit-based Evolutionary Algorithm, Chapter 16
RMD: Correlation-based Mahalanobis Distance, Chapter 14
RMFD: Correlation-based Matched Filter Distance, Chapter 14
ROC: Receiver Operating Characteristics
RSDE: Relative Spectral Discriminatory Entropy, Chapter 2
RSDPB: Relative Spectral Discriminatory ProBability, Chapter 2
RSDPW: Relative Spectral Discriminatory PoWer, Chapter 2
RXD: RX Detector, Chapters 6, 14
SAM: Spectral Angle Mapper, Chapter 2
SCEM: Sum Constrained Energy Minimization, Chapter 11
SCLS: Sum-to-one Constrained Least-Squares, Chapter 3
SD: Subpixel Detection, Chapters 3-7

SID: Spectral Information Divergence, Chapter 2
SIM: Spectral Information Measure, Chapter 2
SSP: Signature Subspace Projection, Chapter 8
TACMPC: Target Abundance-Constrained Mixed Pixel Classification, Chapter 10
TCIM: Target-Constrained Interference-Minimized, Chapter 11
TCIMF: Target-Constrained Interference-Minimized Filter, Chapters 4, 11
TD: Tchebyshev Distance, Chapter 2
TSCMPC: Target signature-Constrained Mixed Pixel Classification, Chapters 11, 12
TSP: Target Subspace Projection, Chapter 8
UCEM: Unsupervised Constrained Energy Minimization, Chapter 11
UFCLS: Unsupervised Fully Constrained Least-Squares, Chapter 10
UNCLS: Unsupervised Nonnegativity Constrained Least-Squares, Chapter 5
UOSP: Unsupervised Orthogonal Subspace Projection, Chapter 13
UTD: Uniform Target Detector, Chapters 6,14
UTGP: Unsupervised Target Generation Process, Chapter 13
UVQ: Unsupervised Vector Quantization, Chapter 5
VD: Virtual Dimensionality, Chapter 17
WTA: Winner-Take-All, Chapter 9
WTACEM: Winner-Take-All Constrained Energy Minimization, Chapter 11
WTAMPCV: Winner-Take-All Mixed-to-Pure Conversion (Converter), Chapter 9

REFERENCES

Adams, J.B. and M.O. Smith (1986), "Spectral mixture modeling: a new analysis of rock and soil types at the Viking lander 1 suite," *J. Geophysical Res.*, vol. 91, no. B8, pp. 8098-8112, July 10, 1986.

Adams, J.B., M.O. Smith and A.R. Gillepie (1989), "Simple models for complex natural surfaces: a strategy for hyperspectral era of remote sensing," *Proc. IEEE Int. Geoscience and Remote Sensing Symposium'89*, pp. 16-21, 1989.

Adams, J.B., M.O. Smith, and A.R. Gillespie (1993), "Image spectroscopy: interpretation based on spectral mixture analysis," *Remote Geochemical Analysis: Elemental and Mineralogical Composition*, edited by C.M. Pieters and P.A. Englert, Cambridge University Press, pp. 145-166, 1993.

Akaike, H. (1974), "A new look at the statistical model identification," *IEEE Trans. Automatica Control,* vol. AC-19, pp. 716-723, 1974.

Althouse, M.L.G. and C.-I Chang (1991), "Chemical vapor detection with a multispectral thermal imager," *Optical Engineering*, vol. 30, no. 11, pp. 1725-1733, July 1991. (invited paper)

Amari, S. (1999), "Natural gradient learning for over-and under-complete bases in ICA," Neural Computations, vol. 11, pp. 1875-1883, 1999.

Anderson, T.W. (1984), *An Introduction to Multivariate Statistical Analysis*, 2nd. ed., New York: John Wiley & Sons, 1984.

Ashton, E.A. and A. Schaum (1998), "Algorithms for the detection of sub-pixel targets in multispectral imagery," *Photogrammetric Engineering and Remote Sensing*, pp. 723-731, July 1998.

Bayliss, J., J.A. Gualtieri and R.F. Cromp (1997), "Analyzing hyperspectral data with independent component analysis," *Proceedings of SPIE*, vol. 3240, pp. 133-143, 1997.

Behrens, R.T. and L.L. Scharf (1994), "Signal processing applications of oblique projections operators," *IEEE Trans. on Signal Processing*, vol. 42, no. 6, pp. 1413-1423, June 1994.

Bell, A.J. and T.J. Sejnowski (1995), "An information-maximization approach to blind separation and blind deconvolution," Neural Computation, vol. 7, pp. 1129-1160, 1995.

Boardman, J.W. (1989), "Inversion of imaging spectrometry data using singular value decomposition," *Proc. IEEE Sym. Geoscience and Remote Sensing*, pp. 2069-2072, 1989.

Boardman, J.W. (1990), "Inversion of high spectral resolution data," *Proc. SPIE*, vol. 1298, pp. 222-233, 1990.

Boardman, J.W. (1993), "Automated spectral unmixing of AVIRIS data using convex geometry concepts," *Summaries, Fourth JPL Airborne Geosci. Workshop*, JPL Publication 93-26, vol. 1, p. 11-14, 1993.

Boardman, J.W. (1994)., "Geometric mixture analysis of imaging spectrometry data," *Int. Geosci. Rem. Sens. Symposium*, vol. 4, pp. 2369-71, 1994.

Boardman, J.W. (1998), "Leveraging the high dimensionality of AVIRIS data for improved sub-pixel target unmixing and rejection of false positive: mixture tuned matched filtering," *Summaries of Seventh Annual JPL Earth Science Workshop*, JPL Publication 98-4, vol. 1, 1998.

Bowles, J., P. Palmadesso, J. Antoniades and M. Baumback (1995), "Use of filter vectors in hyperspectral data analysis," SPIE, vol. 2553, pp. 148-157, 1995.

Bro, R and S.D. Jong (1997), "A fast non-negativity-constrained least squares algorithm," *J. Chemometrics*, vol. 11, pp. 393-401, 1997.

Brumbley, C. (1998), *Kalman Filtering and Subspace Projection Approaches to Multispectral and Hyperspectral Image Classification*, Department of Computer Science and Electrical Engineering, University of Maryland, Baltimore County, MD, May 1998.

Brumbley, C. and C.-I Chang (1999), "An unsupervised vector quantization-based target signature subspace projection approach to classification and detection in unknown background," *Pattern Recognition*, vol. 32, no. 7, pp. 1161-1174, July 1999.

Bullock, M.E., T.J. Patterson and S.R. Fairchild (1994), "Design of optimal transformation for multispectral change detection using projection pursuit," *SPIE*, vol. 2231, pp. 91-102, 1994.

Cardoso, J.-F. and B.H. Laheld (1996), "Equivalent adaptive source separation," *IEEE Trans. Signal Process.*, vol. 44, no. 12, pp. 3017-3030, Dec. 1996.

Chang, C.-I (1998a), "Further results on relationship between spectral unmixing and subspace projection," *IEEE Trans. on Geoscience and Remote Sensing*, vol. 36, no. 3, pp. 1030-1032, May 1998.

Chang, C.-I (1998b), "A study on spectral characteristics for HYDICE data by spectral similarity and discriminability on a single pixel basis," final report, Desert Research Institute, July 1998.

Chang, C.-I (1999), "Least squares error theory for linear mixing problems with mixed pixel classification for hyperspectral imagery," *Recent Research Developments in Optical Engineering*, ed. S.G. Pandalai, vol. 2, 1999, pp. 241-268, Trivandrum, Kerala: Research Signpost, India.

Chang, C.-I (2000), "An information theoretic-based approach to spectral variability, similarity and discriminability for hyperspectral image analysis," *IEEE Trans. on Information Theory*, vol. 46, no. 5, pp. 1927-1932, August 2000.

Chang, C.-I (2002a), "Relationship among orthogonal subspace projection, constrained energy minimization and RX-algorithm," *SPIE Conf. on Algorithms and Technologies for Multispectral, Hyperspectral and Ultraspectral Imagery VIII*, SPIE 4725, Orlando, Florida, 1-5 April, 2002.

Chang, C.-I (2002b), "Target signature-constrained mixed pixel classification for hyperspectral imagery," *IEEE Trans. on Geoscience and Remote Sensing*, vol. 40, no. 2, pp. 1065-1081, May 2002.

Chang, C.-I (2003), An Information-Processed Matched Filter Approach to Hyperspectral Detection and Classification Algorithms, *Computer Vision Beyond the Visible Spectrum* edited by I. Pavlidis and B. Bhanu, Springer-Verlag, 2003.

Chang, C.-I and C. Brumbley (1997), "An orthogonalization target signature space projection approach to image classification in unknown background," *31st Conference on Information Sciences and Systems*, The Johns Hopkins University, pp. 174-178, March, 1997

Chang, C.-I and C. Brumbley (1999a), "A Kalman filtering approach to multispectral image classification and detection of changes in signature abundance," *IEEE Trans. on Geoscience and Remote Sensing*, vol. 37, no. 1, pp. 257-268, January 1999.

Chang, C.-I and C. Brumbley (1999b), "Linear unmixing Kalman filtering approach to signature abundance detection, signature estimation and subpixel classification for remotely sensed images," *IEEE Trans. Aerospace and Electronics Systems*, vol. 37, no 1, pp. 319-330, January 1999.

Chang, C.-I and S.-S. Chiang (2002), "Anomaly detection and classification for hyperspectral imagery," *IEEE Trans. on Geoscience and Remote Sensing*, vol. 40, no. 2, pp. 1314-1325, February 2002.

Chang, C.-I and Q. Du (1999a), "A noise subspace projection approach to determination of intrinsic dimensionality for hyperspectral imagery," *EOS/SPIE Symposium on Remote Sensing, Conference on Image and Signal Processing for Remote Sensing V*, SPIE vol. 3871, Florence, Italy, pp. 34-44, September 20-24, 1999.

Chang, C.-I and Q. Du (1999b), "Interference and noise adjusted principal components analysis," *IEEE Trans. on Geoscience and Remote Sensing*, vol. 37, no. 5, pp. 2387-2396, September 1999.

Chang, C.-I and D. Heinz (2000a), "Constrained subpixel target detection for hyperspectral imagery," *SPIE Conf. on Signal and Data Processing of Small Targets 2000*, Orlando, FL, pp. 35-45, April 2000.

Chang, C.-I and D. Heinz (2000b), "Constrained subpixel detection for remotely sensed images," *IEEE Trans. on Geoscience and Remote Sensing*, vol. 38, no. 3, pp. 1144-1159, May 2000.

Chang, C.-I and H. Ren (1999a), *Computer-Assisted Target Detection and Classification for Hyperspectral Imagery,* June 29 1999.

Chang, C.-I and H. Ren (1999b), "Linearly constrained minimum variance beamforming for target detection and classification in hyperspectral imagery," *IEEE 1999 International Geoscience and Remote Sensing Symp.*, Hamburg, Germany, pp. 1241-1243, 28 June-2 July, 1999.

Chang, C.-I and H. Ren (2000a), "An experiment-based quantitative and comparative analysis of hyperspectral target detection and image classification algorithms," *IEEE Trans on Geoscience and Remote Sensing*, vol. 38, no. 2, pp. 1044-1063, March 2000.

Chang, C.-I and H. Ren (2000b), *Linearly Constrained Minimum Variance Beamforming Methods for Remote Sensing Image Analysis,* Disclosure of invention, University of Maryland, Baltimore County, May, 2000.

Chang, C.-I, T.-L.E. Sun and M.L.G. Althouse (1998a), "An unsupervised interference rejection approach to target detection and classification for hyperspectral imagery," *Optical Engineering*, vol. 37, no. 3, pp. 735-743, March 1998.

Chang, C. I, X. Zhao, M.L.G. Althouse and J.-J. Pan (1998b), "Least squares subspace projection approach to mixed pixel classification in hyperspectral images," *IEEE Trans on Geoscience and Remote Sensing*, vol. 36, no. 3, pp. 898-912, May 1998.

Chang, C.-I, Q. Du, T.S. Sun and M.L.G. Althouse (1999), "A joint band prioritization and band decorrelation approach to band selection for hyperspectral image classification," *IEEE Trans. on Geoscience and Remote Sensing*, vol. 37, no. 6, pp. 2631-2641, November 1999.

Chang, C.-I, J.-M. Liu, B.-C. Chieu, C.-M. Wang, C. S. Lo, P.-C. Chung, H. Ren, C.-W. Yang, D.-J. Ma (2000), "A generalized constrained energy minimization approach to subpixel target detection for multispectral imagery," *Optical Engineering*, vol. 39, no. 5, pp. 1275-1281, May 2000.

Chang, C.-I, H. Ren and S.S. Chiang (2001a), "Real-time processing algorithms for target detection and classification in hyperspectral imagery," *IEEE Trans. on Geoscience and Remote Sensing*, vol. 39, no. 4, pp. 760-768, April 2001.

Chang, C.-I, S.-S. Chiang and I.W. Ginsberg (2001b), "Anomaly detection in hyperspectral imagery," *SPIE Conf. on Geo-Spatial Image and Data Exploitation II*, Orlando, pp. 43-50, Florida, 20-24 April, 2001.

Chang, C.-I, Q. Du, S.-S. Chiang, D. Heinz and I.W. Ginsberg (2001c), "Unsupervised subpixel target detection in hyperspectral imagery," *SPIE Conf. on Algorithms for Multispectral, Hyperspectral and Ultraspectral Imagery VII*, Orlando, Florida, pp. 370-379, 20-24 April, 2001.

Chang, C.-I, H. Ren, Q. Du, S-S. Chiang and A. Ifarraguerri (2001d), "An ROC analysis for subpixel detection," *IEEE 2001 International Geoscience and Remote Sensing Symp.*, Sydney, Australia, July 9-13, 2001.

Chang, C.-I, S.S. Chiang, J.A. Smith and I.W. Ginsberg (2002), "Linear spectral random mixture analysis for hyperspectral imagery," *IEEE Trans. on Geoscience and Remote Sensing*, vol. 40, no. 2, pp. 375-392, February 2002.

Chen, S, C.F.N. Cowan and P.M. Grant (1991), "Orthogonal least squares learning algorithm for radial basis function networks," *IEEE Transaction on Neural Networks*, vol. 2, pp. 302-309, 1991.

Chiang, S.-S. and C.-I Chang (1999), "Target subpixel detection for hyperspectral imagery using projection pursuit," *EOS/SPIE Symposium on Remote Sensing, Conference on Image and Signal Processing for Remote Sensing V*, SPIE vol. 3871, Florence, Italy, pp. 107-115, September 20-24, 1999.

Chiang, S.-S. and C.-I Chang (2001), "Discrimination measures for target classification," *IEEE 2001 International Geoscience and Remote Sensing Symp.*, Sydney, Australia, July 9-13, 2001.

Chiang, S.-S., C.-I Chang and I.W. Ginsberg (2000), "Unsupervised hyperspectral image analysis using independent components analysis," *IEEE 2000 International Geoscience and Remote Sensing Symp.*, Hawaii, USA, July 24-28, 2000.

Chiang, S.-S., C.-I Chang and I.W. Ginsberg (2001), "Unsupervised subpixel target detection for hyperspectral images using projection pursuit," *IEEE Trans. on Geoscience and Remote Sensing*, vol. 39, no. 7, pp. 1380-1391, July 2001.

Cichocki, A. and R. Unbehauen (1993), *Neural Networks for Optimization and Signal Processing*, New York: Wiley, pp. 364-367, 1993.

Comon, P. (1994), "Independent component analysis, A new concept?", *Signal Processing*, vol. 36, pp. 287-314, 1994.

Conese, C. and F. Maselli (1993), "Selection of optimum bands from TM scenes through mutual information analysis," *ISPRS J. of Photogrammetry and Remote Sensing*, vol. 48, no. 3, pp. 2-11, 1993.

Cover and J. Thomas (1991), *Elements of Information Theory*, New York: John Wiley and Sons, 1991.

Craig, M.D. (1994), "Minimum-volume transforms for remotely sensed data," *IEEE Trans. Geosci. Rem. Sens.*, vol. 32, no. 3, pp. 542-552, 1994.

DARPA Spectral Exploitation Workshop (1996), the Defense Advanced Research Projects Agency, Annapolis, MD. July 1-2, 1996.

Dasgupta, D. and Z. Michalewicz (1997), *Evolutionary Algorithms in Engineering Applications*, Springer-Verlag.

Duda, R.O. and P.E. Hart (1973), *Pattern Classification and Scene Analysis*, New York: John Wiley & Sons, 1973.

Du, Q. (2000), *Topics in Hyperspectral Image Analysis*, Department of Computer Science and Electrical Engineering, University of Maryland, Baltimore County, MD, May 2000.

Du, Q. and C.-I Chang (1998), "Radial basis function neural networks approach to hyperspectral image classification," *1998 Conference on Information Science and Systems*, Princeton University, Princeton, NJ, pp. 721-726, March 1998.

Du, Q. and C.-I Chang (1999), "An interference rejection-based radial basis function neural network approach to hyperspectral image classification," *International Joint Conference on Neural Network*, Washington DC, pp. 2698-2703, July 1999.

Du, Q. and C.-I Chang (2000), "A hidden Markov model-based spectral measure for hyperspectral image analysis," *SPIE Conf. Algorithms for Multispectral, Hyperspectral, and Ultraspectral Imagery VI*, Orlando, FL, pp. 375-385, April 2000.

Du, Q. and C.-I Chang (2001a), "A linear constrained distance-based discriminant analysis for hyperspectral image classification," *Pattern Recognition*, vol. 34, no. 2, 2001.

Du, Q. and C.-I Chang (2001b), "An interference subspace projection approach to subpixel target detection," *SPIE Conf. on Algorithms for Multispectral, Hyperspectral and Ultraspectral Imagery VII*, Orlando, Florida, pp. 570-577, 20-24 April, 2001.

Du, Q. and H. Ren (2002), "On relationship between OSP and CEM," *SPIE Conf. on Algorithms for Multispectral, Hyperspectral and Ultraspectral Imagery VIII*, Orlando, Florida, 20-24 April, 2002.

Du, Q., C.-I Chang, D.C. Heinz, M. L.G. Althouse and I.W. Ginsberg (2000), "Hyperspectral image compression for target detection and classification," *IEEE 2000 International Geoscience and Remote Sensing Symp.*, Hawaii, USA, July 24-28, 2000.

Fano, R.M. (1961), *Transmission of Information: A Statistical Theory of Communication*, John Wiley & Sons, N.Y., 1961.

Farrand, W., and J.C. Harsanyi (1997), "Mapping the distribution of mine tailing in the coeur d'Alene river valley, Idaho, through the use of constrained energy minimization technique," *Remote Sensing of Environment*, vol. 59, pp. 64-76, 1997.

Friedman, J.H. and J.W. Tukey (1974), "A projection pursuit algorithm for exploratory data analysis," *IEEE Transactions on Computers*, vol. c-23, no. 9, pp. 881-889, 1974.

Friedman, J.H. (1987) "Exploratory projection pursuit," *Journal of American Statistical Association*, 82, pp. 249-266, 1987.

Frost III, O.L. (1972), "An algorithm for linearly constrained adaptive array processing," *Proc. IEEE*, vol. 60, pp. 926-935, 1972.

Fukunaga, K. (1982), "Intrinsic Dimensionality Extraction", *Classification, Pattern Recognition and Reduction of Dimensionality*, Handbook of Statistics, vol. 2, P.R. Krishnaiah and L.N. Kanal eds., Amsterdam: North-Holland Publishing Company, 1982, pp. 347-360.

Fukunaga, K (1992), *Statistical Pattern Recognition*, 2nd ed., New York: Academic Press, 1992.

Gillespie, A.R., M.O. Smith, J.B. Adams, S.C. Willis, A.F. Fischer, III and D.E. Sabol (1990), "Interpretation of residual images: spectral mixture analysis of

AVIRIS images, Owens valley, California," *Proc. 2nd AVIRIS Workshop*, pp. 243-270, 1990.

Girolami, M (2000), *Self-Organising Neural Networks: Independent Component Analysis and Blind Source Separation*, , Springer-Verlag, 2000.

Goetz, A.F.H. and J.W. Boardman (1989), "Quantitative determination of imaging spectrometer specifications based on spectral mixing models," *Proc. IEEE Int. Geoscience and Remote Sensing Symposium'89*, pp. 1036-1039, 1989.

Golderg, D.E. (1989), *Genetic Algorithms in Search, Optimization and Machine Learning*, Addison-Wesley, 1989.

Golub, G.H. and G.F. Van Loan (1989), *Matrix Computations*, 2nd. ed., John Hopkins University Press, 1989.

Green, A.A. M. Berman, P. Switzer and M.D. Craig (1988), "A transformation for ordering multispectral data in terms of image quality with implications for noise removal," *IEEE Trans. on Geoscience and Remote Sensing*, vol. 26, pp. 65-74, 1988.

Guilfoyle, K., M.L.G. Althouse and C.-I Chang (2001), "A quantitative and comparative analysis of linear and nonlinear spectral mixture models using radial basis function neural networks," *IEEE Trans. on Geoscience and Remote Sensing*, vol. 39, no. 10, pp. 2314-2318, October 2001.

Guilfoyle, K., M.L.G. Althouse, C.-I Chang, "Further results on linear and nonlinear mixture models for analyzing hyperspectral imagery," *SPIE Conf. on Algorithms and Technologies for Multispectral, Hyperspectral and Ultraspectral Imagery VIII*, SPIE 4725, Orlando, Florida, 1-5 April, 2002.

Hapke, B. (1981), "Bidirection reflectance spectrocopy. I. theory," *J. Geophysical Res.*, vol. 86, pp. 3039-3054, 1981.

Harsanyi, J.C. (1993), *Detection and Classification of Subpixel Spectral Signatures in Hyperspectral Image Sequences*, Department of Electrical Engineering, University of Maryland, Baltimore County, MD, August 1993.

Harsanyi, J.C. and C.-I Chang (1994), "Hyperspectral image classification and dimensionality reduction: an orthogonal subspace projection approach," *IEEE Trans. on Geoscience and Remote Sensing*, vol. 32, no. 4, pp. 779-785, July, 1994.

Harsanyi, J.C., W. Farrand and C.-I Chang (1994), "Detection of subpixel spectral signatures in hyperspectral image sequences," *Annual Meeting, Proceedings of American Society of Photogrammetry & Remote Sensing*, Reno, pp. 236-247, 1994.

Harsanyi, J.C., W. Farrand, J. Hejl and C.-I Chang (1994), "Automatic identification of spectral endmembers in hyperspectral image sequences," *International Symposium on Spectral Sensing Research '94* (ISSSR), San Diego, July 10-15, pp. 267-277, 1994.

Haskell, K.H. and R.J. Hanson (1981),"An algorithm for linear least squares problems with equality and nonnegativity constraints generalized," *Mathematical Programming*, vol. 21, pp. 98-118, 1981.

Haykin, S. (1986), *Adaptive Filter Theory*, NJ: Prentice-Hall, 1986, chapter 10.

Heinz, D., C.-I Chang and M.L.G. Althouse (1997), "Fully constrained least squares-based linear unmixing," *IEEE 1999 International Geoscience and Remote Sensing Symp.*, Hamburg, Germany, pp. 1401-1403, 28 June-2 July, 1999.

Heinz, D. and C.-I Chang (2000), *Constraint Least Squares Linear Spectral Mixture Methods for Material Abundance Estimation, Discrimination, Detection, Classification, Identification, Recognition and Quantification, Data Compression and Noise Estimation in Hyperspectral and Multispectral Imagery*, February 14, 2000.

Heinz, D. and C.-I Chang (2001), "Fully constrained least squares linear mixture analysis for material quantification in hyperspectral imagery," *IEEE Trans. on Geoscience and Remote Sensing*, vol. 39, no. 3, pp. 529-545, March 2001.

Heute, A.R. (1986), " Separation of soil-plant spectral mixtures by factor analysis," *Remote Sensing Environment*, vol. 19, pp. 237-251, 1986.

Huber, P.J. (1985), "Projection pursuit," *Annals of Statistics*, vol. 13, no. 2, pp. 435-475, 1985.

HYMSMO (1998), HYperspectral MASINT Support to Military Operations Program, 1998.

Hyvarinen, A. and E. Oja (2000), "Independent component analysis: algorithms and applications," *Neural Networks*, vol. 13, pp. 411-430, 2000.

Ifarraguerri, A. and C.-I Chang (1999), "Hyperspectral image segmentation with convex cones," *IEEE Trans. on Geoscience and Remote Sensing*, vol. 37, no 2, pp. 756-770, March 1999.

Ifarraguerri, A. (2000), *Hyperspectral Image Analysis with Convex Cones and Projection Pursuit*, Department of Computer Science and Electrical Engineering, University of Maryland, Baltimore County, MD, May 2000.

Ifarraguerri, A. and C.-I Chang (2000), "Multispectral and hyperspectral image analysis with projection pursuit," *IEEE Trans. on Geoscience and Remote Sensing*, vol. 38, no. 6, pp. 2529-2538, November 2000.

Jensen, J.R (1996), *Introductory Digital Image Processing: A Remote Sensing Perspective*, second edition, Upper Saddle River, Prentice-Hall, N.J., 1996.

Jenson, S.K. and F.A. Waltz (1979), "Principal components analysis and canonical analysis in remote sensing," *Proc. Amer. Soc. Photogrammetry 45th Ann. Meeting*, pp. 337-348, 1979.

Jia, X. and J.A. Richards (1994), "Efficient maximum likelihood classification for imaging spectrometer data sets," *IEEE Trans. Geoscience and Remote Sensing*, vol. 32, no. 2, pp. 274-281, March 1994.

Jimenez, L.O. and D.A. Landgrebe (1999), "Hyperspectral data analysis and supervised feature reduction via projection pursuit," *IEEE Trans on Geoscience and Remote Sensing*, vol. 37. , no. 6, pp. 2653-2667, Nov. 1999.

Johnson, P., M. Smith, S. Taylor-George and J. Adams (1983), "A semiempirical method for analysis of the reflectance spectra of binary mineral mixtures," *J. Geophysical Res.*, vol. 88, pp. 3557-3561, 1983.

Jones, M.C. and R. Sibson (1987), "What is projection pursuit," *Journal Royal Statistics Society of Association*, vol. 150, part 1, pp. 1-36, 1987.

Juang, B.H. and S. Katagiri (1992), "Discriminative learning for minimum error classification," *IEEE Trans. on Signal Processing*, vol. 40, pp. 3043-3054, 1992.

Karhunen, J. E. Oja, L. Wang, R. Vigário and J. Joutsensalo (1997), "A class of neural networks for independent component analysis", *IEEE Trans. Neural Networks*, vol. 8, no. 3, pp. 486-504, May 1997.

Katsavounides, I., C.C.J. Kuo, Z. Zhang (1994), "A new initialization technique for generalization Lloyd iteration," *IEEE Trans. Signal Process. Letters*, vol. 1, pp. 144-146, 1994.

Kim, B. and D.A. Landgrebe (1991), "Hierarchical classifier design in high-dimensional, numerous class cases," *IEEE Trans. Geoscience and Remote Sensing*, vol. 29, no. 4, pp. 792-800, July 1991.

Kullback, S. (1959), *Information Theory and Statistics*, John Wiley & Sons, N.Y., 1959 or Dover, 1968.

Lawson, C.L. and R.J. Hanson (1995), *Solving Least Squares Problems*, CAM vol. 15, SIAM, Philadelphia, PA, 1995.

Lee, C. and D.A. Landgrebe (1993), "Feature extraction based on decision boundaries," *IEEE Trans. Pattern Analysis Mach. Intell.*, vol. 15, no. 4, pp. 388-400, April, 1993.

Lee, C. and D.A. Landgrebe (1993), "Analyzing high-dimensional multispectral data," *IEEE Trans. Geoscience and Remote Sensing*, vol. 31, no. 4, pp. 792-800, July 1993.

Lee, J.B., A.S. Woodyatt and M. Berman (1990), "Enhancement of high spectral resolution remote sensing data by a noise-adjusted principal components transform," *IEEE Trans. on Geoscience and Remote Sensing*, vol. 28, no. 3, pp. 295-304, May 1990.

Lee, T.W. (1998), *Independent Component Analysis: Theory and Applications*, Boston: Kluwer Academic Publishers, 1998.

Lee, T.W., M.S. Lewicki and M. Girolami (1999), "Bind source separation of more sources than mixtures using overcomplete representations," *IEEE Trans. Signal Process.*, vol. 6, no. 4, pp. 87-90, April 1999.

Linde, Y., A Buzo and R.M. Gray (1980), "An algorithm for vector quantizer design," *IEEE Trans. on Communication*, vol.28, pp. 84-95, 1980.

Lloyd, S. (1982), "Least squares quantization in PCM," *IEEE Transaction on Information Theory*, vol. 28, no. 2, pp. 129-137, 1982.

Malinowski, E.R. (1977a), "Theory of error in factor analysis," *Analytical Chemistry*, vol. 49, pp. 606-612, 1977.

Malinowski, E.R. (1977b), "Determination of the number of factors and experimental error in a data matrix," *Analytical Chemistry*, vol. 49, pp. 612-617, 1977.

Mausel, P.W., W.J. Kramber and J.K. Lee (1990), "Optimum band selection for supervised classification of multispectral data," *Photogrammetric Engineering and Remote Sensing*, vol. 56, no. 1, pp. 55-60, Jan. 1990.

Mavrovouniotis, M.L., Harper, A.M. and Ifarraguerri, A. (1994a), "Classification of pyrolysis mass spectra of biological agents using convex cones," *J. Chemometrics*, vol. 8, pp. 305-33, 1994.

Mavrovouniotis, M.L., Harper, A.M. and Ifarraguerri, A. (1994b), "Convex-cone analysis of the time profiles of pyrolysis mass spectra of biological agents," *U.S. Army Edgewood Research, Development and Engineering Center*, Technical Report ERDEC-CR-130, July 1994.

Mavrovouniotis, M.L., Harper, A.M. and Ifarraguerri, A. (1996), "A method for extracting patterns from pyrolysis mass spectra," *Computer Assisted Analytical Spectroscopy*, Chichester, England: Wiley, pp. 189-240, 1996.

Metz, C.E (1978), "ROC methodology in radiological imaging," *Invest. Radiol.*, vol. 21, pp. 720-723, 1978.

Michalewicz, Z. (1996), *Genetic Algorithms + Data Structures = Evolution Programs*, 3rd. rev. and extended., Springer-Verlag, 1996.

Miller, J.W.V., J.B. Farison and Y. Shin (1992), "Spatially invariant image sequences," *IEEE Trans. Image Process.*, vol. 1, pp. 148-161, 1992.

Murata, N. S. Yoshizawa and S. Amari (1994), "Network information criterion-determining the number of hidden units for an artificial neural network model," *IEEE Trans. Neural Networks*, vol. 5, no. 6, pp. 865-871, Nov. 1994.

Oja, E., J. Karhunen, L. Wang and R. Vigário (1995), "Principal and independent components in neural networks – recent developments", *Proc. VII Italian Workshop Neural Networks WIRN '95, Vietrisul Mare, Italy,* pp. 16–35, May 1995.

Otsu, N (1979), "A threshold selection method from gray-level histograms," *IEEE Trans. System Man Cybernetics*, Vol. SMC-9, No. 1, pp. 62-66, 1979.

Pesses, M.E. (1999), "A least squares-filter vector hybrid approach to hyperspectral subpixel demixing," *IEEE Trans. on Geoscience and Remote Sensing*, vol. 37 no. 2, pp. 846-849, March 1999.

Pal, N.R. and S. K. Pal (1989), "Entropic thresholding," *Signal Processing*, Vol. 16, pp. 97-108, 1989.

Poor, H.V. (1994), *An Introduction to Detection and Estimation Theory*, 2nd. ed., New York: Springer-Verlag, pp. 58-59, 1994.

Rabiner, L. and B.-H. Juang (1993), *Fundamentals of Speech Recognition*, Prentice-Hall, 1993.

Reed, I.S. and X. Yu (1990), "Adaptive multiple-band CFAR detection of an optical pattern with unknown spectral distribution," *IEEE Trans. on Acoustic, Speech and Signal Process.*, vol. 38, no. 10, pp. 1760-1770, Oct. 1990.

Resmini, R.S., M.E. Kappus, W.S. Aldrich, J.C. Harsanyi and M. Anderson (1997), "Mineral mapping with HYperspectral Digital Imagery Collection Experiment (HYDICE) sensor data at Cuprite, Nevada, U.S.A.," *Int. J. Remote Sensing*, vol. 18, no. 17, pp. 1553-1570, 1997.

Ren, H. (1998), *A Comparative Study of Mixed Pixel Classification Versus Pure Pixel Classification for Multi/Hyperspectral Imagery*, Department of Computer Science and Electrical Engineering, University of Maryland, Baltimore County, MD, May 1998.

Ren, H. (2000), *Unsupervised and Generalized Orthogonal Subspace Projection and Constrained Energy Minimization for Target Detection and Classification in Remotely Sensed Imagery*, Department of Computer Science and Electrical Engineering, University of Maryland, Baltimore County, MD, May 2000.

Ren, H. and C.-I Chang (1998), "A computer-aided detection and classification method for concealed targets in hyperspectral imagery," *IEEE 1998 International Geoscience and Remote Sensing Symposium*, Seattle, WA, pp. 1016-1018, July 5-10, 1998.

Ren, H. and C.-I Chang (1999), "A constrained least squares approach to hyperspectral image classification," *1999 Conference on Information Science and Systems*, pp. 551-556, Johns Hopkins University, Baltimore, MD, March 17-19, 1999.

Ren, H. and C.-I Chang (2000a), "A generalized orthogonal subspace projection approach to unsupervised multispectral image classification," *IEEE Trans. on Geoscience and Remote Sensing*, vol. 38, no. 6, pp. 2515-2528, November 2000.

Ren, H. and C.-I Chang (2000b), "Target-constrained interference-minimized approach to subpixel target detection for hyperspectral imagery," *Optical Engineering*, vol. 39, no. 12, pp. 3138-3145, December 2000.

Richards, J.A. (1993), *Remote Sensing Digital Image Analysis*, 2nd ed. Springer Verlag, 1993.

Rissanen, J. (1978), "Modeling by shortest data description," *Automatica*, vol. 14, pp. 465-471, 1978.

Roger, R.E. (1996), "Principal components transform with simple, automatic noise adjustment," *International Journal Remote Sensing*, vol. 17, no. 14, pp. 2719-2727, 1996.

Roger, R.E. and J.F. Arnold (1996), "Reliably estimating the noise in AVIRIS hyperspectral imagers," *Int. J. Remote Sensing*, vol. 17, no. 10, 1951-1962, 1996.

Sabol, D.E., J.B. Adams and M.O. Smith (1992), "Quantitative sub-pixel spectral detection of targets in multispectral images," *J. Geophys. Research*, 97, pp. 2659-2672, 1992.

Sahoo, P.K., S. Soltani, A.K.C. Wong and Y.C. Chen (1988), "A survey of thresholding techniques," *Computer Vision, Graphics and Image Process. (CVGIP)*, vol. 41, pp. 233-260, 1988.

Schalkoff, R. (1992), *Pattern Recognition: Statistical, Structure and Neural Network*, New York: John Wiley and Sons, 1992.

Scharf, L.L. (1991), *Statistical Signal Processing*, MA: Addison-Wesley, 1991.

Schowengerdt, R.A. (1997), *Remote Sensing: Models and Methods for Image Processing*, 2nd ed., Academic Press, 1997.

Schwarz, G. (1978), "Estimating the dimension of a model," *Ann. Stat.*, vol. 6, pp. 461-464, 1978.

Settle, J.J. (1996), "On the relationship between spectral unmixing and subspace projection," *IEEE Trans on Geoscience and Remote Sensing*, vol. 34, no. 4, pp. 1045-1046, July 1996.

Settle, J.J. and N.A. Drake (1993), "Linear mixing and estimation of ground cover proportions," *Int. J. Remote Sensing*, vol. 14, no. 6, pp. 1159-1177, 1993.

Shahshahani, B.M. and D.A. Landgrebe (1994), "The effect of unlabeled samples in reducing the small sample size problem and mitigating the Hugh phenomenon," *IEEE Trans. Geoscience and Remote Sensing*, vol. 32, no. 5, pp. 1087-1095, September 1994.

Shimabukuro, Y.E. (1987), *Shade Images Derived from Linear Mixing Models of Multispectral Measurements of Forested Areas*, Ph.D. dissertation, Department of Forest and Wood Science, Colorado State University, Fort Collins, 1987.

Shimabukuro, Y.E. and J.A. Smith (1991), "The least-squares mixing models to generate fraction images derived from remote sensing multispectral data," *IEEE Trans on Geoscience and Remote Sensing*, vol. 29, pp. 16-20, 1991.

Singer, R.B. and T.B. McCord (1979), "Mars: large scale mixing of bright and dark surface materials and implications for analysis of spectral reflectance," *Proc. Lunar Planet. Sci. Conf. 10th*, pp. 1835-1848, 1979.

Smith, M.O., J.B. Adams and D.E. Sabol (1994), "Spectral mixture analysis-new strategies for the analysis of multispectral data," *Image Spectroscopy-a tool for Environmental Observations* edited by J. Hill and J. Mergier, Brussels and Luxembourg, pp. 125-143, 1994.

Smith, M.O., D.A. Roberts, J. Hill, W. Mehl, B. Hosgood, J. Verdebout, G. Schmuck, C. Koechler and J.B. Adams (1994), "A new approach to quantifying abundances of materials in multispectral images," *Proc. IEEE Int. Geoscience and Remote Sensing Symposium'94*, Pasadena, CA, pp. 2372-2374, 1994.

Soltanian-Zadeh,H., J.P. Windham, D.J. Peck (1996), "Optimal linear transformation for MRI feature extraction," *IEEE Trans. on Medical Imaging*, vol. **15**, pp. 749-767, 1996.

Stark, H. and J. Woods (1994), *Probability, Random Processes, and Estimation for Engineers*, 3rd ed., Prentice-Hall, 2002.

Stearns, S.D., B.E. Wilson and J.R. Peterson (1993), "Dimensionality reduction by optimal band selection for pixel classification of hyperspectral imagery," *Applications of Digital Image Processing XVI*, SPIE, vol. 2028, pp. 118-127, 1993.

Stellman, C.M., G.G. Hazel, F. Bucholtz, J.V. Michalowicz, A. Stocker and W. Scaaf (2000), "Real-time hyperspectral detection and cuing," Optical Engineering, vol. 39, no. 7, pp. 1928-1935, July 2000.

Swets, J.A. and R.M. Pickett (1982), *Evaluation of Diagnostic Systems: Methods from Signal Detection Theory*, Academic Press, 1982.

Szu, H (1999), "Independent component analysis (ICA): an enabling technology for intelligent information/image technology (IIT)," *IEEE Circuit and System Magazine*, vol. 10, no. 4, pp. 14-37, December 1999.

Thai, B. and G. Healey (2002), "Invariant subpixel material detection in hyperspectral imagery," *IEEE Trans. on Geoscience and Remote Sensing*, vol. 40, no. 3, pp. 599-608, March 2002.

Tompkins, S., J.F. Mustard, C.M. Pieters and D.W. Forsyth (1997), "Optimization of targets for spectral mixture analysis," *Remote Sensing of Environment,* vol. 59, pp. 472-489, 1997.

Tu, T.M., C.-H. Chen and C.-I Chang (1997), "*A posteriori* least squares orthogonal subspace projection approach to weak signature extraction and detection," *IEEE Trans. on Geoscience and Remote Sensing*, vol. 35, no. 1, pp. 127-139, January 1997.

Tu, T.M., H.C. Shy, C.-H. Lee and C.-I Chang (1999), "An oblique subspace projection to mixed pixel classification in hyperspectral images," *Pattern Recognition*, vol. 32, no. 8, pp. 1399-1408, August, 1999.

Tu, T.M. (2000), "Unsupervised signature extraction and separation in hyperspectral images: a noise-adjusted fast independent component analysis approach," *Optical Engineering*, vol. 39, no 4, pp. 897-906.

Van Veen, B.D. and K.M. Buckley (1988), "Beamforming: a versatile approach to spatial filtering," *IEEE ASSP Magazine*, pp. 4-24, April 1988.

Wang, C.-M (2002), *Applications of Multispectral Imaging Processing Techniques to Brain Magnetic Resonance Image Classification*, Ph.D. dissertation, Department of Electrical Engineering, National Cheng Kung University, Tainan, Taiwan, June 2002.

Wang, C.-M., S-C. Yang, P.-C. Chung, C.S. Lo, C.-I Chang, C.C. Chen, C.-W. Yang and C.H. Wen (2002a), "Orthogonal subspace projection-based approaches to classification of MR image sequences," *Computerized Medical Imaging and Graphics*, vol. 25, no. 6, pp. 465-476, October 2001.

Wang, C.-M., C.C. Chen, S.-C. Yang, Y.-N. Chung, P.C. Chung, C.W. Yang and C.-I Chang (2002b), "An unsupervised Orthogonal subspace projection approach to MR image classification MR images for classification," *Optical Engineering*, vol. 41, no. 7, pp. 1546-1557, July 2002.

Wang, C.-M., S.-C. Yang, P.C. Chung, C.C. Chen, Y.-N. Chung, C.W. Yang and C.-I Chang (2003), "Detection of spectral signatures in MR images for classification," *IEEE Trans. on Medical Imaging*, vol. TMI-22, no. 1, pp. 50-61, Jan. 2003.

Wang, J and C.-I Chang (2002), "Unsupervised Kalman filter approach to signature estimation for remotely sensed imagery," *International Geoscience and Remote Sensing Symp.*, 24-28, June, Toronto, Canada, 2002.

Ward, C.R., P.J. Hargrave and J.G. McWhirter (1986), "A novel algorithm and architecture for adaptive digital beamforming," *IEEE Trans. Antennas and Propagation*, vol. AP-34, no. 3, pp. 338-346, March 1986.

Wax, M. and T. Kailath (1985), "Detection of signals by information criteria," *IEEE Trans, Acoustic, Speech, and Signal Process.*, vol. ASSP-33, no. 2, pp. 387-392, April, 1985.

Wu, H.T., J.F. Yang and F.K. Chen (1995), "Source number estimators using transformed Gerschgorin radii," *IEEE Trans. on Signal Process.*, vol. 43, no. 6, pp. 1325-1333, June 1995.

Yu, X, I.S. Reed and A.D. Stocker (1993), "Comparative performance analysis of adaptive multispectral detectors," *IEEE Trans. Signal Process.*, vol. 41, no. 8, pp. 2639-2656, Aug. 1993.

Zenzo, S.D., Degloria, R. Bernstein and H.C. Kolsky (1987a), "Gaussian maximum likelihood and contextual classification algorithms for multicrop classification," *IEEE Trans. Geoscience and Remote Sensing*, vol. GE-25, no. 6, pp. 805-814, November 1987.

Zenzo, S.D., R. Bernstein, S.D. Degloria and H.C. Kolsky (1987b), "Gaussian maximum likelihood and contextual classification algorithms for multicrop classification experiments using thematic mapper and multispectral scanner sensor data," *IEEE Trans. Geoscience and Remote Sensing*, vol. GE-25, pp. 815-824, November 1987.

Zhao, X. (1996), *Subspace Projection Approach to Multispectral/Hyperspectral Image Classification Using Linear Mixture Modeling*, Master Thesis, Department of Computer Sciences and Electrical Engineering, University of Maryland Baltimore County, MD, May 1996.

Zhang, X., C.H. Chen (2002), "New independent component analysis method using higher order statistics with applications to remote sensing images," *Optical Engineering*, vol. 41, pp. 1717-1728, July 2002.

INDEX